Radar Array Design Using Optimization Theory

Other related titles:

You may also like

- SBRA533 | Cui | Radar Waveform Design Based on Optimization Theory | 2020
- SBRA526 | Williams | Electronic Scanned Array Design | 2020

We also publish a wide range of books on the following topics:
Computing and Networks
Control, Robotics and Sensors
Electrical Regulations
Electromagnetics and Radar
Energy Engineering
Healthcare Technologies
History and Management of Technology
IET Codes and Guidance
Materials, Circuits and Devices
Model Forms
Nanomaterials and Nanotechnologies
Optics, Photonics and Lasers
Production, Design and Manufacturing
Security
Telecommunications
Transportation

All books are available in print via https://shop.theiet.org or as eBooks via our Digital Library https://digital-library.theiet.org.

SciTech RADAR, SONAR AND NAVIGATION SERIES 566

Radar Array Design Using Optimization Theory

Edited by
Guolong Cui, Junli Liang, Bin Liao, Xiangrong Wang
and Lingjiang Kong

The Institution of Engineering and Technology

About the IET

This book is published by the Institution of Engineering and Technology (The IET).

We inspire, inform, and influence the global engineering community to engineer a better world. As a diverse home across engineering and technology, we share knowledge that helps make better sense of the world, accelerate innovation, and solve the global challenges that matter.

The IET is a not-for-profit organization. The surplus we make from our books is used to support activities and products for the engineering community and promote the positive role of science, engineering, and technology in the world. This includes education resources and outreach, scholarships and awards, events and courses, publications, professional development and mentoring, and advocacy to governments.

To discover more about the IET please visit https://www.theiet.org/.

About IET books

The IET publishes books across many engineering and technology disciplines. Our authors and editors offer fresh perspectives from universities and industry. Within our subject areas, we have several book series steered by editorial boards made up of leading subject experts.

We peer review each book at the proposal stage to ensure the quality and relevance of our publications.

Get involved

If you are interested in becoming an author, editor, series advisor, or peer reviewer, please visit https://www.theiet.org/publishing/publishing-with-iet-books/ or contact author_support@theiet.org.

Discovering our electronic content

All of our books are available online via the IET's Digital Library. Our Digital Library is the home of technical documents, eBooks, conference publications, real-life case studies, and journal articles. To find out more, please visit https://digital-library.theiet.org.

In collaboration with the United Nations and the International Publishers Association, the IET is a Signatory member of the SDG Publishers Compact. The Compact aims to accelerate progress to achieve the Sustainable Development Goals (SDGs) by 2030. Signatories aspire to develop sustainable practices and act as champions of the SDGs during the Decade of Action (2020–30), publishing books and journals that will help inform, develop, and inspire action in that direction.

In line with our sustainable goals, our UK printing partner has FSC accreditation, which is reducing our environmental impact on the planet. We use a print-on-demand model to further reduce our carbon footprint.

Published by The Institution of Engineering and Technology, London, United Kingdom

The Institution of Engineering and Technology (the "**Publisher**") is registered as a Charity in England & Wales (no. 211014) and Scotland (no. SC038698).

First published 2024

The Institution of Engineering and Technology
Futures Place,
Kings Way, Stevenage,
Herts, SG1 2UA,
United Kingdom

www.theiet.org

British Library Cataloguing in Publication Data

A catalogue record for this product is available from the British Library.

ISBN 978-1-83953-933-6 (hardback)
ISBN 978-1-83953-934-3 (PDF)

Typeset in India by MPS Limited

Cover image: Blue Spectrogram by Flavio Coelho/Moment via Getty Images

Contents

Foreword

Radar array design aims at shaping a specific transmit/receive beampattern through optimizing array geometry, transmit waveform, and transmit/receive weight vectors. It has been a hot topic among the scientific and industrial communities due to the advent of modern digital arrays with multiple digital transmit and receive channels, arbitrary digital waveform generators, solid-state transmitters, etc.

The adoption of suitable array beamforming techniques can yield a plethora of benefits, markedly enhancing radar performance across various domains including detection, localization, anti-clutter measures, anti-jamming capabilities, and spectrum coexistence. Transmit and receive beamforming adaptivity are also at the base of a cognitive array radar to implement the perception-action cycle playing a key role in the stimulation of the environment. In such contexts, the radar array beampattern emerges as a critical performance metric necessitating optimization, while accommodating constraints stemming from both internal factors and external information acquired through the perception cycle.

Optimization theory turns out to be useful to address the specific performance metric optimization (objective function) as for instance beampattern integrated sidelobe level or peak sidelobe level (PSL). As a matter of fact, in the past decades there has been a proliferation of innovative techniques for radar array design heavily exploiting mathematical results developed in the context of optimization theory.

This book attempts to provide an overview of some radar beampattern synthesis obtained as the result of an optimization process trying to cover some of the most challenging application fields including phased array beampattern, multiple input-multiple output (MIMO) beampattern in the presence of clutter, jamming, and congested environments. Innovative and sophisticated instruments from the optimization theory such as alternating direction method of multipliers (ADMM), coordinate descent technique, semidefinite relaxation, rank-one matrix decomposition, Lagrange duality theory, fractional programming, convex approximation, Pareto optimization, and machine learning (ML) will be framed in the context of radar array design.

The book provides a rigorous mathematical approach corroborated by a wealth of numerical study cases. Additionally, the book has a cross-disciplinary approach because it tries to exploit cross-fertilization by the recent research and discoveries in optimization theory. The material of the book is organized in 11 chapters each one completed by a very comprehensive list of references.

Precisely, Chapter 1 investigates the sparse array reconfiguration via antenna selection based on ML. First, ML algorithms, specifically support vector machine (SVM) and artificial neural network, are utilized to solve combinatorial antenna selection problems in terms of maximum signal-to-interference-plus-noise ratio. Numerical examples are presented to validate the effectiveness and efficiency of ML algorithms for sparse array design. Moreover, the SVM-based antenna selection is robust against the direction of arrival (DOA) estimate uncertainties. To this end, the sparse array optimization for enhanced DOA estimation as a classification problem is formulated and the convolutional neural network (CNN) structure is devised. The dataset using the Cramer–Rao bound of DOA estimation as an indicator is generated to train the CNN and numerical results are provided to demonstrate the adaptability and effectiveness of the proposed CNN-based sparse receiver design in an arbitrary unknown environment.

Chapter 2 deals with a joint design of subarray layout vector (LV) and weighted coefficient of radiation element to achieve the desired beampattern for a new array architecture. An architecture of thinned array composed of multiple regular subarrays is presented and corresponding subarray layout and beampattern models are established. Then, a new optimization problem is considered, aiming to minimize the PSL and the LV norm under the constraints of power beampattern and subarray layout. The resulting problem is quite challenging to solve due to the highly nonconvex discrete LV and double-sided quadratic constraints. In this respect, the resultant hybrid optimization problem is transformed into a new ADMM form via introducing a series of auxiliary variables, and an approximation algorithm based on ADMM is proposed. In each iteration of the proposed ADMM, the tractable and small-sized subproblems are globally or locally solved in parallel.

In Chapter 3, the problem of reconfigurable array beampattern synthesis based on sensor network consensus computation theory is considered. First, the concept of sensor network consensus computation is introduced and the synthesis problem is recast as a sensor network computation problem via imitating a virtual (conceptual) multi-node sensor network, where each "node" corresponds to an individual beampattern synthesis task. More specifically, these "nodes" with the same number as those of radiated patterns share a set of common excitation magnitudes, which results in a special magnitude-consensused computation problem in the imitated sensor network. Then, via locally computing the excitation in each "node" and exchanging current computation results with neighboring "nodes", all "nodes" obtain the magnitude-consensused excitations after the "network" is stable.

Chapter 4 addresses the problem of array pattern synthesis with the minimization of dynamic range ratio (DRR). Unlike the common DRR control approaches which limit DRR to be less than a certain threshold, the DRR of the excitation vectors while synthesizing the anticipated array pattern is minimized, which results in a new nonconvex and nonlinear optimization problem owing to its fractional objective function and nonconvex constraints. Then, a new algorithm to solve the optimization problem is introduced. More specifically, via introducing auxiliary variables, an equivalent optimization problem is formulated that transforms the fractional objective function into a linear one, decomposes the original optimization problem into several subproblems in each iteration, and simplifies them as either single-variable quadratic unconstrained optimization or least squares problems which are solved efficiently in each iteration.

Chapter 5 proposes a method based on primary-dual iteration to address the array beampattern synthesis problem of minimizing the PSL under the constraints of controlled mainlobe shape and DRR of current excitations. The proposed method can be utilized to synthesize a variety of desired-shaped beampatterns, such as focused beam, flat-top beam, and cosecant square beam as well as null beampattern. Additionally, the proposed method can be applied in arbitrary array configurations, such as uniform linear array, planar array, and circular rings array, even in generic array with non-uniform spacing between array elements.

Chapter 6 delves into the techniques of pattern synthesis through array response control, introducing several distinct methods: the accurate array response control (A2RC) algorithm, the multipoint accurate array response control (MA2RC) algorithm, optimal and precise array response control (OPARC) algorithm, the weight vector orthogonal decomposition (WORD) algorithm, and flexible array response control via oblique projection (FARCOP) algorithm. These algorithms are rooted in the adaptive array theory, aiming to provide precise, optimal, flexible, and robust control over array response levels. Collectively, these methods present a comprehensive and effective framework for pattern synthesis through precise array response control. They offer promising solutions for wireless communications, radar systems, and beyond, enabling fine-tuned control of array responses and pattern synthesis in a wide range of applications.

Chapter 7 proposes a novel wideband beamforming method that can generate a deep notch in the spatial-spectral region of interest (SSRI) for interference suppression, assuming that the system has some prior information on the spectral features and locations of potential interferences. Specifically, a wideband beamforming design that minimizes the beampattern PSL in the SSRI is investigated. Meanwhile, the white noise gain and mainlobe level constraints are imposed to guarantee an acceptable output signal-to-noise ratio gain. To tackle the difficult optimization problem containing non-convex constraints, two competitive iterative optimization algorithms are proposed, referred to as alternating optimization algorithm and convex approximation algorithm, and analyze their performance in terms of convergence, complexity, and null depth.

In Chapter 8, two cases of hybrid beamforming (HBF) design for millimeter wave dual-function radar-communication (DFRC) systems are studied. In the first case, we consider a scenario in which a single-carrier DFRC system communicates with a single multi-antenna user while detecting an extended target in the presence of extended clutters. In the second case, a multi-user and multi-carrier DFRC system is investigated, where this system leverages multicarrier signals to achieve high-accuracy DOA and ensure high-quality communication services. Two state-of-the-art HBF design algorithms are proposed for the two cases under consideration.

Chapter 9 investigates the problem of robust beamforming for the MIMO DFRC system, when there exist communication channel state information (CSI) uncertainties. Two different CSI error models are considered to achieve robustness. In the first model, the communication CSI is assumed to be available with a bounded unknown error. In the second model, it is assumed that the CSI error follows a complex zero-mean Gaussian distribution. Under this model, nonconvex optimization problems are formulated to design robust beamformers, and nonconvex optimization techniques such as semidefinite relaxation (SDR) and ADMM are applied to achieve feasible solutions.

Chapter 10 focuses on the colocated MIMO array. The transmit waveform of the colocated MIMO array can be divided into two categories: orthogonal and partially correlated. When the transmit waveforms are orthogonal, the transmit beampattern is omnidirectional. In this case, the virtual aperture expansion can be realized by signal separation at the receive end, to obtain high resolution. When the transmit waveforms are partially correlated, the transmit beampattern can be optimized to any desired shape and direction. In this case, transmit beampatterns can be designed so that the transmit energy can be reasonably allocated to the target area of interest.

Chapter 11 deals with the design of a virtual low-sidelobe beampattern in distributed phased MIMO radar. An iterative framework with respect to subarray layout and receive weighting coefficients is proposed to minimize beampattern PSL accounting for practical position constraints as well as the mainlobe level restriction. In each iteration, an efficient cyclic optimization algorithm based on the coordinate descent framework is developed to seek the subarray layout by splitting high dimensional layout optimization problem into multiple one-dimensional optimization problems, and the semi-definite relaxation technique and rank one approximation is explored to design weighting coefficients.

Each chapter within this compendium stands as a self-contained exposition, meticulously crafted by esteemed scholars in the field of radar array design and optimization theory. The book emphasizes both theoretical results and practical applications, clearly demonstrating the potential benefits achievable in radar array design through the utilization of modern optimization theory. A common list of symbols (reported at the end of this foreword) and extensive cross-referencing have been realized so that the related material can be easily found and connected, thus significantly augmenting the book's value as a scholarly reference.

This book is intended for systems engineers and their managers within the aerospace and defense sector, technical personnel in procurement agencies and their advisory teams, as well as students pursuing MSc and PhD degrees in signal processing, electrical engineering, and optimization theory. Additionally, it caters to individuals with an interest in the practical applications of optimization theory in the realm of radar array engineering.

The editors express their heartfelt gratitude for the professionalism and camaraderie extended by the collaborating authors who contributed to the writing of select chapters in this book. Their invaluable cooperation has played an instrumental role in shaping and enriching the content of this volume. To our esteemed readers, we eagerly anticipate and welcome your constructive criticisms and insightful comments, which will further enhance the scholarly discourse surrounding these works.

Guolong Cui
Junli Liang
Bin Liao
Xiangrong Wang
Lingjiang Kong

Notation

$\boldsymbol{a}$	column vector (lowercase)
$\boldsymbol{A}$	matrix (upper case)
$a(n)$ or a_n	n-th element of $\boldsymbol{a}$
$\boldsymbol{A}(m,n)$	(m,n)-th entry of $\boldsymbol{A}$
$(\cdot)^T$	transpose operator
$(\cdot)^*$	conjugate operator
$(\cdot)^\dagger$	conjugate transpose operator
$\mathrm{tr}\,(\cdot)$	trace of the square matrix argument
$\mathrm{Rank}\,(\cdot)$	rank of the square matrix argument
$\lambda_{max}(\cdot)$	maximum eigenvalue of the square matrix argument
$\lambda_{min}(\cdot)$	minimum eigenvalue of the square matrix argument
$\mathbf{diag}(\boldsymbol{a})$	N-dimensional diagonal matrix whose i-th diagonal element is $a(i)$, for $i = 1, \ldots, N$, with $\boldsymbol{a} \in \mathbb{C}^N$
$\mathbf{Diag}(\boldsymbol{A})$	N-dimensional column vector whose i-th entry is $\boldsymbol{A}(i,i)$, $i = 1, \ldots, N$
$\mathrm{Range}\,(\boldsymbol{X})$	range span of the column vectors of the matrix $\boldsymbol{X}$
$\boldsymbol{I}$	identity matrix (its size is determined from the context)
0	matrix with zero entries (its size is determined from the context)
$\mathbb{R}^N$	set of N-dimensional vectors of real numbers
$\mathbb{C}^N$	set of N-dimensional vectors of complex numbers
$\mathbb{H}^N$	set of $N \times N$ Hermitian matrices
$\mathbb{Z}$	set of integer numbers
$\mathbb{N}$	set of nature numbers
$\mathbb{R}_{\geq}0$	set of non-negative real numbers
$\mathbb{R}_{<}0$	set of negative real numbers
$\succeq$	for any $\boldsymbol{A} \in \mathbb{H}^N$, $\boldsymbol{A} \succeq \boldsymbol{0}$ means that $\boldsymbol{A}$ is a positive semidefinite matrix
$\succ$	for any $\boldsymbol{A} \in \mathbb{H}^N$, $\boldsymbol{A} \succ \boldsymbol{0}$ means that $\boldsymbol{A}$ is a positive definite matrix
$\\|\boldsymbol{x}\\|$	Euclidean norm of the vector $\boldsymbol{x}$
j	imaginary unit (i.e., $j = \sqrt{-1}$)
i	generic index
$\mathrm{Re}(x)$	real part of the complex number x
$\mathrm{Im}(x)$	imaginary part of the complex number x
$\lvert x \rvert$	modulus of the complex number x
$\arg(x)$	argument of the complex number x
$\arg(\boldsymbol{a})$	vector of the element-wise arguments of $\boldsymbol{a}$
$\mathbb{E}\,[\cdot]$	statistical expectation
$\mathbb{P}[\cdot]$	probability measure

$\odot$	Hadamard product
$\otimes$	Kronecker product
$\star$	convolution operator
$\lfloor \cdot \rfloor$	floor function
$\dot{y}, \frac{\partial y}{\partial x}, \frac{dy}{dx}$	first derivative of y with respect to variable x
$\ddot{y}, \frac{\partial^2 y}{\partial x^2}, \frac{d^2 y}{dx^2}$	second derivative of y with respect to variable x
$\partial h(\boldsymbol{x})$	gradient or subgradient of h to the vector $\boldsymbol{x}$
$v(\mathscr{P})$	optimal value of the optimization problem $\mathscr{P}$
$(\cdot)^\star$	optimized solution
$\mathscr{N}(m, \Sigma)$	circular symmetric complex Gaussian distribution with mean m and covariance matrix Σ

About the editors

Guolong Cui is a professor at the University of Electronic Science and Technology of China (UESTC), Chengdu, China. Between June 2012 and August 2013, he was a post-doctoral researcher at the Department of Electrical and Computer Engineering, Stevens Institute of Technology, Hoboken, NJ, USA. His research interests include cognitive radar, array signal processing, MIMO radar, and through-the-wall radar. He is a senior member of the IEEE.

Junli Liang is currently a professor at the School of Electronics and Information, Northwestern Polytechnical University, Xi'an, China. He received a PhD degree in signal and information processing from the Chinese Academy of Sciences, Beijing, China, in 2007. His research interests include radar signal processing, machine learning, 6G, and its applications. He is a senior member of the IEEE.

Bin Liao is currently a full professor at the College of Electronics and Information Engineering, Shenzhen University, China. He received his PhD degree from the University of Hong Kong in 2013. He was previously a research assistant at the University of Hong Kong. Dr Liao is an associate editor of *IEEE Transactions on Aerospace and Electronic Systems*, *IET Signal Processing*, and *Multidimensional Systems and Signal Processing*.

Xiangrong Wang is currently a professor at the School of Electronic and Information Engineering, Beihang University, Beijing, China. Her research interests include array signal processing, radar signal processing, integrated sensing, and communications. She has published two books and two invited book chapters. She is an associate editor of *IEEE Transactions on Radar Systems* and *Elsevier Digital Signal Processing*. She is the winner of the 2023 IEEE TAES Barry Carlton Award.

Lingjiang Kong is currently a professor at the School of Information and Communication Engineering, University of Electronic Science and Technology of China. He is a senior member of the IEEE. From September 2009 to March 2010, he was a visiting researcher at the University of Florida, Gainesville, FL, USA. His research interests include MIMO radar, through-the-wall radar, and statistical signal processing.

Chapter 1

Machine learning-based antenna selection for sparse array reconfiguration

Xiangrong Wang[1], Weitong Zhai[1] and Xianghua Wang[2]

The sparse array design for adaptive beamforming is usually formulated into combinatorial antenna selection problems, which belong to notorious NP-hard problems. As the commonly deployed convex relaxation algorithms are susceptible to local optima, several trials with different initial points are conducted for the global optima. Moreover, the high computational load of optimization techniques prohibits the real-time adaptive array reconfiguration. In this chapter, we consider to utilize machine learning algorithms, specifically support vector machine (SVM) and convolutional neural network (CNN), for solving combinatorial antenna selection problems.

1.1 Introduction

Spatial filtering is a capability possessed by antenna arrays, extensively applied in radar, sonar, wireless communications, radio astronomy, and satellite navigation, to list a few [1–5]. While the nominal array configuration remains uniform, sparse arrays have recently become fundamental in various sensing systems that involve multi-antenna transmitters and receivers [6–8]. Recent demonstrations have indicated that sparse arrays, comprising a designated number of antennas strategically positioned at optimal subsets of grid locations and interfacing with radio-frequency (RF) front-end receivers, can maintain significant aperture size while simplifying system complexity [9–13]. In essence, the primary objective of sparse array design involves determining the optimal placement of a specified number of sensors to achieve optimum performance. Various optimization metrics result in diverse configurations of sparse arrays [14–23]. An optimal array configuration, from the perspective of signal enhancement and interference suppression, maximizes the signal-to-interference-plus-noise ratio (MaxSINR). From the viewpoint of parameter estimation, such as direction of arrival (DOA), the usual metric of optimality is the estimation accuracy lower bounded by the Cramer–Rao Bound (CRB).

The underlying problem, whether cast as the optimum placement of a given number of antennas or equivalently selecting a subset of antennas to connect with

[1]School of Electronic and Information Engineering, Beihang University, China
[2]School of Intelligent Engineering and Automation, Beijing University of Posts and Telecommunications, China

RF front-end receivers, fits within the framework of antenna selection. With the fundamental goal of reducing hardware costs linked to expensive RF chains, the problem from the antenna selection perspective relies on low-complexity RF switches [13,24]. Formulated into combinatorial optimization problems in terms of the MaxSINR or the CRB, the antenna selection was tackled and solved by convex relaxation algorithms [25–28]. The adaptive array reconfiguration applications may not find optimization techniques suitable due to their high computational complexity. Furthermore, relaxation techniques used to solve non-convex optimization problems are susceptible to local optima. Conducting several runs with different initial points is necessary to search for the global optima, which significantly increases the required computational time.

In this chapter, we investigate the sparse array reconfiguration via antenna selection based on machine learning (ML). In Section 1.2, we propose to utilize machine learning algorithms, specifically support vector machine (SVM) and artificial neural network (ANN), for solving combinatorial antenna selection problems in terms of MaxSINR. Numerical examples are presented to validate the effectiveness and efficiency of machine learning algorithms for sparse array design. Moreover, the SVM-based antenna selection is robust against DOA estimate uncertainties. In Section 1.3, we formulate the sparse array optimization for enhanced DOA estimation as a classification problem and design the convolutional neural network (CNN) structure. We generate the dataset using the CRB of DOA estimation as an indicator to train the CNN and numerical results are provided to demonstrate the adaptability and effectiveness of the proposed CNN based sparse receiver design in arbitrary unknown environment.

1.2 Sparse array beamformer design via machine learning

Recently, the formulation of sparse array design for adaptive beamforming into combinatorial antenna selection problems has highlighted their status as notorious NP-hard problems. As the commonly deployed convex relaxation algorithms are susceptible to local optima, several trials with different initial points are conducted for the global optima. Furthermore, the high computational load of optimization techniques hinders real-time adaptive array reconfiguration. In this section, the utilization of ML algorithms, specifically support vector machine (SVM) [29,30] and artificial neural network (ANN) [31,32], is proposed for achieving optimal sparse array reconfiguration.

Figure 1.1 illustrates the flowchart of the ML-based antenna selection strategy. A large set of training data from all possible scenarios is used to completely train the SVM and ANN. Following training, the first step involves employing a Capon beamformer to sense the operating environment and extract features, such as the number and DOAs of interferences. According to the features provided by the Capon beamformer, the SVM/ANN then determines the status of each antenna (either selected or discarded). The ML-based selection algorithms have demonstrated the capability to rapidly obtain the global optimum solution, making them highly suitable for rapidly

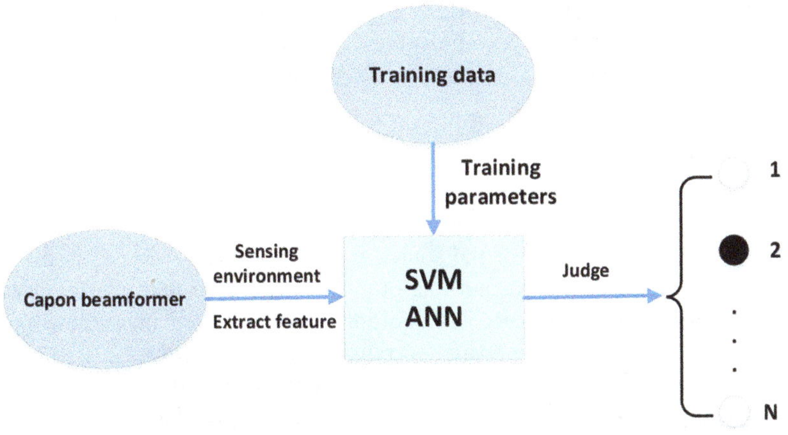

Figure 1.1 Flowchart of machine learning-based antenna selection

changing environments. Furthermore, simulation results also illustrate the robustness of the SVM algorithm against uncertainties in DOA estimates.

1.2.1 Mathematical model

Consider a linear array of N isotropic antennas with positions specified by multiple integer of unit inter-element spacing $p_n d$, $p_n \in \mathbb{N}, n = 1, \ldots, N$. The symbol $\mathbb{N}$ denotes the set of integer numbers. Suppose that a single source is impinging on the array from directions ϕ_s and q strong interfering signals are arriving from directions $\{\phi_1, \ldots, \phi_q\}$. The corresponding steering vectors are

$$\begin{aligned} \mathbf{s} &= [e^{jk_0 p_1 d \cos\phi_s}, \ldots, e^{jk_0 p_N d \cos\phi_s}]^T, \\ \mathbf{v}_l &= [e^{jk_0 p_1 d \cos\phi_l}, \ldots, e^{jk_0 p_N d \cos\phi_l}]^T, \quad l = 1, \ldots, q, \end{aligned} \tag{1.1}$$

where the wavenumber is defined as $k_0 = 2\pi/\lambda$ with λ being the wavelength and ^{T} denotes the transpose operation. The received signal at time instant t is given by

$$\mathbf{x}(t) = s(t)\mathbf{s} + \mathbf{V}\mathbf{v}(t) + \mathbf{n}(t), \tag{1.2}$$

where $\mathbf{V} = [\mathbf{v}_1, \ldots, \mathbf{v}_q]$ is the interference array manifold matrix with the full column rank. In the above equation, $s(t) \in \mathbb{C}$ and $\mathbf{v}(t) \in \mathbb{C}^q$ are, respectively, the statistically independent source and interfering signals, $\mathbf{n}(t) \in \mathbb{C}^N$ denotes the received Gaussian noise vector. The symbol $\mathbb{C}$ denotes the set of complex numbers. The Capon beamformer, which minimizes the output variance while keeping the desired signal distortionless [33], is $\mathbf{w}_c = \mathbf{R}_n^{-1}\mathbf{s}$, where $\mathbf{R}_n = \mathbf{V}\mathbf{C}_v\mathbf{V}^H + \sigma_n^2\mathbf{I}$ is the interference plus noise covariance matrix with $\mathbf{C}_v = E\{\mathbf{v}(t)\mathbf{v}^H(t)\}$ denoting the interference correlation matrix, σ_n^2 the noise power level and ^{H} stands for the Hermitian operation. The output SINR of the Capon beamformer can be expressed as [25,26]

$$\begin{aligned} \text{SINR} &= \sigma_s^2 \mathbf{s}^H \mathbf{R}_n^{-1} \mathbf{s}, \\ &\approx \text{SNR}\{\mathbf{s}^H [\mathbf{I} - \mathbf{V}(\mathbf{V}^H\mathbf{V})^{-1}\mathbf{V}^H]\mathbf{s}\}, \\ &= N\text{SNR}(1 - |\alpha|^2), \end{aligned} \tag{1.3}$$

where σ_s^2 denotes the power of the source signal, and $\text{SNR} = \sigma_s^2/\sigma_n^2$ is the input signal-to-noise ratio. In the second line of (1.3), we assume that interferences are much stronger than white noise, a condition often observed in satellite navigation and radio astronomy. The spatial correlation coefficient (SCC) α is defined as $|\alpha|^2 = (1/N)\mathbf{s}^H\mathbf{V}(\mathbf{V}^H\mathbf{V})^{-1}\mathbf{V}^H\mathbf{s}$. From (1.3), we can observe that the output SINR depends solely on the squared SCC value for a given number of antennas. Therefore, maximizing the output SINR through sparse array design is equivalent to minimizing the SCC value.

1.2.2 Optimization-based sparse array design

The sparse array design can be cast as selecting K out of N candidate antennas placed on uniform grid points with half wavelength spacing. The positions of the K antennas are freely determined by the optimization technique, which in this case is the Capon beamformer. Denote an antenna selection vector $\mathbf{z} = [z_i, i = 1, \ldots, N] \in \{0, 1\}^N$ with "zero" entry for a discarded antenna and "one" entry for a selected one. As mentioned in Section 1.2.1, the optimum sparse array can be configured by minimizing the SCC value, that is, [26]

$$\begin{aligned} &\min_{\mathbf{z}} \log|\mathbf{V}^H \text{diag}(\mathbf{z})\mathbf{V}| - \log|\mathbf{V}_s^H \text{diag}(\mathbf{z})\mathbf{V}_s|, \\ &\text{s.t. } \mathbf{1}^T\mathbf{z} = K, \ \mathbf{z} \in \{0, 1\}^N, \end{aligned} \tag{1.4}$$

where s.t. stands for "subject to," diag($\mathbf{z}$) is a diagonal matrix with the vector $\mathbf{z}$ populating along the diagonal, $\mathbf{V}_s = [\mathbf{V}, \mathbf{s}]$, and $|\cdot|$ denotes the determinant operation. The difficulty of solving the problem in (1.4) is twofold: the non-convexity of the objective function and boolean constraints of the selection variable $\mathbf{z}$. In order to solve the problem, the objective function is approximated by its affine upper bound iteratively and the boolean constraints are relaxed to a box constraint $0 \leq \mathbf{z} \leq 1$ [26]. The antenna selection in the $(k+1)$-th iteration can be formulated based on the solution from the k-th iteration as

$$\begin{aligned} &\min_{\mathbf{z}} \Delta\mathbf{g}^T(\mathbf{z}^{(k)})(\mathbf{z} - \mathbf{z}^{(k)}) - \log|\mathbf{V}_s^H \text{diag}(\mathbf{z})\mathbf{V}_s|, \\ &\text{s.t. } \mathbf{1}^T\mathbf{z} = K, \\ &\quad\quad 0 \leq \mathbf{z} \leq 1, \end{aligned} \tag{1.5}$$

where $\Delta\mathbf{g}(\mathbf{z}^{(k)})$ is the gradient of the concave function $\log|\mathbf{V}^H\text{diag}(\mathbf{z})\mathbf{V}|$ evaluated at the point $\mathbf{z}^{(k)}$. That is,

$$\Delta\mathbf{g} = [\mathbf{v}_{r,i}^H(\mathbf{V}^H\text{diag}(\mathbf{z}^{(k)})\mathbf{V})^{-1}\mathbf{v}_{r,i}, i = 1, \ldots, N]^T, \tag{1.6}$$

with $\mathbf{v}_{r,i}$ denoting the i-th column vector of the matrix $\mathbf{V}^H$. Note that the iterative relaxation in (1.5) is a local heuristic and its performance depends on the initial point $\mathbf{z}^{(0)}$. It is, therefore, typical to initialize the algorithm with several feasible points $\mathbf{z}^{(0)}$ and find the one with the minimum objective value over the different runs. The interior point method can be utilized to solve the optimization problem with a computational complexity of order $O(n^{3.5}L^2)$, where n and L are the number of variables and bitlength, respectively.

1.2.3 Machine learning-based sparse array design

While convex relaxation and optimization are effective in many cases, their susceptibility to local optima and high computational complexity hinder practical implementation. In this section, we explore the utilization of ML algorithms for sparse array design. First, we summarize two principal ML techniques, and then we describe how antenna selection problems are formulated within the framework of ML.

1.2.3.1 Support vector machine

The support vector machine (SVM) is known as the maximum margin classifier, which calculates the optimum hyperplane $\mathbf{u}^T\mathbf{x} + b$ with the maximal margin of separation between the two classes. Given a set of training data $\mathbf{x}_i, i = 1, \ldots, L$ and the corresponding labels $y_i \in \{-1, 1\}, i = 1, \ldots, L$. The hyperplane can be calculated as follows:

$$\begin{aligned} &\min_{\mathbf{u},\mathbf{b}} \frac{1}{2}\|\mathbf{u}\|^2 + C\sum_{l=1}^{L} \varepsilon_l, \\ &\text{s.t. } y_i(\mathbf{u}^T\mathbf{x}_i + b) \geq 1 - \varepsilon_l, l = 1, \ldots, L, \\ &\quad \varepsilon_l \geq 0, l = 1, \ldots, L, \end{aligned} \tag{1.7}$$

where C is a trade-off parameter. The Lagrangian dual of the problem in (1.7) can be expressed as the following quadratic programming form:

$$\begin{aligned} &\max_{\alpha} \frac{1}{2}\alpha^T\mathbf{H}\alpha - \mathbf{1}^T\alpha, \\ &\text{s.t. } \alpha^T\mathbf{y} = 0,\ 0 \leq \alpha \leq C, \end{aligned} \tag{1.8}$$

where $\mathbf{y} = [y_i, i = 1, \ldots, L]^T$ and $\mathbf{H} = \mathbf{G}^T\mathbf{G}$ with $\mathbf{G} = [y_1\mathbf{x}_1, \ldots, y_L\mathbf{x}_L]$. In order to increase the Vapnik-Chervonenkis (VC) dimension of the SVM classifier, which is defined as the maximum number of points that can be labeled in all possible ways, kernel mapping, $\kappa : \mathscr{X} \rightarrow \mathscr{F}$ from data space $\mathscr{X}$ to a dot product feature space $\mathscr{F}$, can be employed. The matrix $\mathbf{H}$ in (1.8) with kernel can be rewritten as

$$\mathbf{H}_{ij} = y_i y_j \kappa(\mathbf{x}_i, \mathbf{x}_j). \tag{1.9}$$

1.2.3.2 Artificial neural network

The structure of the employed artificial neural network (ANN) is shown in Figure 1.2, which comprises five layers, an input layer, three hidden layers, and an output layer.

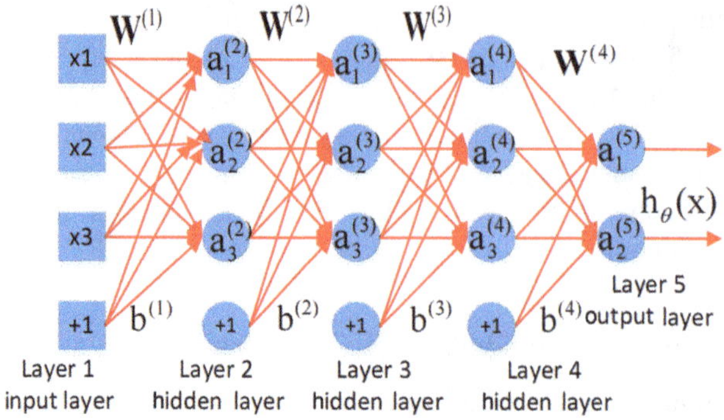

Figure 1.2 The structure of a five-layer artificial neural network

Let n_l indicate the number of layers and s_l denote the number of nodes in layer l. We write $a_k^{(l)}$ to denote the activation of neuron k in layer l and for l = 1, $a_i^{(1)} = x_i$, i.e., the i-th feature of the input data vector $\mathbf{x}$. The circles labeled "+1" are called bias units, and correspond to the intercept term. The neural network has parameters $\theta = (\mathbf{W}^{(1)}, b^{(1)}, \mathbf{W}^{(2)}, b^{(2)}, \mathbf{W}^{(3)}, b^{(3)}, \mathbf{W}^{(4)}, b^{(4)})$, where $\mathbf{W}_{ij}^{(l)}$ denotes the weight associated with the connection between neuron i in layer l and neuron j in layer $l+1$. Thus, $\mathbf{W}^{(1)} \in \mathbb{R}^{s_l \times s_{l+1}}$. Given a fixed setting of the parameters θ, our neural network defines a hypothesis $h_\theta(\mathbf{x})$ that outputs the prediction $\mathbf{a}^{(5)}$. Specifically, the forward propagation that this neural network represents is given by

$$\mathbf{a}^{(l+1)} = g(\mathbf{W}^{(l)}\mathbf{a}^{(l)} + b^{(l)}), l = 1, \ldots, n_l - 1 \tag{1.10}$$

where the activation $g(x)$ is a sigmoid function. The neural network parameters can be trained by backward propagation using batch gradient descent algorithm.

1.2.3.3 Sparse array design using ML

The sparse array design for Capon beamforming aims to detect the source from a specified direction of arrival (DOA) considering the known directions of interferers. Antenna selection can be conceptualized as classification problems, where each antenna is categorized into two classes: "selected" and "discarded." The training data $\mathbf{x}$ and $\mathbf{y}$ under the two strategies can be generated either by enumeration or optimization described in Section 1.2.2 for every possible scenario, characterized by the DOA of the source signal, the number q and DOAs of interferences. Upon completing training, the Capon beamformer can initially sense the environment and extract feature data $\mathbf{x}$. Based on this information, the well-trained machine then determines the status of all antennas in practical applications. The selected antennas constitute the optimal sparse array corresponding to the operating environment. It is noted that the definition of the feature space $\mathscr{X}$ differs between the SVM and the ANN. A detailed description is provided below.

The feature space $\mathscr{X}$ of the SVM is defined as $\mathbf{x}_l = [\phi_s, \phi_1, \ldots, \phi_q]$, then the dimension of the feature space is $q+1$ with the feature value within the range of $[0, 180]$. The classification variable $y_i \in \{-1, 1\}, i = 1, \ldots, N$ with value -1

denoting discarded antenna and 1 selected. We train each antenna separately and obtain the SVM parameters α^n and $b^n, n = 1, \ldots, N$. Denote $\mathbf{x}$ as the sensed electromagnetic environment, we then predict the status of each antenna according to the following formula:

$$\sum_{l=1}^{L} \alpha_l^n y_l \kappa(\mathbf{x}_l, \mathbf{x}) + b^n \begin{cases} > 0 \text{ then } y = 1 \\ < 0 \text{ then } y = -1. \end{cases} \tag{1.11}$$

We employ a Gaussian kernel with bandwidth τ in this chapter, which is defined as

$$\kappa(\mathbf{x}_1, \mathbf{x}_2) = \exp\left(-\frac{\|\mathbf{x}_1 - \mathbf{x}_2\|_2^2}{\tau}\right). \tag{1.12}$$

The Gaussian kernel has a theoretical infinite VC dimension.

For the ANN algorithm, we define the feature space $\mathscr{X}$ as $\mathbf{x}_l = \{0, 1\}^{180\times 1}$ with value 1 indicating a signal arriving from the corresponding direction. Therefore, the dimension of the feature space for the ANN is 180, and the feature vector $\mathbf{x}$ is sparse, containing only "one" entries corresponding to the source and interference arrival angles. The classification variable is defined as $y_i \in \{0, 1\}$ with value 0 denoting discarded antenna and 1 selected. Different from the separate training of SVM, we train all the N antennas at the same time for the ANN and obtain the parameters θ. For a new feature input $\mathbf{x}$, the following classification is made:

$$y = g\left\{\mathbf{W}^{(4)}\left[\mathbf{W}^{(3)}[\mathbf{W}^{(2)}(\mathbf{W}^{(1)}\mathbf{x} + b^{(1)}) + b^{(2)}] + b^{(3)}\right] + b^{(4)}\right\}. \tag{1.13}$$

Randomized initialization of the parameters $\mathbf{W}$ and b is utilized and serves the purpose of symmetry breaking.

1.2.4 Simulations

In this section, we consider $K = 8$ available antennas and $N = 16$ uniformly spaced positions with an inter-element spacing of $d = \lambda/2$.

1.2.4.1 Validation of optimum sparse array

Assume that there are two interfering signals arriving from $\phi_1 = 58°, \phi_2 = 120°$ relative to the endfire direction with INR being 20 dB. A single source is impinging on the array from $\phi_s = 64°$ with SNR being 0 dB. Note that there are totally $C_{16}^8 = 12,870$ different sparse array configurations. We enumerate all the different sparse arrays for Capon beamforming and calculate the output SINR. The results are plotted in Figure 1.3 in ascending order. The structure of the optimum 8-antenna sparse array is presented in Figure 1.4. The difference in the SINR offerings of different sparse arrays is clearly seen in Figure 1.3. Hence, various array configurations significantly influence the performance of the Capon beamformer.

1.2.4.2 Validation of ML algorithms

Consider a signal arriving from the angular range $\Phi_s \in [60° \sim 120°]$ and two strong interfering signals arriving from the angular ranges $\Phi_1 \in [5° \sim 55°]$ and $\Phi_2 \in [125° \sim 175°]$, respectively. The 8-antenna optimum sparse array is constructed from a 16-antenna uniform linear array. Using either enumeration or

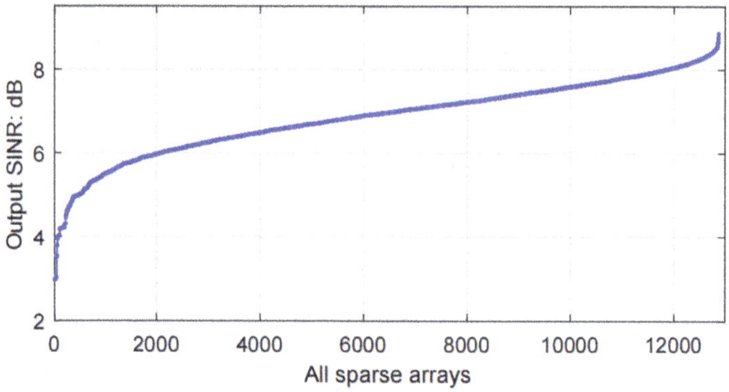

Figure 1.3 Output SINR of all sparse array Capon beamformers

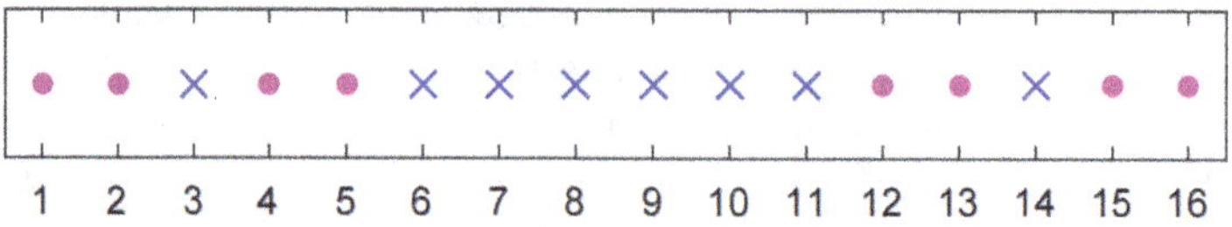

Figure 1.4 The optimum 8-antenna sparse array (a). Filled circles denote selected locations and crosses denote discarded.

optimization-based techniques, we calculate the optimal sparse array for each possible scenario and prepare the feature data according to the description in Section 1.2.3.3. Subsequently, both the SVM and ANN algorithms are trained with this data. Note that there are $N = 16$ independent SVM classifiers with parameters $[\alpha^n, b^n], n = 1, \ldots, N$, while there is only one ANN classifier with one parameter $\theta = [\mathbf{W}^{(1)}, b^{(1)}, \ldots, \mathbf{W}^{(4)}, b^{(4)}]$. The number of neurons for the first, second, and third hidden layers is $s_2 = 25$, $s_3 = 50$, and $s_4 = 25$, respectively. The input and output layers have $s_1 = 180$ and $s_5 = 16$ neurons. The feature space dimension of the SVM classifier is 3, i.e., $\mathbf{x}_i = [\phi_s, \phi_1, \phi_2]$, whereas the feature space dimension of the ANN classifier is 180, i.e., $\mathbf{x}_i \in \{0, 1\}^{180}$ with one entry corresponding to the DOAs of source and interferences. Figure 1.5 illustrates the training data of the first antenna. Clearly, distinguishing the two sets of points in the original 3-dimensional space is impossible. A series of transformations can be performed by both the SVM and ANN to project the data onto a higher dimensional feature space, where the two sets of points can be easily separated.

In Table 1.1, we present a comparison of the classification accuracy and computational time among the optimization, the ML-based method, and the table search. It is evident that both the SVM and ANN algorithms demonstrate significantly higher accuracy than the optimization method. While optimization does not guarantee the global optimum solution, it can yield a satisfactory sub-optimal solution, as demonstrated in [25,26]. Undoubtedly, the table search method always achieves 100 percent accuracy; however, its searching time will increase dramatically with the feature space dimension. Following complete training, the SVM and ANN only require simple matrix multiplication for classification, thus exhibiting much faster

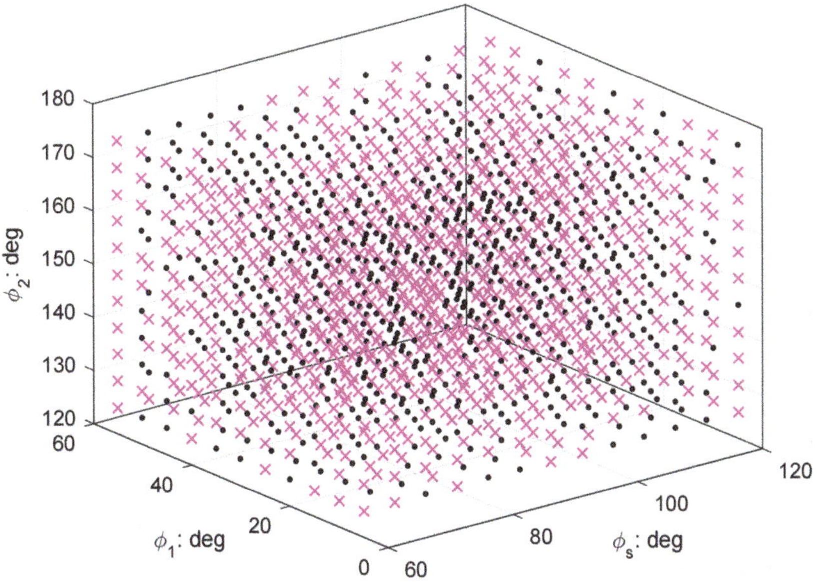

Figure 1.5 The illustration of training data for the first antenna using SVM. The cross denotes the label −1, while the dot denotes the label 1.

Table 1.1 The classification accuracy and computational time of four methods under the UAC and RAC

Method	Accuracy %	Computational time (s)
Opt	64.3	7.17
SVM	100	0.09
ANN	97.08	2.21e-4
Table	100	5.6

computational speed than the other two methods. As the kernel computation is computationally involved, the SVM is slower than the ANN, whereas the training time required by the ANN is much longer than that of the SVM. Adding more complicated hidden layers, such as convolution neurons, can further improve the accuracy of the ANN.

Next, a pragmatic scenario is considered where the estimated DOAs of the source and interferences do not exactly coincide with the training set. This scenario can occur due to possible biases and perturbations in the DOAs caused by platform motion. Assume that the estimated DOAs of the source and interferences are deviated from the training set of integer angles. We set them as $\phi_s = 65.5°$, $\phi_1 = 45.4°$, $\phi_2 = 125.6°$. The SVM is still capable of returning the true optimum sparse array, depicted in Figure 1.6(b). However, the table-search-based method approximates the off-grid angles to $\phi_s = 66°$, $\phi_1 = 45°$, $\phi_2 = 126°$ and returns the

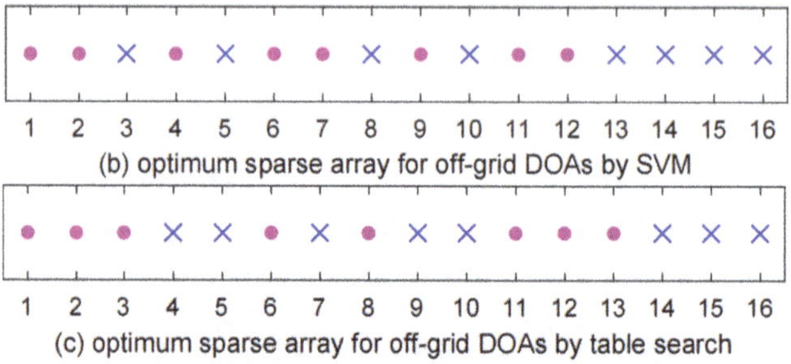

Figure 1.6 The 8-antenna sparse arrays returned by the SVM and table search

corresponding sparse array, as plotted in Figure 1.6(c). The robustness of the SVM-based antenna selection method against DOA estimate uncertainties is clearly demonstrated by the structural difference between the two sparse arrays (b) and (c).

1.3 CNN-based sparse array design for enhanced DOA estimation

In this section, we address the problem of sparse array design to enhance the accuracy of DOA estimation. Given that the accuracy of DOA estimation depends on the sparse array structure, optimizing the configuration of sparse receivers is necessary [12,34]. In the literature, various antenna selection strategies have been employed to optimize the array configuration for enhanced DOA estimation. These include the iterative reweighted l_1 algorithm [35], semi-definite programming (SDP) [36], and heuristic search algorithms [37]. These strategies require prior environmental information and demonstrate high computational complexity, rendering them challenging to respond quickly to dynamic environments [16]. The proposal of environment-independent deep learning (DL)-based sparse array design for DOA estimation under unknown dynamic environments has recently emerged [38]. Once trained with ample environmental data, neural networks can rapidly and accurately adapt to the new environment.

In this section, we suggest designing the sparse array by selecting the optimal receive antennas using CNN. We frame the sparse array design as a classification problem and concentrate on designing the CNN structure. The CRB of DOA estimation serves as an indicator for generating the corresponding dataset. Additionally, we introduce a new labeling strategy that significantly reduces the number of classes and enhances classification accuracy. The proposed CNN-based method can achieve precise DOA estimation for multi-target sensing in the presence of multipath under an unknown environment.

1.3.1 Mathematical model

Consider an M-element ULA receiver and there are K radar targets to be sensed. Then, the received signal at time instant t is given by

$$\mathbf{y}(t) = \mathbf{A}\mathbf{s}(t) + \mathbf{n}(t), \tag{1.14}$$

where $\mathbf{s}(t) \in \mathbb{C}^{K\times 1}$ is the reflection echoes of K targets, $\mathbf{A}_k \in \mathbb{C}^{M\times q_k}$ is defined as $\mathbf{A}_k = [\mathbf{a}(\theta_{k,1}), \ldots, \mathbf{a}(\theta_{k,q_k})] \in \mathbb{C}^{M\times q_k}$, and q_k is the number of multipaths from the k-th radar target to the receiver. The steering vector $\mathbf{a}(\theta_{k,i})$ is given as follows:

$$\mathbf{a}(\theta_{k,i}) = [1, e^{j\frac{2\pi}{\lambda}d\sin\theta_{k,i}}, \ldots, e^{j\frac{2\pi}{\lambda}(M-1)d\sin\theta_{k,i}}]^T, \tag{1.15}$$

where λ is the carrier wavelength, d is the inter-element spacing of the receiver, and $\theta_{k,i}$ is the incident angle of the k's target's i-th path. Matrix $\mathbf{B}$ is defined by

$$\mathbf{B} = \begin{bmatrix} \boldsymbol{\beta}_1 & \mathbf{0}_{q_1\times 1} & \cdots & \mathbf{0}_{q_1\times 1} \\ \mathbf{0}_{q_2\times 1} & \boldsymbol{\beta}_2 & \cdots & \mathbf{0}_{q_2\times 1} \\ \cdots & \cdots & \cdots & \cdots \\ \mathbf{0}_{q_K\times 1} & \mathbf{0}_{q_K\times 1} & \cdots & \boldsymbol{\beta}_K \end{bmatrix}, \tag{1.16}$$

where $\boldsymbol{\beta}_k \in \mathbb{C}^{q_k\times 1}$ is defined as $\boldsymbol{\beta}_k = [\beta_{k,1}, \ldots, \beta_{k,q_k}]^T$ with $\beta_{k,i}$ representing the propagation coefficient of the k-th target's i-th path.

Based on (1.14), if we select L antennas out of the M-element ULA to form a sparse receiver, then the signal received by the selected L antennas time instant t is given by

$$\widetilde{\mathbf{y}}(t) = \widetilde{\mathbf{A}}\mathbf{B}\mathbf{s}(t) + \widetilde{\mathbf{n}}(t) = \widetilde{\mathbf{A}}\bar{\mathbf{s}}(t) + \widetilde{\mathbf{n}}(t), \tag{1.17}$$

where matrix $\widetilde{\mathbf{A}} = [\widetilde{\mathbf{A}}_1, \ldots, \widetilde{\mathbf{A}}_K] \in \mathbb{C}^{L\times Q}$ is composed of the sparse steering vectors of all incident signals (including multipath signals), Q is the number of all incident signals which is defined as $Q = q_1 + \cdots + q_K$, $\widetilde{\mathbf{A}}_k = [\widetilde{\mathbf{a}}(\theta_{k,1}), \ldots, \widetilde{\mathbf{a}}(\theta_{k,q_k})]$ is composed of the L rows of $\mathbf{A}_k$ corresponding to the selected sparse elements, $\widetilde{\mathbf{a}}(\theta_{k,i})$ is the corresponding sparse steering vector selected from $\mathbf{a}(\theta_{k,i})$, and $\widetilde{\mathbf{n}}(t)$ is the zero-mean adaptive white Gaussian noise vector with covariance $\sigma_n^2\mathbf{I}_L$. Based on the definition of $\mathbf{B}$, $\bar{\mathbf{s}}(t)$ is the Q-dimensional incident signal vector defined as $\bar{\mathbf{s}}(t) = \mathbf{B}\mathbf{s}(t)$.

For the received signal $\widetilde{\mathbf{y}}(t)$, the data covariance matrix $\mathbf{R}_y$ is given by

$$\widetilde{\mathbf{R}}_y = \mathbb{E}\{\widetilde{\mathbf{y}}(t)\widetilde{\mathbf{y}}^H(t)\} = \widetilde{\mathbf{A}}\mathbf{R}_s\widetilde{\mathbf{A}}^H + \sigma_n^2\mathbf{I}_L, \tag{1.18}$$

where $\mathbb{E}\{\cdot\}$ represents the operation of taking expectation and $\mathbf{R}_s = \mathbb{E}\{\bar{\mathbf{s}}(t)\bar{\mathbf{s}}^H(t)\}$. If we assume that the signals of different communication users are independent, then $\mathbf{R}_s$ can be rewritten as

$$\mathbf{R}_s = \mathbf{B}\mathbb{E}\{\mathbf{s}(t)\mathbf{s}^H(t)\}\mathbf{B}^H = \mathbf{B}\mathbf{P}_s\mathbf{B}^H, \tag{1.19}$$

where $\mathbf{P}_s = \mathbb{E}\{\mathbf{s}(t)\mathbf{s}^H(t)\}$ is the reflection correlation matrix with the echoes' power of K target, i.e., P_k, $k = 1, \ldots, K$, populating the diagonal.

In practice, the data covariance matrix can be estimated by taking J discrete data samples, which is given by

$$\hat{\widetilde{\mathbf{R}}}_y = \frac{1}{J}\sum_{i=1}^{J}\widetilde{\mathbf{y}}(i)\widetilde{\mathbf{y}}^H(i), \tag{1.20}$$

where $\widetilde{\mathbf{y}}(i)$ is the i-th discrete sample of $\widetilde{\mathbf{y}}(t)$.

DOA estimation can then be conducted following the construction of $\hat{\widetilde{\mathbf{R}}}_y$. In this chapter, we employ the CRB to assess the accuracy of DOA estimation. The CRB represents the lower bound of estimation variance of any unbiased estimator and is widely used to evaluate sensing performance. In the multi-source scenario, the CRBs of estimated DOAs correspond to the diagonal elements of the inverse Fisher information matrix, symbolized as **F**. We denote the DOAs to be estimated as $\{\omega_1, \ldots, \omega_Q\}$ where ω_q is defined as $\omega_q = \frac{2\pi}{\lambda} d\sin\theta_q$ (θ_q is the incident angle of the q-th signal). Based on the signal model given in (1.17), the element of the i-th row and j-th column in matrix **F** can be calculated by

$$[\mathbf{F}]_{i,j} = J \times \text{trace}\left\{ \widetilde{\mathbf{R}}_y^{-1} \frac{\partial \widetilde{\mathbf{R}}_y}{\partial \omega_i} \widetilde{\mathbf{R}}_y^{-1} \frac{\partial \widetilde{\mathbf{R}}_y}{\partial \omega_j} \right\}. \tag{1.21}$$

Substituting (1.18), the CRB can be simplified as

$$\text{CRB} = \frac{\sigma_n^2}{2J}\{\text{Re}[\{\widetilde{\mathbf{D}}^H(\mathbf{I} - \widetilde{\mathbf{A}}(\widetilde{\mathbf{A}}^H\widetilde{\mathbf{A}})^{-1}\widetilde{\mathbf{A}}^H)\widetilde{\mathbf{D}}\}\mathbf{B}\mathbf{P}_s\mathbf{B}^H\}]\}^{-1}, \tag{1.22}$$

where $\odot$ represents the element-wise product and $\widetilde{\mathbf{D}} = [\widetilde{\mathbf{d}}_1, \ldots, \widetilde{\mathbf{d}}_Q]$ with vector $\widetilde{\mathbf{d}}_q$, $q = 1, \ldots, Q$ being defined as the first-order derivative of $\widetilde{\mathbf{a}}(\theta_q)$ with respect to ω_q, that is,

$$\widetilde{\mathbf{d}}_q = \frac{\partial \widetilde{\mathbf{a}}(\theta_q)}{\partial \omega_q}. \tag{1.23}$$

1.3.2 CNN for antenna selection

Equation (1.22) suggests that the accuracy of DOA estimation depends on the configuration of the sparse array, necessitating the selection of appropriate elements to construct the optimal sparse array. In this section, we propose a CNN-based network for designing the sparse receiver. Details regarding the generation of training data and the network structure of CNN are provided below.

1.3.2.1 Training data design for antenna selection

In this subsection, we outline the method for generating the training dataset. The training samples are employed to train the parameters of the CNN, encompassing input data and corresponding labels. Without loss of generality, we assume that the direction of the radar echoes remains constant for a certain period of time. Consequently, the optimal sparse array corresponding to the current environment remains unchanged for this duration. From this perspective, we can optimize the configuration of the sparse array based on the received data from the full M-element ULA and subsequently perform DOA estimation using the selected sparse array.

For the antenna selection problem, we use the data covariance matrix $\hat{\mathbf{R}}_y$ generated by all M receive antennas as the input of CNN, which can be estimated in maximum likelihood by J groups of discrete samples per (1.20). Utilizing a complementary subarray switching strategy [39], the M-dimensional received signal of the full ULA can be obtained. Additionally, leveraging the properties of the Toeplitz matrix further simplifies the data collection process. For more details, interested readers can consult [40].

We treat antenna selection as a classification problem, where each possible configuration of the sparse array is classified into a distinct category. Theoretically, there are $U = C_M^L$ different classes for selecting L from M receive antennas to construct the sparse array. We employ the trace of the CRB matrix, as presented in (1.22), as the evaluation metric for the DOA estimation accuracy of various sparse arrays. Within a specific environment, we define the evaluation function for the u-th sparse array as follows:

$$f(\boldsymbol{\Theta}, \mathbf{B}, u) = \text{trace}\{\text{CRB}\}, \tag{1.24}$$

where $\boldsymbol{\Theta}$ is the set composed of DOAs of all incident signals.

After establishing the training data and classification metrics, we need to address how to label the training samples. Given that the optimal sparse array minimizes the value of the evaluation function, the index of the sparse array that minimizes $f(\boldsymbol{\Theta}, \mathbf{B}, \iota_u)$ serves as the label for the current training sample. Directly considering all possible configurations of sparse arrays as the set of labels would result in a significant increase in the total number of classes U with the element number M, which would be detrimental to the classification accuracy. Indeed, the number of configurations that can be selected is less than U, as certain sparse configurations may be impossible to select (for instance, antennas at the two ends are easier to select than those in the internal positions). We define a new label set Γ_1 consisting of the indexes of all sparse arrays that can be selected. Suppose the length of Γ_1 is U_1, then Γ_1 is defined as

$$\Gamma_1 = \{\iota_1, \iota_2, \ldots, \iota_{U_1}\}, \tag{1.25}$$

where ι_u, $(u = 1, \ldots, U_1)$ is the class of sparse arrays that provide the highest DOA estimation accuracy under the scenario $(\boldsymbol{\Theta}, \mathbf{B})$ and is given by

$$\iota_u = \arg \min_{u=1,\ldots,U} f(\boldsymbol{\Theta}, \mathbf{B}, u). \tag{1.26}$$

To better illustrate the difference between U and U_1, we provide an example without considering multipath signals, and the results are given in Table 1.2. It can be seen from Table 1.2 that U_1 is smaller than U, especially when $Q = 1, 2$. However, with the increase in the number of incident signals, the value of U_1 also becomes very large, rendering it impractical to simplify the label set. Consequently, to ensure the accuracy of classification, we opt to refine the labeling criteria of the training samples further, aiming to reduce the number of classes. This is outlined as follows:

Table 1.2 Number of classes under different labeling strategies

L \ Q		$L = 3$	$L = 4$	$L = 5$	$L = 6$	$L = 7$	$L = 8$
	U	120	210	252	210	120	45
$Q = 1$	U_1	2	1	2	1	2	1
	U_2	1	1	1	1	1	1
$Q = 2$	U_1	35	13	14	12	13	7
	U_2	6	5	3	3	3	3
$Q = 3$	U_1	120	194	167	122	86	40
	U_2	47	34	31	23	16	9

In a given scenario, since there may be multiple sparse arrays that demonstrate good DOA estimation performance, selecting any one of them can ensure satisfactory results. Assuming that the number of all possible scenarios is T, for the τ-th scenario, we take the first C sparse arrays with the best DOA estimation accuracy as the candidate set, which is defined as Υ_τ. Any candidate in Υ_τ can be selected to label the τ-th scenario. We define the set of label classes for all T scenarios as Γ_2. And our goal is to find a labeling strategy that can minimize the number of elements in Γ_2. In this regard, we propose a new labeling strategy that can minimize the number of elements in Γ_2 as much as possible. First, we select the sparse array with the highest frequency appearing in all T sets (Υ_τ, $\tau = 1, \ldots, T$), denoted as ℓ_1. We utilize ℓ_1 to label all the scenarios, where the candidate set contains ℓ_1, and take ℓ_1 as the first element of Γ_2. Second, excluding the scenarios marked as ℓ_1, we select the sparse array that is most selected in the other scenarios and mark it as ℓ_2. Similarly, we label the corresponding scenarios with ℓ_2 and take it as the second element of the Γ_2. Following the above process, we can classify all scenarios and obtain the set of labels Γ_2. Assuming that the number of elements in Γ_2 is U_2, then Γ_2 can be represented as $\Gamma_2 = \{\ell_1, \ldots, \ell_{U_2}\}$. This process is illustrated more intuitively in Algorithm 1.1.

In Table 1.2, we provide simulations of the number of classes under different scenarios utilizing the new label strategy. We take $C = 10$, and the results show that compared with U and U_1, U_2 is much more smaller, which indicates the new labeling strategy can significantly reduce the number of classes.

After defining the input covariance matrix and introducing the corresponding labeling strategy, we label all T scenarios and generate P groups of training covariance matrices for each scenario. We record the pair $\{\hat{\mathbf{R}}_y^{(\tau,p)}, z^{(\tau)}\}$ (where $z^{(\tau)} \in \Gamma_2$ is the label of the τ-th scenario) for the p-th realization under the τ-th scenario as a group of training data. The steps for generating all training data are summarized in Algorithm 1.2.

1.3.2.2 The structure of CNN

In this subsection, we outline the structure of the proposed CNN for sparse receiver design. The CNN classifier consists of 10 layers, as illustrated in Figure 1.7. The first

---

Algorithm 1.1 The new labeling strategy

Initialization: $\Gamma_2 = \oslash$, $i = 1$, $\bar{\Omega} = \{1, \ldots, T\}$.
Input: C, T, Υ_τ ($\tau = 1, \ldots, T$).
Output: Γ_2.
repeat

1. Find the sparse array with the highest frequency among all unlabeled sets Υ_τ, $\tau \in \bar{\Omega}$, denoted as ℓ_i.
2. Label the scenario whose selectable set contains the ℓ_i-th sparse array as ℓ_i.
3. $\Gamma_2 = \Gamma_2 \cup \ell_i$.
4. Remove the labeled scenarios from $\bar{\Omega}$ and obtain the new unlabeled set.
5. $i = i + 1$.

until $\bar{\Omega} = \oslash$

Algorithm 1.2 Generation of training data

Input: Total number of receive antennas M, number of antennas to be selected L, number of scenarios T, number of snapshots J, number of realizations under each scenario P and selectable parameter C.

Output: Training data $\{\hat{\mathbf{R}}_y^{(\tau,p)}, z^{(\tau)}\}$ and training dataset.

1. Generate the candidate sets Υ_τ $(\tau = 1, \ldots, T)$ for all scenarios.
2. Label the scenarios according to **Algorithm 1.1**.

for the τ-th scenario, $\tau = 1, \ldots, T$ **do**

- The label of the τ-th scenario $z^{(\tau)}$.

for p = 1, ..., P **do**

1. Generate J snapshots of data received by the full receive array $\mathbf{y}(i)$ $(i = 1, \ldots, J)$, where the noise follows the Gaussian distribution and the signals of interest are generated according to the preset SNRs.
2. Calculate the sample data covariance matrix, $\hat{\mathbf{R}}_y^{(\tau,p)}$.
3. Obtain the training data $\{\hat{\mathbf{R}}_y^{(\tau,p)}, z^{(\tau)}\}$.

end for

end for

3. Construct the set of all training data,

$$\left\{\{\hat{\mathbf{R}}_y^{(1,1)}, z^{(1)}\}; \{\hat{\mathbf{R}}_y^{(1,2)}, z^{(1)}\}; \ldots; \{\hat{\mathbf{R}}_y^{(T,P)}, z^{(T)}\}\right\},$$

whose size is $T \times P$.

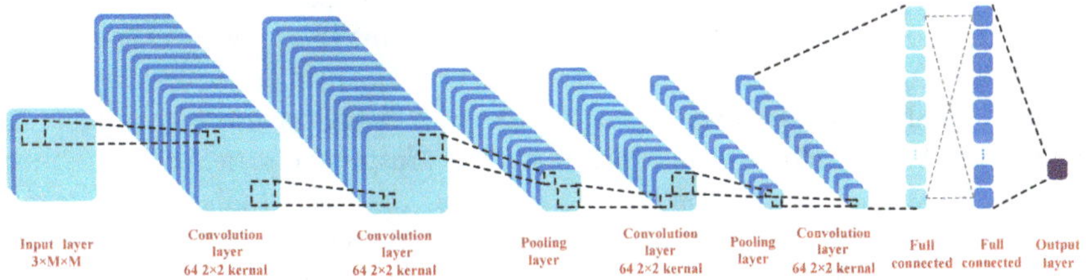

Figure 1.7 Structure of the proposed CNN for sparse receiver design

layer is the input layer, which consists of three real-valued channels, each with the size of $M \times M$. The input real-valued matrix, which is defined as $\mathbf{X} \in \mathbb{R}^{M\times M\times 3}$, is constructed as follows: the inputs of the first and second channels are defined as the real part and the imaginary part of the covariance matrix $\hat{\mathbf{R}}_y^{(\tau,p)}$, respectively, i.e., $[\mathbf{X}]_{:,:,1} = \text{Re}\{\hat{\mathbf{R}}_y^{(\tau,p)}\}$ and $[\mathbf{X}]_{:,:,2} = \text{Im}\{\hat{\mathbf{R}}_y^{(\tau,p)}\}$; that of the third channel is the angle of $\hat{\mathbf{R}}_y^{(\tau,p)}$, that is $[\mathbf{X}]_{:,:,3} = \angle\{\hat{\mathbf{R}}_y^{(\tau,p)}\}$. The second, third, fifth, and seventh layers consist of convolutional layers with 64 filters of size 2×2. The fourth and sixth layers are pooling layers, employed to reduce the dimension by 2. The eighth and ninth layers are fully connected layers, each comprising 1024 units. The final layer is the output layer, responsible for yielding the best sparse receiver by classifying the input data. Additionally, after each convolutional layer and fully connected layer, a rectified linear unit (ReLU) serves as the activation function for neurons.

1.3.3 Numerical results

In this section, we assess the effectiveness of the proposed CNN for sparse receiver design through numerical simulations.

Consider a ULA radar receiver composed of $M = 10$ elements with an inter-element spacing of half wavelength. We aim to select $L = 5$ antennas, and we are examining the angular interval of the incident signal, which spans from $0°$ to $90°$. The generation process of the training data is provided below. We divide the observation interval evenly into 90 grids with a grid size of $1°$ $(0°, 1°, \ldots, 89°)$. Assuming that the maximum number of signals incident to the receiver is Q, the total number of possible scenarios can be calculated as $T = C_{90}^{1} + \cdots + C_{90}^{Q}$. For each scenario, we consider $P = 100$ realizations according to Algorithm 1.2 and thus generate a total of $T \times 100$ groups of training data. Regarding the training of the CNN, we allocate 60% of the training data as the training dataset, another 20% as the validation dataset, and the remaining 20% as the testing dataset. We set the maximum number of incident signals Q as 2, 3, and 4, respectively, and generate the corresponding training data to train the CNN. The classification accuracy is provided in Table 1.3. As depicted in Table 1.3, the proposed CNN demonstrates commendable classification performance across various environmental conditions, achieving a probability of over 90% to select the optimal sparse receiver. As Q increases, reflecting more complex environmental conditions, there is a slight impact on the classification accuracy. However, this effect can be mitigated by augmenting the training data.

Next, we assess the DOA estimation accuracy of the sparse receiver selected by the CNN. We examine three scenarios for this evaluation. In Scenario 1, there are two incident signals ($Q = 2$) whose DOAs are $20°$ and $30°$. In Scenario 2, $Q = 3$ and the three incident signals come from $10°$, $30°$, and $50°$, respectively. In Scenario 3, there are four signals ($Q = 4$) coming from $15°$, $35°$, $50°$, and $80°$. In these three scenarios, we optimize the configurations of the sparse receiver using both the proposed CNN framework and the iterative reweighted approach proposed in [35]. The results are illustrated in Figure 1.8. Furthermore, we randomly select two sparse receivers under each scenario and compare the DOA estimation performance of the four sparse receivers. We utilize the high-resolution MUSIC algorithm to estimate the DOAs of the incident signals and conduct 10,000 Monte Carlo runs for statistical performance analysis. The curves depicting estimation mean square errors (MSEs) versus SNR

Table 1.3 Classification accuracy of training, validation, and testing datasets under different Q values

	Training dataset	**Validation dataset**	**Testing dataset**
Q=2	98.9%	97.5%	96.1%
Q=3	96.3%	95.3%	94.7%
Q=4	92.6%	91.9%	90.1%

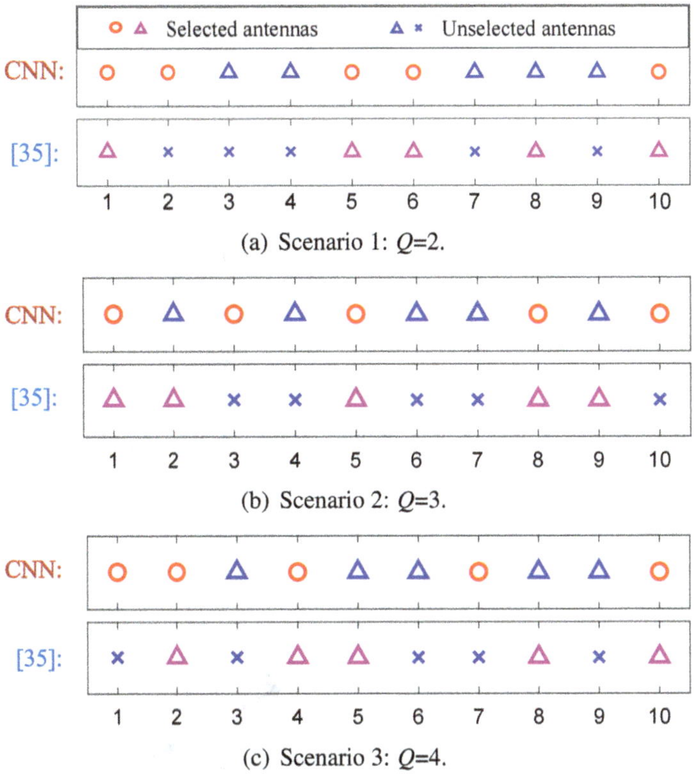

(a) Scenario 1: Q=2.

(b) Scenario 2: Q=3.

(c) Scenario 3: Q=4.

Figure 1.8 The configurations of selected sparse receivers via CNN and the approach in [35] under three scenarios

of incident signals using four sparse receivers under three scenarios are plotted in Figure 1.9. In this figure, the red dashed line represents the CRB of the globally optimal sparse receiver obtained via the exhaustive method. It is evident from the results that the DOA estimation accuracy of the sparse receiver selected by the proposed CNN surpasses that of the other three sparse receivers. In addition, the CNN-based method exhibits a greater likelihood of discovering the configuration that approaches the global optimal solution compared to the approach proposed in [35]. The numerical results affirm that the proposed CNN can select a sparse receiver with superior DOA estimation performance under various scenarios.

1.4 Conclusion

In this chapter, we explored machine learning-based antenna selection for sparse array reconfiguration.

In Section 1.2, we introduced the utilization of two principal machine learning algorithms, namely support vector machine and artificial neural network, to address

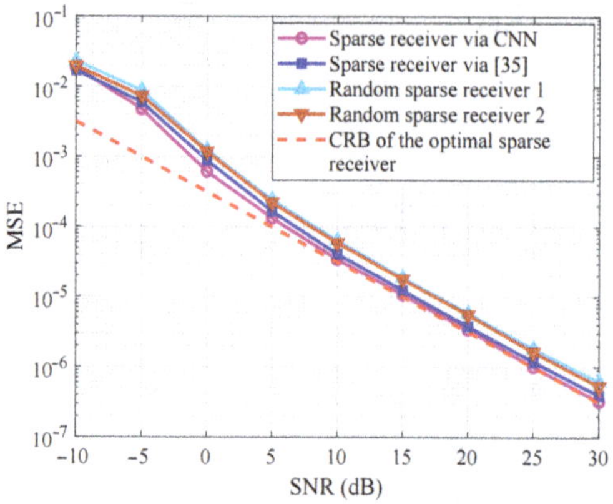

(a) Scenario 1: Q=2.

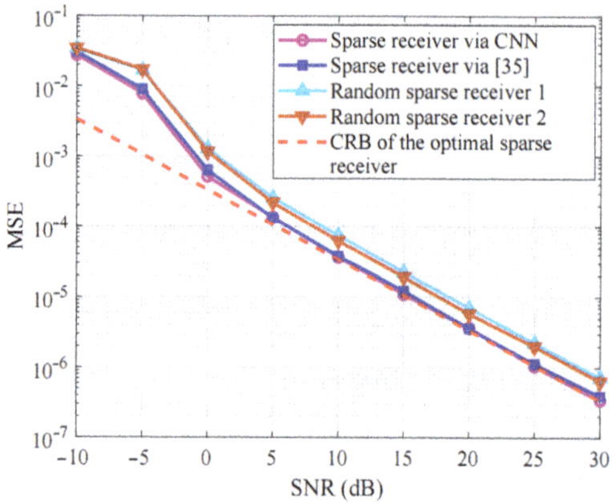

(b) Scenario 2: Q=3.

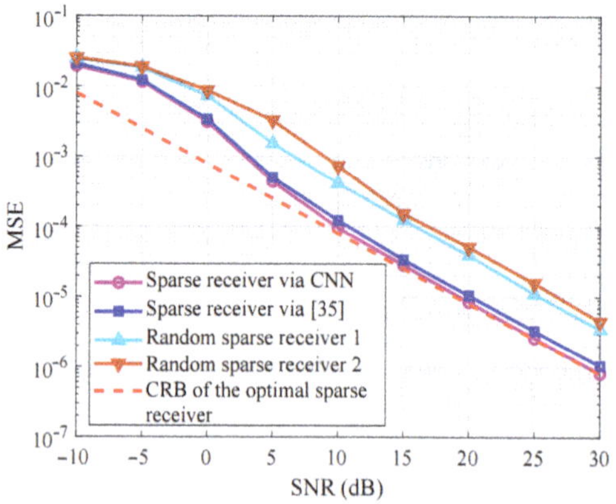

(c) Scenario 3: Q=4.

Figure 1.9 The curves of MSE versus SNR for different sparse receivers under three different scenarios combinatorial antenna selection problems for optimal sparse array design. Numerical examples showcased the efficacy of machine learning algorithms for optimal sparse array design. Their high accuracy, rapid computational speed, and resilience against DOA uncertainties underscored the ML algorithms as a desirable solution for adaptive sparse array reconfiguration in dynamic environments.

In Section 1.3, we introduced a CNN-based sparse receiver design approach to enhance DOA estimation. We structured the CNN to optimize the configuration of the sparse receiver and generated the dataset using the CRB as a criterion for training. By proposing a new labeling strategy, we reduced the number of classes, thereby enhancing the classification efficiency of the CNN. Numerical results illustrated that the proposed method efficiently designed the optimal sparse receiver for improved DOA estimation under unknown environments.

References

[1] L. E. Brennan and L. S. Reed, "Theory of adaptive radar," *IEEE Transactions on Aerospace and Electronic Systems*, vol. AES-9, no. 2, pp. 237–252, 1973.

[2] R. L. Fante and J. J. Vaccaro, "Wideband cancellation of interference in a GPS receive array," *IEEE Transactions on Aerospace and Electronic Systems*, vol. 36, no. 2, pp. 549–564, 2000.

[3] A. J. van der Veen, A. Leshem, and A. J. Boonstra, "Signal processing for radio astronomical arrays," in *Sensor Array and Multichannel Signal Processing Workshop Proceedings, 2004*, July 2004, pp. 1–10.

[4] R. Compton, "An adaptive array in a spread-spectrum communication system," *Proceedings of the IEEE*, vol. 66, no. 3, pp. 289–298, 1978.

[5] D. Johnson and S. DeGraaf, "Improving the resolution of bearing in passive sonar arrays by eigenvalue analysis," *IEEE Transactions on Acoustics, Speech, and Signal Processing*, vol. 30, no. 4, pp. 638–647, 1982.

[6] Y. Selén, H. Tullberg, and J. Kronander, "Sensor selection for cooperative spectrum sensing," in *2008 3rd IEEE Symposium on New Frontiers in Dynamic Spectrum Access Networks. DySPAN 2008*. Piscataway, NJ: IEEE; 2008, pp. 1–11.

[7] X. Shen and P. Varshney, "Sensor selection based on generalized information gain for target tracking in large sensor networks," *IEEE Transactions on Signal Processing*, vol. 62, no. 2, pp. 363–375, 2014.

[8] M. Hawes and W. Liu, "Sparse array design for wideband beamforming with reduced complexity in tapped delay-lines," *IEEE/ACM Transactions on Audio, Speech, and Language Processing*, vol. 22, no. 8, pp. 1236–1247, 2014.

[9] M. Wax and Y. Anu, "Performance analysis of the minimum variance beamformer," *IEEE Transactions on Signal Processing*, vol. 44, no. 4, pp. 928–937, 1996.

[10] A. Gorokhov, D. A. Gore, and A. J. Paulraj, "Receive antenna selection for MIMO spatial multiplexing: theory and algorithms," *IEEE Transactions on Signal Processing*, vol. 51, no. 11, pp. 2796–2807, 2003.

[11] A. Molisch and M. Win, "MIMO systems with antenna selection," *Microwave Magazine, IEEE*, vol. 5, no. 1, pp. 46–56, 2004.
[12] M. G. Amin, X. Wang, Y. D. Zhang, F. Ahmad, and E. Aboutanios, "Sparse arrays and sampling for interference mitigation and DOA estimation in GNSS," *Proceedings of the IEEE*, vol. 104, no. 6, pp. 1302–1317, 2016.
[13] A. G. Rodriguez, C. Masouros, and P. Rulikowski, "Efficient large scale antenna selection by partial switching connectivity," in *2017 42th IEEE International Conference on Acoustics, Speech and Signal Processing*, Mar 2017.
[14] P. Vaidyanathan and P. Pal, "Sparse sensing with co-prime samplers and arrays," *IEEE Transactions on Signal Processing*, vol. 59, no. 2, pp. 573–586, 2011.
[15] P. Pal and P. Vaidyanathan, "Nested arrays in two dimensions, part II: Application in two dimensional array processing," *IEEE Transactions on Signal Processing*, vol. 60, no. 9, pp. 4706–4718, 2012.
[16] X. Wang, E. Aboutanios, and M. G. Amin, "Adaptive array thinning for enhanced DOA estimation," *IEEE Signal Processing Letters*, vol. 22, no. 7, pp. 799–803, 2015.
[17] H. Gazzah and S. Marcos, "Cramer-Rao bounds for antenna array design," *IEEE Transactions on Signal Processing*, vol. 54, no. 1, pp. 336–345, 2006.
[18] H. Gazzah and K. Abed-Meraim, "Optimum ambiguity-free directional and omnidirectional planar antenna arrays for DOA estimation," *IEEE Transactions on Signal Processing*, vol. 57, no. 10, pp. 3942–3953, 2009.
[19] D. Bajovic, B. Sinopoli, and J. Xavier, "Sensor selection for event detection in wireless sensor networks," *IEEE Transactions on Signal Processing*, vol. 59, no. 10, pp. 4938–4953, 2011.
[20] A. Bertrand, J. Szurley, P. Ruckebusch, I. Moerman, and M. Moonen, "Efficient calculation of sensor utility and sensor removal in wireless sensor networks for adaptive signal estimation and beamforming," *IEEE Transactions on Signal Processing*, vol. 60, no. 11, pp. 5857–5869, 2012.
[21] O. Mehanna, N. Sidiropoulos, and G. Giannakis, "Joint multicast beamforming and antenna selection," *IEEE Transactions on Signal Processing*, vol. 61, no. 10, pp. 2660–2674, 2013.
[22] N. K. Dhingra, M. R. Jovanovic, and Z. Q. Luo, "An ADMM algorithm for optimal sensor and actuator selection," in *53rd IEEE Conference on Decision and Control*, Dec 2014, pp. 4039–4044.
[23] J. Ranieri, A. Chebira, and M. Vetterli, "Near-optimal sensor placement for linear inverse problems," *IEEE Transactions on Signal Processing*, vol. 62, no. 5, pp. 1135–1146, 2014.
[24] N. B. Mehta, S. Kashyap, and A. F. Molisch, "Antenna selection in LTE: from motivation to specification," *IEEE Communications Magazine*, vol. 50, no. 10, pp. 144–150, 2012.
[25] X. Wang, E. Aboutanios, M. Trinkle, and M. G. Amin, "Reconfigurable adaptive array beamforming by antenna selection," *IEEE Transactions on Signal Processing*, vol. 62, no. 9, pp. 2385–2396, 2014.

[26] X. Wang, E. Aboutanios, and M. G. Amin, "Slow radar target detection in heterogeneous clutter using thinned space-time adaptive processing," *IET Radar, Sonar & Navigation*, vol. 10, no. 4, pp. 726–734, 2015.
[27] X. Wang, M. Amin, X. Wang, and X. Cao, "Sparse array quiescent beamformer design combining adaptive and deterministic constraints," *IEEE Transactions on Antennas and Propagation*, vol. 65, no. 11, pp. 5808–5818, 2017.
[28] X. Wang, M. Amin, and X. Cao, "Analysis and design of optimum sparse array configurations for adaptive beamforming," *IEEE Transactions on Signal Processing*, vol. 66, no. 2, pp. 340–351, 2018.
[29] C. J. Burges, "A tutorial on support vector machines for pattern recognition," *Data Mining and Knowledge Discovery*, vol. 2, no. 2, pp. 121–167, 1998.
[30] M. A. Hearst, S. T. Dumais, E. Osuna, J. Platt, and B. Scholkopf, "Support vector machines," *IEEE Intelligent Systems and Their Applications*, vol. 13, no. 4, pp. 18–28, 1998.
[31] D. W. Patterson, *Artificial neural networks: theory and applications*. Prentice Hall PTR, 1998.
[32] X. Yao, "Evolving artificial neural networks," *Proceedings of the IEEE*, vol. 87, no. 9, pp. 1423–1447, 1999.
[33] J. Capon, "High-resolution frequency-wavenumber spectrum analysis," *Proceedings of the IEEE*, vol. 57, no. 8, pp. 1408–1418, 1969.
[34] E. Aboutanios, H. Nosrati, and X. Wang, "Online antenna selection for enhanced doa estimation," in *ICASSP 2021 – 2021 IEEE International Conference on Acoustics, Speech and Signal Processing (ICASSP)*, 2021, pp. 8468–8472.
[35] X. Wang, W. Zhai, X. Zhang, X. Wang, and M. G. Amin, "Enhanced automotive sensing assisted by joint communication and cognitive sparse MIMO radar," *IEEE Transactions on Aerospace and Electronic Systems*, vol. 59, no. 5, pp. 4782–4799, 2023.
[36] P. Chen, Z. Yang, Z. Chen, and Z. Guo, "Reconfigurable intelligent surface aided sparse DOA estimation method with non-ULA," *IEEE Signal Processing Letters*, vol. 28, pp. 2023–2027, 2021.
[37] B. Errasti-Alcala and R. Fernandez-Recio, "Performance analysis of metaheuristic approaches for single-snapshot DOA estimation," *IEEE Antennas and Wireless Propagation Letters*, vol. 12, pp. 166–169, 2013.
[38] D. Hu, Y. Zhang, L. He, and J. Wu, "Low-complexity deep-learning-based DOA estimation for hybrid massive MIMO systems with uniform circular arrays," *IEEE Wireless Communications Letters*, vol. 9, no. 1, pp. 83–86, 2020.
[39] X. Wang, M. S. Greco, and F. Gini, "Adaptive sparse array beamformer design by regularized complementary antenna switching," *IEEE Transactions on Signal Processing*, vol. 69, pp. 2302–2315, 2021.
[40] W. Zhai, X. Wang, S. A. Hamza, and M. G. Amin, "Cognitive-driven optimization of sparse array transceiver for MIMO radar beamforming," in *2021 IEEE Radar Conference (RadarConf21)*, 2021, pp. 1–6.

Chapter 2
Beampattern synthesis via the constrained subarray layout optimization

Xianxiang Yu[1], Lifang Feng[2], Guolong Cui[1] and Lingjiang Kong[1]

2.1 Introduction

Beampatterns synthesis involves designing the physical configuration and the weighting coefficients (WCs) of radiating elements (REs) or subarrays to achieve a specified array beampattern, which can significantly enhance the efficiency of antenna radiation or reception [1,2]. Therefore, it has remained a significant and enduring subject in radar [1] and wireless communications [2] research.

Traditional methods of beampattern synthesis typically concentrate on the configuration of REs and the formulation of WCs. Common approaches include convex optimization [3–6], the matrix pencil method [7–11], array response control algorithms [12–15], group sparse optimization (GSO) [16], iterative soft-thresholding-based optimization [17,18], and penalty function approaches within the alternating direction method of multipliers (ADMM) framework [19–21]. Other heuristic optimization techniques include particle swarm optimization [22], genetic algorithms [23], ant colony algorithms [24], and differential evolution algorithms [25]. Notably, the ADMM algorithm is highly favored for its distributed structure, convergence properties, and computational efficiency [26], making it extensively applied in beampattern synthesis [19–21] and radar waveform design [27–29].

With advancements in subarray technology, new studies on beampattern synthesis for arrays composed of multiple subarrays (ACMS) are emerging [30–47]. The design of ACMS aims to optimize the balance between performance and cost [30]. Typically, each subarray is a phased array comprising multiple REs, and for ACMS, it is unnecessary to equip every RE with a transmit/receive (TR) component and a data acquisition channel. This setup is only needed for each subarray, reducing overall system costs. Additionally, when ACMS is paired with thinned array techniques, the count of REs (or subarrays) and data channels can be further reduced to decrease expenses [30]. Moreover, each subarray might contain an equal number

[1]School of Information and Communication Engineering, University of Electronic Science and Technology of China, China
[2]College of Electronics and Information Engineering, Shenzhen University, China

of REs, which simplifies the modular design of large-aperture arrays [31,39]. Alternatively, having subarrays with varying numbers of REs allows for innovative array structures, such as the clustered phased array (also known as the subarrayed phased array) [30,31,34,37,44–46].

Similar to the traditional approach, the optimal beam performance for ACMS hinges on the strategic design of subarray layouts and WCs. Methods such as compressed sensing [31,32], genetic algorithms [33,35,41], contiguous partition techniques [36,44,46], and K-means clustering [34,37,45] have been employed mainly to tackle irregular subarray layouts, while the combined optimization of regular subarray layouts and WCs has been less frequently addressed in the literature.

Additionally, among the synthesis techniques discussed previously, there exists a norm minimization approach centered around weighting coefficients (WCs), which is extensively applied in optimizing thinned arrays [16,19,32]. However, this model struggles to enforce constraints related to the layout of subarrays (or REs). Furthermore, employing this approach tends to introduce a sparsity operation that can diminish the performance of the resulting beampattern.

In this chapter, we introduce a novel norm minimization model that is based on the layout vector. We then formulate an approximate algorithm using ADMM framework. This approach is designed to facilitate the beampattern synthesis for ACMRS, leveraging the robust convergence properties of ADMM.

Our key contributions are summarized as follows:

- We introduce a novel architecture for thinned ACMRS, along with the associated subarray layout and beampattern models. Unlike the subarrayed phased arrays that consist of irregular subarrays [30], our model is designed to facilitate beampattern synthesis for both thinned ACMRS and conventional thinned arrays.
- We develop a new norm minimization approach based on the layout vector. This model is an improvement over the norm minimization model based on WC [16, 19,32] because it more effectively enforces layout constraints. Building on this, we define a hybrid optimization challenge that involves the discrete layout vector and WCs, aiming to minimize the peak sidelobe level (PSL) and the norm of the layout vector under the constraints of power beampattern and discrete subarray layout.
- We propose a new approximation algorithm based on alternating direction method of multipliers (AA-ADMM) to solve the resulting highly non-convex hybrid optimization problem. We resort to the ADMM framework to decompose the resultant problem into three tractable optimization sub-problems and solve them iteratively. Finally, numerical simulation and full-wave simulation are provided to assess the effectiveness and reliability of the proposed method.

The remainder of this chapter is organized as follows. Section 2.2 describes the modeling of a thinned ACMRS. Section 2.3 outlines the formulation of the beampattern synthesis problem. Section 2.4 introduces the newly developed AA-ADMM method to solve the complex optimization problem. Section 2.5 presents various synthesis results to assess the performance of our proposed method. Finally, Section 2.6 concludes with remarks and directions for future research.

2.2 Array model

In this section, we construct a thinned array that consists of N_H regular subarrays, as depicted in Figure 2.1(a). These subarrays are arranged in a linear grid array containing M ($M > N_H$) grids, as illustrated in Figure 2.1(b). with each subarray occupying a single grid, as demonstrated in Figure 2.1(b) and (c). All subarrays utilize a uniform linear array structure, comprising M_H REs. The distance between consecutive REs is $d = d_0/M_H$ where d_0 represents the size of each grid. The WC of each RE can be adjusted individually. For instance, in Figure 2.1, the values are $N_H = 5$, $M = 11$ and $M_H = 4$.

Within these M grids, certain grids are reserved and not filled by subarrays. These reserved grids form contiguous sequences of P grids, and we define them as Reserved Successive Grid (RSG), as illustrated in Figure 2.1(b) and (c). Next, we consider N_L such RSGs, with each RSG containing P grids. It is clear that the values of M, N_H, N_L, and P must meet the relationship $N_H + N_L P = M$. For instance, in Figure 2.1, we have $N_L = 3$ and $P = 2$. Some observations about the array structure and the RSGs are important to note.

- The aim of the proposed array architecture is to fulfill the operational needs of a modular dual-band shared-aperture array. First, using regular subarrays as the fundamental unit enhances the modular design of large-scale arrays. Second, these reserved successive grids (RSGs) can be utilized to install subarrays for other frequency bands, making the entire array a dual-band shared-aperture system [48–50]. The configuration and number of subarrays within the linear grid

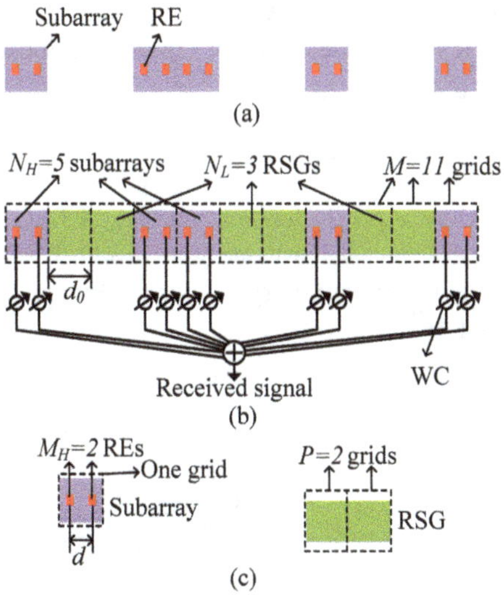

Figure 2.1 Array architecture example: (a) thinned array, (b) subarray layout rules, and (c) subarray and RSG sizes

array can be adjusted flexibly based on specific requirements. This concept is influenced by the phased array radar SEA FIRE 500 [51].

- The use of the thinned array architecture helps decrease the number of REs and T/R modules, thereby reducing costs.
- When the band of subarrays placed in RSGs is lower than that occupied in other grids, P is typically more than 1. This is due to the fact that subarrays operating at lower frequencies generally have larger dimensions [1,2], making $P > 1$ a reasonable choice.
- If the number of RSGs is 0, the whole array degenerates into a uniform linear array (ULA) [1].

Next, we will derive the model for the subarray layout and the formula for the beampattern of this array.

2.2.1 Subarray layout model

We initially introduce a layout vector (LV), $\mathbf{b} \in \mathbb{R}^{M\times 1}$, defined as

$$\mathbf{b} = [b_1, b_2, \ldots, b_M]^{\mathrm{T}}, b_m \in \{0, 1\}, \tag{2.1}$$

where $m = 1, 2, \ldots, M$. Here, $b_m = 0$ indicates that the mth grid is filled by a subarray, whereas $b_m = 1$ implies that the mth grid is part of an RSG. For example, the LV in Figure 2.1 is $\mathbf{b} = [0, 1, 1, 0, 0, 1, 1, 0, 1, 1, 0]^{\mathrm{T}}$. Furthermore, the total number of subarrays must satisfy $\sum_{m=1}^{M} (1 - b_m) = N_H$.

To impose constraints on $\mathbf{b}$, we first identify the candidate RSGs (CRSGs) as the successive P grids that might be RSGs. For a grid array with M grids, there are $\hat{M}$ ($\hat{M} = M - P + 1$) CRSGs, as illustrated in Figure 2.2. Subsequently, we define a layout auxiliary vector, $\mathbf{e} \in \mathbb{R}^{\hat{M}\times 1}$, as

$$\mathbf{e} = [e_1, e_2, \ldots, e_{\hat{M}}]^{\mathrm{T}}, e_{\hat{m}} \in \{0, 1\}, \tag{2.2}$$

where $\hat{m} = 1, \ldots, \hat{M}$. Here $e_{\hat{m}} = 1$ denotes that $\hat{m}$th CRSG is designated as an RSG and $e_{\hat{m}} = 0$ indicates that at least one grid in $\hat{m}$th CRSG is filled by a subarray. For example, the layout auxiliary vector in Figure 2.1 can be given by $\mathbf{e} = [0, 1, 0, 0, 0, 1, 0, 0, 1, 0]^{\mathrm{T}}$.

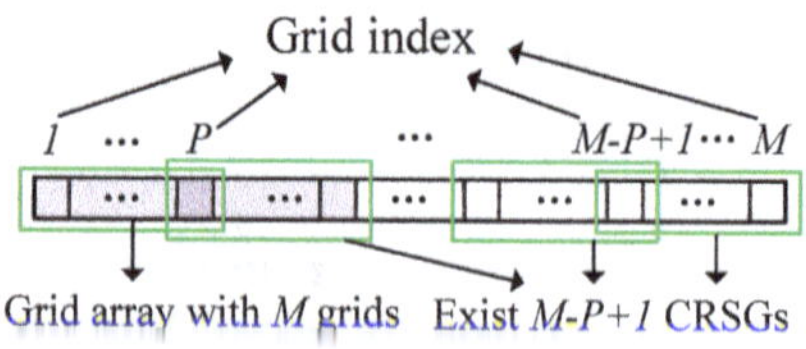

Figure 2.2 $\hat{M}$ CRSGs in grid array with M grids

In line with the definitions above, the element $e_{\hat{m}}$ of $\mathbf{e}$ is

$$e_{\hat{m}} = \prod_{r=0}^{P-1} b_{\hat{m}+r}, \tag{2.3}$$

where $\hat{m} = 1, \ldots, \hat{M}$. In any adjacent P CRSGs, a maximum of one CRSG may be designated as an RSG. Therefore, the variables, $e_{\hat{m}}$ ($\hat{m} = 1, \ldots, \hat{M}$) must fulfill

$$\sum_{x=\hat{x}-P+1}^{\hat{x}} e_x = 1 \text{ or } 0, \tag{2.4}$$

where $\hat{x} = P, \ldots, \hat{M}$. Additionally, the total number of RSGs should satisfy $\sum_{c=1}^{\hat{M}} e_{\hat{m}} = N_L$.

2.2.2 *Thinned array beampattern*

The beampattern $f(\theta)$ of the proposed array is described by

$$f(\theta) = \mathbf{w}^{\dagger}\mathbf{C}[\mathbf{g}(\theta) \odot \tilde{\mathbf{a}}(\theta)] = \mathbf{w}^{\dagger}\mathbf{a}(\theta), \tag{2.5}$$

where $\mathbf{w} = [w_1, \ldots, w_{MM_H}]^{\mathrm{T}}$, with w_m as the WC of the mth RE ($m = 1, \ldots, MM_H$). The array vector $\mathbf{a}(\theta) = \mathbf{C}[\mathbf{g}(\theta) \odot \tilde{\mathbf{a}}(\theta)]$, with $\mathbf{C}$ as the mutual coupling matrix (MCM) and $[\mathbf{C}]_{m,n}$ as the mutual coupling coefficient from the mth RE to the nth element. The vector $\mathbf{g}(\theta) = [g_1(\theta), \ldots, g_M(\theta)]^{\mathrm{T}}$, where $g_m(\theta)$ is the beampattern of the mth RE. The steering vector $\tilde{\mathbf{a}}(\theta)$ is

$$\tilde{\mathbf{a}}(\theta) = [e^{j\frac{2\pi d_1 \sin(\theta)}{\lambda}}, e^{j\frac{2\pi d_2 \sin(\theta)}{\lambda}}, \ldots, e^{j\frac{2\pi d_{MM_H} \sin(\theta)}{\lambda}}]^{\mathrm{T}}, \tag{2.6}$$

where λ represents the wavelength. The distance d_m denotes between the mth RE and the first RE is $d_m = (m-1)d$. Concerning (2.5), several notes are important:

- In the thinned array beampattern model, the WC (i.e., $\mathbf{w}$) is influenced by LV (i.e., $\mathbf{b}$), and we will detail their interrelationship in the subsequent section.
- $\mathbf{C}$ is the mutual coupling matrix of entire array (i.e., $\mathbf{b} = 0$) without any weighting (i.e., $\mathbf{w} = \mathbf{i}_{MM_H}$), and it can be derived from experimental methods or full-wave simulations [52]. This matrix is assumed known in this chapter.
- To ensure a modular design, each RE maintains the same beampattern, i.e., $g_1(\theta) = g_2(\theta) = \cdots = g_M(\theta)$. Furthermore, to lower PSL in the real array's beampattern, we can utilize the isotropic RE model (i.e., $g_1(\theta) = 1$) during the synthesis.

Overall, the beampattern of the entire thinned array is influenced by both the layout vector (including the layout and number of subarrays) and the WC. Next, we will examine how to achieve beampattern synthesis for the proposed array by simultaneously designing the LV and WC.

2.3 Problem formulation

In this section, we will formulate a multi-constrained optimization problem with respect to beampattern synthesis.

2.3.1 *Power beampattern constraints*

(1) Mainlobe:

$$\gamma - \delta \leq |\mathbf{w}^{\dagger}\mathbf{a}(\theta)|^2 \leq \gamma + \delta,\ \theta \in \Phi_m, \tag{2.7}$$

where Φ_m is the discrete angle set at mainlobe region. γ and δ are the desired mainlobe power level and the maximum allowable power fluctuation, respectively.

(2) Sidelobe:

$$|\mathbf{w}^{\dagger}\mathbf{a}(\theta)|^2 \leq t_H, \theta \in \Phi_s, t_H \leq \hat{t}_H, \tag{2.8}$$

where t_H is the desired sidelobe power level. Φ_s is the discrete angle set at sidelobe region and $\hat{t}_H$ is a preset value.

2.3.2 *Subarray layout constraints*

To constrain subarray layout by using WC, we propose a joint constraint of WC and LV which is

$$\mathbf{w} = \beta \odot [(\mathbf{i}_M - \mathbf{b}) \otimes \mathbf{i}_{M_H}], \tag{2.9}$$

where $\beta \in \mathbb{C}^{MM_H \times 1}$ and the equation also gives a coupling relationship between WC and LV. Besides, the constraints for the number of subarrays is

$$N_l \leq M - \mathbf{i}_M^{\mathrm{T}}\mathbf{b} \leq N_u, \tag{2.10}$$

where N_l and N_u are the lower and upper bounds of the number of subarrays, respectively.

2.3.3 *Optimization problem*

The beampattern synthesis can be formulated as the following optimization problem:

$$\begin{aligned}
\min_{\mathbf{b},\beta}\quad & t_H + \alpha\|\mathbf{i}_M - \mathbf{b}\|_1 \\
\text{s.t.}\quad & \gamma - \delta \leq |\mathbf{w}^{\dagger}\mathbf{a}(\theta)|^2 \leq \gamma + \delta,\ \theta \in \Phi_m, \\
& |\mathbf{w}^{\dagger}\mathbf{a}(\theta)|^2 \leq t_H, \theta \in \Phi_s, t_H \leq \hat{t}_H, \\
& \mathbf{w} = \beta \odot [(\mathbf{i}_M - \mathbf{b}) \otimes \mathbf{i}_{M_H}], \\
& N_l \leq M - \mathbf{i}_M^{\mathrm{T}}\mathbf{b} \leq N_u, \\
& e_{\hat{m}} = \prod_{r=0}^{P-1} b_{\hat{m}+r},\ \hat{m} = 1, \ldots, \hat{M}, \\
& \sum_{x=\hat{x}-P+1}^{\hat{x}} \hat{e}_x = 1 \text{ or } 0,\ \hat{x} = P, \ldots, \hat{M}, \\
& b_m = 1 \text{ or } 0,\ m = 1, \ldots, M, \\
& e_{\hat{m}} = 1 \text{ or } 0,\ \hat{m} = 1, \ldots, \hat{M},
\end{aligned} \tag{2.11}$$

where α is a trade-off factor between t_H and $\|\mathbf{i}_M - \mathbf{b}\|_1$. For the modeled problem, some remarks are necessary:

- This problem aims to minimize the number of subarrays and the PSL of the beampattern by designing $\mathbf{b}$ and β simultaneously, under the constraints of the power pattern, WC, and LV.
- If $N_l = N_u$, this problem also represents the beampattern synthesis of a thinned array with a fixed number of subarrays, making the trade-off factor α inactive.
- $\|\mathbf{i}_M - \mathbf{b}\|_1$ in the cost function is a norm model based on LV, differing from the norm model based on WC [16,19,32]. It better facilitates the constraint on array layout (e.g., (2.1)–(2.4)), whereas for the norm model based on WC, it is more challenging.

To avoid the continuous multiplication operation of the fifth constraint in problem (2.11), we present the following Proposition 1.

Proposition 1. *Suppose that* $\Gamma = [\gamma_1, \ldots, \gamma_M]^{\mathrm{T}} \in \mathbb{R}^{M\times\hat{M}}$, where $\gamma_m \in \mathbb{R}^{\hat{M}\times 1}$ is the mth column vector of Γ. *When b_m meets (2.1) and $e_{\hat{m}}$ satisfies (2.2), (2.3), and (2.4) simultaneously, $\boldsymbol{b}$ is*

$$\mathbf{b} = \Gamma\mathbf{e}, \tag{2.12}$$

where γ_m can be described by (2.13) and $m = 1, \ldots, M$.

$$\gamma_m = \begin{cases} \sum\limits_{\hat{m}=m-P+1}^{m} \mathbf{q}_{\hat{m}}, & P \le m \le \hat{M}, \\ \sum\limits_{\hat{m}=1}^{m} \mathbf{q}_{\hat{m}}, & 1 \le m < P, \\ \sum\limits_{\hat{m}=m-P+1}^{\hat{M}} \mathbf{q}_{\hat{m}}, & \hat{M} < m \le M, \end{cases} \tag{2.13}$$

where $\boldsymbol{q}_{\hat{m}} \in \mathbb{R}^{\hat{M}\times 1}$. The $\hat{m}$th element of $\boldsymbol{q}_{\hat{m}}$ is 1 and others are 0, i.e., it is given by $\mathbf{q}_{\hat{m}} = [0, \ldots, 0, \underbrace{1}_{m\text{th element}}, 0, \ldots, 0]^{\mathrm{T}}$, where $m = 1, \ldots, \hat{M}$.

Proof. Based on (2.1), (2.2), (2.3), we obtain the following equation:

$$b_m = \begin{cases} \prod\limits_{\hat{m}=m-P+1}^{m} e_{\hat{m}}, & P \le m \le \hat{M}, \\ \prod\limits_{\hat{m}=1}^{m} e_{\hat{m}}, & 1 \le m < P, \\ \prod\limits_{\hat{m}=m-P+1}^{\hat{M}} e_{\hat{m}}, & \hat{M} < m \le M, \end{cases} \tag{2.14}$$

where $m = 1, 2, \ldots, M$. This equation indicates that the state of each grid can be determined by all CRSGs associated with it.

Given that at most one CRSG in the adjacent P CRSGs can be designated as an RSG (i.e., (2.4) holds), b_m can be further expressed as

$$b_m = \begin{cases} \sum_{\hat{m}=m-P+1}^{m} e_{\hat{m}}, & P \leq m \leq \hat{M}, \\ \sum_{\hat{m}=1}^{m} e_{\hat{m}}, & 1 \leq m < P, \\ \sum_{\hat{m}=m-P+1}^{\hat{M}} e_{\hat{m}}, & \hat{M} < m \leq M, \end{cases} \\ = \gamma_m^{\mathrm{T}} \mathbf{e} \tag{2.15}$$

where γ_m is defined by (2.13) and $m = 1, 2, \ldots, M$. Consequently, $\mathbf{b}$ can be represented by (2.12), proving Proposition 1.

By substituting $\mathbf{b} = \Gamma\mathbf{e}$ into (2.9), we derive

$$\mathbf{w} = \beta \odot [(\mathbf{i}_M - \mathbf{b}) \otimes \mathbf{i}_{M_H}] = \beta \odot (\mathbf{a}_e - \mathbf{B}_e\mathbf{e}), \tag{2.16}$$

where $\mathbf{a}_e = \mathbf{E}_e\mathbf{i}_M$, $\mathbf{B}_e = \mathbf{E}_e\Gamma$, and $\mathbf{E}_e = \mathbf{E} \otimes \mathbf{i}_{M_H}$. Here $\mathbf{E}$ is an $M \times M$ dimensional identity matrix. Therefore, problem (2.11) can be equivalently transformed into

$$\begin{aligned} \min_{\mathbf{e},\beta} \quad & t_H + \alpha\|\mathbf{i}_M - \Gamma\mathbf{e}\|_1 \\ \text{s.t.} \quad & \gamma - \delta \leq |[\beta \odot (\mathbf{a}_e - \mathbf{B}_e\mathbf{e})]^\dagger \mathbf{a}_m|^2 \leq \gamma + \delta, \\ & |[\beta \odot (\mathbf{a}_e - \mathbf{B}_e\mathbf{e})]^\dagger \mathbf{a}_s|^2 \leq t_H, \ t_H \leq \hat{t}_H, \\ & N_l \leq M - \mathbf{i}_M^{\mathrm{T}}\Gamma\mathbf{e} \leq N_u, \\ & \hat{\mathbf{q}}_{\hat{x}}^{\mathrm{T}}\mathbf{e} = 1 \text{ or } 0, \ \hat{x} = P, \ldots, \hat{M}, \\ & e_{\hat{m}} = 1 \text{ or } 0, \ \hat{m} = 1, \ldots, \hat{M}, \end{aligned} \tag{2.17}$$

where $\mathbf{a}_m = \mathbf{a}(\theta_m)$ $(\theta_m \in \Phi_m, \ m = 1, \ldots, M_0)$ and $\mathbf{a}_s = \mathbf{a}(\theta_s)$ $(\theta_s \in \Phi_s, \ s = 1, \ldots, S_0)$. Here M_0 and S_0 are the number of sampling points at the mainlobe and sidelobe regions, respectively. $\hat{\mathbf{q}}_{\hat{x}} = \sum_{\hat{m}=\hat{x}-P+1}^{\hat{x}} \mathbf{q}_{\hat{m}}$ and the fourth constraint of problem (2.17) represents the vector form of equation (2.3).

Finally, both $\mathbf{e}$ and β can be determined by solving problem (2.17). The values of $\mathbf{b}$ and $\mathbf{w}$ can be obtained using (2.12) and (2.9), respectively.

2.4 Optimization method

In this section, we develop a novel approximation algorithm using the AA-ADMM to address the optimization problem outlined in (2.17).

Given that the first constraint in problem (2.17) is a two-sided quadratic constraint and the fourth and fifth constraints are discrete combinatorial constraints, problem (2.17) is classified as a highly non-convex hybrid optimization problem, which is typically challenging to solve directly. To facilitate the solution process, we first introduce a set of auxiliary variables to decouple the optimization variables from the two-sided quadratic constraint. Then, leveraging the ADMM framework, we alternate between solving for the continuous variable β and the discrete variable $\mathbf{e}$. To accurately determine $\mathbf{e}$, we devise an approximate algorithm (AA) and integrate it within the ADMM framework.

We introduce two new auxiliary variables, $\mathbf{y} = [y_1, \ldots, y_{M_0}]$ and $\mathbf{z} = [z_1, \ldots, z_{S_0}]$, and reformulate the first and second constraints of problem (2.17) into linear forms involving β or $\mathbf{e}$. Consequently, problem (2.17) can be equivalently transformed into:

$$\begin{aligned} \min_{\mathbf{e},\beta,y_m,z_s} \quad & t_H + \alpha\|\mathbf{i}_M - \Gamma\mathbf{e}\|_1 \\ \text{s.t.} \quad & y_m = [\beta \odot (\mathbf{a}_e - \mathbf{B}_e\mathbf{e})]^\dagger \mathbf{a}_m,\ m = 1, \ldots, M_0, \\ & z_s = [\beta \odot (\mathbf{a}_e - \mathbf{B}_e\mathbf{e})]^\dagger \mathbf{a}_s,\ s = 1, \ldots, S_0, \\ & \gamma - \delta \le |y_m|^2 \le \gamma + \delta,\ \ m = 1, \ldots, M_0, \\ & |z_s|^2 \le t_H,\ s = 1, \ldots, S_0,\ t_H \le \hat{t}_H, \\ & N_l \le M - \mathbf{i}_M^{\mathrm{T}}\Gamma\mathbf{e} \le N_u, \\ & \hat{\mathbf{q}}_{\hat{x}}^{\mathrm{T}}\mathbf{e} = 1 \text{ or } 0,\ \hat{x} = P, \ldots, \hat{M}, \\ & e_{\hat{m}} = 1 \text{ or } 0,\ \hat{m} = 1, \ldots, \hat{M}. \end{aligned} \tag{2.18}$$

where $\mathbf{A}_m = [\mathbf{a}_1, \ldots, \mathbf{a}_{M_0}]$ and $\mathbf{A}_s = [\mathbf{a}_1, \ldots, \mathbf{a}_{S_0}]$. $\kappa = [\kappa_1, \ldots, \kappa_{M_0}]^{\mathrm{T}} \in \mathbb{C}^{M_0}$ and $\iota = [\iota_1, \ldots, \iota_{S_0}]^{\mathrm{T}} \in \mathbb{C}^{S_0}$ are dual variables. The mainlobe and sidelobe residuals are $\varepsilon_m = \|\mathbf{y}^{\mathrm{T}} - \mathbf{A}_m^\dagger[\beta \odot (\mathbf{a}_e - \mathbf{B}_e\mathbf{e})]\|_\infty$ and $\varepsilon_s = \|\mathbf{z}^{\mathrm{T}} - \mathbf{A}_s^\dagger[\beta \odot (\mathbf{a}_e - \mathbf{B}_e\mathbf{e})]\|_\infty$, respectively. The penalty factors for these residuals are ρ_1 and ρ_2, respectively.

At the kth iteration, suppose the optimization variables are β^k, $\mathbf{y}^k$, t_H^k, $\mathbf{z}^k$, $\mathbf{e}^k$, κ^k, and ι^k, respectively. The variables for the $k+1$th iteration, β^{k+1}, $\mathbf{y}^{k+1}$, t_H^{k+1}, $\mathbf{z}^{k+1}$, $\mathbf{e}^{k+1}$, κ^{k+1}, and ι^{k+1} at the $(k+1)$th can be determined by the following equations:

$$\beta^{k+1} := \arg\min_{\beta} \mathbb{L}(t_H^k, \mathbf{e}^k, \beta, \mathbf{y}^k, \mathbf{z}^k, \kappa^k, \iota^k), \tag{2.19}$$

$$\begin{aligned} \mathbf{y}^{k+1} &:= \arg\min_{\mathbf{y}} \mathbb{L}(t_H^k, \mathbf{e}^k, \beta^{k+1}, \mathbf{y}, \mathbf{z}^k, \kappa^k, \iota^k), \\ & \quad \text{s.t. } \gamma - \delta \le |y_m|^2 \le \gamma + \delta, m = 1, \ldots, M_0, \end{aligned} \tag{2.20}$$

$$\begin{aligned} \{t_H^{k+1}, \mathbf{z}^{k+1}\} &:= \arg\min_{t_H,\mathbf{z}} \mathbb{L}(t_H, \mathbf{e}^k, \beta^{k+1}, \mathbf{y}^k, \mathbf{z}^k, \kappa^k, \iota^k), \\ & \quad \text{s.t. } |z_s|^2 \le t_H,\ t_H \le \hat{t}_H, s = 1, \ldots, S_0, \end{aligned} \tag{2.21}$$

$$\begin{aligned} \mathbf{e}^{k+1} := \arg\min_{\mathbf{e}} \ & \mathbb{L}(t_H^k, \mathbf{e}, \beta^{k+1}, \mathbf{y}^k, \mathbf{z}^k, \kappa^k, \iota^k), \\ \text{s.t. } & N_l \le M - \mathbf{i}_M^{\mathrm{T}}\Gamma\mathbf{e} \le N_u, \\ & \hat{\mathbf{q}}_{\hat{x}}^{\mathrm{T}}\mathbf{e} = 1 \text{ or } 0,\ \hat{x} = P, \ldots, \hat{M}, \\ & e_{\hat{m}} = 1 \text{ or } 0,\ \hat{m} = 1, \ldots, \hat{M}, \end{aligned} \tag{2.22}$$

$$\kappa^{k+1} := \kappa^k + \rho_1(\mathbf{y}^{k+1} - \mathbf{A}_m^\dagger[\beta^{k+1} \odot (\mathbf{a}_e - \mathbf{B}_e\mathbf{e}^{k+1})]), \tag{2.23}$$

$$\iota^{k+1} := \iota^k + \rho_2(\mathbf{z}^{k+1} - \mathbf{A}_s^\dagger[\beta^{k+1} \odot (\mathbf{a}_e - \mathbf{B}_e\mathbf{e}^{k+1})]). \tag{2.24}$$

Next, we will focus on optimizing the sub-problems in (2.19)–(2.22).

2.4.1 Update β^{k+1}

Disregarding terms independent of β, (2.19) simplifies to

$$\mathbb{L}(t_H^k, \mathbf{e}^k, \beta, \mathbf{y}^k, \mathbf{z}^k, \kappa^k, \iota^k) = \|\mathbf{A}_\beta \beta - \mathbf{b}_\beta\|_2^2 + C_\beta, \tag{2.25}$$

where $\mathbf{A}_\beta = [\sqrt{\frac{\rho_1}{2}}(\mathbf{A}_m^\dagger \mathbf{D}_1)^{\mathrm{T}}, \sqrt{\frac{\rho_2}{2}}(\mathbf{A}_s^\dagger \mathbf{D}_1)^{\mathrm{T}}]^{\mathrm{T}}$, $\mathbf{b}_\beta = [\sqrt{\frac{\rho_1}{2}}(\mathbf{y}^k + \frac{\kappa^k}{\rho_1})^{\mathrm{T}}, \sqrt{\frac{\rho_1}{2}}(\mathbf{z}^k + \frac{\iota^k}{\rho_2})^{\mathrm{T}}]^\dagger$ and $\mathbf{D}_1 = \mathrm{diag}(\mathbf{a}_e - \mathbf{B}_e\mathbf{e}^k)$. Here diag($\cdot$) returns a square diagonal matrix with the elements of vector $\mathbf{a}_e - \mathbf{B}_e\mathbf{e}^k$ on the main diagonal. C_β is a constant independent of β.

The optimization problem in (2.19) reduces to

$$\min_\beta \|\mathbf{A}_\beta^\dagger \beta - \mathbf{b}_\beta\|_2^2, \tag{2.26}$$

with its closed-form solution given by

$$\beta^{k+1} = (\mathbf{A}_\beta \mathbf{A}_\beta^\dagger)^{-1}\mathbf{A}_\beta \mathbf{b}_\beta. \tag{2.27}$$

2.4.2 Update $\mathbf{y}^{k+1}$

Neglecting terms independent of $\mathbf{y}$, the cost function in the optimization problem (2.20) simplifies to

$$\mathbb{L}(t_H^k, \mathbf{e}^k, \beta^{k+1}, \mathbf{y}, \mathbf{z}^k, \kappa^k, \iota^k) = \|\mathbf{y} - \mathbf{y}^c\|_2^2 + C_y, \tag{2.28}$$

where $\mathbf{y}^c = \mathbf{A}_m^\dagger[\beta^{k+1} \odot (\mathbf{a}_e - \mathbf{B}_e\mathbf{e}^k)] - \frac{\kappa^k}{\rho_1}$ and C_y is a constant independent of $\mathbf{y}$.

Expanding the norm term in (2.28), the optimization problem in (2.20) simplifies to

$$\begin{aligned} &\min_{\mathbf{y}} \sum_{m=1}^{M_0} (y_m - y_m^c)^2 \\ &\text{s.t. } \gamma - \delta \le |y_m|^2 \le \gamma + \delta, \ m = 1, \ldots, M_0, \end{aligned} \tag{2.29}$$

where the closed-form solution, y_m^{k+1}, is [27]

$$y_m^{k+1} = \begin{cases} \sqrt{\gamma + \delta}\dfrac{y_m^c}{|y_m^c|}, & \text{if } |y_m^c| \ge \sqrt{\gamma + \delta}, \\ \sqrt{\gamma - \delta}\dfrac{y_{m,}^c}{|y_{m,}^c|}, & \text{if } |y_m^c| \le \sqrt{\gamma - \delta}, \\ y_m^c, & \text{otherwise}, \end{cases} \tag{2.30}$$

and $m = 1, \ldots, M_0$.

2.4.3 Update t_H, $\mathbf{z}^{k+1}$

Similar to subsection B, the cost function in the optimization problem (2.21) reduces to

$$\mathbb{L}(t_H^k, \mathbf{e}^k, \beta^{k+1}, \mathbf{y}, \mathbf{z}^k, \kappa^k, \iota^k) = t_H + \|\mathbf{z} - \mathbf{z}^c\|_2^2 + C_z, \tag{2.31}$$

where $\mathbf{z}^c = \mathbf{A}_s^{\dagger}[\beta^{k+1} \odot (\mathbf{a}_e - \mathbf{B}_e\mathbf{e}^k)] - \frac{\kappa^k}{\rho_1}$ and C_z is a constant independent of t_H and $\mathbf{z}$.

Expanding the norm term in (2.31), the optimization problem in (2.21) simplifies to

$$\begin{aligned} &\min_{t_H,\mathbf{z}} t_H + \sum_{s=1}^{S_O} (z_s - z_s^c)^2, \\ &\text{s.t. } |z_s|^2 \leq t_H, t_H \leq \hat{t}_H, \; s = 1, \ldots, S_O. \end{aligned} \tag{2.32}$$

First, suppose the solution, t_H^{k+1}, at $(k+1)$th iteration has been obtained, then $\mathbf{z}^{k+1}$ is

$$\begin{aligned} &\min_{\mathbf{z}} \sum_{s=1}^{S_O} (z_s - z_s^c)^2, \\ &\text{s.t. } |z_s|^2 \leq t_H^{k+1}, s = 1, \ldots, S_O, \end{aligned} \tag{2.33}$$

thus, $\mathbf{z}^{k+1}$ is

$$z_s^{k+1} = \begin{cases} \sqrt{t_H^{k+1}} \dfrac{z_s^c}{|z_s^c|}, & \text{if } |z_s^c| \geq \sqrt{t_H^{k+1}}, \\ z_s^c, & \text{otherwise,} \end{cases} \tag{2.34}$$

where $s = 1, \ldots, S_O$.

Substituting (2.34) into the cost function in problem (2.32) yields a new optimization problem with respect to the single variable t_H,

$$\begin{aligned} &\min_{t_H} t_H + \frac{\rho}{2} \sum_{s=1}^{S_O} \varpi_s \left(\sqrt{t_H} - |z_s^c|\right)^2 \\ &\text{s.t. } t_H \leq \hat{t}_H, \end{aligned} \tag{2.35}$$

where $\varpi_s = 1$ if $|z_s^c| > \sqrt{t_H}$, otherwise, $\varpi_s = 0$.

To solve problem (2.35), we first sort $|z_s^c|$ $(s = 1, 2, \ldots, S_O)$ in ascending order to obtain a new sequence, $r_0, r_1, \ldots, r_U$, which satisfy $r_0 < r_1 < \cdots < r_U$ with $U \leq S_O$ (removing repetitions). On this basis, we define the following U index sets, Ω_u $(u = 1, 2, \ldots, U)$, which is

$$\Omega_u = \{s | z_s^c \geq r_{u-1}, s \in \{1, 2, \ldots, S_O\}\}, \tag{2.36}$$

especially when $u = 1$, $\Omega_1 = \{1, 2, \ldots, S_O\}$.

Next, we define the following functions $f_u(t_H)$ $(u = 1, 2, \ldots, U)$,

$$f_u(t_H) = \begin{cases} t_H + \dfrac{\rho}{2} \displaystyle\sum_{s \in \Omega_u} (\sqrt{t_H} - |z_s^c|)^2, \sqrt{t_H} \in [r_{u-1}, r_u], \\ t_H + \dfrac{\rho}{2} \displaystyle\sum_{s \in \Omega_1} (\sqrt{t_H} - |z_s^c|)^2, \sqrt{t_H} < r_0, u = 0, \end{cases} \tag{2.37}$$

where $t_H \leq \hat{t}_H$. Equation (2.37) is equivalent to the piecewise form of the cost function in the optimization problem (2.35).

Algorithm 2.1 Update t_H^{k+1}

Input: $\hat{t}_H$, $|z_s^c|$ for $s = 1, 2, \ldots, S_0$
Output: t_H^{k+1}

1: Sort $|z_s^c|$ $(s = 1, 2, \ldots, S_0)$ in ascending order to obtain a new sequence, $r_0, r_1, \ldots, r_U$.
2: Obtain U index sets, Ω_u $(u = 1, 2, \ldots, U)$, by (2.36);
3: If $\sqrt{\hat{t}_H} \le r_0$, update t_H^{k+1} by (2.41) and (2.40);
4: If $\sqrt{\hat{t}_H} \ge r_U$, update t_H^{k+1} by (2.42)–(2.44);
5: If $r_0 < \sqrt{\hat{t}_H} < r_U$, update t_H^{k+1} by (2.45), (2.43), and (2.44).

Supposing $v = \sqrt{t_H}$ $(0 \le v \le \sqrt{\hat{t}_H})$, then the new functions, $g_u(v) = f_u(t_H)$ $(u = 1, 2, \ldots, U)$, is

$$g_u(v) = \begin{cases} \tilde{a}_u(v - \tilde{b}_u)^2 + \tilde{c}_u, v \in [r_{u-1}, r_u], \\ \tilde{a}_0(v - \tilde{b}_0)^2 + \tilde{c}_0, v \le r_0, u = 0, \end{cases} \tag{2.38}$$

where $\tilde{a}_u$, $\tilde{c}_u$, $\tilde{a}_0$, and $\tilde{c}_0$ are the constants independent of s. In addition,

$$\tilde{b}_u = \frac{\rho \sum\limits_{s \in \Omega_u} |z_s^c|}{2 + \rho \sum\limits_{s \in \Omega_u} 1}, \tag{2.39}$$

and

$$\tilde{b}_0 = \tilde{b}_1 = \frac{\rho}{2 + \rho S_0} \sum_{s \in \Omega_1} |z_s^c|. \tag{2.40}$$

According to the size relation of $\sqrt{\hat{t}_H}$, r_0, and r_U, the solution, t_H^{k+1}, to problems (2.35) or (2.32) can be obtained through several cases:

Case 1: if $\sqrt{\hat{t}_H} \le r_0$, $\varpi_s = 1$ $(s = 1, 2, \ldots, S_0)$. Thus, problem (2.35) can be equivalent to the following problem:

$$\begin{aligned} t_H^{k+1} &= \arg\min_{t_H} t_H + \frac{\rho}{2} \sum_{s \in \Omega_1} (\sqrt{t_H} - |z_s^c|)^2, \\ &= \{\arg\min_{v} \tilde{a}_0(v - \tilde{b}_0)^2 + \tilde{c}_0\}^2, \\ &= \begin{cases} \hat{t}_H, & \text{if } \sqrt{\hat{t}_H} < \tilde{b}_0, \\ \tilde{b}_0^2, & \text{otherwise}, \end{cases} \end{aligned} \tag{2.41}$$

where $\tilde{b}_0$ is given by (2.40).

Case 2: if $\sqrt{\hat{t}_H} \ge r_U$, $\varpi_s = 1$ $(s \in \Omega_u)$, otherwise, $\varpi_s = 0$ $(s \in \Omega_u)$. Thus, problem (2.35) can be equivalent to

$$t_H^{k+1} = \arg\min_{t_u} \{f_u(t_u) | u = 1, \ldots, U\}, \tag{2.42}$$

where t_u $(u = 2, 3, \ldots, U)$ is given by (2.43) and t_1 is given by (2.44).

$$\begin{aligned} t_u &= \arg\min_t \{f_u(t) | r_{u-1} \le \sqrt{t} \le r_u\} \\ &= \{\arg\min_v \tilde{a}_u \left(v - \tilde{b}_u\right)^2 + \tilde{c}_u\}^2 \\ &= \begin{cases} r_{u-1}^2, & \text{if } r_{u-1} > \tilde{b}_u, \\ r_u^2, & \text{if } r_u < \tilde{b}_u, \\ \tilde{b}_u^2, & \text{otherwise,} \end{cases} \end{aligned} \tag{2.43}$$

where $\tilde{b}_u$ $(u = 2, 3, \ldots, U)$ is given by (2.39).

$$t_1 = \begin{cases} r_1^2, & \text{if } r_1 < \tilde{b}_1, \\ \tilde{b}_1^2, & \text{otherwise.} \end{cases} \tag{2.44}$$

Case 3: $r_0 < \sqrt{\hat{t}_H} < r_U$. Sorting $r_0, r_1, \ldots, r_U, \sqrt{\hat{t}_H}$ in ascending order yields a new sequence, $r_0, r_1, \ldots, r_{\hat{U}}$, which meets $r_0 < r_1 < \cdots < r_{\hat{U}} = \sqrt{\hat{t}_H}$ (removing terms greater than $\sqrt{\hat{t}_H}$). Then, problem (2.35) can be equivalent to

$$t_H^{k+1} = \arg\min_{t_u} \{f_u(t_u) | u = 1, \ldots, \hat{U}\}, \tag{2.45}$$

where t_u $(u = 2, 3, \ldots, \hat{U})$ and t_1 are given by (2.43) and (2.44), respectively.

Finally, the update of t_H^{k+1} is summarized in Algorithm 1.

2.4.4 Update $\mathbf{e}^{k+1}$

Disregarding terms independent of $\mathbf{e}$, the cost function in (2.22) reduces to

$$\begin{aligned} &\mathbb{L}(t_H^{k+1}, \mathbf{e}, \beta^{k+1}, \mathbf{y}^{k+1}, \mathbf{z}^{k+1}, \kappa^k, \iota^k) \\ &= \|\mathbf{A}_e\mathbf{e} - \mathbf{b}_e\|_2^2 - \alpha \mathbf{i}_M^{\mathrm{T}} \Gamma \mathbf{e} + C_{e1} \\ &= \mathbf{e}^\dagger \mathbf{A}_e^\dagger \mathbf{A}_e \mathbf{e} - \Re\{2(\mathbf{b}_e \mathbf{A}_e)^\dagger\}\mathbf{e} + C_{e2}, \end{aligned} \tag{2.46}$$

where $\mathbf{A}_e = [\sqrt{\frac{\rho_1}{2}}(\mathbf{A}_m^\dagger \mathbf{D}_2 \mathbf{B}_e)^{\mathrm{T}}, \sqrt{\frac{\rho_2}{2}}(\mathbf{A}_s^\dagger \mathbf{D}_2 \mathbf{B}_e)^{\mathrm{T}}]^{\mathrm{T}}$ and $\mathbf{b}_e = [\sqrt{\frac{\rho_1}{2}}(\mathbf{y}^{k+1} + \frac{\kappa^k}{\rho_1} - \mathbf{A}_m^\dagger \mathbf{D}_2 \mathbf{a}_e)^{\mathrm{T}}, \sqrt{\frac{\rho_2}{2}}(\mathbf{z}^{k+1} + \frac{\iota^k}{\rho_2} - \mathbf{A}_s^\dagger \mathbf{D}_2 \mathbf{a}_e)^{\mathrm{T}}]^\dagger$, with $\mathbf{D}_2 = \mathrm{diag}(\beta^{k+1})$. C_{e1} and C_{e2} are two constants independent of $\mathbf{e}$.

The optimization problem in (2.22) simplifies to

$$\begin{aligned} &\min_{\mathbf{e}} \mathbf{e}^\dagger \mathbf{A}_e^\dagger \mathbf{A}_e \mathbf{e} - \Re\{2\mathbf{b}_e^\dagger \mathbf{A}_e + \alpha \mathbf{i}_M^{\mathrm{T}} \Gamma\}\mathbf{e} \\ &\text{s.t. } N_l \le M - \mathbf{i}_M^{\mathrm{T}} \Gamma \mathbf{e} \le N_u, \\ &\quad \hat{\mathbf{q}}_{\hat{x}}^{\mathrm{T}} \mathbf{e} = 1 \text{ or } 0, \ \hat{x} = P, \ldots, \hat{M}, \\ &\quad e_{\hat{m}} = 1 \text{ or } 0, \ \hat{m} = 1, \ldots, \hat{M}, \end{aligned} \tag{2.47}$$

where problem (2.47) is a combinatorial optimization problem with multiple constraints and is generally very hard to solve. We develop an AA to tackle it.

Algorithm 2.2 AA for updating $\mathbf{e}^{k+1}$

Input: N_l, N_u, α, t_H^{k+1}, β^{k+1}, $\mathbf{y}^{k+1}$, $\mathbf{z}^{k+1}$, κ^k, ι^k
Output: $\mathbf{e}^{k+1}$;

1: Compute $\tilde{\mathbf{e}}$ by solving problem (2.48);
2: Compute $\hat{\mathbf{e}} = [\hat{\mathbf{E}}, 0\hat{\mathbf{E}}]^{\mathrm{T}}\tilde{\mathbf{e}}$;
3: Construct the candidate solution set Ω by (2.49);
4: Update $\mathbf{e}^{k+1}$ by solving problem (2.50).

Firstly, we replace the Boolean constraints $e_{\hat{m}} = 1$ or 0 with the convex constraints $0 \leq e_{\hat{m}} \leq 1$ and $\hat{\mathbf{q}}_{\hat{x}}^{\mathrm{T}}\mathbf{e} = 1$ or 0 with $0 \leq \hat{\mathbf{q}}_{\hat{x}}^{\mathrm{T}}\mathbf{e} \leq 1$ to obtain a relaxed solution of the antenna selection optimization problem.

Then, we transform problem (2.47) into the real domain for further analysis as

$$\begin{aligned} &\min_{\tilde{\mathbf{e}}} \tilde{\mathbf{e}}^{\mathrm{T}}\mathbf{R}_0\tilde{\mathbf{e}} - \mathbf{h}_0^{\mathrm{T}}\tilde{\mathbf{e}} \\ &\text{s.t. } M - N_u \leq \mathbf{h}_1^{\mathrm{T}}\tilde{\mathbf{e}} \leq M - N_l, \\ &\quad 0_{\hat{M}+P-1} \preceq \mathbf{R}_1\tilde{\mathbf{e}} \preceq \mathbf{i}_{\hat{M}+P-1}, \\ &\quad 0_{\hat{M}} \preceq \tilde{\mathbf{e}} \preceq \mathbf{i}_{\hat{M}},\ \mathbf{R}_2^{\mathrm{T}}\tilde{\mathbf{e}} = 0_{\hat{M}}, \end{aligned} \tag{2.48}$$

where $\tilde{\mathbf{e}} = [\Re\{\mathbf{e}\}^{\mathrm{T}}, \Im\{\mathbf{e}^{\mathrm{T}}\}]^{\mathrm{T}}$,
$\mathbf{h}_0 = [\Re\{2\mathbf{b}_e^{\dagger}\mathbf{A}_e + \alpha\mathbf{i}_M^{\mathrm{T}}\Gamma\}, 0_{M_1}^{\mathrm{T}}]^{\mathrm{T}}$, $\mathbf{h}_1 = [\mathbf{i}_{M_1}^{\mathrm{T}}, 0_{M_1}^{\mathrm{T}}]^{\mathrm{T}}$,
$\mathbf{R}_0 = \begin{bmatrix} \Re\{\mathbf{A}_e^{\dagger}\mathbf{A}_e\}, -\Im\{\mathbf{A}_e^{\dagger}\mathbf{A}_e\} \\ \Im\{\mathbf{A}_e^{\dagger}\mathbf{A}_e\}, \Re\{\mathbf{A}_e^{\dagger}\mathbf{A}_e\} \end{bmatrix}$, $\mathbf{R}_1 = [\hat{\mathbf{q}}_P, \ldots, \hat{\mathbf{q}}_{\hat{M}}]^{\mathrm{T}}$ and $\mathbf{R}_2 = [0\hat{\mathbf{E}}, \hat{\mathbf{E}}]$ with $\hat{\mathbf{E}}$ denoting a $\hat{M} \times \hat{M}$ dimensional unit matrix. Problem (2.48) is a quadratic convex optimization problem that can be solved using the CVX toolkit.

After solving problem (2.48), we can obtain $\hat{\mathbf{e}}$ by computing $\hat{\mathbf{e}} = [\hat{\mathbf{E}}, 0\hat{\mathbf{E}}]^{\mathrm{T}}\tilde{\mathbf{e}}$. Next, we construct a candidate solution set, Ω, to problem (2.47), which is

$$\Omega = \{\mathbf{e}_x | x = M - N_u, M - N_u + 1, \ldots, M - N_l\} \tag{2.49}$$

with $\mathbf{e}_x$ obtained by setting the maximum x elements of $\hat{\mathbf{e}}$ to 1 and the rest to 0. Thus, problem (2.47) can be approximated as

$$\begin{aligned} &\min_{\mathbf{e}} \mathbf{e}^{\dagger}\mathbf{A}_e^{\dagger}\mathbf{A}_e\mathbf{e} - \Re\{2\mathbf{b}_e^{\dagger}\mathbf{A}_e + \alpha\mathbf{i}_M^{\mathrm{T}}\Gamma\}\mathbf{e}, \\ &\text{s.t. } \mathbf{e} \in \Omega, \end{aligned} \tag{2.50}$$

where the problem can be solved by the direct search method. Finally, the AA for updating $\mathbf{e}^{k+1}$ is summarized in Algorithm 2.

After updating β^{k+1}, $\mathbf{y}^{k+1}$, t_H^{k+1}, $\mathbf{z}^{k+1}$, and $\mathbf{e}^{k+1}$, we update κ^{k+1} and ι^{k+1} by (2.23) and (2.24), respectively. The iteration termination criteria of the proposed AA-ADMM is given by

$$\varepsilon_m \leq \varepsilon_1 \text{ and } \varepsilon_s \leq \varepsilon_2, \tag{2.51}$$

where ε_1 and ε_2 represent the maximum tolerances for the mainlobe and sidelobe residuals, respectively, referred to as the mainlobe termination residual (MTR) and

Algorithm 2.3 AA-ADMM for beampattern synthesis

Input: γ, δ, $\hat{t}_H$, N_l, N_u, M, P, Φ_m, Φ_s, M_0, S_0, α, ρ, ε_1, ε_2
Output: $\mathbf{b}$, $\mathbf{w}$;

Randomly initialize β^1, $\mathbf{y}^1$, $\mathbf{z}^1$, $\mathbf{e}^1$, κ^1, ι^1;
For $k = 1, 2, \ldots, K$
  Update β^{k+1} by (2.27),
  Update $\mathbf{y}^{k+1}$ by (2.30),
  Update t_H^{k+1} by Algorithm 1,
  Update $\mathbf{z}^{k+1}$ by (2.34),
  Update $\mathbf{e}^{k+1}$ by Algorithm 2,
  Update κ^{k+1} by using (2.23),
  Update ι^{k+1} by using (2.24),
  If $\varepsilon_m \leq \varepsilon_1$ and $\varepsilon_s \leq \varepsilon_2$;
    break;
  EndIf
EndFor
Output $\mathbf{b} = \Gamma \mathbf{e}^{k+1}$,
Output $\mathbf{w} = \beta^{k+1} \odot [(\mathbf{i}_M - \mathbf{b}) \otimes \mathbf{i}_{M_H}]$.

sidelobe termination residual (STR). Finally, the iterative procedure of the proposed AA-ADMM is summarized in Algorithm 3.

2.5 Simulation results

In this section, we will evaluate the effectiveness and reliability of the proposed method as described in subsections V-A, V-B, and V-C. To ensure a broad applicability, we present synthesis cases that include both a single mainlobe and multiple mainlobes. Additionally, we benchmark our method against the group sparse optimization (GSO) approach, as detailed in [16]. The GSO method is recognized as the cutting-edge in thinned array synthesis and shares the most similarities with our approach. For the array architecture introduced in Section II, the beampattern synthesis is typically conducted offline. Therefore, it is unnecessary to consider the computational time in our analysis. More importantly, we focus on the number of subarrays (or REs), the PSL, adherence to layout rules, and the reliability of our proposed method. Furthermore, all numerical and full-wave simulations are conducted using MATLAB® 2016a (on a standard PC equipped with an Intel Core i5 8400U processor and 8 GB RAM) and HFSS 2018 (on a Dell PowerEdge R940xa with 1 TB RAM), respectively.

2.5.1 Beampattern synthesis with single mainlobe

In this subsection, we use the synthesis of a single mainlobe beampattern as an example to examine the performance of our proposed method.

The fundamental parameters for the array and the AA-ADMM are established as follows: The grid count is $M = 40$ with each subarray comprising a single RE, that is, $M_H = 1$. The number of subarrays is fixed at $N_l = N_u = 22$, which directly constrains the number of REs to 22. Each RSG includes $P = 2$ grids. The desired power level of the mainlobe is $\gamma = 1$, with a maximum allowable deviation of $\delta = 0.1$. The desired power level range of mainlobe is $[-0.4576 \text{ dB}, 0.4139 \text{ dB}]$. The trade-off factor is set as $\alpha = 0.01$. The designated PSL is preset as $\hat{t}_H = 10^{-23/10}$ (i.e., -23 dB). The mainlobe angle region is set as $\Phi_m = [-20°, 20°]$, the sidelobe angle region is set as $\Phi_s = [-90°, -25°] \cup [25°, 90°]$, and the angle sampling interval is set as $0.1°$. The thresholds MTR and STR are set as $\varepsilon_1 = 10^{-4}$ (i.e., -80 dB) and $\varepsilon_2 = 10^{-4}$ (i.e., -80 dB), respectively. The penalty factors are $\rho_1 = 10$ and $\rho_2 = 50$, respectively. Besides, to achieve a fair comparison, both the isotropic REs and the ideal mutual coupling characteristics are considered, i.e., $g_1(\theta) = 1$ and $\mathbf{C} = \mathbf{E}$, which is because the real array model is not considered in GSO. The beampattern parameters (including mainlobe and sidelobe levels, mainlobe power fluctuation, and the angle regions for both) and the number of REs for the GSO are set identical to those for the AA-ADMM. The additional parameters for the GSO are detailed in Section II-E of [16] to ensure the method's convergence and accuracy.

During each iteration of AA-ADMM, we check if the mainlobe and sidelobe residuals (ε_m and ε_s, respectively) meet the conditions $\varepsilon_m \le \varepsilon_1$ and $\varepsilon_s \le \varepsilon_2$. If both conditions are satisfied simultaneously, it indicates that the AA-ADMM has converged and the iteration stops. Otherwise, it suggests that the AA-ADMM has not yet converged, prompting a move to the next iteration.

Figure 2.3 presents the convergence trajectories for both the mainlobe and sidelobe residuals (represented in logarithmic form as ε_m and ε_s, respectively) as a function of the iteration count. This graph illustrates that the proposed AA-ADMM method converges within polynomial time based on the specified termination criteria. In Figure 2.3, the mainlobe and sidelobe residuals, depicted by the red dotted line and the blue dash-dotted line respectively, both satisfy the termination criteria (i.e., $\varepsilon_m = -313.1 \text{ dB} \le -80 \text{ dB}$ and $\varepsilon_s = -79.13 \text{ dB} \approx -80 \text{ dB}$) by the 106th iteration, which takes 37.24 s to complete. Despite some minor fluctuations during the iterations, the overall trend of the residuals is downward, with a notable rapid decrease after a few iterations. It is important to note that the sidelobe residual at convergence (-79.13 dB) in Figure 2.3 is slightly above the pre-set STR (-80 dB). This discrepancy is due to the rounding mechanisms used in the software during the convergence evaluation process.

In Figure 2.4(a)–(c), "AA-ADMM, 22 REs" indicates that a thinned array with 22 REs has been synthesized using the AA-ADMM method. "GSO, full" denotes the full array before any sparsification, synthesized using the GSO method. The sparsification process involves removing REs with the smallest absolute values of $|\beta_m|$ from the full array. Therefore, "GSO, 22 REs" signifies that a thinned array with 22 REs has been achieved using the GSO method and subsequent sparsification.

Figure 2.4(a) illustrates that the PSLs for "AA-ADMM, 22 REs," "GSO, full," and "GSO, 22 REs" are −23.27 dB, −22.88 dB, and −20.27 dB, respectively. This indicates that both the AA-ADMM and GSO methods can meet the

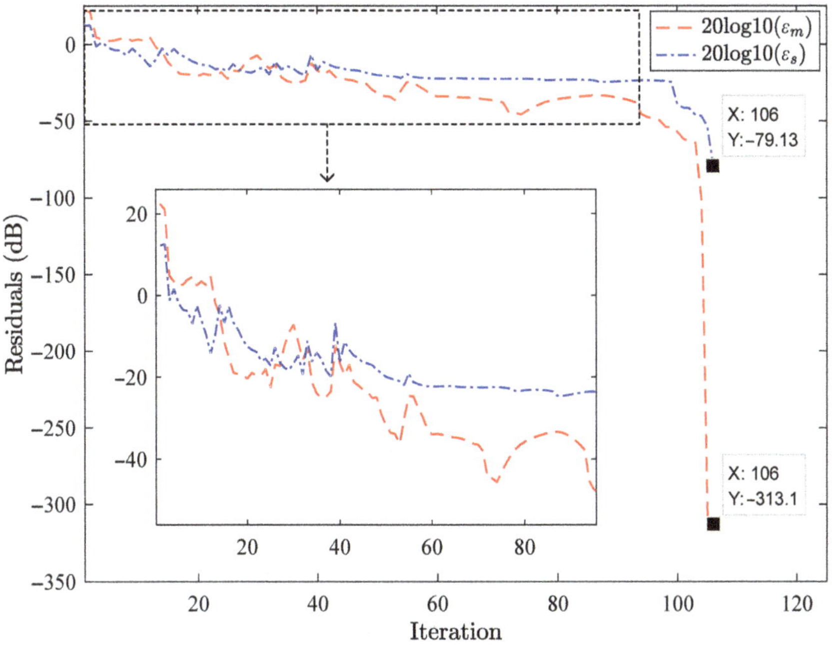

Figure 2.3 Mainlobe or sidelobe residuals (ε_m or ε_s in logarithmic form, respectively) versus iteration number

beampattern synthesis criteria under a single mainlobe constraint, with AA-ADMM demonstrating superior sidelobe suppression compared to GSO.

The reasons for this are outlined as follows: (1) A comparison between the "AA-ADMM, 22 REs" and "GSO, full" curves shows that the PSL of the thinned array produced by AA-ADMM is lower than that of the full array from GSO. This occurs because AA-ADMM actively minimizes the PSL while adhering to beampattern constraints, whereas GSO focuses on optimizing the beampattern within the specified constraints. (2) A further comparison among "AA-ADMM, 22 REs," "GSO, full," and "GSO, 22 REs" reveals that the beampattern of the thinned array synthesized using the AA-ADMM approach is more precise than that obtained via GSO.

For these methods, AA-ADMM employs a norm minimization model based on LV, allowing direct synthesis of the thinned array without additional sparse operations. In contrast, GSO uses a norm minimization model based on WC, necessitating sparse operations to achieve a thinned array. Consequently, this sparse operation degrades the PSL of the thinned array by 2.61 dB compared to the full array.

Figure 2.4(b) shows that the thinned array is directly derived using AA-ADMM. Figure 2.4(c) illustrates the process by which the thinned array is obtained from GSO through sparse operations. These operations remove REs with non-zero $|\beta_m|$, resulting in an increased PSL for the thinned array.

Figure 2.4(d) and (e) depicts the layout results based on LVs in Figure 2.4(b) and (c), respectively. The former demonstrates that the thinned array synthesized by

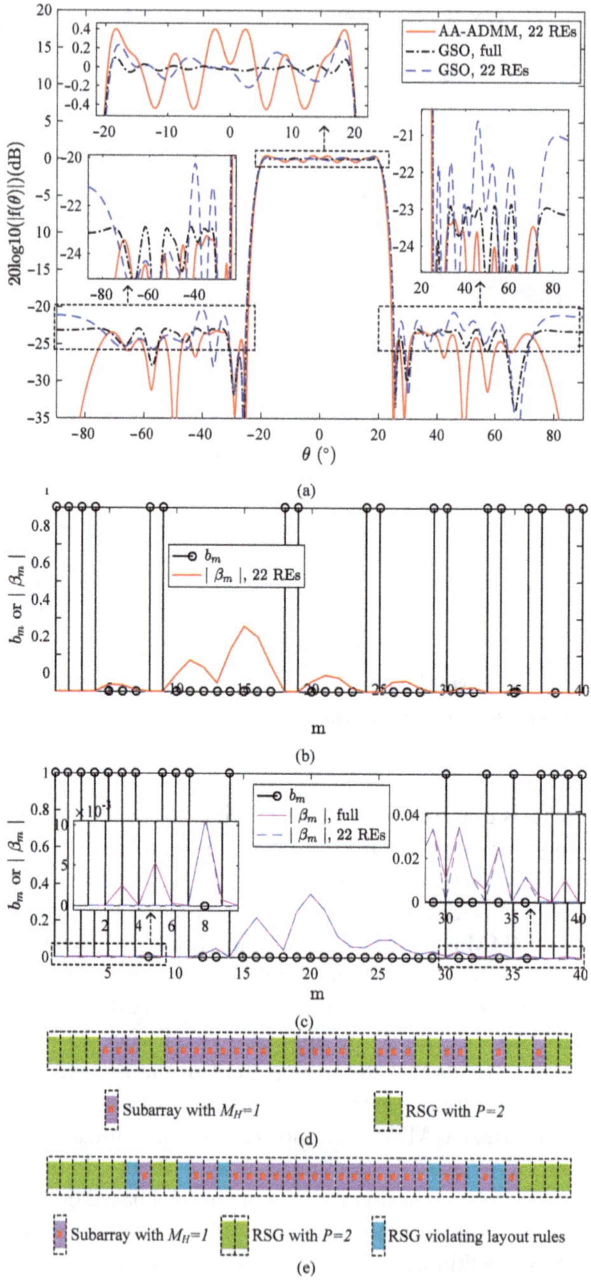

Figure 2.4 Single mainlobe pattern synthesis results: (a) beampatterns with single mainlobe, (b) b_m and $|\beta_m|$ obtained by AA-ADMM, (c) b_m and $|\beta_m|$ obtained by GSO, (d) layout result of (b), and (e) layout result of (c)

AA-ADMM fully complies with the subarray layout rules, whereas the latter indicates that the GSO-synthesized thinned array breaches these rules due to the occurrence of RSG with $P \neq 2$ (i.e., RSG violating layout rules in Figure 2.4(e)). Moreover, both methods result in 22 subarrays (or REs), representing 55% of the full array.

Additionally, the beampattern synthesis results (including b_m, $|\beta_m|$, and $\measuredangle\beta_m$) produced by AA-ADMM and GSO are presented in Table 2.1 to assist readers in reproducing or comparing the findings.

2.5.2 Beampattern synthesis with multiple mainlobes

In this subsection, we use the synthesis of a multiple mainlobe beampattern as an example to examine the performance of our proposed method.

The fundamental parameters for the array and the AA-ADMM are established as follows:

The basic parameters of array and AA-ADMM are set as follows. The grid count is $M = 20$, with each subarray containing $M_H = 2$ REs, meaning the full array has a total of 40 REs. The number of subarrays is bounded above and below by $N_u = 13$ and $N_l = 9$, respectively. Each RSG contains $P = 2$ grids. The PSL is preset as $\hat{t}_H = 10^{-20/10}$. The mainlobe angle region is $\Phi_m = [-30°, -15°] \cup [20°, 35°]$, while sidelobe angle region is $\Phi_s = [-90°, -35°] \cup [-10°, 15°] \cup [-40°, 90°]$. The target beampattern parameters (including mainlobe and sidelobe levels, mainlobe power fluctuation, and the angle regions for both) and the number of REs for the GSO are set identical to those for the AA-ADMM. Additional parameters for the GSO are detailed in Section II-E of [16] to ensure the method's convergence and accuracy.

Figure 2.5(a)–(c) illustrates that, under multiple mainlobe constraints, the proposed synthesis method achieves a thinned array with fewer REs and improved sidelobe performance compared to GSO.

Figure 2.5(a) shows that the PSLs for "AA-ADMM, 24 REs," "GSO, full," "GSO, 35 REs," and "GSO, 34 REs" are −20.28 dB, −19.89 dB, −19.89 dB, and −19.12 dB, respectively. This also indicates that to meet the beampattern constraints, AA-ADMM requires 24 REs (60% of the full array), whereas GSO requires 35 REs (87.5% of the full array), meaning the proposed method saves 11 REs compared to GSO. Additionally, it is evident that the proposed method attains a lower PSL than both the predefined value and the results (including the PSLs of both the thinned and full arrays) synthesized by GSO.

Figure 2.5(b) and (c) presents the layout results based on AA-ADMM and GSO, respectively. The former shows that the thinned array synthesized by AA-ADMM adheres completely to the subarray layout rules. The latter, however, reveals that the thinned array synthesized by GSO breaches these rules due to the presence of subarrays with $M_H \neq 2$ and RSGs with $P \neq 2$ (i.e., RE violating layout rules in Figure 2.5(c)).

Moreover, Figure 2.5(d) and (e) depicts the amplitudes and phases of the WC synthesized by AA-ADMM and GSO, respectively.

Table 2.1 Single mainlobe beampattern synthesis results

m	AA-ADMM			GSO		
	b_m	$\|\beta_m\|$	$\measuredangle\beta_m$(rad)	b_m	$\|\beta_m\|$	$\measuredangle\beta_m$(rad)
1	1	0	0	1	0	0
2	1	0	0	1	0	0
3	1	0	0	1	0	0
4	1	0	0	1	0	0
5	0	0.0379	−3.1395	1	0	0
6	0	0.0329	−3.1408	1	0	0
7	0	0.0086	3.1343	1	0	0
8	1	0	0	0	0.0108	−2.1982
9	1	0	0	1	0	0
10	0	0.0909	−3.14	1	0	0
11	0	0.1693	−3.1402	1	0	0
12	0	0.1313	−3.14	0	0.0111	−2.4185
13	0	0.0463	0	0	0.047	−2.2943
14	0	0.228	0.0011	1	0	0
15	0	0.3562	0.0011	0	0.1184	0.9145
16	0	0.2975	0.0011	0	0.2137	0.9266
17	0	0.1384	0.0009	0	0.1256	0.754
18	1	0	0	0	0.0386	−1.5333
19	1	0	0	0	0.25	−1.9796
20	0	0.0537	0.0008	0	0.3451	−2.0381
21	0	0.0886	0.0007	0	0.262	−1.9569
22	0	0.0754	0.0005	0	0.1151	−1.6904
23	0	0.0202	0.0012	0	0.058	−1.0146
24	1	0	0	0	0.0539	−1.0518
25	1	0	0	0	0.0946	−1.7415
26	0	0.0537	0.0004	0	0.0959	−1.8108
27	0	0.0579	0.0004	0	0.0455	−1.4746
28	0	0.0201	−0.0014	0	0.0193	−0.1624
29	1	0	0	0	0.0334	0.2109
30	1	0	0	1	0	0
31	0	0.02	0.0026	0	0.0342	−1.7938
32	0	0.0231	0.002	0	0.0109	−1.528
33	1	0	0	1	0	0
34	1	0	0	0	0.0249	0.7331
35	0	0.0093	3.1398	1	0	0
36	1	0	0	0	0.0114	−1.8436
37	1	0	0	1	0	0
38	0	0.0047	−0.0094	1	0	0
39	1	0	0	1	0	0
40	1	0	0	1	0	0

2.5.3 Reliability analysis

Typically, the beampattern generated by full-wave simulation closely approximates that of an actual manufactured array, and it can serve as the benchmark pattern for

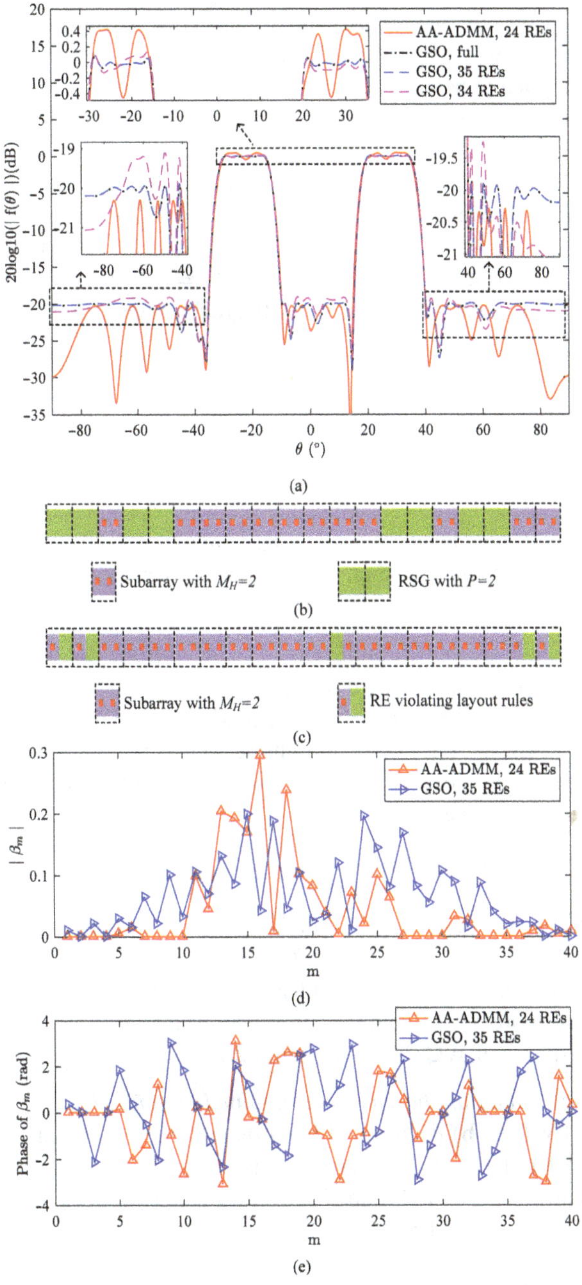

Figure 2.5 Multiple mainlobes pattern synthesis results: (a) beampattern with double mainlobes, (b) layout result of AA-ADMM, (c) layout result of GSO, (d) m versus $|\beta_m|$, and (e) m versus $\measuredangle\beta_m$

evaluating the quality of theoretical simulations. Thus, in the subsequent sections, we will employ full-wave simulation to investigate the reliability of the proposed method when applied to beampattern synthesis for real arrays.

The fundamental parameters for the array and the AA-ADMM method are established as follows: The grid count is $M = 23$, with each subarray comprising $M_H = 2$ REs, meaning the full array has a total of 46 REs. The number of subarrays is bounded above and below by $N_u = 13$ and $N_l = 9$, respectively. Each RSG includes $P = 1$ grids. The PSL is preset as $\hat{t}_H = 10^{-21/10}$. The mainlobe and sidelobe angle regions are $\Phi_m = [0°, 30°]$ and $\Phi_s = [-90°, -5°] \cup [-35°, 90°]$, respectively. The elements of the MCM for the real array are detailed in [52] by the following equation:

$$[\mathbf{C}]_{m,n} = \begin{cases} 1, & m = n, \\ -\dfrac{[\mathbf{Z}]_{m,n}}{Z_L}, & m \neq n, \end{cases} \tag{2.52}$$

where $[\mathbf{Z}]_{m,n}$ represents the mutual impedance between the mth and nth antennas, and Z_L denotes the load impedance at the antenna terminal. All other parameters remain as described in subsection A. The process for verifying reliability is outlined below:

Step 1: modeling real array.
We use the HFSS 2018 tool to build a real ULA with the 46 same REs working in the 5G frequency band, as shown in Figure 2.6(e). Then we can obtain both the mutual impedance $\mathbf{Z}$ and the load impedance $Z_L = 49.1376 - j10$. Consequently, $\mathbf{C}$ is computed using (2.52), with its amplitudes presented in Figure 2.6(a) and (b). Additionally, the amplitudes and phases of the beampattern for the REs are shown in Figure 2.6(c) and (d), respectively.

Step 2: theoretical simulation based on the proposed method.
By incorporating the obtained $\mathbf{C}$ into (2.5), we use the proposed method to determine the synthesized beampattern and the thinned array configuration, shown by the red dotted curve in Figure 2.7(a) and Figure 2.7(b), respectively. As indicated in Figure 2.7(b), the synthesized array contains 24 REs, which constitutes 52.1% 52.1% of the full array. Furthermore, the LV and the amplitudes of the WC are displayed in Figure 2.7(d), with the phase of WC shown in Figure 2.7(e).

Step 3: full-wave simulation based on the obtained LV and WC.
Following the LV (i.e., $\mathbf{b}$) determined from theoretical simulation, we remove the REs with $b_m = 1$ in the actual full array (illustrated in Figure 2.6(e)) to derive the actual thinned array (shown in Figure 2.7(c)). Subsequently, we conduct a full-wave simulation to acquire the real beampattern, depicted by the blue dash-dotted curve in Figure 2.7(a).

Figure 2.7(a) demonstrates a strong agreement between the theoretical simulation and the full-wave simulation. Within the mainlobe angle region, the ripple of the beampattern from the full-wave simulation is only marginally degraded by 0.32 dB, while in the sidelobe angle region, the PSL of the beampattern from the full-wave simulation is worsened by just 1.28 dB (the PSLs of the beampatterns obtained by theoretical and full-wave simulations are −20.17 dB and −21.45 dB, respectively).

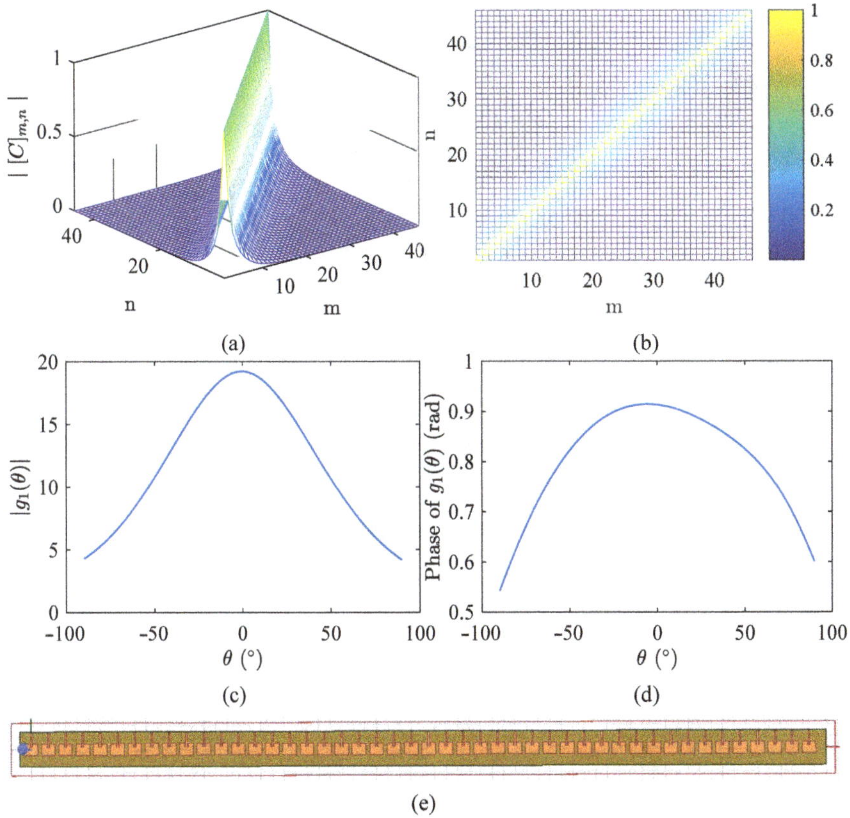

Figure 2.6 Real REs and array model: (a) magnitude of $[\mathbf{C}]_{m,n}$*, (b) top view of (a), (c) magnitude of* $g_1(\theta)$*, (d) phase of* $g_1(\theta)$*, and (e) real full array*

Despite this, the use of the isotropic RE model in beampattern synthesis enables the full-wave simulation based on LV and WC, provided by the proposed method, to achieve enhanced sidelobe performance compared to the theoretical simulation, except at a $-8.8°$ direction. Therefore, we assert that the proposed method is highly reliable for beampattern synthesis applications in real arrays.

2.6 Conclusion

This chapter has concentrated on solving the problem of synthesizing thinned array beampatterns with multiple constraints. Initially, we modeled a thinned ACMRS and set up a beampattern synthesis problem that focuses on minimizing the PSL and the number of subarrays (or REs), subject to the constraints of power beampattern and subarray layout. We then introduced an ADMM approach, specifically an AA-ADMM, to tackle this problem. To evaluate the effectiveness and reliability of our

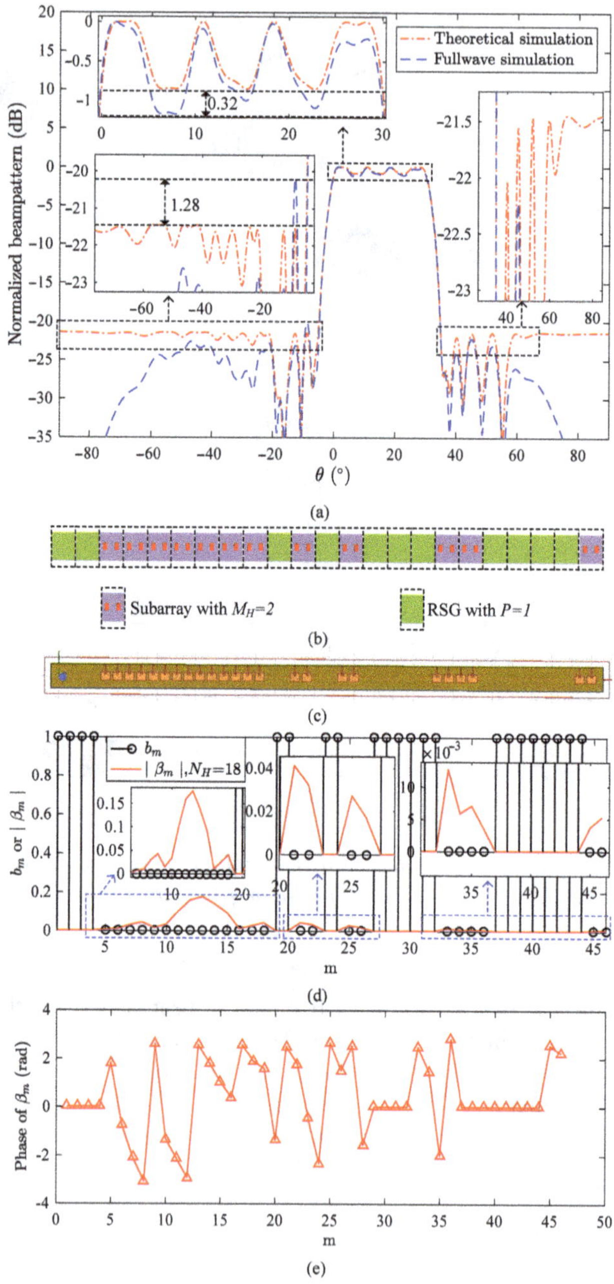

Figure 2.7 Theoretical simulation and full-wave simulation results: (a) synthesized beampattern, (b) subarray layout results, (c) real thinned array based on LV, (d) b_m or $|\beta_m|$ versus m, and (e) $\angle\beta_m$ versus m

proposed method, we provided results from both numerical and full-wave simulations. The key findings are as follows: (1) The proposed method can realize the beampattern synthesis for the thinned ACMRS or the traditionally thinned array under the given constraints. (2) Compared with the state-of-the-art GSO [16], the proposed method can achieve lower PSL and thinner (i.e., fewer REs or subarrays) array conforming layout rules. (3) The proposed method has good reliability in the application of beampattern synthesis for real array.

Possible future research tracks might concern the joint design of transmitting waveform and subarray layout for beampattern synthesis with constant modulus constraint [53].

References

[1] Skolnik MI. *Radar Handbook* (2nd edition). New York: McGraw-Hill; 1991.

[2] Godara LC. *Handbook of Antennas in Wireless Communications*. Boca Raton, FL: CRC Press; 2001.

[3] Boyd S and Vandenberghe L. *Convex Optimization*. Cambridge: Cambridge University Press; 2004.

[4] Nai SE, Ser W, Yu ZL, *et al.* Beampattern synthesis for linear and planar arrays with antenna selection by convex optimization. *IEEE Transactions on Antennas and Propagation*, 2010;58(12):3923–3930.

[5] D'Urso M, Prisco G, and Tumolo RM. Maximally sparse, steerable, and non-superdirective array antennas via convex optimizations. *IEEE Transactions on Antennas and Propagation*. 2016;64(9):3840–3849.

[6] Fuchs B. Synthesis of sparse arrays with focused or shaped beampattern via sequential convex optimizations. *IEEE Transactions on Antennas and Propagation*. 2012;60(7):3499–3503.

[7] Liu Y, Nie Z, and Liu QH. Reducing the number of elements in a linear antenna array by the matrix pencil method. *IEEE Transactions on Antennas and Propagation*. 2008;56(9):2955–2962.

[8] Liu Y, Liu QH, and Nie Z. Reducing the number of elements in multiple-pattern linear arrays by the extended matrix pencil methods. *IEEE Transactions on Antennas and Propagation*. 2014;62(2):652–660.

[9] Shen H and Wang B. An effective method for synthesizing multiple-pattern linear arrays with a reduced number of antenna elements. *IEEE Transactions on Antennas and Propagation*. 2017;65(5):2358–2366.

[10] Liu Y, Liu QH, and Nie Z. Reducing the number of elements in the synthesis of shaped-beam patterns by the forward-backward matrix pencil method. *IEEE Transactions on Antennas and Propagation*. 2010;58(2):604–608.

[11] Liu Y, Zhang L, Zhu C, *et al.* Synthesis of nonuniformly spaced linear arrays with frequency-invariant patterns by the generalized matrix pencil methods. *IEEE Transactions on Antennas and Propagation*. 2015;63(4):1614–1625.

[12] Zhang X, He Z, Liao B, *et al.* A^2RC: An accurate array response control algorithm for pattern synthesis. *IEEE Transactions on Signal Processing*. 2017;65(7):1810–1824.

[13] Zhang X, He Z, Liao B, *et al.* Pattern synthesis with multipoint accurate array response control. *IEEE Transactions on Antennas and Propagation*, 2017;65(8):4075–4088.

[14] Zhang X, He Z, Xia X, *et al.* OPARC: optimal and precise array response control algorithm part I: fundamentals. *IEEE Transactions on Signal Processing*. 2019;67(3):652–667.

[15] Zhang X, He Z, Xia X, *et al.* OPARC: optimal and precise array response control algorithm part II: multi-points and applications. *IEEE Transactions on Signal Processing*. 2019;67(3):668–683.

[16] Wang X, Aboutanios E, and Amin MG. Thinned array beampattern synthesis by iterative soft-thresholding-based optimization algorithms. *IEEE Transactions on Antennas and Propagation*. 2014;62(12):6102–6113.

[17] Wang X, Hassanien A, and Amin MG. Sparse Transmit array design for dual-function radar communications by antenna selection. *Digital Signal Processing*. 2018;83:223–234.

[18] Wang X, Amin M, Wang X, *et al.* Sparse array quiescent beamformer design combining adaptive and deterministic constraints. *IEEE Transactions on Antennas and Propagation*, 2017;65(11):5808–5818.

[19] Liang J, Zhang X, So HC, *et al.* Sparse array beampattern synthesis via alternating direction method of multipliers. *IEEE Transactions on Antennas and Propagation*, 2018;66(5):2333–2345.

[20] Fan X, Liang J, Zhang Y, *et al.* Shaped power pattern synthesis with minimization of dynamic range ratio. *IEEE Transactions on Antennas and Propagation*, 2019;67(5):3067–3078.

[21] Gemechu AY, Cui G, Yu X, *et al.* Beampattern synthesis with sidelobe control and applications. *IEEE Transactions on Antennas and Propagation*, 2020;68(1):297–310.

[22] Yu X, Cui G, Yang S, *et al.* Coherent unambiguous transmit for sparse linear array with geography constraint. *IET Radar Sonar and Navigation*. 201;11(2):386–393.

[23] Chen K, Yun X, He Z, *et al.* Synthesis of sparse planar arrays using modified real genetic algorithm. *IEEE Transactions on Antennas and Propagation*. 2007;55(4):1067–1073.

[24] Quevedo-Teruel O and Rajo-Iglesias E. Ant colony optimization in thinned array synthesis with minimum sidelobe level. *IEEE Antennas and Wireless Propagation Letters*. 2006;5:349–352.

[25] Cui C, Jiao Y, Zhang L, *et al.* Synthesis of subarrayed monopluse arrays with contiguous elements using a DE algorithm. *IEEE Transactions on Antennas and Propagation*. 2017;65(8):4340–4345.

[26] Huang K and Sidiropoulos ND. Consensus-ADMM for general quadratically constrained quadratic programming. *IEEE Transactions on Signal Processing*. 2016;64(20):5297–5310.

[27] Fan W, Liang J, and Li J. Constant modulus MIMO radar waveform design with minimum peak sidelobe transmit beampattern. *IEEE Transactions on Signal Processing*. 2018;66(16):4207–4222.

[28] Yu X, Cui G, Yang J, *et al.* Wideband MIMO radar waveform design. *IEEE Transactions on Signal Processing*. 2019;67(13):3487–3501.

[29] Yu X, Cui G, Yang J, *et al.* MIMO radar transmit-receive design for moving target detection in signal-dependent clutter. *IEEE Transactions on Vehicular Technology*. 2020;69(1):522–536.

[30] Rocca P, Oliveri G, Mailloux RJ, *et al.* Unconventional phased array architectures and design methodologies – a review. *Proceedings of the IEEE*. 2016;104(3):544–560.

[31] Oliveri G, Salucci M, and Massa A. Synthesis of modular contiguously clustered linear arrays through a sparseness-regularized solver. *IEEE Transactions on Antennas and Propagation*. 2016;64(10):4277–4287.

[32] Zhao X, Yang Q, and Zhang Y. Synthesis of minimally subarrayed linear arrays via compressed sensing method. *IEEE Antennas and Wireless Propagation Letters*. 2019;18(3).

[33] Anselmi N, Poli L, Rocca P, *et al.* Design of simplified array layouts for preliminary experimental testing and validation of large AESAs. *IEEE Transactions on Antennas and Propagation*. 2018;66(12):6906–6920.

[34] Li X, Duan B, and Song L. Design of clustered planar arrays for microwave wireless power transmission. *IEEE Transactions on Antennas and Propagation*. 2019;67(1):606–611.

[35] Epcacan E and Ciloglu T. A hybrid nonlinear method for array thinning. *IEEE Transactions on Antennas and Propagation*. 2018;66(5):2318–2325.

[36] Rocca P, Hannan MA, Poli L, *et al.* Optimal phase-matching strategy for beam scanning of sub-arrayed phased arrays. *IEEE Transactions on Antennas and Propagation*. 2019;67(2):951–959.

[37] Manica L, Rocca P, Oliveri G, *et al.* Synthesis of multi-beam sub-arrayed antennas through an excitation matching strategy. *IEEE Transactions on Antennas and Propagation*. 2011;59(2):482–492.

[38] Lopez P, Rodriguez JA, Ares F, *et al.* Subarray weighting for the difference patterns of monopulse antennas: joint optimization of subarray configurations and weights. *IEEE Transactions on Antennas and Propagation*, 2001;49(11):1606–1608.

[39] Dong W, Xu Z, Liu X, *et al.* Modular subarrayed phased-array design by means of iterative convex relaxation optimization. *IEEE Antennas and Wireless Propagation Letters*. 2019;18(3):447–451.

[40] Bianchi D, Genovesi S, and Monorchio A. Randomly overlapped subarrays for reduced sidelobes in angle-limited scan arrays. *IEEE Antennas and Wireless Propagation Letters*. 2017;16:1969–1972.

[41] Rocca P, Mailloux RJ, and Toso G. GA-based optimization of irregular subarray layouts for wideband phased arrays design, *IEEE Antennas and Wireless Propagation Letters*. 2015;14:131–134.

[42] Oliveri G. Multibeam antenna arrays with common subarray layouts. *IEEE Antennas and Wireless Propagation Letters*. 2010;9:1190–1193.

[43] Xiong Z, Xu Z, and Chen S. Subarray partition in array antenna based on the algorithm X. *IEEE Antennas and Wireless Propagation Letters*. 2013;12: 906–909.

[44] Rocca P and D'Urso M Poli L. Advanced strategy for large antenna array design with subarray-only amplitude and phase control. *IEEE Antennas and Wireless Propagation Letters*. 2014;13:91–94.

[45] Yang X, Xi W, Sun Y, *et al.* Optimization of subarray partition for large planar phased array radar based on weighted K-means clustering method. *IEEE Journal of Selected Topics in Signal Processing*. 2015;9(8):1460–1468.

[46] Manica L, Rocca P, Pastorino M, *et al.* Boresight slope optimization of subarrayed linear arrays through the contiguous partition method. *IEEE Antennas and Wireless Propagation Letters*. 2009;8:253–257.

[47] Xiong Z, Xu Z, and Xiao S. Beamforming properties and design of the phased arrays in terms of irregular subarrays. *IET Microwaves Antennas & Propagation*. 2015;9(4):369–379.

[48] Kwon G, Park J, Kim D, *et al.* Optimization of a shared-aperture dual-band transmitting/receiving array antenna for radar applications. *IEEE Transactions on Antennas and Propagation*, 2017;65(12):7038–7051.

[49] Chen Y and Vaughan RG. Dual-polarized *L*-band and single-polarized *X*-band shared-aperture SAR array. *IEEE Transactions on Antennas and Propagation*. 2018;66(7):3391–3400.

[50] Zhang JF, Cheng YJ, Ding YR, *et al.* A dual-band shared-aperture antenna with large frequency ratio, high aperture reuse efficiency, and high channel isolation. *IEEE Transactions on Antennas and Propagation*. 2019;67(2): 853–860.

[51] SEA FIRE 500. Available from: https://www.youtube.com/watch?v=1JTLLssrKAg [Accessed 5 Oct 2019].

[52] Zhang T and Ser W. Robust Beampattern synthesis for antenna arrays with mutual coupling effect. *IEEE Transactions on Antennas and Propagation*. 2011;59(8):2889–2895.

[53] Yu X, Cui G, Kong L, *et al.* Constrained waveform design for colocated MIMO radar with uncertain steering matrices. *IEEE Transactions on Aerospace and Electronic Systems*. 2019;55(3):356–370.

Chapter 3

Reconfigurable array beampattern synthesis via conceptual sensor network modeling and computation

Junli Liang[1], Xuhui Fan[2], Xuan Zhang[3], Guoyang Yu[4] and Hing Cheung So[5]

This section explores the issue of reconfigurable array beampattern synthesis using sensor network consensus computation theory. Initially, we apply the principles of sensor network consensus computation to frame the synthesis challenge as a sensor network computation issue by simulating a virtual (conceptual) multi-node sensor network, where each "node" represents an individual beampattern synthesis task. Specifically, these "nodes" corresponding in number to the radiated patterns, share a common set of excitation magnitudes, leading to a unique magnitude-consensused computation challenge within the simulated sensor network. Subsequently, through local excitation computation within each "node" and exchanging computation results with neighboring "nodes," all "nodes" achieve magnitude-consensused excitations once the "network" stabilizes.

3.1 Introduction

Reconfigurable arrays of antennas can create various emitted patterns, allowing them to serve various functions. This versatility makes them valuable in air traffic control systems, satellite communications, and wireless communications [1]. This feature enables the use of a single reconfigurable array for multiple tasks by switching between patterns, reducing costs and the size of antenna systems [2]. Specifically, synthesizing array patterns by manipulating excitation phases is a cost-effective and straightforward approach. This method ensures that the magnitude of each array element remains constant during pattern switches, enabling precise control over phase adjustments. As a result, it is crucial for a reconfigurable array with phase control

[1]School of Electronics and Information, Northwestern Polytechnical University, China
[2]School of Communication and Information Engineering, Xi'an University of Science and Technology, China
[3]School of Electronic and Information Engineering, Beihang University, China
[4]Information Processing Technology Research Department, Xi'an Institute of Electromechanical Information Technology, China
[5]Department of Electrical Engineering, City University of Hong Kong, China

to create multiple excitation vectors where elements with the same indexes share identical magnitudes.

In recent years, numerous techniques have been created to create radiation patterns in reconfigurable arrays controlled by phase. For instance, [1,2] utilize a method known as successive projection to tackle the issue of pattern synthesis with phase-only control, while in [3,4] this method is paired with the quasi-Newton BFGS method [5] to address the power synthesis problem for conformal arrays. Nevertheless, the alternating projection methods between nonlinear sets and non-convex sets in [1–4] rely heavily on the selection of the starting point to minimize the risk of encountering trap points that are distant from the optimal solution. Additionally, in [6], individual patterns are synthesized separately to produce beampatterns for phase-only reconfigurable linear arrays, although the resulting solution is only in close proximity to the optimal solution. More recently, the utilization of the SemiDefinite Relaxation (SDR) technique has emerged in array synthesis. For instance, in [7], SDR is implemented in the synthesis of array patterns with phase-only control, however, the complexity of this synthesis approach in [7] is substantial and is not suitable for reconfiguring multiple patterns or beam scanning, since the SDR technique increases the dimension of the original problem twofold and relaxes the rank-1 constraint to achieve an approximately feasible solution. Furthermore, [8] employs a customized version of the Woodward–Lawson technique to craft multiple patterns for linear arrays whilst preserving a predefined excitation magnitude distribution. Despite this, it fails to utilize the complete degree of freedom (DoF) as it optimizes solely phases and not a combination of joint excitation magnitudes and phases. While genetic algorithms [9], particle swarm [10], and simulated annealing [11] are able to identify the optimal solution globally, the computational expense grows substantially with larger problem sizes.

Our recent studies have focused on solving the beampattern synthesis challenges using sparse antenna elements while ensuring a minimum dynamic range ratio [12,13]. However, this article presents a fresh perspective inspired by the successful applications of distributed consensus computation in sensor networks [14–21]. The approach aims to create reconfigurable patterns through phase-only control and develop a unique array synthesis technique that enables precise control of radiation power patterns while meeting given mask constraints rapidly by mimicking consensus computation in sensor networks to determine a common excitation amplitude distribution. Additionally, this novel method is adaptable for uniform amplitude distribution arrays. In contrast to existing phase-only methods discussed in prior works [22–24], this chapter focuses on minimizing the shared excitation amplitude norm to enable controlled excitations, particularly for uniform amplitude cases. Moreover, unlike conventional approaches where magnitudes are predetermined or fixed to achieve desired phase configurations in array synthesis as mentioned in previous studies [22–24], this chapter addresses a minimization problem of constant excitation magnitude to reduce power consumption.

The following chapter is structured as follows: Section 3.2 formulates the problem of reconfigurable array synthesis, aiming to minimize the norm of excitation vectors with phase-only control. In Section 3.3, a new approach is introduced to emulate sensor network consensus computation, offering a redefined synthesis problem and its corresponding solution. The theoretical analysis on computational complexity and convergence is also discussed. Section 3.4 showcases numerical examples, while Section 3.5 concludes the chapter.

3.2 Problem formulation

Let's consider a collection of N antenna elements positioned at coordinates $p_n \in \mathbb{R}^{2\times 1}$ for $n = 1, \ldots, N$. To simplify, we will focus on one-dimensional pattern synthesis in this section, as extending to two dimensions is simple. When pointing in the direction θ, the nth element emits a pattern $g_n(\theta)$, and the far field pattern $f(\theta)$ emitted by the array is denoted as follows:

$$f(\theta) = \boldsymbol{w}^H a(\theta), \tag{3.1}$$

where $\boldsymbol{w} \in \mathbb{C}^{N\times 1}$ is the complex excitation vector, and $\boldsymbol{a}(\theta) = [g_1(\theta)e^{j\frac{2\pi}{\lambda}\boldsymbol{p}_1^T\boldsymbol{v}(\theta)} \cdots g_N(\theta)e^{j\frac{2\pi}{\lambda}\boldsymbol{p}_N^T\boldsymbol{v}(\theta)}]^T$ is the steering vector associated with $\boldsymbol{v}(\theta)$ being the unit vector in the direction θ.

The power pattern of the array is then given by

$$|f(\theta)|^2 = |\boldsymbol{w}^H\boldsymbol{a}(\theta)|^2. \tag{3.2}$$

For the synthesis of focused beams, the power pattern of the array is usually required to satisfy the following constraints [7]

$$\mathscr{CF} \begin{cases} |f(\theta_0)|^2 \geq 1 \\ |f(\theta)|^2 \leq \rho(\theta),\ \forall\, \theta \in \Theta_{sl}, \end{cases} \tag{3.3}$$

where the main beam radiated by the array points to the direction θ_0 while the sidelobes Θ_{sl} are limited by a given upper bound $\rho(\theta)$.

Accordingly, the synthesis of shaped beams generally desires to yield a power pattern imposed by the following constraints [7]:

$$\mathscr{CS} \begin{cases} L(\theta) \leq |f(\theta)|^2 \leq U(\theta),\ \forall\, \theta \in \Theta_{ml} \\ |f(\theta)|^2 \leq \varepsilon(\theta),\ \forall\, \theta \in \Theta_{sl}, \end{cases} \tag{3.4}$$

where the main lobes Θ_{ml} are constrained in the specified the lower bound $L(\theta)$ and the upper bound $U(\theta)$, respectively, with sidelobes Θ_{sl} below an envelope $\varepsilon(\theta)$.

The problem of synthesizing reconfigurable arrays aims to find K excitation vectors. The excitation elements with identical indexes share common magnitudes and produce K distinct power patterns that satisfy mask $\mathscr{C}_k$, $k = 1, \ldots, K$. This problem can be formulated as follows:

$$\text{find } \boldsymbol{w}_k \begin{cases} f(\boldsymbol{w}_k;\theta) \in \mathscr{C}_k,\quad k = 1, \ldots, K. & \text{(3.5a)} \\ |w_{1n}| = |w_{2n}| = \cdots = |w_{Kn}|, n = 1, \ldots, N, & \text{(3.5b)} \end{cases}$$

In (3.5a), the mask $\mathscr{C}_k$ specifies the limitations on radiation power pattern, such as $\mathscr{CF}$ and $\mathscr{CS}$ for creating a focused or shaped beam. The magnitude of the nth element of excitation vector k, denoted as $|w_{kn}|$, is also considered. Condition (3.5a) enforces phase-only control on the nth excitation element with consistent magnitudes across all K excitation vectors.

Instead of aiming for a viable solution, this section focuses on determining the excitation vectors $\boldsymbol{w}_k, k = 1, \ldots, K$, by solving the subsequent optimization problems:

$$\begin{aligned} &\min_{\boldsymbol{w}_k} \ \frac{1}{K}\sum_{k=1}^{K} \|\boldsymbol{w}_k\|_2^2 \\ &\text{s.t.}\, f(\boldsymbol{w}_k;\theta) \in \mathscr{C}_k, \ \ k = 1, \ldots, K \\ &\quad |w_{1n}| = |w_{2n}| = \cdots = |w_{Kn}|, n = 1, \ldots, N, \end{aligned} \tag{3.6}$$

Here, utilizing vector norm minimization as the objective function is aimed at reducing power consumption, preventing significant fluctuations in excitations, and indirectly limiting the dynamic range ratio (DRR) of excitations.

3.3 Proposed method

3.3.1 Sensor network modeling

In recent times, significant attention has been given to distributed consensus computation in sensor networks [14–21], illustrated in Figure 3.1(a). The primary goal is to leverage the collaborative efforts of sensor nodes to achieve consensus estimation or to carry out large-scale tasks involving sensing, signal processing, and communication. For instance, the sensor network with K nodes depicted in Figure 3.1(a) follows a loop topology defined by an undirected graph $G(\Pi, E)$. Here, the set of nodes is denoted by $\Pi = \{1, 2, \ldots, K\}$ (represented as black circles), and the edges E (illustrated as black straight lines with arrows) signify the communication links

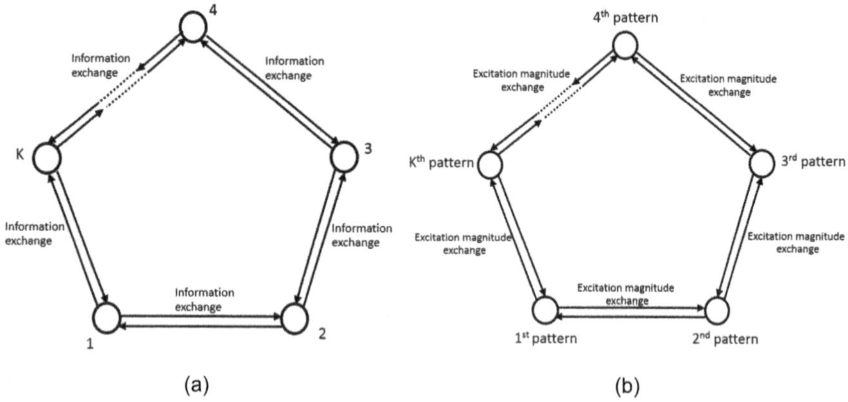

Figure 3.1 Reconfigurable beampattern synthesis modeling in a virtual (conceptual) sensor network. Black circles: nodes. Black straight lines: connections among network nodes. (a) Practical sensor network and (b) reconfigurable pattern synthesis in a conceptual sensor network.

between adjacent nodes [25–27]. Each node within this network possesses the capability to store local data, conduct computations, and communicate with neighboring nodes. Owing to its distinctive attributes, distributed consensus computation has found applications in various domains such as dictionary learning [16], dimensionality reduction [17], distributed classification [18], adaptive filtering [19], distributed detection [20], and distributed estimation [21], among others.

Please be advised that the primary objective of reconfigurable array beampattern synthesis is to identify K sets of excitation vectors with identical excitation magnitudes to generate K distinct radiated patterns. When it comes to determining consistent excitation magnitudes, reconfigurable array beampattern synthesis can be viewed as a specialized consensus calculation issue. Drawing inspiration from sensor network consensus computation, this chapter proposes the use of distributed consensus computation techniques to address the reconfigurable array beampattern synthesis challenge by simulating a virtual sensor network, illustrated in Figure 3.1(b), where each virtual "node" corresponds to an individual beampattern synthesis task. In this context, these "nodes" align with the number of radiated patterns and share equivalent excitation magnitudes. Hence, this chapter tackles a more complex magnitude-consensus computation challenge compared to the traditional consensus computation issue [14–18,20,21,28].

To streamline the calculation of distributed consensus, auxiliary variables denoted as $\boldsymbol{v}_k$ for $k = 1, \ldots, K$ are introduced, totaling K copies. These auxiliary variables are designed as part of the consensus constraints, where each variable $\boldsymbol{w}_k$ is equated to its corresponding $\boldsymbol{v}_k$. The original constraint (3.5a) in (3.6) is then replaced by magnitude constraints for the local neighboring variables within the virtual K-node sensor network. This entails ensuring that the magnitudes of $\boldsymbol{w}_k$ are equal to the magnitudes of $\boldsymbol{v}_{k'}$. Consequently, we transform the problem in (3.6) into a functionally equivalent form as follows:

$$\begin{aligned}
&\min_{\boldsymbol{w}_k, \boldsymbol{v}_k} \frac{1}{K} \sum_{k=1}^{K} \boldsymbol{w}_k^H \boldsymbol{w}_k \\
&\text{s.t.}\, f(\boldsymbol{w}_k; \theta) \in \mathscr{C}_k, \quad k = 1, \ldots, K \\
&\quad \boldsymbol{w}_k = \boldsymbol{v}_k \\
&\quad |\boldsymbol{w}_k| = |\boldsymbol{v}_{k'}|, k' \in \mathrm{Ne}(k),
\end{aligned} \tag{3.7}$$

where $\mathrm{Ne}(k)$ stands for the neighbor set of the kth conceptual "node" for $k = 1, \ldots, K$.

Clearly, (3.7) includes multiple limitations on the identical vector variable $\boldsymbol{w}_k$ (such as $\mathscr{C}_k$ and $|\boldsymbol{w}_k|$ with the absolute value operation), creating challenges in addressing (3.7). To simplify the calculation, we incorporate the agreement restriction $\boldsymbol{w}_k = z_k$ into (3.7) and separate $\boldsymbol{w}_k$ from the absolute value operation, resulting in the subsequent equivalent expression:

$$\min_{\boldsymbol{w}_k, \boldsymbol{v}_k, z_k} \frac{1}{K} \sum_{k=1}^{K} \boldsymbol{w}_k^H \boldsymbol{w}_k$$

$$\text{s.t.}\, f(\boldsymbol{w}_k; \theta) \in \mathscr{C}_k, \quad k = 1, \ldots, K \tag{3.8a}$$

$$\boldsymbol{w}_k = \boldsymbol{v}_k \tag{3.8b}$$

$$|z_k| = |\boldsymbol{v}_{k'}|, k' \in \mathrm{Ne}(k) \tag{3.8c}$$

$$\boldsymbol{w}_k = z_k. \tag{3.8d}$$

By utilizing the virtual sensor network structure depicted in Figure 3.1 and the transformation from (3.6) to (3.8), we have represented the issue of reconfigurable array synthesis as a unique distributed consensus calculation problem within a hypothetical sensor network consisting of K nodes. In the subsequent section, we will establish the appropriate method for sensor network computation to resolve the consensus calculation problem outlined in (3.8).

3.3.2 Sensor network computation

In general, there exist three types of techniques for distributed computation, specifically diffusion [29], gossip tactics [30], and the alternating direction method of multipliers (ADMM) [28,31–33]. With the motivation arising from the advantageous combination of dual decomposition and augmented Lagrangian methods leading to superior convergence properties, this chapter explores the utilization of ADMM to address the distributed consensus computation challenge defined in (3.8).

First, we construct the augmented Lagrangian for (3.8) as follows:

$$\begin{aligned}
&\mathscr{L}_1\left(\boldsymbol{w}_k, z_k, \boldsymbol{\lambda}_k, \boldsymbol{v}_k, \boldsymbol{\lambda}_{k,k}, \boldsymbol{\lambda}_{k,k'}\right) \\
&= \frac{1}{K}\sum_{k=1}^{K} \boldsymbol{w}_k^H \boldsymbol{w}_k + \sum_{k=1}^{K}\left(\boldsymbol{\lambda}_{k,k}^H(\boldsymbol{w}_k - \boldsymbol{v}_k) + \frac{\rho_1}{2}\|\boldsymbol{w}_k - \boldsymbol{v}_k\|_2^2\right) \\
&+ \sum_{k=1}^{K}\sum_{k'\in \mathrm{Ne}(k)}\left(\boldsymbol{\lambda}_{k,k'}^H(|z_k| - |\boldsymbol{v}_{k'}|) + \frac{\rho_2}{2}\||z_k| - |\boldsymbol{v}_{k'}|\|_2^2\right) \\
&+ \sum_{k=1}^{K}\left(\boldsymbol{\lambda}_k^H(\boldsymbol{w}_k - z_k) + \frac{\rho_3}{2}\|\boldsymbol{w}_k - z_k\|_2^2\right),
\end{aligned} \tag{3.9}$$

where $\rho_1 > 0$, $\rho_2 > 0$, and $\rho_3 > 0$ are the augmented Lagrangian parameters and act as penalty parameters, and $\boldsymbol{\lambda}_{k,k} \in \mathbb{C}^{N\times 1}$, $\boldsymbol{\lambda}_{k,k'} \in \mathbb{C}^{N\times 1}$, and $\boldsymbol{\lambda}_k \in \mathbb{C}^{N\times 1}$ are the Lagrange multipliers corresponding to the constraints $\boldsymbol{w}_k = \boldsymbol{v}_k$, $|z_k| = |\boldsymbol{v}_{k'}|$, and $\boldsymbol{w}_k = z_k$ for $k = 1, \ldots, K$, $k' \in \mathrm{Ne}(k)$, respectively.

After completing the square, (3.9) can be rewritten as the following scaled dual variables form [28]:

$$\begin{aligned}
&\mathscr{L}_1\left(\boldsymbol{w}_k, z_k, \boldsymbol{\mu}_k, \boldsymbol{v}_k, u_{k,k}, u_{k,k'}\right) \\
&= \frac{1}{K}\sum_{k=1}^{K} \boldsymbol{w}_k^H \boldsymbol{w}_k + \frac{\rho_1}{2}\sum_{k=1}^{K}\left(\|\boldsymbol{w}_k - \boldsymbol{v}_k + u_{k,k}\|_2^2 - \|u_{k,k}\|_2^2\right) \\
&+ \frac{\rho_2}{2}\sum_{k=1}^{K}\sum_{k'\in \mathrm{Ne}(k)}\left(\||z_k| - |\boldsymbol{v}_{k'}| + u_{k,k'}\|_2^2 - \|u_{k,k'}\|_2^2\right) \\
&+ \frac{\rho_3}{2}\sum_{k=1}^{K}\left(\|\boldsymbol{w}_k - z_k + \boldsymbol{\mu}_k\|_2^2 - \|\boldsymbol{\mu}_k\|_2^2\right),
\end{aligned} \tag{3.10}$$

where $u_{k,k} = (1/\rho_1)\boldsymbol{\lambda}_{k,k}$, $u_{k,k'} = (1/\rho_2)\boldsymbol{\lambda}_{k,k'}$ and $\boldsymbol{\mu}_k = (1/\rho_3)\boldsymbol{\lambda}_k$ are the scaled dual variables for $k = 1, \ldots, K$ and $k' \in \mathrm{Ne}(k)$.

By implementing the approach of iteratively reducing the augmented Lagrangian $\mathscr{L}_1$ in relation to the variables $\{\mathbf{w}_k, z_k, \boldsymbol{\mu}_k, \mathbf{v}_k, u_{k,k}, u_{k,k'}\}$ and the connectivity between the conceptual node variable and its neighboring nodes (illustrated in Figure 3.2), we derive the step-by-step distributed consensus computation process outlined in Table 3.1, with t representing the iteration count.

Through Steps 1–6 outlined in Table 3.1, it can be observed that (i) the set of variables $\{\psi_k\} = \{\mathbf{w}_k, z_k, \boldsymbol{\mu}_k, \mathbf{v}_k, u_{k,k}, u_{k,k'}\}$ can be computed independently by each kth "node," leading to the parallel computation of K groups of local information $\{\psi_k\}$ by all K "nodes"; (ii) each "node" is responsible for synthesizing a specific pattern, as depicted in Figure 3.2, where K patterns are synthesized individually by the K "nodes" and the computed results are shared with neighboring "nodes" to facilitate achieving a consensus on excitation magnitudes. For instance, "Node 1" carries out the synthesis computation for the first pattern while updating $\{z_1^{(t+1)}, \mathbf{v}_1^{(t+1)}, u_{1,2}^{(t+1)}, u_{1,K}^{(t+1)}\}$, and subsequently transmits the results to its neighboring "nodes," namely the 2nd and Kth "nodes." In parallel, "Node 1" receives the updated results $\{z_2^{(t+1)}, \mathbf{v}_2^{(t+1)}, u_{2,1}^{(t+1)}\}$ and $\{z_K^{(t+1)}, \mathbf{v}_K^{(t+1)}, u_{K,1}^{(t+1)}\}$ from "Nodes 2" and "K," respectively, to prepare for the next round of local computation.

Then, we particularly discuss the local computation in Steps 1, 2, and 4.

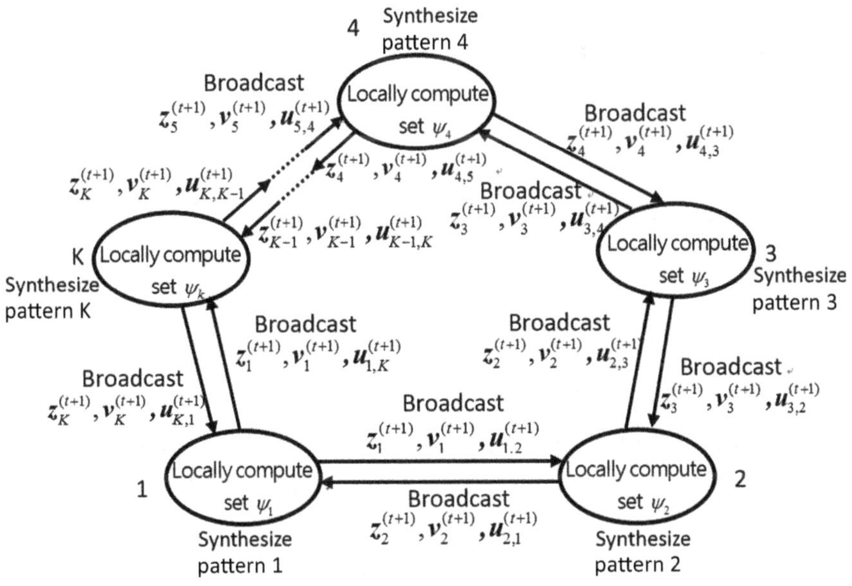

Figure 3.2 The parameter passing process on reconfigurable array synthesis via distributed computation in a conceptual K-node network

Table 3.1 The iterative steps of distributed consensus computation

Step 1	Locally compute $w_k^{(t+1)}$ in the kth conceptual "node" via solving $\min \mathcal{L}_1\left(w_k, z_k^{(t)}, \mu_k^{(t)}, v_k^{(t)}, u_{k,k}^{(t)}, u_{k,k'}^{(t)}\right)$ s.t. $f(w_k;\theta) \in \mathscr{C}_k$;
Step 2	Locally compute $z_k^{(t+1)}$ in the kth conceptual "node" via solving $\min \mathcal{L}_1\left(w_k^{(t+1)}, z_k, \mu_k^{(t)}, v_k^{(t)}, u_{k,k}^{(t)}, u_{k,k'}^{(t)}\right)$ and then broadcast it to the neighboring "nodes" to achieve information exchange purposes;
Step 3	Locally update $\mu_k^{(t+1)}$ in the kth "node" for $k = 1, \ldots, K$: $\mu_k^{(t+1)} = \mu_k^{(t)} + w_k^{(t+1)} - z_k^{(t+1)}$;
Step 4	Locally compute $v_k^{(t+1)}$ in the kth conceptual "node" via solving $\min \mathcal{L}_1\left(w_k^{(t+1)}, z_k^{(t+1)}, \mu_k^{(t+1)}, v_k, u_{k,k}^{(t)}, u_{k,k'}^{(t)}\right)$ and then broadcast it to the neighboring "nodes" to achieve information exchange;
Step 5	Locally update $u_{k,k}^{(t+1)}$ in the kth "node" for $k = 1, \ldots, K$: $u_{k,k}^{(t+1)} = u_{k,k}^{(t)} + w_k^{(t+1)} - v_k^{(t+1)}$;
Step 6	Locally update $u_{k,k'}^{(t+1)}$ in the kth "node" for $k' \in \text{Ne}(k)$: $u_{k,k'}^{(t+1)} = u_{k,k'}^{(t)} + \lvert z_k^{(t+1)}\rvert - \lvert v_{k'}^{(t+1)}\rvert$, and then broadcast it to the neighboring "nodes" to achieve information exchange purposes.

1) Local computation of $w_k^{(t+1)}$ with known $\{z_k^{(t)}, \mu_k^{(t)}, v_k^{(t)}, u_{k,k}^{(t)}, u_{k,k'}^{(t)}\}$ in Step 1, we implement the optimization computation to obtain $w_k^{(t+1)}$ via solving

$$\begin{aligned}
&\min_{w_k} \mathcal{L}_1\left(w_k, z_k^{(t)}, \mu_k^{(t)}, v_k^{(t)}, u_{k,k}^{(t)}, u_{k,k'}^{(t)}\right) \\
&= \min_{w_k} \frac{1}{K}\sum_{k=1}^{K} w_k^H w_k + \frac{\rho_1}{2}\sum_{k=1}^{K} \|w_k - v_k^{(t)} + u_{k,k}^{(t)}\|_2^2 \\
&\quad + \frac{\rho_3}{2}\sum_{k=1}^{K}\left(\|w_k - z_k^{(t)} + \mu_k^{(t)}\|_2^2\right) \\
&\text{s.t. } f(w_k;\theta) \in \mathscr{C}_k, \quad k = 1, \ldots, K,
\end{aligned} \tag{3.11}$$

which implies that the K vector variables $\{w_k\}_{k=1}^K$ can be determined in a distributed fashion by the K "nodes" due to the fact that (3.11) may be divided into K local subproblems across K "nodes" as follows:

$$\begin{aligned}
&\min_{w_k} \frac{1}{K} w_k^H w_k + \frac{\rho_1}{2}\|w_k - x_k\|_2^2 + \frac{\rho_3}{2}\|w_k - \tilde{w}_k\|_2^2 \\
&\text{s.t. } f(w_k;\theta) \in \mathscr{C}_k,
\end{aligned} \tag{3.12}$$

where $x_k = v_k^{(t)} - u_{k,k}^{(t)}$ and $\tilde{w}_k = z_k^{(t)} - \mu_k^{(t)}$.

Note that (3.12) shows a special synthesis problem of the kth "node" with the added penalty terms $\frac{\rho_1}{2}\|w_k - x_k\|_2^2$ and $\frac{\rho_3}{2}\|w_k - \tilde{w}_k\|_2^2$ yielded from the information exchange to assist to achieve the consensus purposes. After removing constant items, we rewrite (3.12) as follows:

$$\min_{w_k} \; w_k^H A w_k + b^H w_k + w_k^H b$$
$$\text{s.t.}\, f(w_k;\theta) \in \mathscr{C}_k, \tag{3.13}$$

where

$$A = \left(\frac{1}{K} + \frac{\rho_1}{2} + \frac{\rho_3}{2}\right) I_N, b = -\frac{\rho_1}{2} x_k - \frac{\rho_3}{2}\tilde{w}_k. \tag{3.14}$$

For the sake of generality, we discuss three concrete cases of the constraint set $\mathscr{C}_k$ in (3.13) as follows:

- When $\mathscr{C}_k$ denotes $\mathscr{CF}$, the corresponding conceptual network is used for synthesizing the focused beam, which is described in Appendix A in detail;
- When $\mathscr{C}_k$ stands for $\mathscr{CS}$, the conceptual network is used for synthesizing shaped beam, which is described in Appendix B in detail;
- Otherwise, all nodes yield focused beams with different main beam directions, including unconstrained and constrained beam scanning, which are described in Appendix C in detail.

2) Local computation of $z_k^{(t+1)}$ with known $\{w_k^{(t+1)}, \mu_k^{(t)}, v_k^{(t)}, u_{k,k}^{(t)}, u_{k,k'}^{(t)}\}$ in Step 2, we implement the optimization computation to obtain $z_k^{(t+1)}$ via solving

$$\{z_k^{(t+1)}\} = \arg\min_{z_k} \mathscr{L}_1\left(w_k^{(t+1)}, z_k, \mu_k^{(t)}, v_k^{(t)}, u_{k,k}^{(t)}, u_{k,k'}^{(t)}\right) \tag{3.15}$$

which reduces into the following simple form via ignoring irrelevant constant items:

$$\min_{z_k} \frac{\rho_2}{2} \sum_{k' \in \mathrm{Ne}(k)} \||z_k| - y_{k'}\|_2^2 + \frac{\rho_3}{2}\|z_k - \tilde{z}_k\|_2^2, \tag{3.16}$$

where $y_{k'} = |v_{k'}^{(t)}| - u_{k,k'}^{(t)}$ and $\tilde{z}_k = w_k^{(t+1)} + \mu_k^{(t)}$.

It is simple to confirm that the quantity $\|z_k - \tilde{z}_k\|_2^2$ reaches its minimum when the direction of z_k is the same as that of $\tilde{z}_k$, indicating that we only have to ascertain the magnitudes of the components of the complex vector z_k as the phases can be derived directly from the components of $\tilde{z}_k$, i.e.,

$$\angle z_k^{(t+1)} = \angle \tilde{z}_k. \tag{3.17}$$

Then, substituting $\angle z_k = \angle \tilde{z}_k$ into (3.16) yields the following least-squares problem:

$$\min_{|z_k|} \frac{\rho_2}{2} \sum_{k' \in \mathrm{Ne}(k)} \||z_k| - y_{k'}\|_2^2 + \frac{\rho_3}{2}\||z_k| - |\tilde{z}_k|\|_2^2, \tag{3.18}$$

the solution to which is easily derived as follows:

$$|z_k^{(t+1)}| = \max\left(\frac{\rho_3|\tilde{z}_k| + \rho_2 \sum_{k'\in \mathrm{Ne}(k)} y_{k'}}{\rho_2 \mathrm{card}(k) + \rho_3}, 0\right), \tag{3.19}$$

In this case, the count of neighboring nodes of the kth "node" is denoted as card(k), and a vector is formed with components of x that are above 0 using max(x, 0). Thus, the creation of $z_k^{(t+1)}$ involves the utilization of the angles in (3.17) and the magnitudes in (3.19).

3) To calculate $\mathbf{v}_k^{(t+1)}$ *locally with respect to* $\{\mathbf{w}_k^{(t+1)}, z_k^{(t+1)}, \boldsymbol{\mu}_k^{(t+1)}, \mathbf{v}_k, u_{k,k}^{(t)}, u_{k,k'}^{(t)}\}$ *in Step 3, an optimization procedure is performed to find* $\mathbf{v}_k^{(t+1)}$ *by solving*

$$\min_{\mathbf{v}_k} \mathscr{L}_1\left(\mathbf{w}_k^{(t+1)}, z_k^{(t+1)}, \boldsymbol{\mu}_k^{(t+1)}, \mathbf{v}_k, u_{k,k}^{(t)}, u_{k,k'}^{(t)}\right)$$

$$= \min_{\mathbf{v}_k} \frac{\rho_1}{2}\sum_{k=1}^{K} \|\mathbf{v}_k - \boldsymbol{\alpha}_k\|_2^2 + \frac{\rho_2}{2}\sum_{k=1}^{K}\sum_{k'\in \mathrm{Ne}(k)} \||\mathbf{v}_{k'}| - \boldsymbol{\beta}_k\|_2^2 \tag{3.20}$$

where $\boldsymbol{\alpha}_k = \mathbf{w}_k^{(t+1)} + u_{k,k}^{(t)}$ and $\boldsymbol{\beta}_k = |z_k^{(t+1)}| + u_{k,k'}^{(t)}$.

Given that the undirected graph demonstrates symmetry, meaning if $k' \in \mathrm{Ne}(k)$, then $k \in \mathrm{Ne}(k')$ [30,34], we can rephrase (3.20) as follows:

$$\min_{\mathbf{v}_k} \frac{\rho_1}{2}\sum_{k=1}^{K} \|\mathbf{v}_k - \boldsymbol{\alpha}_k\|_2^2 + \frac{\rho_2}{2}\sum_{k=1}^{K}\sum_{k'\in \mathrm{Ne}(k)} \||\mathbf{v}_k| - \boldsymbol{\beta}_{k'}\|_2^2, \tag{3.21}$$

where $\boldsymbol{\beta}_{k'} = |z_{k'}^{(t+1)}| + u_{k',k}^{(t)}$.

It can be confirmed that the best $\mathbf{v}_k$ in (3.21) shares the same phase as $\boldsymbol{\alpha}_k$, i.e.,

$$\angle \mathbf{v}_k^{(t+1)} = \angle \boldsymbol{\alpha}_k. \tag{3.22}$$

Just like (3.15)–(3.19), the magnitudes of $|\mathbf{v}_k^{(t+1)}|$ can be easily calculated by substituting (3.22) into (3.21) as follows:

$$|\mathbf{v}_k^{(t+1)}| = \max\left(\frac{\rho_1|\boldsymbol{\alpha}_k| + \rho_2 \sum_{k'\in \mathrm{Ne}(k)} \boldsymbol{\beta}_{k'}}{\rho_2 \mathrm{card}(k) + \rho_1}, 0\right). \tag{3.23}$$

Thus, $\mathbf{v}_k^{(t+1)}$ can be constructed from the phases in (3.22) and amplitudes in (3.23).

Once the data circulates within all the K "nodes" in a virtual (abstract) sensor network, each of the excitation vectors $\{\mathbf{w}_k\}_{k=1}^K$ will possess a shared set of magnitudes once the network reaches a stable state. The stability of the network is determined by comparing whether t is equal to the maximum iteration count T_m or if the consensus errors from (3.8b), (3.8c), and (3.8d) are below 10^{-5}. Based on the aforementioned analysis and calculations, Algorithm 3.1 introduces the VIrtual SEnsor Network COmputation (VISENCO) technique for synthesizing reconfigurable array patterns.

Algorithm 3.1 VISENCO method for reconfigurable array beampattern synthesis

---

Initialization: set $\{w_k^{(0)}, z_k^{(0)}, u_k^{(0)}, v_k^{(0)}, u_{k,k}^{(0)}, u_{k,k'}^{(0)}\}$ and T_m;

Construct a virtual (conceptual) sensor network with K nodes, where each "node" corresponds to an individual synthesis task with given pattern type (focused beam, shaped beam) and corresponding constraint or mask parameters.

for $t = 0, \ldots, T_m$

Each "node" performs its own local computation, and then transmits the temporal computation results in each iteration to its "neighbors" and receives the neighbors' results to achieve information exchange purpose as shown in the following K cycles:

for $k = 1, \ldots, K$

locally compute $w_k^{(t+1)}$,$z_k^{(t+1)}$,$\mu_k^{(t+1)}$, where the kth "node" carries out the synthesis task of the kth pattern, and broadcasts the temporal computation result $z_k^{(t+1)}$ to its "neighbors".

locally compute $v_k^{(t+1)}$ from (3.22) and (3.23), where the kth "node" receives $z_{k'}^{(t+1)}, k' \in \mathrm{Ne}(k)$ from its "neighbors" and then $v_k^{(t+1)}$ is forwarded to its "neighbors".

locally compute $u_{k,k}^{(t+1)}, u_{k,k'}^{(t+1)}$ from Step 5 and Step 6, and $u_{k,k'}^{(t+1)}, k' \in \mathrm{Ne}(k)$ is broadcasted to its "neighbors."

end

end for $t = T_m$ or the maximal consensus errors of constraint (3.8b), (3.8c) and (3.8d) are all less than 10^{-5}.

Output the optimal $w_k^\star,\ k = 1, \ldots, K$.

3.4 Extension to uniform amplitude array synthesis

In real-world applications, engineers are more interested in synthesizing a uniform amplitude array because it can be easily controlled by active antenna amplifiers, resulting in maximum array efficiency. This type of array maintains a consistent excitation vector magnitude, denoted as $|w_{k,n}| = c, \forall n, k$, where c represents a specific constant value. To achieve this, the VISENCO method can be expanded to generate K patterns by optimizing K excitations with a constant magnitude c and adjusting phases. This can be accomplished by solving the modified problem outlined in (3.6) with the additional restriction $|w_{kn}| = c$ for $k = 1, \ldots, K, n = 1, \ldots, N$.

Please note that the cost function of the resulting problem is aimed at minimizing the square magnitude c^2 (for example, $\frac{1}{K}\sum_{k=1}^{K}\|w_k\|_2^2 = Nc^2$). This is in contrast to the previous approach outlined in [22–24], where the fixed magnitude c was used to determine phases and element locations without considering beam scanning. In this current method, however, the magnitude c is considered an unknown variable to be determined. This chapter leverages the Degrees of Freedom (DoF) associated with the unknown magnitude c to address a more challenging issue of minimizing magnitude to reduce power consumption and generate multiple patterns (such as beam scanning).

We adopt the methodology from the VISENCO method and construct the distributed consensus optimization problem in a similar manner to (3.8), with the exception of imposing additional constraints on z_k:

$$\min_{w_k, v_k, z_k, c} \; c^2$$

$$\text{s.t.}\; f(w_k;\theta) \in \mathscr{C}_k, \tag{3.24a}$$

$$w_k = v_k \tag{3.24b}$$

$$|z_k| = |v_{k'}|, k' \in \text{Ne}(k) \tag{3.24c}$$

$$w_k = z_k \tag{3.24d}$$

$$|z_k| = c, \;\; k = 1, \ldots, K. \tag{3.24e}$$

Since z_k is a copy of w_k from (3.24d), w_k has the same magnitude as z_k. Similarly, Step 1–Step 6 are employed to solve (3.24). However, the only to be modified is the local computation of z_k in (3.18) from Step 2 :

$$\min_{|z_k|, c} \frac{\rho_2}{2} \sum_{k' \in \text{Ne}(k)} \| |z_k| - y_{k'} \|_2^2 + \frac{\rho_3}{2} \| |z_k| - |\tilde{z}_k| \|_2^2$$

$$\text{s.t.}\; |z_{k,n}| = c, \tag{3.25}$$

the solution to which is easily given as follows:

$$|z_{k,n}^{(t+1)}| = \max\left(\frac{\rho_3 \sum_{n=1}^{N} |\tilde{z}_{k,n}| + \rho_2 \sum_{k' \in \text{Ne}(k)} \sum_{n=1}^{N} y_{k',n}}{\rho_2 N \text{card}(k) + \rho_3 N}, 0 \right), \forall n. \tag{3.26}$$

At the conclusion, the substitution of (3.19) with (3.26) in Step 2 is utilized to develop the modified VISENCO technique for uniform amplitude array synthesis.

3.5 Numerical examples

Within this segment, illustrations of adaptable array synthesis with phase-only regulation are introduced to showcase the potential and efficiency of the proposed approach. To maintain a general approach, emphasis is placed on the synthesis of linear arrays where each antenna is treated as an isotropic element, denoted as $g_n(\theta) = 1$. For all experiments, the initialization of our method involves the utilization of the predefined set $\{w_k^{(0)}, z_k^{(0)}, u_k^{(0)}, v_k^{(0)}, u_{k,k}^{(0)}, u_{k,k'}^{(0)}\}$ with all elements set to 0, and it is terminated either at the maximum iteration count of $T_m = 1000$ or when the largest consensus error among the K nodes regarding the auxiliary variables and the excitation vectors (AVEV), specifically $\max_n |w_n - v_n|$, $\max_n |w_n - v_n'|$, and $\max_n |w_n - z_n|$, is below 10^{-5}. All experimental procedures are carried out on our PC equipped with an Intel i7-6700 CPU operating on a 64-bit system and 16GB RAM.

3.5.1 Focused beam-shaped beam synthesis

Consider a linear array containing 21 isotropic elements spaced at half-wavelength intervals. By using phase-only control, the array can alternate between emitting a focused beam with sidelobe levels below −26.5 dB for $|\sin(\theta)| \geq 0.16$ and a shaped beam with main lobe ripple of ± 0.45 dB over $|\sin(\theta)| \leq 0.2$ and sidelobe levels below −24.5 dB for $|\sin(\theta)| \geq 0.35$.

Please be advised that the array reconfigures two patterns, prompting the creation of a virtual loop-topology sensor network equipped with a duo of nodes. Our approach involves using the parameters $\rho_1 = 100, \rho_2 = 40, \rho_3 = 100$ to formulate two distinct configurations of the reconfigurable array, while juxtaposed with the methodology outlined in [7]. The outcomes of the array synthesis are graphically depicted in Figure 3.3, where the blue and red lines signify the focused and shaped beams achieved, respectively, with the green dashed lines denoting the mask constraint. The results displayed in Figure 3.3(a)–(c) correspond to the synthesized patterns, showcasing common excitation magnitudes and varying phases, respectively, through the approach delineated in [7], whereas Figure 3.3(d)–(g) showcases the corresponding synthesis outcomes utilizing our proposed technique. A graphical representation elucidating the maximal consensus AVEV error across K "nodes" over numerous iterations is showcased in Figure 3.3(g), revealing the convergence of our method following 524 iterations with AVEV errors well below 10^{-5} (−100 dB). To determine the effectiveness of the two methodologies, we compute the consensus error between the distinct excitation vector magnitudes (DEVM), denoted as

$$\text{DEVM error=norm}(|\boldsymbol{w}_i| - |\boldsymbol{w}_j|), \tag{3.27}$$

where $\boldsymbol{w}_i$ and $\boldsymbol{w}_j$ denote the excitation vectors aligned with different patterns. The DEVM error associated with the [7] method stands at 1.7721×10^{-3}, contrasting with our method's significantly lower error rate of 3.9785×10^{-5}, ensuring precise phase-only control. Additionally, the excitation vector norm for each pattern derived from [7] is calculated at 0.51, marginally higher than our method's norm of 0.507. In terms of the dynamic range ratio (DRR) of the excitations, quantified as [1],

$$\text{DRR}(\boldsymbol{w}) = \frac{\max_n |w_n|}{\min_n |w_n|}, \tag{3.28}$$

our ratio stands at 8.5 compared to that of [7] is 8.8, emphasizing our method's capability in achieving smaller DRR for the excitations. The proposed method and reference [7] have running durations of 45 seconds and 315 seconds, respectively. This disparity is due to the faster convergence speed inherited by the proposed method from ADMM, while the SDR technique employed in [7] increases the dimensions of the initial problem to twice its size.

3.5.2 Shaped beam-shaped beam synthesis

In the second experiment, two additional types of configurations are examined for integration into a linear array consisting of 32 isotropic components spaced at intervals of one-third wavelength. The array emits the cosecant square beam and shaped

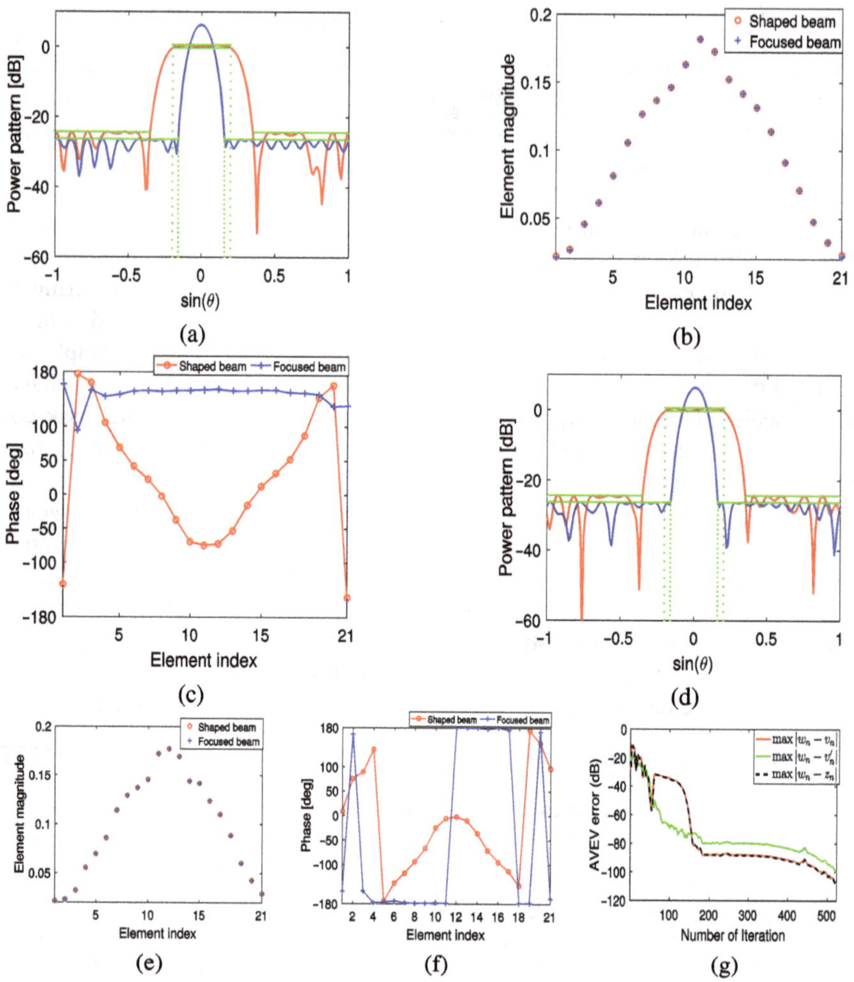

Figure 3.3 Synthesis of reconfigurable array by phase-controlled: panels (a), (b), and (c) generated by the method in [7] show the corresponding radiation power pattern, excitation magnitudes, and phases; panels (d), (e), (f), and (g) are obtained by the proposed method with a conceptual two-node loop-topology sensor network.

beam through phase-only manipulation. The cosecant square beam exhibits a main-lobe variation of ± 0.5 dB within the range of $0.1 \leq \sin(\theta) \leq 0.5$, with sidelobes attenuated below -22 dB for $-1 \leq \sin(\theta) \leq 0$ and $0.55 \leq \sin(\theta) \leq 1$. On the other hand, the shaped beam displays a variation of ± 0.25 dB within $|\sin(\theta)| \leq 0.2$ and achieves sidelobes below -20 dB for $|\sin(\theta)| \geq 0.3$.

Utilizing $\rho_1 = 80$, $\rho_2 = 200$, $\rho_3 = 80$ in our methodology, the synthesis outcomes produced by the approach proposed in [7] and our novel method are depicted in Figure 3.4(a)–(c) and (d)–(g), respectively. The DEVM inaccuracies for the

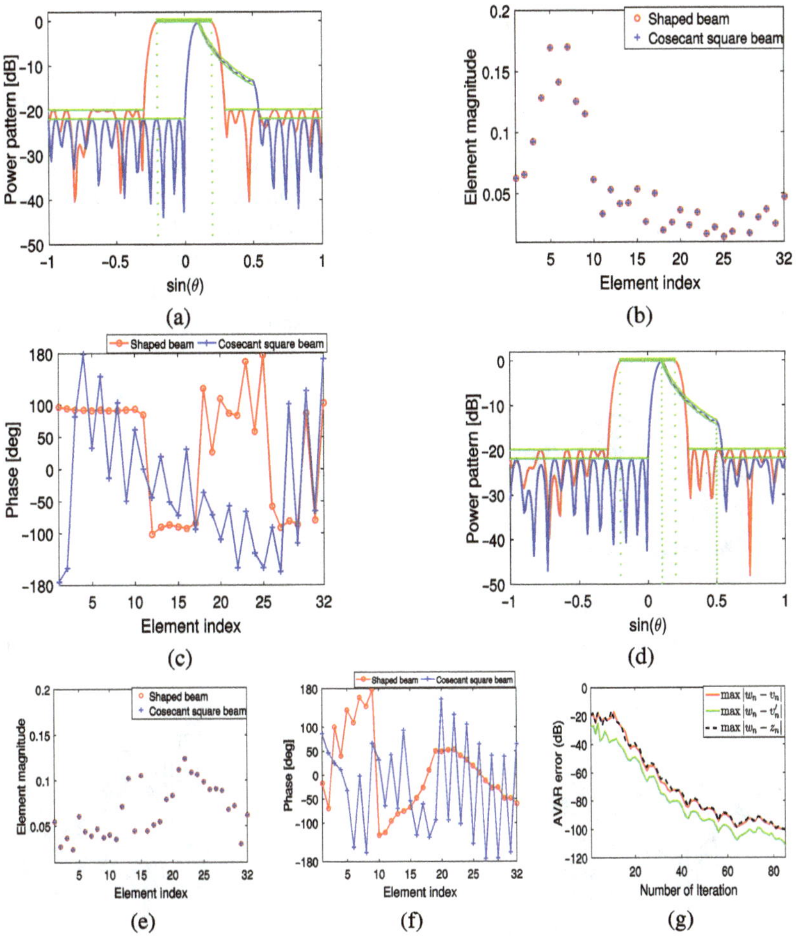

Figure 3.4 *Synthesis of reconfigurable array by phase-controlled: panels (a), (b), and (c) correspond to the radiation power pattern, excitation magnitudes, and phases generated by the method [7]; panels (d), (e), (f), and (g) are obtained by the proposed method with a conceptual two-node loop-topology sensor network.*

methodologies mentioned are computed as 4.2818×10^{-12} and 3.2253×10^{-5} for [7] and our method, respectively, along with corresponding execution times of 1141 and 12 seconds. Significantly, our approach reaches convergence after 85 iterations, as illustrated in Figure 3.4(g). Notably, despite fulfilling the phase-only stipulations, our method is highlighted for producing excitations with a reduced DRR value of 5.3 compared to the DRR of 12.3 from the technique described in [7]. Moreover, our method achieves a smaller excitation vector norm of 0.40, outperforming the norm of 0.41 in [7].

3.5.3 Multiple beam scanning

In the third test, a linear array of $N = 21$ elements spread with half-wavelength intervals is used, and the scanning range of the beams covers 16 beams within the range of $[-15°, 15°]$ at intervals of 2°. Each beam exhibits a sidelobe level lower than −25 dB, and a transition zone of 12° is present on both sides of every scanning angle. The research focuses on the iterative projection technique [1] for generating multiple beam scans for the array, aimed at minimizing the Euclidean distance from the specified power pattern collections through a projection process that reconciles the convex and non-convex sets. Results depicted in Figure 3.5(a)–(c) require 61s of CPU time. However, it is evident that [1] falls short of completely meeting the

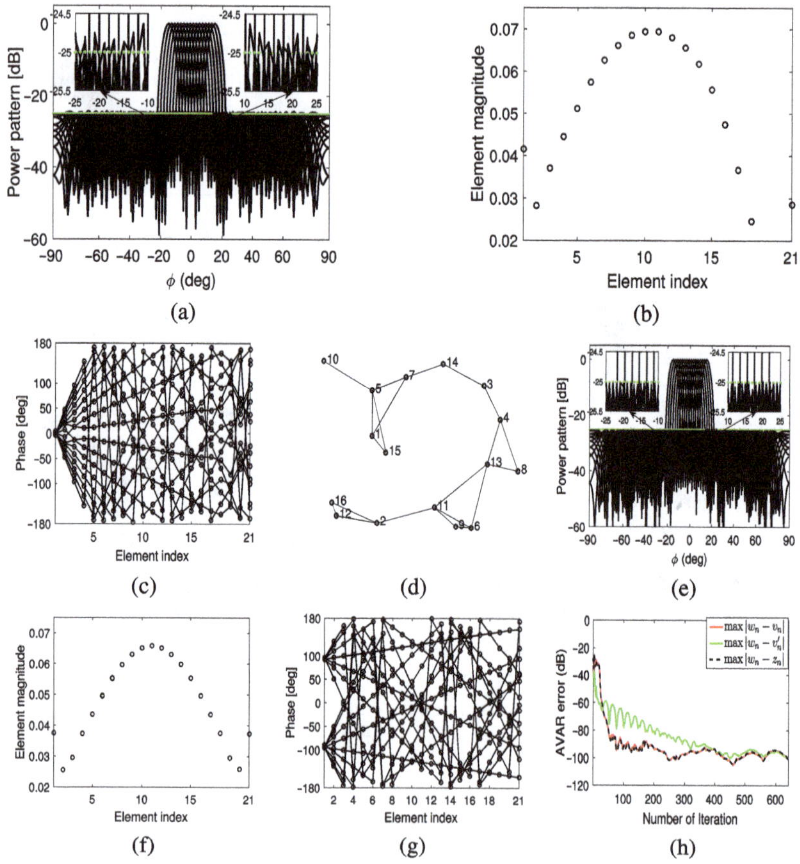

Figure 3.5 Beam scanning for linear array of $N = 21$ elements with phase-only control shifting 16 beams at interval 2° from −15° to 15°: panels (a), (b), and (c) generated by the method [1] show the corresponding global views of the synthesized patterns by overlapping 16 beams, excitation magnitudes, and phases, panels (e), (f), (g), and (h) are obtained by the proposed method with the conceptual sensor network of Figure 5(d).

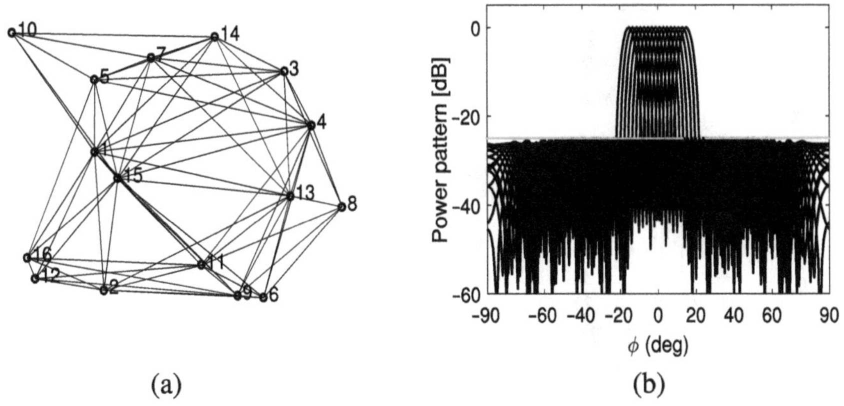

Figure 3.6 (a) Synthesis of 16 beams scanning by the proposed method with (b) the conceptual sensor network of panel (a)

Table 3.2 Performance of different virtual sensor networks with varying neighbors

	DRR	Maximal DEVM error	Excitations norm	CPU time [s]
Figure 5(d)	2.56	8.8020×10^{-5}	0.2270	19 (641 iterations)
Figure 6(a)	2.55	3.9724×10^{-5}	0.2270	20 (554 iterations)

requirements outlined in the pattern due to inaccuracies in pattern matching. Utilizing a virtual 16-node sensor network as illustrated in Figure 3.5(d), we apply our proposed method with $\rho_1 = 90, \rho_2 = 40, \rho_3 = 90$ to achieve 16 beam scans, yielding the outcomes displayed in Figure 3.5(e)–(h) at a cost of only 19s of CPU time. Figure 3.5(e) clearly demonstrates our method's adeptness in precisely controlling the radiation power pattern to adhere to the prescribed mask by optimizing the pattern exclusively according to the mask constraints. Furthermore, Figure 3.5(h) showcases the convergence of our method after 641 iterations, surpassing the speed of [1]. In addition, the DRR from [1] amounts to 10.3, contrasting with our technique's rate of 2.5. Notably, the maximum DEVM errors from the 16 patterns derived from both our approach and [1] are 8.8020×10^{-5} and 4.076×10^{-17}, respectively, signifying that both approaches have met the identical magnitude criteria. The excitation vector norm achieved through our strategy is 0.2270, marginally inferior to the value attained by [1], standing at 0.23.

Next, an examination of the impact of varying neighbor configurations of each "node" on the effectiveness of the proposed approach is conducted. The network nodes illustrated in Figure 3.6(a) exhibit a higher number of neighboring connections compared to those in Figure 3.5(d), facilitating the synthesis of 16 scanning beams showcased in Figure 3.6(b). Evaluating different virtual (conceptual) sensor network topologies with diverse neighbor structures on the operational efficiency of our approach, as presented in Table 3.2, reveals that slight performance variations

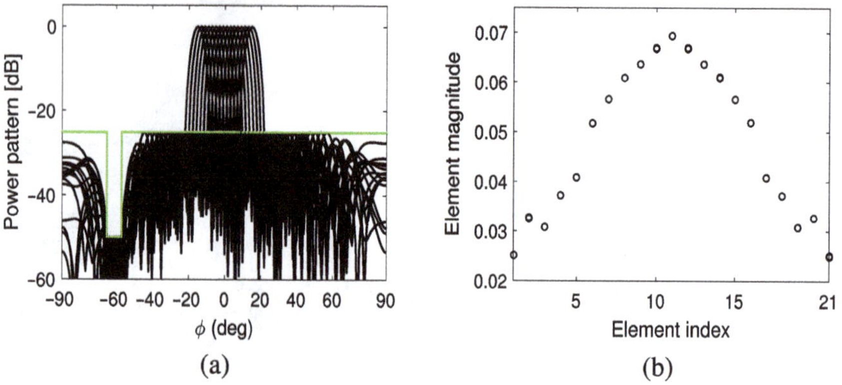

Figure 3.7 (a) Synthesis of constrained beam scanning by the proposed method with (b) the conceptual sensor network of Figure 3.5(d)

exist between two conceptual network layouts featuring varying numbers of adjacent nodes when generating patterns through phase-only control. Notably, the conceptual network denoted in Figure 3.6(a) with more neighbors requires fewer iterations to achieve convergence relative to that in Figure 3.5(d). However, it is essential to note that the increase in network edges entails additional constraints on neighboring variables, offering no time-saving advantage.

Alternatively, an exploration of constrained beam scanning using identical experimental conditions as Figure 3.5, except for the imposition of a null zone of $[-65^\circ, -57^\circ]$ and a null level of -50 dB, as depicted in Figure 3.7(a) and (b), further underscores the efficacy of our method in achieving beam scanning while adhering to null constraints.

3.5.4 Uniform amplitude array synthesis for beam scanning

In the fourth experiment, a modified VISENCO approach utilizing (3.26) in Step 2 is applied to create a linear array with uniform amplitude consisting of $N = 21$ elements spaced at half-wavelength intervals. The scan range for beam coverage includes 16 beams ranging from -15° to 15° at intervals of 2°. The sidelobe levels for each beam, whether unconstrained or constrained, are limited to -16 dB and -15 dB, respectively, with a transitional zone of 12° on either side of each scan angle. Unlike previous studies [22–24] that did not address the scenario of unknown magnitude and beam scanning in a uniform amplitude array, our focus is solely on the modified VISENCO technique for designing patterns on a conceptual 16-node sensor network as depicted in Figure 3.5(d). The synthesis outcome is illustrated in Figure 3.8, revealing that achieving uniform amplitude imposes more stringent requirements compared to a general reconfigurable array with a standard set of excitation magnitudes as seen in Figure 3.5(f) and 3.7(b). Consequently, Figure 3.8 exhibits higher sidelobe levels than Figures 3.5–3.7 due to reduced degrees of freedom. Notably, the minimized magnitudes for unconstrained and constrained

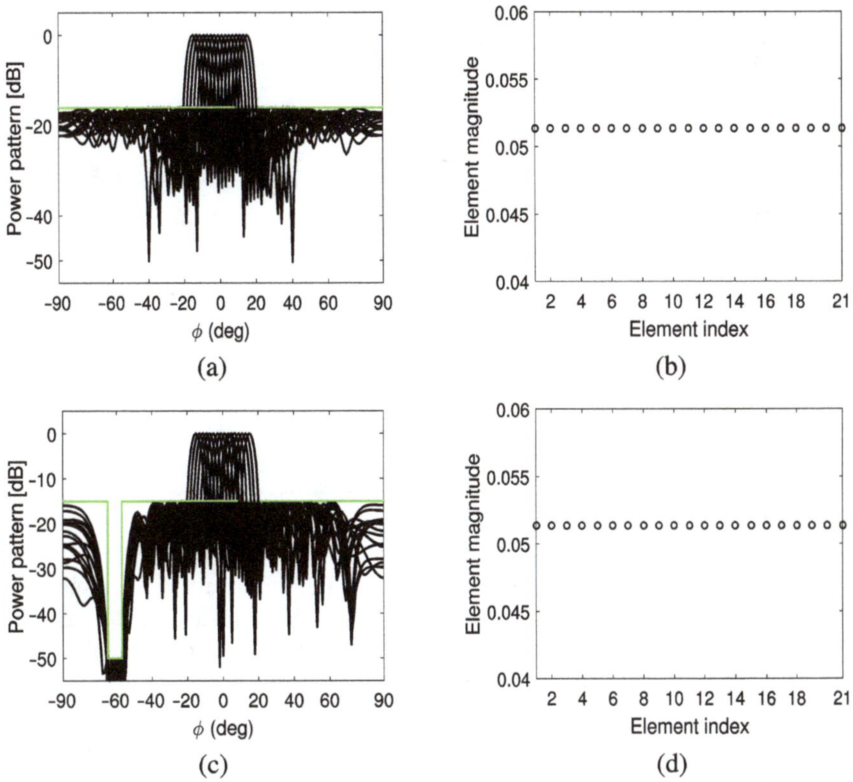

Figure 3.8 Synthesis of 16 beams in the uniform amplitude array: panels (a) and (b) depict the unconstrained beam scanning; panels (c) and (d) depict the constrained beam scanning with the null zone $[-65°, -57°]$ and the null level -50 dB.

beam scanning are nearly identical, approximately 0.05134 and 0.05138 as demonstrated in Figure 3.8(b) and (d). When null constraints are introduced in Figure 3.8(c), resulting in a reduction of degrees of freedom, there is a corresponding 1 dB increase in the sidelobe region compared to Figure 3.8(a).

3.5.5 Practical performance of the proposed method simulated with Ansys HFSS

In the final experiment, investigating the practical performance of the proposed method in array synthesis begins with a study on the case where each antenna is isotropic without any coupling effects. For this scenario, $g_n(\theta)$ is uniform across antennas $n = 1, \ldots, N$, with $g_n(\theta)$ representing the radiation pattern of a half-wave dipole operating at 3 GHz (e.g., wavelength $\lambda = 100$ mm) with a length and radius of 0.48 λ and 0.005λ, respectively, as depicted in Figure 3.9(a). The initial experiment involves 21 half-wave dipoles spaced at 0.5λ to test the reconfiguration of

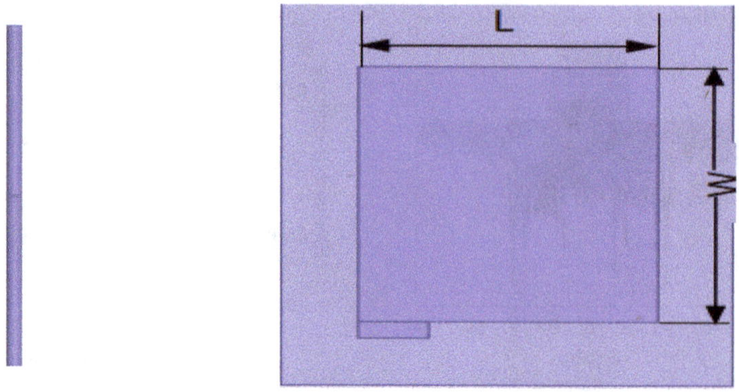

(a) (b)

Figure 3.9 Antenna structure: (a) half-wave dipole and (b) microstrip patch

a focused-shaped beam using the 3-D full-wave electromagnetic software Ansys High-Frequency Simulation Software (HFSS) as illustrated in Figure 3.10(a). Similarly, Figure 3.10(b) demonstrates the utilization of 32 half-wave dipoles spaced at one-third of λ to reconfigure the shaped beam with a common amplitude excitation in the second experiment. Moreover, the performance evaluation of multiple beam scanning is carried out using both common and uniform amplitude excitations in Figure 3.10(c)–(f), showcasing scan directions of -15°, -1°, and 15° for simplicity. It is observed that the proposed method achieves satisfactory patterns through phase control with minimal loss in actual patterns, validating its effectiveness.

The inevitable coupling effect is observed in practical scenarios. Another commonly used microstrip patch antenna, operating at 16 GHz with dimensions of 5.25 mm in length and 8.2 mm in width, includes a dielectric layer with a thickness of 0.5 mm, as illustrated in Figure 3.9(b). To consider the coupling effect between the patches, a linear array consisting of 21 patches is simulated along the X-axis in the HFSS environment to calculate the steering vector $\boldsymbol{a}(\theta)$. The proposed techniques are employed to generate the desired radiation patterns. In this case, 16 beams are synthesized using a combination of common amplitude excitation and uniform amplitude excitation, as depicted in Figure 3.11. A comparison between Figure 3.11 with mutual coupling and Figures 3.5–3.8 without mutual coupling, also including Figure 3.10, reveals that all excitation amplitudes are lower compared to the case without coupling, possibly influenced by the individual patch pattern $g_n(\theta)$. Moreover, our methods demonstrate effectiveness even in the presence of mutual coupling to achieve desired radiation patterns with phase-only control, emphasizing the practical utility of the proposed techniques.

3.6 Conclusion

In this chapter, a new approach is introduced for generating array patterns through the controlling phase only. The synthesis problem is treated as a distributed computation

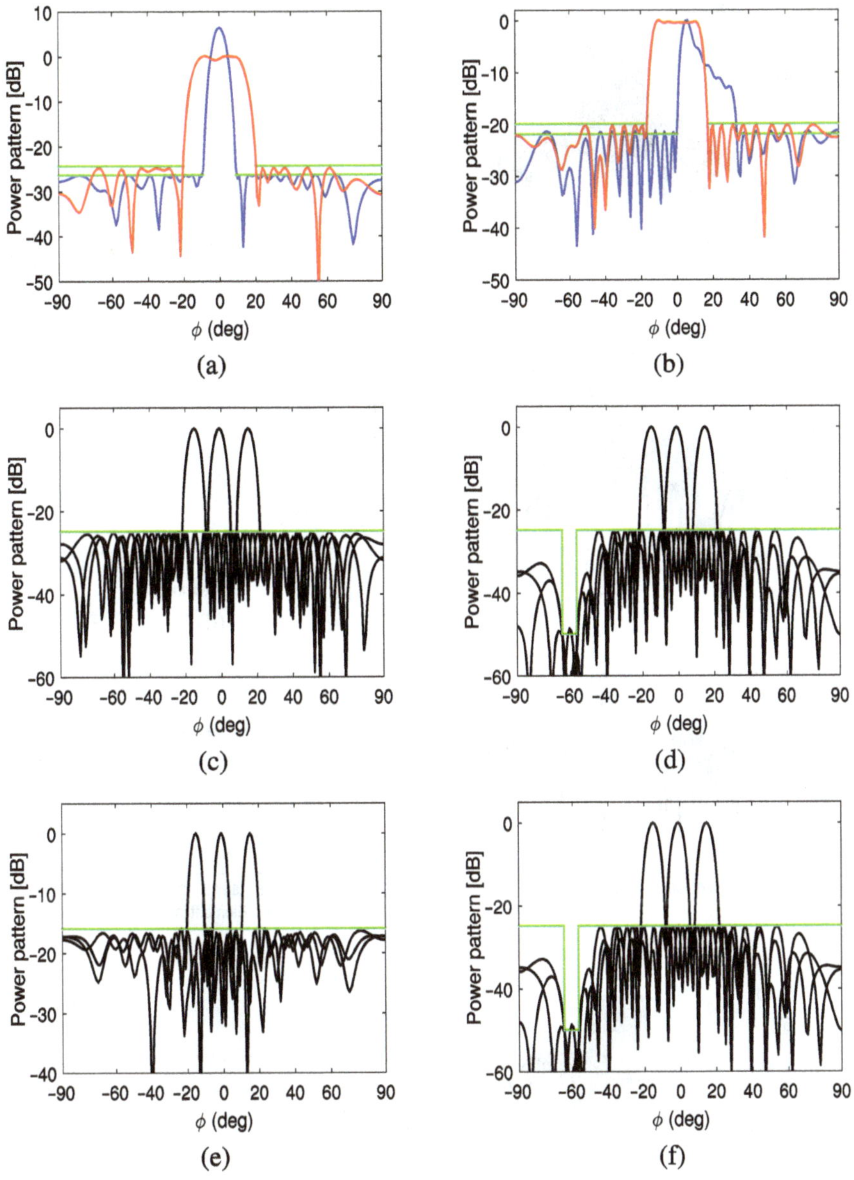

Figure 3.10 Synthesis of a linear array composed of half-wave dipoles working at 3 GHz via using Ansys HFSS: panels (a) and (b) show the reconfigured patterns; panels (c) and (d) show the beam scanning from a set of common excitations; panels (e) and (f) show the beam scanning from a uniform amplitude excitation.

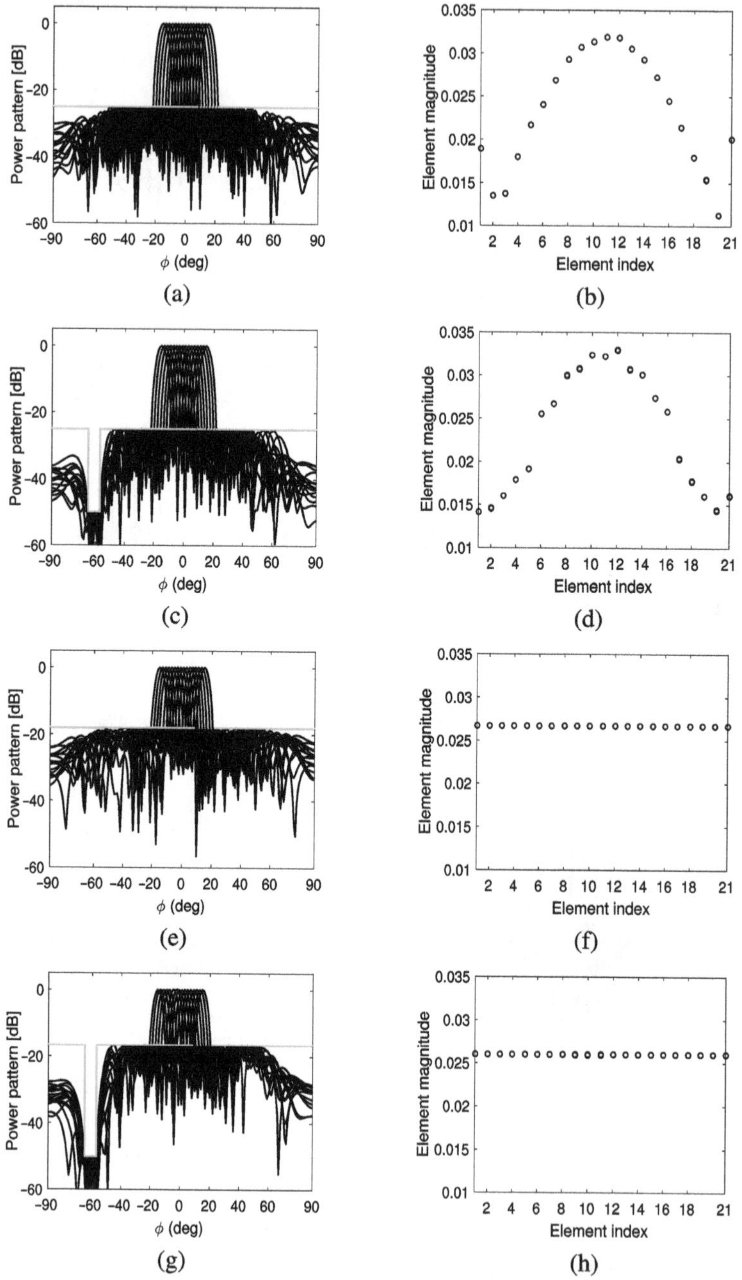

Figure 3.11 Synthesis of a linear array composed of 21 patches via using Ansys HFSS: panels (a)–(d) beam scanning from a set of common amplitude excitation; panels (e)–(h) beam scanning from a uniform amplitude excitation.

challenge within a sensor network concept, with the number of nodes representing the patterns to be generated. By leveraging communication between neighboring nodes, a consensus is reached on the excitation levels for each element within the array. Additionally, this method is expanded to include uniform amplitude array synthesis. Results from numerical simulations demonstrate the effectiveness and practicality of this technique, outperforming other methods in terms of both execution time and Dynamic Range Ratio (DRR).

Appendix A: Focused beam synthesis

When denoting $\mathscr{C}_k$ as $\mathscr{C}\mathscr{F}$, the related network is utilized for the synthesis of a focused beam, specifically, (3.13) represents the subsequent optimization problem:

$$\begin{aligned}
&\min_{\boldsymbol{w}_k} \ \boldsymbol{w}_k^H A \boldsymbol{w}_k + b^H \boldsymbol{w}_k + \boldsymbol{w}_k^H b \\
&\text{s.t.} \ |\boldsymbol{w}_k^H \boldsymbol{a}(\theta_0)|^2 \geq 1 \\
&\qquad |\boldsymbol{w}_k^H \boldsymbol{a}(\theta_s)|^2 \leq \rho(\theta_s), \ \forall \theta_s \in \Theta_{sl},
\end{aligned} \tag{A.1}$$

Here, the sidelobe region Θ_{sl} is discretized into S equally spaced angles, namely, $\{\theta_1, \ldots, \theta_S\}$.

To separate $\boldsymbol{w}_k$ from the constraints, two auxiliary variables p and $\hat{s}_s, s = 1, \ldots, S$ are introduced, generating consensus constraints $p = \boldsymbol{w}_k^H \boldsymbol{a}(\theta_0)$ and $\hat{s}_s = \boldsymbol{w}_k^H \boldsymbol{a}(\theta_s)$. Consequently, (A.1) can be equivalently represented as follows:

$$\begin{aligned}
&\min_{\boldsymbol{w}_k} \ \boldsymbol{w}_k^H A \boldsymbol{w}_k + b^H \boldsymbol{w}_k + \boldsymbol{w}_k^H b \\
&\text{s.t.} \ |p|^2 \geq 1 \\
&\qquad |\hat{s}_s|^2 \leq \rho(\theta_s), \ s = 1, \ldots, S \\
&\qquad p = \boldsymbol{w}_k^H \boldsymbol{a}(\theta_0) \\
&\qquad \hat{s}_s = \boldsymbol{w}_k^H \boldsymbol{a}(\theta_s).
\end{aligned} \tag{A.2}$$

Based on (A.2), we then formulate the augmented Lagrangian:

$$\begin{aligned}
&\mathscr{L}_2\left(\boldsymbol{w}_k, p, \hat{s}_s, \zeta, \vartheta_s\right) \\
&= \boldsymbol{w}_k^H A \boldsymbol{w}_k + b^H \boldsymbol{w}_k + \boldsymbol{w}_k^H b + \frac{\rho}{2}|p - \boldsymbol{w}_k^H \boldsymbol{a}(\theta_0) + \zeta|^2 \\
&\quad - |\zeta|^2 + \frac{\rho}{2}\sum_{s=1}^{S}\left(|\hat{s}_s - \boldsymbol{w}_k^H a(\theta_s) + \vartheta_s|^2 - |\vartheta_s|^2\right) \\
&\text{s.t.} \ |p|^2 \geq 1 \\
&\qquad |\hat{s}_s|^2 \leq \rho(\theta_s), s = 1, \ldots, S,
\end{aligned} \tag{A.3}$$

where ζ and ϑ_s denote scaled dual variables, and $\rho > 0$ denotes the Lagrangian parameter.

Now we utilize Steps 1–4 to solve (A.3) in an iterative fashion.

Step 1: By utilizing the provided $\{\boldsymbol{w}_k^{(t)}, \hat{s}_s^{(t)}, \zeta^{(t)}, \vartheta_s^{(t)}\}$, we determine the sequence $\{p^{(t+1)}\}$ by ignoring immaterial terms, from

$$
\begin{aligned}
\{p^{(t+1)}\} &= \arg\min_{p} \ \mathscr{L}_2\left(\boldsymbol{w}_k^{(t)}, p, \hat{s}_s^{(t)}, \zeta^{(t)}, \vartheta_s^{(t)}\right) \\
&= \arg\min_{p} \ |p - \boldsymbol{w}_k^{(t)H}\boldsymbol{a}(\theta_0) + \zeta^{(t)}|^2 \\
&\text{s.t. } |p|^2 \geq 1,
\end{aligned} \tag{A.4}
$$

which is actually a projection of the scalar $\boldsymbol{w}_k^{(t)H}\boldsymbol{a}(\theta_0) + \zeta^{(t)}$ onto the constraint set, yielding a closed-form solution denoted as follows:

$$
p^{(t+1)} = \begin{cases} \tilde{p}, & \text{if } \ |\tilde{p}| \geq 1 \\ e^{j\angle\tilde{p}}, & \text{otherwise} \end{cases} \tag{A.5}
$$

where $\tilde{p} = \boldsymbol{w}_k^{(t)H}\boldsymbol{a}(\theta_0) - \zeta^{(t)}$.

Step 2: Compute $\{\hat{s}_s^{(t+1)}\}$ using the provided $\{\boldsymbol{w}_k^{(t)}, p^{(t+1)}, \zeta^{(t)}, \vartheta_s^{(t)}\}$ values, after omitting constant elements, from

$$
\begin{aligned}
\{\hat{s}_s^{(t+1)}\} &= \arg\min_{S_s} \ \mathscr{L}_2\left(\boldsymbol{w}_k^{(t)}, p^{(t+1)}, \hat{s}_s, \zeta^{(t)}, \vartheta_s^{(t)}\right) \\
&= \arg\min_{\hat{s}_s} \ \sum_{s=1}^{S}\left(|\hat{s}_s - \boldsymbol{w}_k^{(t)H} a(\theta_s) + \vartheta_s^{(t)}|^2\right) \\
&\text{s.t. } |\hat{s}_s|^2 \leq \rho(\theta_s).
\end{aligned} \tag{A.6}
$$

Likewise, (A.6) can be divided into S smaller problems, which can each be resolved through a basic projection process as described:

$$
\hat{s}_s^{(t+1)} = \begin{cases} \sqrt{\rho(\theta_s)}e^{j\angle\tilde{s}_s}, & \text{if } |\tilde{s}_s| \geq \sqrt{\rho(\theta_s)} \\ \tilde{s}_s, & \text{otherwise} \end{cases} \tag{A.7}
$$

where $\tilde{s}_s = \boldsymbol{w}_k^{(t)H} a(\theta_s) - \vartheta_s^{(t)}$.

Step 3: Find the set $\{\boldsymbol{w}_k^{(t+1)}\}$ using the provided $\{p^{(t+1)}, \hat{s}_s^{(t+1)}, \zeta^{(t)}, \vartheta_s^{(t)}\}$ derived from the simplified expression:

$$
\begin{aligned}
&\min_{\boldsymbol{w}_k} \ \boldsymbol{w}_k^H A \boldsymbol{w}_k + b^H \boldsymbol{w}_k + \boldsymbol{w}_k^H b + \frac{\rho}{2}|\boldsymbol{w}_k^H \boldsymbol{a}(\theta_0) - c|^2 \\
&+ \frac{\rho}{2}\sum_{s=1}^{S}\left(|\boldsymbol{w}_k^H \boldsymbol{a}(\theta_s) - d_s|^2\right),
\end{aligned} \tag{A.8}
$$

where $c = p^{(t+1)} + \zeta^{(t)}, d_s = \hat{s}_s^{(t+1)} + \vartheta_s^{(t)}$, and setting the derivative of (A.8) with respect to $\boldsymbol{w}_k^*$ to zeros yields $\tilde{A}\boldsymbol{w}_k = \tilde{b}$, where

$$
\tilde{A} = A + \frac{\rho}{2}\boldsymbol{a}(\theta_0)\boldsymbol{a}(\theta_0)^H + \frac{\rho}{2}\sum_{s=1}^{S}\boldsymbol{a}(\theta_s)\boldsymbol{a}(\theta_s)^H, \tag{A.9}
$$

$$
\tilde{b} = \frac{\rho}{2}\boldsymbol{a}(\theta_0)c^* + \frac{\rho}{2}\sum_{s=1}^{S}\boldsymbol{a}(\theta_s)d_s^* - b, \tag{A.10}
$$

and since $\tilde{A}$ is invertible, $\boldsymbol{w}_k^{(t+1)} = \tilde{A}^{-1}\tilde{b}$.

Step 4: Update the scaled dual variables $\{\zeta^{(t+1)}, \vartheta_s^{(t+1)}\}$ by

$$\zeta^{(t+1)} = \zeta^{(t)} + p^{(t+1)} - \boldsymbol{w}_k^{(t+1)H}\boldsymbol{a}(\theta_0) \tag{A.11}$$

$$\vartheta_s^{(t+1)} = \vartheta_s^{(t)} + \hat{s}_s^{(t+1)} - \boldsymbol{w}_k^{(t+1)H}\boldsymbol{a}(\theta_s), \tag{A.12}$$

for $s = 1, \ldots, S$.

Repeat Steps 1–4 until a preset maximum iteration number T or the residuals of consensus constraints $p = \boldsymbol{w}_k^H\boldsymbol{a}(\theta_0)$ and $\hat{s}_s = \boldsymbol{w}_k^H\boldsymbol{a}(\theta_s)$ are less than 10^{-5}.

Appendix B: Shaped beam synthesis

When $\mathscr{C}_k$ represents the $\mathscr{CS}$, the network is utilized for creating shaped beam patterns. In other words, the optimization problem

$$\begin{aligned}
&\min_{\boldsymbol{w}_k} \ \boldsymbol{w}_k^H A \boldsymbol{w}_k + b^H \boldsymbol{w}_k + \boldsymbol{w}_k^H b \\
&\text{s.t. } L(\theta_m) \le |\boldsymbol{w}_k^H \boldsymbol{a}(\theta_m)|^2 \le U(\theta_m), \ \forall \theta_m \in \Theta_{ml} \\
&\quad |\boldsymbol{w}_k^H \boldsymbol{a}(\bar{\theta}_s))|^2 \le \varepsilon(\bar{\theta}_s), \ \forall \bar{\theta}_s \in \Theta_{sl},
\end{aligned} \tag{B.1}$$

corresponding to (3.13) involves discretizing the main beam region Θ_{ml} and the side-lobe region Θ_{sl} into equally spaced angles, namely $\{\theta_1, \ldots, \theta_M\}$ and $\{\bar{\theta}_1, \ldots, \bar{\theta}_S\}$, respectively. It is evident that the non-convex nature of (B.1) arises due to the dual constraints imposed on quadratic terms concerning $\boldsymbol{w}_k$. Similarly, to disentangle $\boldsymbol{w}_k$ from the mainlobe and sidelobe constraints, we introduce two auxiliary variables q_m and r_s, $m = 1, \ldots, M$, $s = 1, \ldots, S$, and establish consensus constraints $q_m = \boldsymbol{w}_k^H \boldsymbol{a}(\theta_m)$ and $r_s = \boldsymbol{w}_k^H \boldsymbol{a}(\bar{\theta}_s)$. Subsequently, (B.1) is rephrased as

$$\begin{aligned}
&\min_{\boldsymbol{w}_k} \ \boldsymbol{w}_k^H A \boldsymbol{w}_k + b^H \boldsymbol{w}_k + \boldsymbol{w}_k^H b \\
&\text{s.t. } L(\theta_m) \le |q_m|^2 \le U(\theta_m), \ m = 1, \ldots, M \\
&\quad |r_s|^2 \le \varepsilon(\bar{\theta}_s), \ s = 1, \ldots, S \\
&\quad q_m = \boldsymbol{w}_k^H \boldsymbol{a}(\theta_m), \quad m = 1, \ldots, M \\
&\quad r_s = \boldsymbol{w}_k^H \boldsymbol{a}(\bar{\theta}_s), \quad s = 1, \ldots, S.
\end{aligned} \tag{B.2}$$

Based on (B.2), we construct another augmented Lagrangian as follows:

$$\begin{aligned}
&\mathscr{L}_3\left(\boldsymbol{w}_k, q_m, r_s, \eta_m, \xi_s\right) \\
&= \boldsymbol{w}_k^H A \boldsymbol{w}_k + b^H \boldsymbol{w}_k + \boldsymbol{w}_k^H b + \frac{\rho}{2}\sum_{m=1}^{M}\left(|q_m - \boldsymbol{w}_k^H \boldsymbol{a}(\theta_m)\right. \\
&\left. + \eta_m|^2 - |\eta_m|^2\right) + \frac{\rho}{2}\sum_{s=1}^{S}\left(|r_s - \boldsymbol{w}_k^H a(\bar{\theta}_s) + \xi_s|^2 - |\xi_s|^2\right) \\
&\text{s.t. } L(\theta_m) \le |q_m|^2 \le U(\theta_m), \ m = 1, \ldots, M \\
&\quad |r_s|^2 \le \varepsilon(\bar{\theta}_s), \ s = 1, \ldots, S,
\end{aligned} \tag{B.3}$$

where η_m and ξ_s are scaled dual variables for $m = 1, \ldots, M$ and $s = 1, \ldots, S$, and $\rho > 0$ is the Lagrangian parameter.

Utilizing the identical iterative approach as Steps 1–4 referred to as Algorithm 3.2, we address the (B.2) by resolving the issue in the following manner.

Step 1: Determine $\{q_m^{(t+1)}\}$ with the given $\{w_k^{(t)}, r_s^{(t)}, \eta_m^{(t)}, \xi_s^{(t)}\}$ via solving

$$\begin{aligned}
&\min_{q_m} \mathscr{L}_3\left(\boldsymbol{w}_k^{(t)}, q_m, r_s^{(t)}, \eta_m^{(t)}, \xi_s^{(t)}\}\right)\\
&= \min_{q_m} \sum_{m=1}^{M} |q_m - \boldsymbol{w}_k^{(t)H}\boldsymbol{a}(\theta_m) + \eta_m^{(t)}|^2\\
&\text{s.t. } L(\theta_m) \leq |q_m|^2 \leq U(\theta_m),\ m = 1, \ldots, M,
\end{aligned} \tag{B.4}$$

which amounts to a projection problem with double-side constraints for M subproblems, and thus the solution to which is given by

$$q_m^{(t+1)} = \begin{cases} \sqrt{U(\theta_m)}e^{j\angle\tilde{q}_m}, & \text{if } |\tilde{q}_m| \geq \sqrt{U(\theta_m)} \\ \sqrt{L(\theta_m)}e^{j\angle\tilde{q}_m}, & \text{if } |\tilde{q}_m| \leq \sqrt{L(\theta_m)} \\ \tilde{q}_m, & \text{otherwise} \end{cases} \tag{B.5}$$

for $m = 1, \ldots, M$, and $\tilde{q}_m = \boldsymbol{w}_k^{(t)H}\boldsymbol{a}(\theta_m) - \eta_m^{(t)}$.

Step 2: Determine $\{r_s^{(t+1)}\}$ with the given $\{w_k^{(t)}, q_m^{(t+1)}, r_s, \eta_m^{(t)}, \xi_s^{(t)}\}$ via solving

$$\begin{aligned}
&\min_{r_s} \mathscr{L}_3\left(\boldsymbol{w}_k^{(t)}, q_m^{(t+1)}, r_s, \eta_m^{(t)}, \xi_s^{(t)}\right)\\
&= \min_{r_s} \sum_{s+1}^{S} |r_s - \boldsymbol{w}_k^{(t)H}\boldsymbol{a}(\bar{\theta}_s) + \xi_s^{(t)}|^2\\
&\text{s.t. } |r_s|^2 \leq \varepsilon(\bar{\theta}_s),\ s = 1, \ldots, S.
\end{aligned} \tag{B.6}$$

From (A.6) to (A.7), we can easily obtain $r_s^{(t+1)}$ as follows:

$$r_s^{(t+1)} = \begin{cases} \sqrt{\varepsilon(\bar{\theta}_s)}e^{j\angle\tilde{r}_s}, & \text{if } |\tilde{r}_s| \geq \sqrt{\varepsilon(\bar{\theta}_s)} \\ \tilde{r}_s, & \text{otherwise} \end{cases} \tag{B.7}$$

for $s = 1, \ldots, S$, and $\tilde{r}_s = \boldsymbol{w}_k^{(t)H}\boldsymbol{a}(\bar{\theta}_s) - \xi_s^{(t)}$.

Step 3: Determine $\{\boldsymbol{w}_k^{(t+1)}\}$ with the given $\{\boldsymbol{w}_k^{(t)}, q_m^{(t+1)}, r_s^{(t+1)}, \eta_m^{(t)}, \xi_s^{(t)}\}$ via solving

$$\begin{aligned}
&\min_{\boldsymbol{w}_k} \boldsymbol{w}_k^H A \boldsymbol{w}_k + b^H \boldsymbol{w}_k + \boldsymbol{w}_k^H b + \frac{\rho}{2}\sum_{m=1}^{M}\left(|\boldsymbol{w}_k^H \boldsymbol{a}(\theta_m)\right.\\
&\left. - e_m|^2\right) + \frac{\rho}{2}\sum_{s=1}^{S}\left(|\boldsymbol{w}_k^H \boldsymbol{a}(\bar{\theta}_s) - g_s|^2\right)
\end{aligned} \tag{B.8}$$

where $e_m = q_m^{(t+1)} + \eta_m$, $g_s = r_s^{(t+1)} + \xi_s$.

Differentiating (B.8) with respect to $\boldsymbol{w}_k^*$ and equating the result to zeros yields $\boldsymbol{w}_k^{(t+1)} = \hat{A}^{-1}\hat{b}$, where

$$\hat{A} = A + \frac{\rho}{2}\sum_{m=1}^{M} \boldsymbol{a}(\theta_m)\boldsymbol{a}(\theta_m)^H + \frac{\rho}{2}\sum_{s=1}^{S} \boldsymbol{a}(\bar{\theta}_s)\boldsymbol{a}(\bar{\theta}_s)^H, \tag{B.9}$$

$$\hat{b} = \frac{\rho}{2}\sum_{m=1}^{M} \boldsymbol{a}(\theta_m)e_m^* + \frac{\rho}{2}\sum_{s=1}^{S} \boldsymbol{a}(\bar{\theta}_s)g_s^* - b\,. \tag{B.10}$$

Step 4: Update the dual variables $\{\eta^{(t+1)}, \xi_s^{(t+1)}\}$ as follows:

$$\eta_m^{(t+1)} = \eta^{(t)} + q_m^{(t+1)} - \boldsymbol{w}_k^{(t+1)H}\boldsymbol{a}(\theta_m) \tag{B.11}$$

$$\xi_s^{(t+1)} = \xi_s^{(t)} + r_s^{(t+1)} - \boldsymbol{w}_k^{(t+1)H}\boldsymbol{a}(\bar{\theta}_s) \tag{B.12}$$

for $m = 1, \ldots, M$ and $s = 1, \ldots, S$.
Repeat Steps 1–4 until a preset maximum iteration number T or the residuals of consensus constraints $q_m = \boldsymbol{w}_k^H\boldsymbol{a}(\theta_m)$ and $r_s = \boldsymbol{w}_k^H\boldsymbol{a}(\bar{\theta}_s)$ are less than 10^{-5}.

Appendix C: Beam scanning synthesis

To address the beam scanning synthesis task, the focused beams produced by different nodes have varying main beam directions, making it a unique case of reconfigurable array synthesis. The key to solving this lies in replacing (A.5) with $p^{(t+1)} = e^{j\angle\tilde{p}}$, where $\tilde{p} = \boldsymbol{w}_k^{(t)H}\boldsymbol{a}(\theta_k) - \zeta^{(t)}$. Here, θ_k represents the main beam direction of the kth focused beam, and by following the steps outlined in Appendix A, each focused beam can be synthesized effectively. Adding to this, the consideration of beam scanning synthesis with a null zone brings about a new dimension to the task. With the null zone denoted as Θ_I, it is segmented uniformly into $\{\theta_1, \ldots, \theta_I\}$ and constrained by a specific threshold level $\sigma(\theta_i)$, where $|\boldsymbol{w}_k^H\boldsymbol{a}(\theta_i)|^2 \leq \sigma(\theta_i)$ for $i = 1, \ldots, I$. This constraint introduces a new level of complexity to the synthesis process, requiring a more nuanced approach to ensure the successful generation of focused beams with desired characteristics.

For a concise overview of beam scanning, we can consider the problem formulation (3.13) as follows. This problem is denoted as follows:

$$\begin{aligned}
&\min_{\boldsymbol{w}_k}\ \boldsymbol{w}_k^H A\boldsymbol{w}_k + b^H\boldsymbol{w}_k + \boldsymbol{w}_k^H b\\
&\text{s.t. } |\boldsymbol{w}_k^H\boldsymbol{a}(\theta_0)|^2 = 1\\
&\quad |\boldsymbol{w}_k^H\boldsymbol{a}(\theta_s)|^2 \leq \rho(\theta_s),\ \theta_s \in \Theta_{sl}\\
&\quad |\boldsymbol{w}_k^H\boldsymbol{a}(\theta_i)|^2 \leq \sigma(\theta_i),\ i \in \Theta_I,
\end{aligned} \tag{C.1}$$

which is non-convex in nature. To address this, we can apply a similar approach by introducing three additional variables p, s_s, h_i and defining consensus constraints such as $p = \mathbf{w}_k^H \boldsymbol{a}(\theta_0)$, $\hat{s}_s = \mathbf{w}_k^H \boldsymbol{a}(\theta_s), s = 1, \ldots, S$, and $h_i = |\mathbf{w}_k^H \boldsymbol{a}(\theta_i)|^2 \leq \sigma(\theta_i)$, $i = 1, \ldots, I$. Subsequently, we can construct an augmented Lagrangian function:

$$
\begin{aligned}
&\mathscr{L}_4 \left(\mathbf{w}_k, p, \hat{s}_s, h_i, \zeta, \vartheta_s, \delta_i\right) \\
&= \mathbf{w}_k^H A \mathbf{w}_k + b^H \mathbf{w}_k + \mathbf{w}_k^H b + \frac{\rho}{2}|p - \mathbf{w}_k^H \boldsymbol{a}(\theta_0) + \zeta|^2 \\
&\quad - |\zeta|^2 + \frac{\rho}{2}\sum_{s=1}^{S} \left(|\hat{s}_s - \mathbf{w}_k^H a(\theta_s) + \vartheta_s|^2 - |\vartheta_s|^2\right) \\
&\quad + \frac{\rho}{2}\sum_{i=1}^{I} \left(|h_i - \mathbf{w}_k^H a(\theta_i) + \delta_i|^2 - |\delta_i|^2\right) \\
&\text{s.t. } |p|^2 = 1 \\
&\quad |\hat{s}_s|^2 \leq \rho(\theta_s), s = 1, \ldots, S, \\
&\quad |h_i|^2 \leq \sigma(\theta_i), i = 1, \ldots, I, \\
&\quad p = \mathbf{w}_k^H \boldsymbol{a}(\theta_0) \\
&\quad \hat{s}_s = \mathbf{w}_k^H \boldsymbol{a}(\theta_s), s = 1, \ldots, S, \\
&\quad h_i = |\mathbf{w}_k^H \boldsymbol{a}(\theta_i)|^2,\ i = 1, \ldots, I,
\end{aligned}
\tag{C.2}
$$

incorporating scaled dual variables ζ, ϑ_s, δ_i, and a Lagrangian parameter $\rho > 0$. By utilizing alternating update procedures outlined in Appendix A, we can effectively tackle the optimization problem presented in (C.2).

References

[1] Vescovo R. Reconfigurability and beam scanning with phase-only control for antenna arrays. *IEEE Transactions on Antennas and Propagation*. 2008;56(6):1555–1565.

[2] Bucci OM, Mazzarella G, and Panariello G. Reconfigurable arrays by phase-only control. *IEEE Transactions on Antennas and Propagation*. 1991;39(7):919–925.

[3] Bucci OM and D'Elia G. Power synthesis of reconfigurable conformal arrays with phase-only control. *IEE Proceedings – Microwaves, Antennas and Propagation*. 1998;145(1):131–136.

[4] Bucci OM, Capozzoli A, and D'Elia G. Power pattern synthesis of reconfigurable conformal arrays with near-field constraints. *IEEE Transactions on Antennas and Propagation*. 2004;52(1):132–141.

[5] Luenberger DG and Ye Y. *Linear and Nonlinear Programming*. Springer; 1984.

[6] Morabito AF, Massa A, Rocca P, *et al*. An effective approach to the synthesis of phase-only reconfigurable linear arrays. *IEEE Transactions on Antennas and Propagation*. 2012;60(8):3622–3631.

[7] Fuchs B. Application of convex relaxation to array synthesis problems. *IEEE Transactions on Antennas and Propagation*. 2014;62(2):634–640.

[8] Durr M, Trastoy A, and Ares F. Multiple-pattern linear antenna arrays with single prefixed amplitude distributions: modified Woodward–Lawson synthesis. *Electronics Letters*. 2000;36(16):1345–1346.

[9] Mahanti GK, Chakraborty A, and Das S. Phase-only and amplitude-phase only synthesis of dual-beam pattern linear antenna arrays using floating-point genetic algorithms. *Progress in Electromagnetics Research*. 2007;68: 247–259.

[10] Gies D and Rahmat-Samii Y. Particle swarm optimization for reconfigurable phase-differentiated array design. *Microwave & Optical Technology Letters*. 2003;38(3):168–175.

[11] Díaz X, Rodríguez JA, Ares F, *et al.* Design of phase-differentiated multiple-pattern antenna arrays. *Microwave & Optical Technology Letters*. 2015;26(1):52–53.

[12] Liang J, Zhang X, So HC, *et al.* Sparse array beampattern synthesis via alternating direction method of multipliers. *IEEE Transactions on Antennas and Propagation*. 2018;66(5):2333–2345.

[13] Fan X, Liang J, Zhang Y, *et al.* Shaped power pattern synthesis with minimization of dynamic range ratio. *IEEE Transactions on Antennas and Propagation*. 2019;67(5):3067–3078.

[14] Shi X, Cao J, and Huang W. Distributed parametric consensus optimization with an application to model predictive consensus problem. *IEEE Transactions on Cybernetics*. 2018;48(7):2024–2035.

[15] Kajiyama Y, Hayashi N, and Takai S. Distributed subgradient method with edge-based event-triggered communication. *IEEE Transactions on Automatic Control*. 2018;63(7):2248–2255.

[16] Liang J, Zhang M, Zeng X, *et al.* Distributed dictionary learning for sparse representation in sensor networks. *IEEE Transactions on Image Processing*. 2014;23(6):2528–2541.

[17] Liang J, Yu G, Chen B, *et al.* Decentralized dimensionality reduction for distributed tensor data across sensor networks. *IEEE Transactions on Neural Networks and Learning Systems*. 2016;27(11):2174–2186.

[18] Kokiopoulou E and Frossard P. Distributed classification of multiple observation sets by consensus. *IEEE Transactions on Signal Processing*. 2011;59(1):104–114.

[19] Cavalcante RLG and Mulgrew B. Adaptive filter algorithms for accelerated discrete-time consensus. *IEEE Transactions on Signal Processing*. 2010;58(3):1049–1058.

[20] Nurellari E, McLernon D, and Ghogho M. Distributed two-step quantized fusion rules via consensus algorithm for distributed detection in wireless sensor networks. *IEEE Transactions on Signal and Information Processing over Networks*. 2016;2(3):321–335.

[21] Chen C, Zhu S, Guan X, *et al. Distributed Consensus Estimation of Wireless Sensor Networks*. Springer; 2014.

[22] Fuchs B, Skrivervik A, and Mosig JR. Synthesis of uniform amplitude focused beam arrays. *IEEE Antennas and Wireless Propagation Letters*. 2012;11:1178–1181.

[23] Bucci OM, D'Urso M, Isernia T, *et al.* Deterministic synthesis of uniform amplitude sparse arrays via new density taper techniques. *IEEE Transactions on Antennas and Propagation*. 2010;58(6):1949–1958.

[24] Morabito AF, Isernia T, and Di Donato L. Optimal synthesis of phase-only reconfigurable linear sparse arrays having uniform-amplitude excitations. *Progress in Electromagnetics Research*. 2012;124:405–423.

[25] Tavli B. A survey of visual sensor network platforms. *Multimedia Tools & Applications*. 2012;60(3):689–726.

[26] Akyildiz IF, Su W, Sankarasubramaniam Y, *et al.* Wireless sensor networks: a survey. In: *International Conference on Advanced Information Networking & Applications Workshops*; 2009.

[27] Bhanu B, Ravishankar CV, Roy-Chowdhury AK, *et al. Distributed Video Sensor Networks*. Springer Science & Business Media; 2011.

[28] Boyd S, Parikh N, Chu E, *et al.* Distributed optimization and statistical learning via the alternating direction method of multipliers. *Foundations and Trends® in Machine Learning*. 2011;3(1):1–122.

[29] Chen J and Sayed AH. Diffusion adaptation strategies for distributed optimization and learning over networks. *IEEE Transactions on Signal Processing*. 2012;60(8):4289–4305.

[30] Dimakis AG, Kar S, Moura JMF, *et al.* Gossip algorithms for distributed signal processing. *Proceedings of the IEEE*. 2010;98(11):1847–1864.

[31] Majzoobi L, Lahouti F, and Shah-Mansouri V. Analysis of distributed ADMM algorithm for consensus optimization in presence of node error. *IEEE Transactions on Signal Processing*. 2019;67(7):1774–1784.

[32] Ling Q, Liu Y, Shi W, *et al.* Weighted ADMM for fast decentralized network optimization. *IEEE Transactions on Signal Processing*. 2016;64(22):5930–5942.

[33] Erseghe T, Zennaro D, Dall'Anese E, *et al.* Fast consensus by the alternating direction multipliers method. *IEEE Transactions on Signal Processing*. 2011;59(11):5523–5537.

[34] Olfati-Saber R, Fax JA, and Murray RM. Consensus and cooperation in networked multi-agent systems. *Proceedings of the IEEE*. 2007;95(1):215–233.

Chapter 4

Shaped power pattern synthesis with minimization of dynamic range ratio

Junli Liang[1], Xuhui Fan[2], Yuanhang Zhang[3], Hing Cheung So[4] and Xiaozhe Zhao[1]

In this chapter, we address the issue of array pattern synthesis with a focus on minimizing the dynamic range ratio (DRR). Unlike common approaches that simply limit DRR to a specific threshold, our goal is to reduce the DRR of excitation vectors while creating the desired array pattern. This unique approach results in a new optimization problem characterized by a fractional objective function and nonconvex constraints. To tackle this challenge, we propose a novel algorithm that effectively solves the optimization problem. By introducing auxiliary variables, we formulate an equivalent optimization problem that converts the fractional objective function into a linear one. This allows us to decompose the original problem into subproblems that can be efficiently solved in each iteration. Through this approach, we simplify the subproblems into single-variable quadratic unconstrained optimization or least-squares problems, enabling us to achieve our goal of minimizing DRR in array pattern synthesis.

4.1 Introduction

The objective of synthesizing array patterns with minimized dynamic range ratio (DRR) of excitations involves designing the antenna array excitations to achieve a specified power pattern while managing the DRR of the excitations. This challenge holds significant importance in radar, remote sensing, and wireless communication systems [1], because low DRR excitations aid in controlling mutual coupling among adjacent array elements, decreasing output power loss, and streamlining the feeding network design [2,3]. Numerous techniques exist for shaping radiation patterns, including methods for controlling both amplitude and phase [4,5], controlling only amplitude [6], and controlling only phase [7,8]. While these methods can synthesize

[1]School of Electronics and Information, Northwestern Polytechnical University, China
[2]School of Communication and Information Engineering, Xi'an University of Science and Technology, China
[3]Beijing Huahang Radio Measurement Institute, Beijing, China
[4]Department of Electrical Engineering, City University of Hong Kong, China

diverse radiation patterns suitable for different scenarios, they do not offer DRR control, leading to significant DRR values.

In addressing this challenge, various approaches have been outlined in [2,3,9–16], which can be categorized into three groups. The methodology described in [9,10] involves eliminating array elements with very low amplitudes. Although this is a straightforward approach, it may not effectively restrict the DRR to the target threshold and could introduce pattern distortion due to the reduction in the number of array elements [2]. Another approach, discussed in [2,11–13], specifies the desired DRR threshold and ensures that the optimized DRR does not exceed it. In [2], a projection-based iterative algorithm is proposed for far-field pattern synthesis with DRR reduction, including a consideration of the consistency between null constraints and DRR constraints. The algorithm presented in [11] has been expanded to address the issue of reconfigurable array synthesis, incorporating both DRR and near-field reduction simultaneously. In the work of [12], a method based on projections is employed to create nulls while enforcing an upper limit on the DRR of the excitations for circular arrays. The approach outlined in [13] formulates the synthesis of a flat-top radiation pattern with a specified DRR as a semidefinite relaxation (SDR) problem [17], which is then tackled using the CVX toolbox [18]. Despite these advancements, in scenarios where the user sets a small DRR threshold, the resulting pattern may not meet all desired criteria, whereas setting it too high can lead to increased cost and complexity of the feeding network, failing to adhere to modern attenuator constraints [19]. Therefore, determining an appropriate DRR threshold may require some trial and error. To mitigate the risk of a large DRR, the method proposed in [14–16] introduces a cost function that combines the amplitude difference between the actual and desired radiation patterns with the DRR, subsequently utilizing genetic algorithms [14], fast iterative algorithms [15], or penalty methods [16] for optimization. It is crucial to carefully balance these components in the cost function to achieve both a satisfactory radiation pattern and a minimal DRR. Emphasizing one over the other may result in either a satisfactory pattern with a large DRR or a deviation from the desired radiation pattern. Additionally, in the study by Fuchs [3], the issue of radiation pattern synthesis involving DRR minimization is presented as an SDR problem. This problem is addressed by disregarding the nonconvex rank-1 constraint, leading to a solution that may not always be optimal due to necessary approximations. This is particularly evident when dealing with flat-top beampattern synthesis.

In summary, the methods mentioned above do not directly produce a customized power pattern that meets all radiation requirements while minimizing DRR. This prompts the development of an optimization problem aimed at achieving the lowest DRR for the excitations while ensuring the desired power patterns are achieved.

The subsequent sections are structured as follows: Section 4.2 outlines the formulation of the beampattern synthesis problem with concurrent DRR minimization. The solution to this problem, along with a theoretical analysis of computational complexity and convergence, is presented in Section 4.3. Section 4.4 includes numerical results comparing the proposed algorithm with existing methods. Last, Section 4.5 contains the concluding remarks.

4.2 Problem formulation

Consider a linear array with N antenna elements located at known locations $x_n \in \mathbb{R}$, $n = 1, 2, \cdots, N$. The far-field power pattern of the array in the direction θ measured from the array broadside is given by

$$|\bar{\boldsymbol{w}}^H \boldsymbol{a}(\theta)|^2, \tag{4.1}$$

where $\bar{\boldsymbol{w}} = [\bar{w}_1, \cdots, \bar{w}_N]^T$ is the complex excitation vector, $\bar{w}_n$ is the complex excitation of the nth element for $n = 1, \ldots, N$, $\boldsymbol{a}(\theta) = [g_1(\theta)e^{j\frac{2\pi x_1 \sin\theta}{\lambda}}, \cdots, g_N(\theta)e^{j\frac{2\pi x_N \sin\theta}{\lambda}}]^T$ is the steering vector, $g_n(\theta)$ is the pattern radiated by the nth element, and λ is the wavelength of the transmitted signal.

In general, the DRR of the excitation vector $\bar{\boldsymbol{w}}$ is defined as [2]:

$$\text{DRR}(\bar{\boldsymbol{w}}) = \frac{\max\limits_{1 \le n \le N} \{|\bar{w}_n|\}}{\min\limits_{1 \le n \le N} \{|\bar{w}_n|\}}. \tag{4.2}$$

Motivated by practical considerations regarding the cost and complexity of designing the feeding network [2,3], the beampattern synthesis problem with DRR reduction aims to minimize DRR while shaping the power pattern. To achieve this goal, a new optimization problem (4.3) is formulated to synthesize a tailored power pattern and reduce DRR by optimizing excitations.

$$\begin{aligned} &\min_{\bar{\boldsymbol{w}}} \ \frac{\max\limits_{1 \le n \le N} |\bar{w}_n|}{\min\limits_{1 \le n \le N} |\bar{w}_n|} \\ &\text{s.t.} \ L(\theta_m) \le |\bar{\boldsymbol{w}}^H \boldsymbol{a}(\theta_m)|^2 \le U(\theta_m), \ m = 1, \cdots, M, \\ &|\bar{\boldsymbol{w}}^H \boldsymbol{a}(\bar{\theta}_s)|^2 \le \eta(\bar{\theta}_s), \ s = 1, \cdots, S, \end{aligned} \tag{4.3}$$

In this formulation, $\theta_{m m=1}^{\ M}$, $L(\theta_m)$, and $U(\theta_m)$ represent angle grids and bound functions for the desired magnitude response in the mainlobe area. Similarly, $\{\bar{\theta}_s\}_{s=1}^S$ and $\eta(\bar{\theta}_s)$ represent angle grids and bound functions for the desired magnitude response in the sidelobe area.

When focusing on a specific pattern, constraints in (4.3) can be adjusted as the following constraints while ensuring the objective function remains intact:

$$\begin{cases} |\bar{\boldsymbol{w}}^H \boldsymbol{a}(\theta_0)|^2 \ge 1, \\ |\bar{\boldsymbol{w}}^H \boldsymbol{a}(\bar{\theta}_s)|^2 \le \eta(\bar{\theta}_s), \ s = 1, \cdots, S. \end{cases} \tag{4.4}$$

Furthermore, one can add the null constraints to (4.3) to suppress the strong interferences:

$$\begin{cases} L(\theta_m) \le |\bar{\boldsymbol{w}}^H \boldsymbol{a}(\theta_m)|^2 \le U(\theta_m), \ m = 1, \cdots, M, \\ |\bar{\boldsymbol{w}}^H \boldsymbol{a}(\bar{\theta}_s)|^2 \le \eta(\bar{\theta}_s), \ s = 1, \cdots, S, \\ |\bar{\boldsymbol{w}}^H \boldsymbol{a}(\tilde{\theta}_l)|^2 \le \iota(\tilde{\theta}_l), \ l = 1, \cdots, L, \end{cases} \tag{4.5}$$

where $\{\tilde{\theta}_l\}_{l=1}^L$ and ι denote the L angle grids and the depth within the null region, respectively.

It is important to note that (4.4) and (4.5) are special cases of (4.3), and the solution is derived for (4.3). The ultimate goal is to determine the minimal DRR of excitations to meet both mainlobe and sidelobe level requirements.

4.3 Proposed algorithm

The primary concern in (4.3) pertains to its significant nonconvexity and nonlinearity due to the fractional objective function and nonconvex constraints it contains. To address this issue, a practical approach has been devised for resolving (4.3).

For ease of notation, we denote $U(\theta_m)$ as U, $L(\theta_m)$ as L, and $\eta(\bar{\theta}_s)$ as η in the ensuing calculations.

By substituting $\hat{p}$ and $\hat{q}$ for $\max\limits_{1\leq n\leq N}\{|w_n|\}$ and $\min\limits_{1\leq n\leq N}\{|w_n|\}$ respectively, and defining $\boldsymbol{e}_n \in \mathbb{R}^{N\times 1}$ as a vector with a value of 1 at the nth position and 0 elsewhere, we can express (4.3) as follows:

$$\begin{aligned} \min_{\bar{\boldsymbol{w}},\hat{p},\hat{q}} \quad & \frac{\hat{p}}{\hat{q}} \\ \text{s.t.} \quad & \hat{q} \leq |\bar{\boldsymbol{w}}^H \boldsymbol{e}_n| \leq \hat{p},\ n = 1,\cdots,N, \\ & L \leq |\bar{\boldsymbol{w}}^H \boldsymbol{a}(\theta_m)|^2 \leq U,\ m = 1,\cdots,M, \\ & |\bar{\boldsymbol{w}}^H \boldsymbol{a}(\bar{\theta}_s)|^2 \leq \eta,\ s = 1,\cdots,S. \end{aligned} \tag{4.6}$$

All the variables $\hat{p}$, $\hat{q}$, and $\bar{\boldsymbol{w}}$ are normalized with $\hat{q}$ and become p, 1, and $\boldsymbol{w}$. Similarly, constants L, U, and η are normalized with $\hat{q}^2$, and become $\frac{L}{\hat{q}^2}$, $\frac{U}{\hat{q}^2}$, and $\frac{\eta}{\hat{q}^2}$. Let us simplify $\frac{1}{\hat{q}^2}$ as β. With this transformation, we can rewrite (4.6) with a fractional objective function as an optimization problem with a linear objective function:

$$\begin{aligned} \min_{\boldsymbol{w},p,\beta} \quad & p \\ \text{s.t.} \quad & 1 \leq |\boldsymbol{w}^H \boldsymbol{e}_n| \leq p,\ \ n = 1,\cdots,N, \\ & L\beta \leq |\boldsymbol{w}^H \boldsymbol{a}(\theta_m)|^2 \leq U\beta,\ m = 1,\cdots,M, \\ & |\boldsymbol{w}^H \boldsymbol{a}(\bar{\theta}_s)|^2 \leq \eta\beta,\ s = 1,\cdots,S. \end{aligned} \tag{4.7}$$

By introducing new auxiliary variables $\{y_n\}$, $\{z_m\}$, and $\{v_s\}$, along with new equality constraints $y_n = \boldsymbol{w}^H \boldsymbol{e}_n$, $z_m = \boldsymbol{w}^H \boldsymbol{a}(\theta_m)$, and $v_s = \boldsymbol{w}^H \boldsymbol{a}(\bar{\theta}_s)$, (4.7) can be written as

$$\begin{aligned} \min_{\boldsymbol{w},p,\beta,y_n,z_m,v_s} \quad & p \\ \text{s.t.} \quad & y_n = \boldsymbol{w}^H \boldsymbol{e}_n, 1 \leq |y_n| \leq p,\ \ n = 1,\cdots,N, \\ & z_m = \boldsymbol{w}^H \boldsymbol{a}(\theta_m), L\beta \leq |z_m|^2 \leq U\beta,\ m = 1,\cdots,M, \\ & v_s = \boldsymbol{w}^H \boldsymbol{a}(\bar{\theta}_s), |v_s|^2 \leq \eta\beta,\ s = 1,\cdots,S. \end{aligned} \tag{4.8}$$

Note that (4.8) is actually equivalent to (4.3), and thus we can obtain the solution to (4.3) by solving (4.8).

In the context of (4.8), the pairs of variables $\{\boldsymbol{w}, p\}$ and $\{\boldsymbol{w}, \beta\}$ are decoupled, leading to the variable $\boldsymbol{w}$ being extracted from the complex nonconvex constraints. This extraction results in the inequality constraints $1 \leq |y_n| \leq p$, $L\beta \leq |z_m|^2 \leq U\beta$, and $|v_s|^2 \leq \eta\beta$ playing crucial roles in determining y_n, z_m, and v_s, respectively. To address this, we can leverage the alternating direction method of multipliers (ADMM) [20–25], known for its exceptional decomposability and convergence properties [20–22], to tackle (4.8). A detailed approach to resolving (4.8) is outlined below. To begin, we employ the ADMM technique to formulate the scaled-form augmented Lagrangian function [22], represented as follows:

$$L_\rho(\boldsymbol{w}, p, y_n, z_m, v_s, \kappa_n, \mu_m, \gamma_s) = p + \frac{\rho}{2}\sum_{n=1}^{N}(|y_n - \boldsymbol{w}^H\boldsymbol{e}_n + \kappa_n|^2 - |\kappa_n|^2) \tag{4.9}$$

$$+ \frac{\rho}{2}\sum_{m=1}^{M}(|z_m - \boldsymbol{w}^H\boldsymbol{a}(\theta_m) + \mu_m|^2 - |\mu_m|^2) + \frac{\rho}{2}\sum_{s=1}^{S}(|v_s - \boldsymbol{w}^H\boldsymbol{a}(\bar{\theta}_s) + \gamma_s|^2 - |\gamma_s|^2),$$

where κ_n, μ_m, and γ_s are scaled dual variables, and ρ is the user-defined step size [22].

Secondly, the variables $\{\boldsymbol{w}, p, \beta, y_n, z_m, v_s, \kappa_n, \mu_m, \gamma_s\}$ are determined alternately and iteratively, that is, minimizing (4.10) by subdividing each iteration into three parts:

$$\begin{aligned} \min_{\boldsymbol{w},p,\beta,y_n,z_m,v_s,\kappa_n,\mu_m,\gamma_s} \quad & L_\rho(\boldsymbol{w}, p, y_n, z_m, v_s, \kappa_n, \mu_m, \gamma_s) \\ \text{s.t.} \quad & 1 \leq |y_n| \leq p, \ n = 1, \cdots, N, \\ & L\beta \leq |z_m|^2 \leq U\beta, \ m = 1, \cdots, M, \\ & |v_s|^2 \leq \eta\beta, \ s = 1, \cdots, S. \end{aligned} \tag{4.10}$$

Step 1: Given $\{\boldsymbol{w}^{(t)}, \kappa_n^{(t)}, \mu_m^{(t)}, \gamma_s^{(t)}\}$, $\{p^{(t+1)}, \beta^{(t+1)}, y_n^{(t+1)}, z_m^{(t+1)}, v_s^{(t+1)}\}$ are updated by solving the following optimization problem:

$$\begin{aligned} \min_{p,\beta,y_n,z_m,v_s} \quad & L_\rho(\boldsymbol{w}^{(t)}, p, y_n, z_m, v_s, \kappa_n^{(t)}, \mu_m^{(t)}, \gamma_s^{(t)}) \\ \text{s.t.} \quad & 1 \leq |y_n| \leq p, \ n = 1, \cdots, N, \\ & L\beta \leq |z_m|^2 \leq U\beta, \ m = 1, \cdots, M, \\ & |v_s|^2 \leq \eta\beta, \ s = 1, \cdots, S, \end{aligned} \tag{4.11}$$

where t is the iteration index.

Note that the variables $\{\beta, \{v_s\}_{s=1}^S, \{z_m\}_{m=1}^M\}$ are irrelevant to the variables $\{p, \{y_n\}_{n=1}^N\}$ and hence (4.11) can be divided into the following two subproblems:

$$\begin{aligned} \min_{\beta,z_m,v_s} \quad & \sum_{m=1}^{M}|z_m - \bar{z}_m^{(t)}|^2 + \sum_{s=1}^{S}|v_s - \bar{v}_s^{(t)}|^2 \\ \text{s.t.} \quad & L\beta \leq |z_m|^2 \leq U\beta, \ \ m = 1, \cdots, M, \\ & |v_s|^2 \leq \eta\beta, \beta \in [\beta_L, \beta_U], \ \ s = 1, \cdots, S, \end{aligned} \tag{4.12}$$

and

$$\begin{aligned}
&\min_{p,y_n} \; p + \frac{\rho}{2}\sum_{n=1}^{N} |y_n - \bar{y}_n^{(t)}|^2 \\
&\text{s.t.} \quad 1 \le |y_n| \le p, p \in [p_L, p_U], \;\; n = 1, \cdots, N,
\end{aligned} \tag{4.13}$$

where $\bar{y}_n^{(t)} = \boldsymbol{w}^{(t)^H}\boldsymbol{e}_n - \kappa_n^{(t)}$, $\bar{z}_m^{(t)} = \boldsymbol{w}^{(t)^H}\boldsymbol{a}(\theta_m) - \mu_m^{(t)}$ and $\bar{v}_s^{(t)} = \boldsymbol{w}^{(t)^H}\boldsymbol{a}(\bar{\theta}_s) - \gamma_s^{(t)}$ for $n = 1, \cdots, N$, $m = 1, \cdots, M$, and $s = 1, \cdots, S$, respectively. In addition, β_L and β_U are the predefined lower and upper bounds of the variable β. Similarly, the lower and upper bounds of the variable p are set as p_L and p_U, respectively. It is worth mentioning that the value of the variable p is no less than 1 according to the constraints in (4.13).

We notice that once the variables β and p are provided, for $m = 1, \cdots, M$, $s = 1, \cdots, S$, and $n = 1, \cdots, N$, the optimal z_m, v_s and y_n are given by

$$z_m^{(t+1)} = \begin{cases} \sqrt{U\beta}\dfrac{\bar{z}_m^{(t)}}{|\bar{z}_m^{(t)}|}, & \text{if } |\bar{z}_m^{(t)}| \ge \sqrt{U\beta} \\ \sqrt{L\beta}\dfrac{\bar{z}_m^{(t)}}{|\bar{z}_m^{(t)}|}, & \text{if } |\bar{z}_m^{(t)}| \le \sqrt{L\beta} \\ \bar{z}_m^{(t)}, & \text{otherwise} \end{cases} \tag{4.14}$$

$$v_s^{(t+1)} = \begin{cases} \sqrt{\eta\beta}\dfrac{\bar{v}_s^{(t)}}{|\bar{v}_s^{(t)}|}, & \text{if } |\bar{v}_s^{(t)}| \ge \sqrt{\eta\beta} \\ \bar{v}_s^{(t)}, & \text{otherwise} \end{cases} \tag{4.15}$$

$$y_n^{(t+1)} = \begin{cases} p\dfrac{\bar{y}_n^{(t)}}{|\bar{y}_n^{(t)}|}, & \text{if } |\bar{y}_n^{(t)}| \ge p \\ \dfrac{\bar{y}_n^{(t)}}{|\bar{y}_n^{(t)}|}, & \text{if } |\bar{y}_n^{(t)}| \le 1 \\ \bar{y}_n^{(t)}, & \text{otherwise} \end{cases} \tag{4.16}$$

First, for the first subproblem shown in (4.12), by inserting (4.14) and (4.15) into (4.12), we attain an unconstrained optimization problem of variable β:

$$\begin{aligned}
&\min_{\beta} \sum_{s=1}^{S} Q_{3s}^{(t)}(\sqrt{\eta\beta} - |\bar{v}_s^{(t)}|)^2 \\
&+ \sum_{m=1}^{M} (Q_{1m}^{(t)}(\sqrt{U\beta} - |\bar{z}_m^{(t)}|)^2 + Q_{2m}^{(t)}(\sqrt{L\beta} - |\bar{z}_m^{(t)}|)^2),
\end{aligned} \tag{4.17}$$

where unit-step functions $Q_{1m}^{(t)}$, $Q_{2m}^{(t)}$, and $Q_{3s}^{(t)}$ [23] are defined as follows:

$$Q_{1m}^{(t)} = \begin{cases} 1, \text{if } |\bar{z}_m^{(t)}| \ge \sqrt{U\beta} \\ 0, \text{otherwise} \end{cases} \tag{4.18}$$

$$Q_{2m}^{(t)} = \begin{cases} 1, \text{if } |\bar{z}_m^{(t)}| \leq \sqrt{L\beta} \\ 0, \text{otherwise} \end{cases} \tag{4.19}$$

$$Q_{3s}^{(t)} = \begin{cases} 1, \text{if } |\bar{v}_s^{(t)}| \geq \sqrt{\eta\beta} \\ 0, \text{otherwise} \end{cases} \tag{4.20}$$

Note that the objective function in (4.17) is a piecewise function, and thus we choose $\{\frac{|\bar{z}_1^{(t)}|^2}{L}, \cdots, \frac{|\bar{z}_M^{(t)}|^2}{L}, \frac{|\bar{z}_1^{(t)}|^2}{U}, \cdots, \frac{|\bar{z}_M^{(t)}|^2}{U}, \frac{|\bar{v}_1^{(t)}|^2}{\eta}, \cdots, \frac{|\bar{v}_S^{(t)}|^2}{\eta}\}$ as reasonable turning points, whose ascending order is denoted by $\{\beta_1^{(t)}, \cdots, \beta_K^{(t)}\}$. When $\beta \in [\beta_{k-1}^{(t)}, \beta_k^{(t)}]$, $k \in \{1, \cdots, K\}$, (4.17) can be expressed as a simple quadratic function of variable $\sqrt{\beta}$:

$$\min_{\beta} F_{\beta}(\sqrt{\beta}), \tag{4.21}$$

where

$$\begin{aligned} F_{\beta}(\sqrt{\beta}) &= A_k\beta + B_k\sqrt{\beta} + C_k, \\ A_k &= \sum_{m=1}^{M} (Q_{1m}^{(t)} U + Q_{2m}^{(t)} L) + \sum_{s=1}^{S} Q_{3s}^{(t)} \eta, \\ B_k &= -2\sum_{s=1}^{S} Q_{3s}^{(t)} \sqrt{\eta} |\bar{v}_s^{(t)}| - 2\sum_{m=1}^{M} (Q_{1m}^{(t)} \sqrt{U} + Q_{2m}^{(t)} \sqrt{L}) |\bar{z}_m^{(t)}|, \\ C_k &= \sum_{m=1}^{M} (Q_{1m}^{(t)} + Q_{2m}^{(t)}) |\bar{z}_m^{(t)}|^2 + \sum_{s=1}^{S} Q_{3s}^{(t)} |\bar{v}_s^{(t)}|^2. \end{aligned}$$

whose solution is given by

$$\hat{\beta}_k = \begin{cases} \beta_k^{(t)}, & \text{if } \dfrac{B_k^2}{4A_k^2} \geq \beta_k^{(t)} \\ \beta_{k-1}^{(t)}, & \text{if } \dfrac{B_k^2}{4A_k^2} \leq \beta_{k-1}^{(t)} \\ \dfrac{B_k^2}{4A_k^2}, & \text{otherwise} \end{cases} \tag{4.22}$$

Thus, by substituting $\hat{\beta}_k$ into (4.21) as indicated in Appendix A, the value of (4.17) is determined. After selecting the smallest $\bar{F}_{\beta}$ among the minimal values of K segments and using its corresponding optimal value $\hat{\beta}_k$, $\beta^{(t+1)}$ is calculated as

$$\beta^{(t+1)} = \arg\min_{\hat{\beta}_k} \left\{ F_{\beta}(\sqrt{\hat{\beta}_k}), k = 1, \cdots, K \right\}. \tag{4.23}$$

Upon inserting $\beta^{(t+1)}$ into (4.14) and (4.15), the updated values of $z_m^{(t+1)}$ and $v_s^{(t+1)}$ are obtained.

Solving the second subproblem presented in (4.13), which mirrors (4.12), results in the solution for (4.13), i.e., $p^{(t+1)}$ and $y_n^{(t+1)}$ detailed in Appendix B.

It is important to note that both (4.12) and (4.13) are tackled by converting the intricate constrained optimization problem into a simpler single-variable quadratic unconstrained one, leading to closed-form solutions for the variables in (4.11).

Step 2: Update $\boldsymbol{w}^{(t+1)}$ by solving the following optimization problem with all variables fixed except $\boldsymbol{w}$:

$$\min_{\boldsymbol{w}} L_\rho(\boldsymbol{w}, p^{(t+1)}, y_n^{(t+1)}, z_m^{(t+1)}, v_s^{(t+1)}, \kappa_n^{(t)}, \mu_m^{(t)}, \gamma_s^{(t)}). \tag{4.24}$$

By neglecting the constants, defining $\boldsymbol{y} = [y_1, y_2, \cdots, y_N]^T$, $\boldsymbol{z} = [z_1, z_2, \cdots, z_M]^T$, and $\boldsymbol{v} = [v_1, v_2, \cdots, v_S]^T$ as the introduced auxiliary variable vectors, and rewriting $\boldsymbol{\kappa} = [\kappa_1, \kappa_2, \cdots, \kappa_N]^T$, $\boldsymbol{\mu} = [\mu_1, \mu_2, \cdots, \mu_M]^T$, and $\boldsymbol{\gamma} = [\gamma_1, \gamma_2, \cdots, \gamma_S]^T$ as the Lagrange multiplier vectors, (4.24) is reformulated as a least squares problem:

$$\min_{\boldsymbol{w}} \ \boldsymbol{w}^H \boldsymbol{R} \boldsymbol{w} + \boldsymbol{w}^H \boldsymbol{b} + \boldsymbol{b}^H \boldsymbol{w}, \tag{4.25}$$

where $\boldsymbol{b} = -(\boldsymbol{y} + \boldsymbol{\kappa})^* - \boldsymbol{G}(\boldsymbol{z} + \boldsymbol{\mu})^* - \tilde{\boldsymbol{J}}(\boldsymbol{v} + \boldsymbol{\gamma})^*$, $\boldsymbol{R} = \boldsymbol{I}_N + \boldsymbol{G}\boldsymbol{G}^H + \tilde{\boldsymbol{J}}\tilde{\boldsymbol{J}}^H$, $\boldsymbol{G} = [\boldsymbol{a}(\theta_1), \cdots, \boldsymbol{a}(\theta_M)]$, and $\tilde{\boldsymbol{J}} = [\boldsymbol{a}(\bar{\theta}_1), \cdots, \boldsymbol{a}(\bar{\theta}_S)]$.

Setting the derivative of the objective function in (4.25) with respect to the variable $\boldsymbol{w}$ to zero yields the updating expression of $\boldsymbol{w}^{(t+1)}$:

$$\boldsymbol{w}^{(t+1)} = -\boldsymbol{R}^{-1}\boldsymbol{b}. \tag{4.26}$$

Note that (4.24) is converted into a least squares problem (4.25) so that $\boldsymbol{w}$ has a closed-form solution.

Step 3: Update $\{\kappa_n^{(t+1)}, \mu_m^{(t+1)}, \gamma_s^{(t+1)}\}$ using $\{\boldsymbol{w}^{(t+1)}, p^{(t+1)}, \beta^{(t+1)}, y_n^{(t+1)}, z_m^{(t+1)}, v_s^{(t+1)}\}$ from

$$\kappa_n^{(t+1)} = \kappa_n^{(t)} + y_n^{(t+1)} - w_n^{(t+1)*}, n = 1, \cdots, N, \tag{4.27}$$

$$\mu_m^{(t+1)} = \mu_m^{(t)} + z_m^{(t+1)} - \boldsymbol{w}^{(t+1)^H} \boldsymbol{a}(\theta_m), m = 1, \cdots, M, \tag{4.28}$$

$$\gamma_s^{(t+1)} = \gamma_s^{(t)} + v_s^{(t+1)} - \boldsymbol{w}^{(t+1)^H} \boldsymbol{a}(\bar{\theta}_s), s = 1, \cdots, S. \tag{4.29}$$

Steps 1 to 3 are executed repeatedly until a prescribed maximum iteration number $\tilde{T}$ or all $\max_n |y_n - w_n^*| \leq \varepsilon$, $\max_m |z_m - \boldsymbol{w}^H \boldsymbol{a}(\theta_m)| \leq \varepsilon$, and $\max_s |v_s - \boldsymbol{w}^H \boldsymbol{a}(\bar{\theta}_s)| \leq \varepsilon$ are reached. Then, with the obtained $\boldsymbol{w}$ and β, we normalize $\boldsymbol{w}$ with $\sqrt{\beta}$ and attain the original variable $\bar{\boldsymbol{w}}$.

For clarity, the steps for solving (4.8) are summarized in Algorithm 1.

4.3.1 Computational complexity

In terms of computational complexity, our focus is on the multiplications involved in one iteration. In Algorithm 1, Steps 1, 2, and 3 necessitate flops of $\mathcal{O}(3S + 4M + 2N)$, $\mathcal{O}(N^3 + N^2)$, and $\mathcal{O}(SN + MN)$, respectively. This means that the proposed technique has a complexity of $\mathcal{O}(3S + 4M +$

Algorithm 1 Shaped power pattern synthesis with minimization of DRR

Input:

Initialization: $\kappa_n^{(0)}, \mu_m^{(0)}, \gamma_s^{(0)}, \boldsymbol{w}^{(0)}$;

$t = 0$

while $t \leq \bar{T}$ or ($\max_n |y_n - w_n^*| > \varepsilon$, $\max_m |z_m - \boldsymbol{w}^H \boldsymbol{a}(\theta_m)| > \varepsilon$, or $\max_s |v_s - \boldsymbol{w}^H \boldsymbol{a}(\bar{\theta}_s)| > \varepsilon$) **do**

  $t = t + 1$;

  Determine $p^{(t+1)}$ using (B.7);

  Determine $y_n^{(t+1)}$ using $p^{(t+1)}$ and (4.16);

  Determine $\beta^{(t+1)}$ using (4.18);

  Determine $z_m^{(t+1)}$ and $v_s^{(t+1)}$ using $\beta^{(t+1)}$, (4.14) and (4.15);

  Obtain $\boldsymbol{w}^{(t+1)}$ using (4.26);

  Update $\kappa_n^{(t+1)}$, $\mu_m^{(t+1)}$, $\gamma_s^{(t+1)}$ using (4.27), (4.28), and (4.29);

end while

Obtain $\bar{\boldsymbol{w}}$ via normalizing $\boldsymbol{w}$ with $\sqrt{\beta}$;

Compute DRR using (4.2);

Output:

Optimal weight vector $\bar{\boldsymbol{w}}$, DRR.

$N^2 + N^3 + (S + M + 2)N)$ in one iteration. In comparison, SDR-based techniques in previous studies require significantly more flops. For example, methods presented in [3,13], and [16] demand $\mathcal{O}(N^4(M+S)^{2.5})$ flops and $\mathcal{O}(6N^3 + (6M + 6S + 4)N^2) + 4N(M+S))$ flops, respectively. It is evident that our ADMM scheme showcases a lower complexity than these alternative methods, highlighting its computational efficiency.

4.3.2 *Convergence analysis*

In this section, we examine the convergence of the approach put forth in this study. While the ADMM method is typically used for convex optimization problems, the nonconvex optimization problem addressed in this chapter makes the application of standard ADMM convergence results inappropriate. As highlighted by Tsinos [27] and Wen [28], Theorem 1 offers a convergence analysis of the algorithm proposed in this study, under certain reasonable assumptions. This analysis confirms that the implementation of Steps 1, 2, and 3 can lead to the identification of a locally optimal solution for the specific optimization problem discussed in Equation (4.8).

Theorem 1 [27,28]: Define $\zeta^{(t)} = (\boldsymbol{y}^{(t)}, \boldsymbol{z}^{(t)}, \boldsymbol{v}^{(t)}, \boldsymbol{w}^{(t)}, p^{(t)}, \beta^{(t)}, \boldsymbol{\kappa}^{(t)}, \boldsymbol{\mu}^{(t)}, \boldsymbol{\gamma}^{(t)})$ as a sequence obtained by Steps 1, 2, and 3 with $\rho > 0$, and suppose that the multiplier sequences $(\boldsymbol{\kappa}^{(t)}, \boldsymbol{\mu}^{(t)}, \boldsymbol{\gamma}^{(t)})$ are bounded and satisfy $\lim_{t\to\infty} \kappa^{(t+1)} - \kappa^{(t)} = \mathbf{0}_N$, $\lim_{t\to\infty} \mu^{(t+1)} - \mu^{(t)} = \mathbf{0}_M$, and $\lim_{t\to\infty} \gamma^{(t+1)} - \gamma^{(t)} = \mathbf{0}_S$. Then, the sequence $\{\zeta^{(t)}\}$ converges to a limit point $\zeta^\star = (\boldsymbol{y}^\star, \boldsymbol{z}^\star, \boldsymbol{v}^\star, \boldsymbol{w}^\star, p^\star, \beta^\star, \boldsymbol{\kappa}^\star, \boldsymbol{\mu}^\star, \boldsymbol{\gamma}^\star)$ which is a locally optimal solution to (4.8).

Proof. For simplicity, we rewrite (4.27), (4.28), and (4.29) as vectors, namely, $\boldsymbol{\kappa}^{(t+1)} = \boldsymbol{\kappa}^{(t)} + \boldsymbol{y}^{(t+1)} - \boldsymbol{w}^{(t+1)*}$, $\boldsymbol{\mu}^{(t+1)} = \boldsymbol{\mu}^{(t)} + \boldsymbol{z}^{(t+1)} - \boldsymbol{G}^T \boldsymbol{w}^{(t+1)*}$, and $\gamma^{(t+1)} =$

$\gamma^{(t)} + \mathbf{v}^{(t+1)} - \tilde{\boldsymbol{J}}^T \boldsymbol{w}^{(t+1)*}$. Since $\lim_{t\to\infty} \kappa^{(t+1)} - \kappa^{(t)} = \mathbf{0}_N$, $\lim_{t\to\infty} \mu^{(t+1)} - \mu^{(t)} = \mathbf{0}_M$, $\lim_{t\to\infty} \gamma^{(t+1)} - \gamma^{(t)} = \mathbf{0}_S$, and $\rho > 0$, the following statements hold:

$$\lim_{t\to\infty} \mathbf{y}^{(t+1)} - \mathbf{w}^{(t+1)*} = \mathbf{0}_N, \tag{4.30}$$

$$\lim_{t\to\infty} \mathbf{z}^{(t+1)} - \mathbf{G}^T\mathbf{w}^{(t+1)*} = \mathbf{0}_M, \tag{4.31}$$

$$\lim_{t\to\infty} \mathbf{v}^{(t+1)} - \tilde{\mathbf{J}}^T\mathbf{w}^{(t+1)*} = \mathbf{0}_S. \tag{4.32}$$

Due to $p \in [1, p_U]$, the sequence $\{\boldsymbol{y}^{(t)}\}$ generated by (4.16) is bounded. Following from (4.30), the boundedness of $\{\boldsymbol{y}^{(t)}\}$ and the inequality $||\boldsymbol{y}^{(t+1)}|| \le ||\boldsymbol{y}^{(t+1)} - \boldsymbol{w}^{(t+1)*}|| + ||\boldsymbol{w}^{(t+1)*}||$, we obtain that the sequence $\{\boldsymbol{w}^{(t)}\}$ is also bounded. Hence, there exists a sequence $(\boldsymbol{y}^{(t)}, \boldsymbol{w}^{(t)})$ such that $\boldsymbol{y}^{(t)}$ and $\boldsymbol{w}^{(t)}$ converge to $\boldsymbol{y}^\star$ and $\boldsymbol{w}^\star$, respectively, and

$$\mathbf{0}_N = \lim_{t\to\infty} \mathbf{y}^{(t)} - \mathbf{w}^{(t)*} = \mathbf{y}^\star - (\mathbf{w}^\star)^*. \tag{4.33}$$

Observing that $||\boldsymbol{G}^T\boldsymbol{w}^{(t+1)*}||_2 \le ||\boldsymbol{w}^{(t+1)}||_2||\boldsymbol{G}||_F$, $||\tilde{\boldsymbol{J}}^T\boldsymbol{w}^{(t+1)*}||_2 \le ||\boldsymbol{w}^{(t+1)}||_2 ||\tilde{\boldsymbol{J}}||_F$ ($||\boldsymbol{G}||_F$ and $||\tilde{\boldsymbol{J}}||_F$ are constants), $||\boldsymbol{v}^{(t+1)}|| \le ||\boldsymbol{v}^{(t+1)} - \tilde{\boldsymbol{J}}^T\boldsymbol{w}^{(t+1)*}|| + ||\tilde{\boldsymbol{J}}\boldsymbol{w}^{(t+1)*}||$, and $||\boldsymbol{z}^{(t+1)}|| \le ||\boldsymbol{z}^{(t+1)} - \boldsymbol{G}^T\boldsymbol{w}^{(t+1)*}|| + ||\boldsymbol{G}^T\boldsymbol{w}^{(t+1)*}||$, the sequences $\boldsymbol{z}^{(t)}$ and $\boldsymbol{v}^{(t)}$ are thus bounded following from (4.31), (4.32), and the boundness of $\boldsymbol{w}^{(t)}$. Therefore, the sequence $(\boldsymbol{z}^{(t)}, \boldsymbol{v}^{(t)})$ converges to a limit point $(\boldsymbol{z}^\star, \boldsymbol{v}^\star)$, and

$$\mathbf{0}_M = \lim_{t\to\infty} \mathbf{z}^{(t)} - \mathbf{G}^T\mathbf{w}^{(t)*} = \mathbf{z}^\star - \mathbf{G}^T(\mathbf{w}^\star)^*, \tag{4.34}$$

$$0_S = \lim_{t\to\infty} \mathbf{v}^{(t)} - \tilde{\mathbf{J}}^T\mathbf{w}^{(t)*} = \mathbf{v}^\star - \tilde{\mathbf{J}}^T(\mathbf{w}^\star)^*. \tag{4.35}$$

Additionally, the variable β is bounded because it belongs to the region $[\beta_L, \beta_U]$. Therefore, the sequence $\boldsymbol{\zeta}^{(t)}$ converges to a limit point $\boldsymbol{\zeta}^\star$ which is a local optimum to (4.8).

This completes the proof of Theorem 1. □

According to [28], under certain mild assumptions, the convergence analysis of ADMM for nonconvex optimization problems might not be flawless, but it ensures the dependability of the algorithm. The proposed algorithm's convergence analysis is elaborated in Theorem 1, assuming that the sequences of multipliers $(\boldsymbol{\kappa}^{(t)}, \boldsymbol{\mu}^{(t)}, \boldsymbol{\gamma}^{(t)})$ are bounded and satisfy $\lim_{t\to\infty} \kappa^{(t+1)} - \kappa^{(t)} = \mathbf{0}_N$, $\lim_{t\to\infty} \mu^{(t+1)} - \mu^{(t)} = \mathbf{0}_M$, and $\lim_{t\to\infty} \gamma^{(t+1)} - \gamma^{(t)} = \mathbf{0}_S$.

4.4 Numerical examples

In this chapter, we focus on the study of the synthesis problem related to one-dimensional (1-D) power patterns. The methodology developed here can be seamlessly applied to tackle the 2-D power pattern synthesis by substituting the angle θ in (4.1) with the pair of azimuth and elevation angles (ϑ, φ) in the 2-D scenario. To assess the effectiveness of our algorithm in generating diverse patterns while minimizing the DRR, we carry out numerical tests. The algorithm execution ceases either when the error margin ε reaches 10^{-8} or when the maximum number of iterations $T = 25000$ is completed.

4.4.1 Flat-top pattern synthesis

To begin with, we conduct an initial experiment utilizing a uniform linear array (ULA) with an inter-element spacing of 0.5λ to synthesize a commonly shaped pattern with a broad mainlobe requirement, essential for achieving wide coverage in remote sensing applications [16]. For comparative analysis, we adopt the same array settings and specifications as those examined in a previous study by [13]. The specific configurations entail: $N = 20$, $L = -0.1$ dB, $U = 0.1$ dB covering the range $[-40°, 40°]$, and $\eta = -25$ dB in the interval $[-90°, -50°] \cup [50°, 90°]$. All antennas are assumed to be isotropic elements, characterized by $g_n(\theta) = 1$ for all $n = 1, 2, \ldots, N$. Primarily, we employ Experiment 1 to investigate the impact of the initial starting point, parameter ρ, and sampling interval on the performance of the algorithm.

4.4.1.1 Exp.1-1: CPU time, DRR, SLL, and RMLL versus starting point

In order to demonstrate the impact of the initial position on the performance of the suggested algorithm, a series of 1000 Monte Carlo experiments were conducted. The starting point was randomly chosen for each trial, with $\rho = 20$ and a sampling interval of 1°. The statistical findings are as follows: the dynamic range ratio (DRR) varies between 2.7 and 829.77, with a success rate of achieving the desired pattern requirements standing at 63.5%. It should be noted that 36.5% of synthesized patterns do not fully meet the desired criteria. Nonetheless, the deviation from the desired requirements is minimal, with the obtained radiation pattern's amplitude exceeding by at most 0.15 dB in the sidelobe region and staying within ± 0.14 dB in the mainlobe region [29]. The average CPU time for these simulations, performed on a PC with an Intel i7-6700 CPU and 16 GB RAM utilizing MATLAB® (version R2010b, 64bit), was 533.76s. Figure 4.1 displays 1000 synthesized patterns. Consequently, we can infer that the DRR is significantly influenced by the starting point, whereas the dependence of the synthesized radiation patterns on the initial position is relatively minor [29]. Therefore, selecting an appropriate starting point is crucial for reducing the DRR in the proposed algorithm. Similar to prior work by [16], to minimize the DRR during the synthesis of the desired radiation pattern, the initial phases of $\boldsymbol{w}$ were set to follow a square law distribution:

$$\angle \boldsymbol{w} = (n - 0.5N)^2 \pi / 150, \quad n = 1, 2, \cdots, N. \tag{4.36}$$

4.4.1.2 Exp.1-2: CPU time, DRR, SLL, and RMLL versus ρ

In order to comprehensively analyze the impact of the parameter ρ on the performance of the proposed algorithm, we have generated plots illustrating the optimized DRR, SLL (sidelobe level), RMLL (range of mainlobe level), and CPU time versus ρ as shown in Figure 4.2. The initial phases of $\boldsymbol{w}$ have been set utilizing a square

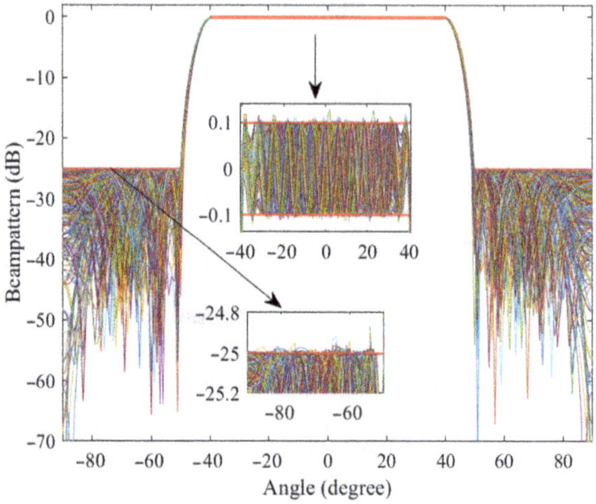

Figure 4.1 Synthesized beampatterns for Exp.1-1

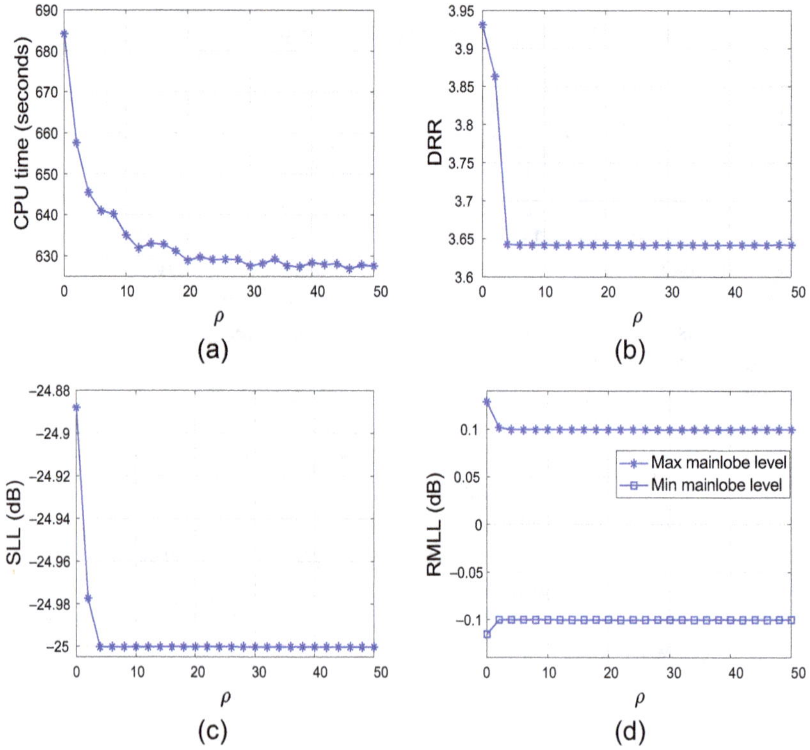

Figure 4.2 (a) CPU time versus ρ; (b) DRR versus ρ; (c) SLL versus ρ; and (d) RMLL versus ρ for Exp.1-2.

law distribution. Our findings can be synthesized as follows. Initially, for values of ρ greater than or equal to 4, the DRR, SLL, and RMLL remain unaffected by changes in ρ. Furthermore, for ρ values below 20, there is a notable decrease in CPU time as ρ increases. However, beyond $\rho \geq 20$, CPU time stabilizes and shows minimal sensitivity to changes in ρ. This suggests that while a higher value of ρ can expedite the execution of the proposed algorithm, it does not significantly impact the CPU time once it reaches a certain threshold. Therefore, based on our analysis, we have selected $\rho = 20$ as the optimal parameter value.

4.4.1.3 Exp.1-3: CPU time, DRR, SLL, and RMLL versus sampling interval

In order to demonstrate the impact of the sampling interval (referred to as S+M, representing the number of samples) on the effectiveness of the algorithm under consideration, we present the results of DRR, SLL, RMLL, and CPU time versus the sampling interval in Figure 4.3. Our experiment involved initializing the phases of

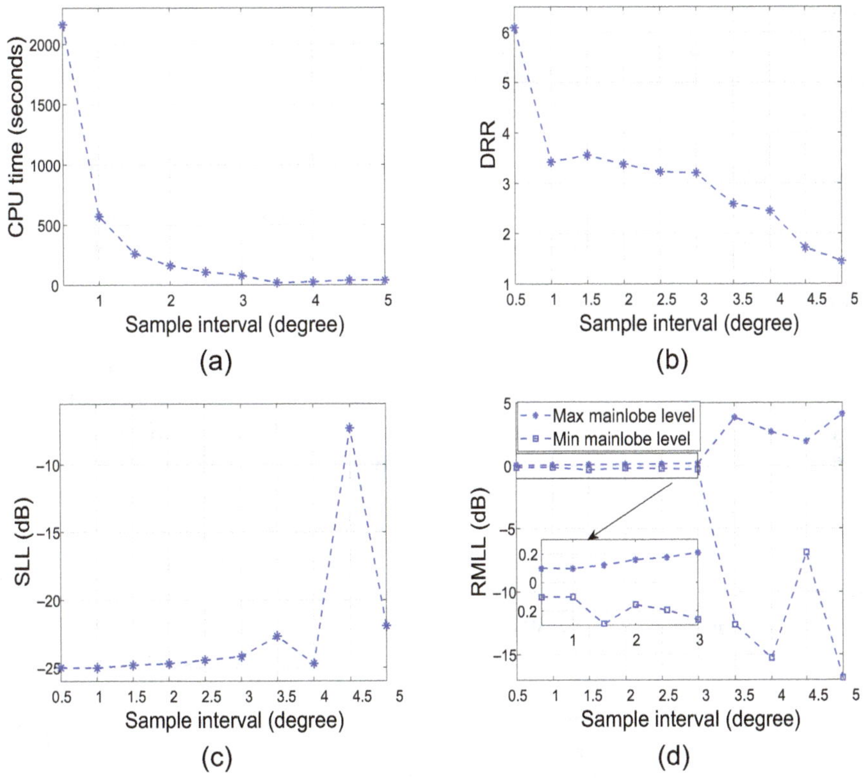

Figure 4.3 (a) CPU time versus sampling interval; (b) DRR versus sampling interval; (c) SLL versus sampling interval; and (d) RMLL versus sampling interval for Exp.1-3.

$\boldsymbol{w}$ according to a square law distribution, with ρ set at 20. The key findings are as follows: (i) The CPU time decreases as the number of samples increases, aligning with our theoretical analysis of computational complexity; (ii) DRR exhibits a gradual decline with increasing $S+M$. One possible explanation for this trend is that a larger sampling interval necessitates controlling fewer angles, thereby enhancing the flexibility of the algorithm; (iii) The radiation pattern synthesized from Figure 4.3(c)–(d) deviates more from the desired mainlobe and sidelobe specifications as the sampling interval expands, particularly beyond 3°. This suggests that a sampling interval greater than 1° lacks the necessary density to deliver precise control over radiation patterns. Hence, taking into account both the CPU time and algorithm performance, we opt for a sampling interval of 1° within the specified angle range.

Based on empirical studies, we set the initial phases of $\boldsymbol{w}$ as the square law distribution, sampling interval as 1°, and $\rho = 20$ in the following experiments.

4.4.1.4 Exp.1-4: comparison with existing algorithms

The array patterns and distributions of excitation amplitude for the proposed algorithm, [3,13], and [16] are displayed in Figures 4.4(a)–(c). Additionally, Table 4.1 provides the values corresponding to DRR, SLL, RMLL, and CPU time. It is important to note that the prescribed DRR value in [13] is represented as PDRR in Table 4.1 and Figure 4.4. For a clearer evaluation of the algorithms in the mainlobe area, Figure 4.4(b) showcases an enlarged version of the mainlobe region depicted in Figure 4.4(a).

Based on our results, we found that: (i) the power patterns generated by the suggested approach and [16] fully meet the specified radiation requirements; (ii) the actual power pattern meets the radiation requirements when the PDRR in [13] is 6. However, if the PDRR is 3.64, the radiation pattern synthesized exceeds the set limits in the mainlobe area, as depicted in Figure 4.4(b). Consequently, if the user-defined DRR threshold is too small, [13] may not deliver satisfactory beampatterns; (iii) the array pattern produced by [3] varies significantly and surpasses the desired limits in the mainlobe region, indicating that [3] is not suitable for flat-top beampattern synthesis; and iv) as shown in Table 4.1, the proposed algorithm achieves a lower DRR compared to other approaches. Although it is more computationally efficient than [13], it is less computationally efficient than [3] and [16].

In summary, even though the proposed algorithm is not the most computationally efficient, it achieves the lowest DRR among the comparison methods while maintaining satisfactory power patterns.

To assess the influence of iteration count on the convergence efficiency of the new approach, we analyze the variation of errors $\max_n|\mathbf{y}_n - \mathbf{w}_n^*|$, $\max_m|\mathbf{z}_m - \mathbf{w}^H\mathbf{a}(\theta_m)|$, and $\max_s|\mathbf{v}_s - \mathbf{w}^H\mathbf{a}(\bar{\theta}_s)|$ between the auxiliary variables and angular responses (AVAR) as illustrated in Figure 4.4(d), utilizing the theoretical findings from the convergence analysis (4.33), (4.34), and (4.35). Observing that the AVAR error diminishes as the number of iterations increases, with a gradual decrease in rate after reaching 5000 iterations, indicates the strong convergence capabilities of the proposed algorithm.

Table 4.1 Performance of different schemes in Exp.1-4

	DRR	SLL (dB)	RMLL (dB)	CPU time (s)
Proposed	3.64	−25	(−0.1, 0.1)	629.54
[13](PDRR=6)	6	−25	(−0.1, 0.1)	1063.70
[13](PDRR=3.64)	4.07	−25.03	(−0.16, 0.1)	1162.82
[3]	6.55	−25.19	(−90, −0.1)	4.57
[16]	4.76	−26.07	(−0.09, 0.09)	65.59

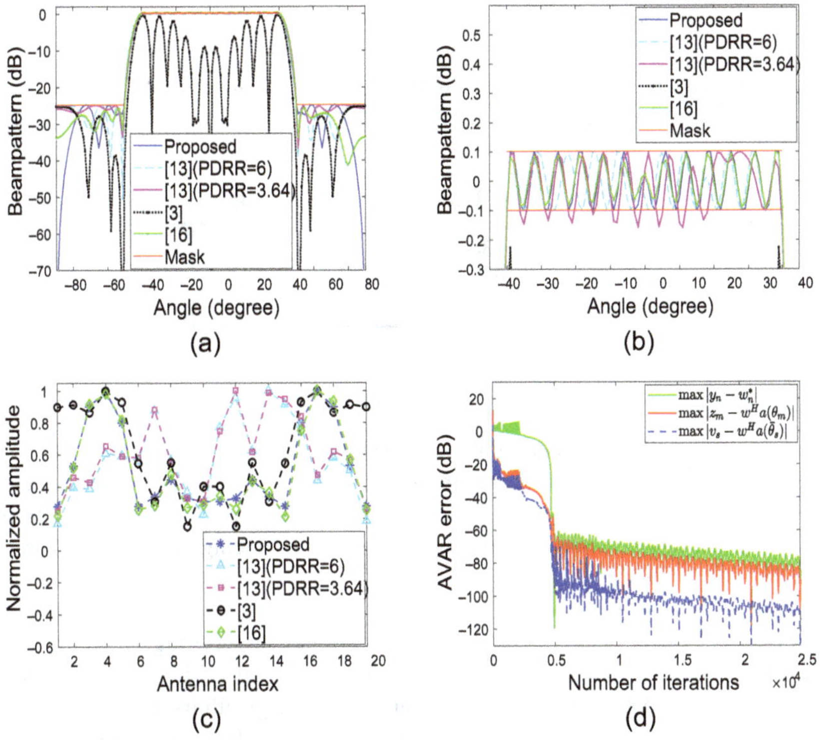

Figure 4.4 (a) Synthesized beampatterns; (b) zoomed mainlobe subregion; (c) normalized excitation magnitudes; and (d) AVAR error versus iteration number for Exp.1-4.

Initially, a comparison is conducted between our proposed approach and the DRR threshold specification technique introduced by [13], showcasing that user-defined DRR thresholds may not consistently align with current attenuator constraints [19]. Subsequent experiments juxtapose our scheme with those outlined in [3] and [16], each of which incorporates DRR management within an objective function for optimization, differing from the DRR threshold approach in [13].

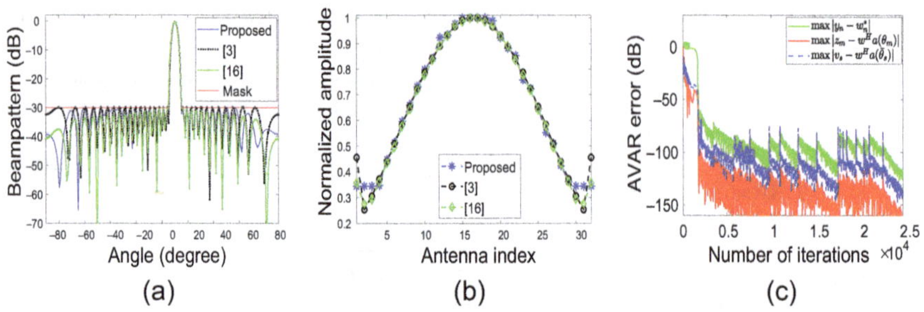

Figure 4.5 (a) Synthesized beampatterns; (b) normalized excitation magnitudes; and (c) AVAR error versus iteration number for synthesizing focused beampattern.

Table 4.2 Performance of different schemes in synthesizing focused beampattern

	DRR	SLL (dB)	RMLL (dB)	CPU time (s)
Proposed	2.89	−30	0	372.41
[3]	3.95	−30	0	6.39
[16]	3.61	−30.26	0	55.40

4.4.2 Focused beampattern synthesis

Given the application of a focused pattern to automatically track targets in radar systems [16], the next step involves creating a focused pattern with a maximum sidelobe level of −30 dB within the range $[-90°, 5°] \cup [15°, 90°]$. To achieve this, we utilize a ULA consisting of 32 sensors with an inter-element spacing of 0.5λ to generate a focused pattern towards the direction of $\theta = 10°$. These antennas match those used in synthesizing a flat-top Pattern.

Figure 4.5 displays the desired power patterns and the resulting excitation amplitude distributions, with detailed numerical results provided in Table 4.2. Analysis of these patterns and values demonstrates that while all algorithms meet the specified requirements in terms of radiation patterns, the proposed method stands out with the lowest DRR compared to other algorithms. Additionally, we assess the AVAR errors and illustrate them in Figure 4.5(c). Despite significant fluctuations after 5000 iterations, the AVAR errors remain below −80 dB, meeting the beampattern design criteria for various practical applications.

4.4.3 Flat-top beampattern synthesis with notching

In this study, we utilized the identical array configuration that was previously used to create a focused pattern. The third case demonstrates the capability of the new algorithm to generate a flat-top pattern with a null to suppress interference. To achieve this goal, we included an extra condition specifying that the null depth should be below −60 dB within the angular range $[35°, 40°]$. Moreover, the shaped power pattern should exhibit a variation of ± 0.1 dB within the mainlobe range $[-10°, 10°]$, and the highest sidelobe level should be -30 dB within the range $[-90°, -15°] \cup [15°, 34°] \cup [41°, 90°]$.

The resulting patterns and excitation amplitude distributions are displayed in Figure 4.6, while Table 4.3 presents the computed values for DRR, SLL, RMLL, NLL (null depth), and CPU time. As shown in Figure 4.6(a), the mainlobe level of the pattern generated by [3] exceeds the specified limits in the mainlobe region, although it meets the requirements for sidelobes and null regions. Both the proposed approach and [16] achieve the desired patterns. However, when compared to [16], the new algorithm performs better in achieving low DRR, as demonstrated in Table III. Hence, our proposed method achieves the lowest DRR among the three approaches while meeting the desired power pattern constraints. Similar to the flat-top pattern synthesis, the variation of the AVAR error with the number of iterations is depicted in Figure 4.6(c). It is evident that all AVAR errors decrease rapidly after the initial 15000 iterations and fall below -99.93 dB when the maximum iteration number $\tilde{T}$ is reached, satisfying the design requirements for the power pattern.

4.4.4 *Multibeam synthesis with notching*

In contrast to the synthesis of flat-top patterns, the fourth experimental setup focuses on a scenario necessitating three main lobes along with two narrow null zones

Table 4.3 Performance of different schemes in flat-top pattern synthesis with notching

	DRR	SLL (dB)	RMLL (dB)	NLL (dB)	CPU time (s)
Proposed	5.86	−30	(−0.1, 0.1)	−60	445.56
[3]	10.81	−37.38	(−24.55, −0.46)	−71.28	7.81
[16]	12.84	−30.85	(−0.1, 0.1)	−60	167.21

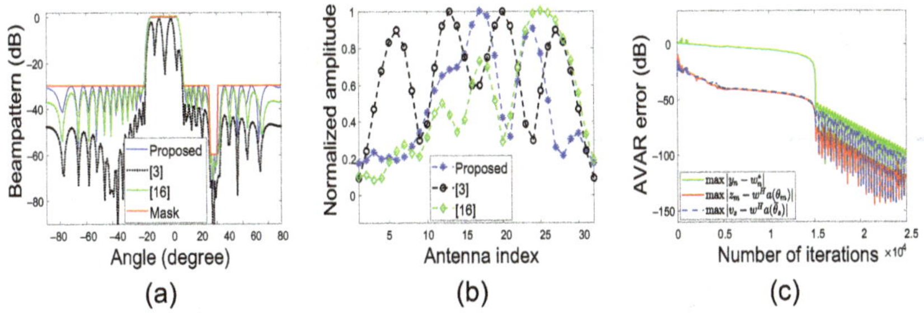

Figure 4.6 (a) Synthesized beampatterns; (b) normalized excitation magnitudes; and (c) AVAR error versus iteration number for synthesizing flat-top pattern with notching.

oriented towards the jamming signals aimed at mitigating interference in communication systems. In this case, the identical array used for producing a focused pattern is taken into account. The objective radiation pattern stipulates a maximum sidelobe level of -22 dB within the specified range $[-90°, -45°] \cup [-35°, -21°] \cup [-18°, -5°] \cup [5°, 18°] \cup [21°, 35°] \cup [45°, 90°]$, coupled with a null depth of -80 dB within the interference region $[-20°, -19°] \cup [19°, 20°]$. The peaks of the three main lobes are situated at $\theta = -40°$, $\theta = 0°$, and $\theta = 40°$. Additionally, an assessment of the impact of mutual coupling on the efficacy of the proposed technique is conducted. As per [30–33], a symmetric Toeplitz matrix with bandwidth constraints is utilized to model the mutual coupling effect among elements of the linear array. In this context, we adopt $\boldsymbol{C} = \text{toeplitz}[1, c_1, c_2, c_3, 0, \cdots, 0]_{1\times N}$ where $c_1 = 0.3527 + 0.4584i$, $c_2 = 0.1618 - 0.2853i$, and $c_3 = 0.0927 - 0.1167i$ [30–33]. This indicates that each sensor is influenced solely by its three adjacent sensors. With mutual coupling taken into consideration, the resultant steering vector is given by $\hat{\ }(\theta) = \boldsymbol{C}\ (\theta)$.

Figure 4.7 illustrates the plotted distributions of excitation amplitude and patterns, while the numerical outcomes can be found in Table 4.4, where the performance of the proposed algorithm under mutual coupling is denoted as ProposedWMC. Analysis of Figure 4.7(a) reveals that all investigated strategies can generate patterns featuring two nulls and three mainlobes. Table 4.4 validates that in comparison to alternative methods, the excitations produced by the proposed scheme exhibit lower DRR while maintaining the desired radiation patterns. It is important to note that the radiation requirements are successfully met even in the presence of mutual coupling, which has minimal impact on the DRRs of the excitations. Moreover, Figure 4.7(c) and (d) presents the variation of AVAR error throughout iterations, both with and without consideration of mutual coupling, showcasing the commendable convergence performance of the proposed algorithm in both scenarios.

4.4.5 Power pattern synthesis for uniform rectangular array

The fifth instance examines a regular rectangular array (RRA) with $N = 11 \times 11$ isotropic elements where the spacing between elements is $\lambda/2$. The array is situated

Table 4.4 Performance of different schemes in synthesizing multibeam with notching

	DRR	SLL (dB)	RMLL (dB)	NLL (dB)	CPU time(s)
Proposed	3.57	−22	0	−80.72	313.57
ProposedWMC	3.86	−22	0	−80.68	312.57
[3]	14.80	−22	0	−80.13	6.04
[16]	9.28	−22.03	0	−91.9	147.37

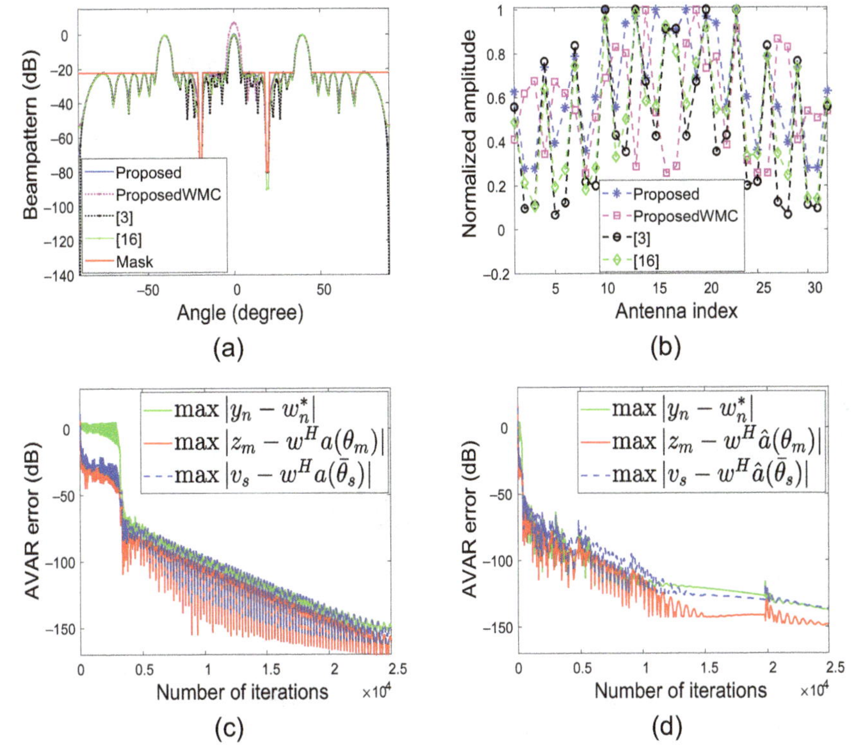

Figure 4.7 *(a) Synthesized beampatterns; (b) normalized excitation magnitudes; (c) AVAR error versus iteration number without the mutual coupling; and (d) AVAR error versus iteration number with the mutual coupling for synthesizing multibeam with notching.*

in the $x-y$ plane of the Cartesian coordinate system. When it comes to planar arrays, our outcomes are contrasted with those derived from convex optimization [10]. In order to ensure that the 3D power pattern synthesized by [10] meets all specified radiation criteria, we decrease the DRR by eliminating array elements with amplitudes below 10^{-3}. The resulting radiation patterns are graphed in the $u_x - u_y$ axes. The primary lobe sector is defined by $u_x^2 + u_y^2 \leq 0.2^2$ with a fluctuation of ± 0.5 dB. The secondary lobe sector is described by $u_x^2 + u_y^2 \geq 0.4^2$ with a maximum secondary lobe level of -25 dB. The sampling step is 0.04 for both the u_x and u_y directions. The experimental outcomes are displayed in Figure 4.8 and Table 4.5. Given that the radiation power patterns in the u_y-direction closely resemble those in the u_x-direction, only the beam patterns in the u_y-direction are illustrated.

It is noticeable that while both the suggested method and [10] achieve satisfactory radiation patterns for RRA, the DRR of the excitations produced by the former is notably lower than that of [10].

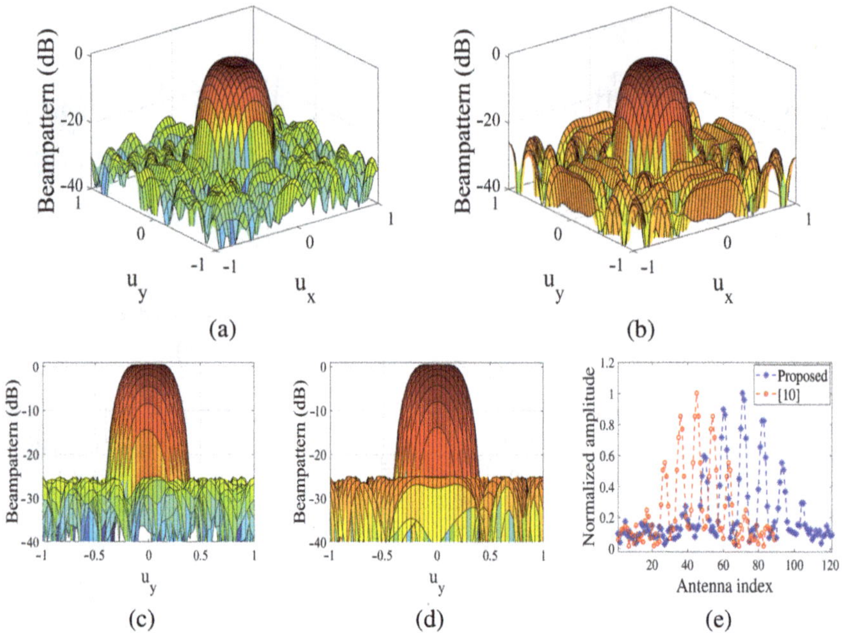

Figure 4.8 (a) Synthesized 3D power pattern by the proposed algorithm; (b) synthesized 3D power pattern by [10]; (c) view of power pattern in u_y-directions of (a); (d) view of power pattern in u_y-directions of (b); and (e) normalized excitation magnitudes for URA.

Table 4.5 Performance of different schemes for URA

	DRR	SLL	RMLL (dB)	CPU time (s)
Proposed	37.39	−25.18	(−0.5, 0.5)	741.40
[10]	65.79	−25	(−0.5, 0.5)	2022.70

4.4.6 Power pattern synthesis for uniform circular array

The example considers a circular array with a radius of $R = 3.2\lambda$, consisting of 80 isotropic elements positioned in the $x - y$ plane of the Cartesian coordinate system. The objective pattern exhibits a peak at $u_x = 0$ and $u_y = 0$, with a maximum sidelobe level of -10 dB within the sidelobe region $u_x^2 + u_y^2 \geq 0.3^2$. All other parameters remain consistent with the fifth example. The numerical results can be observed in Figure 4.9(a)–(e) and Table 4.6. The three-dimensional radiation patterns achieved using the proposed algorithm and [10] meet all specified radiation criteria for the circular array. However, our approach yields a significantly lower DRR value.

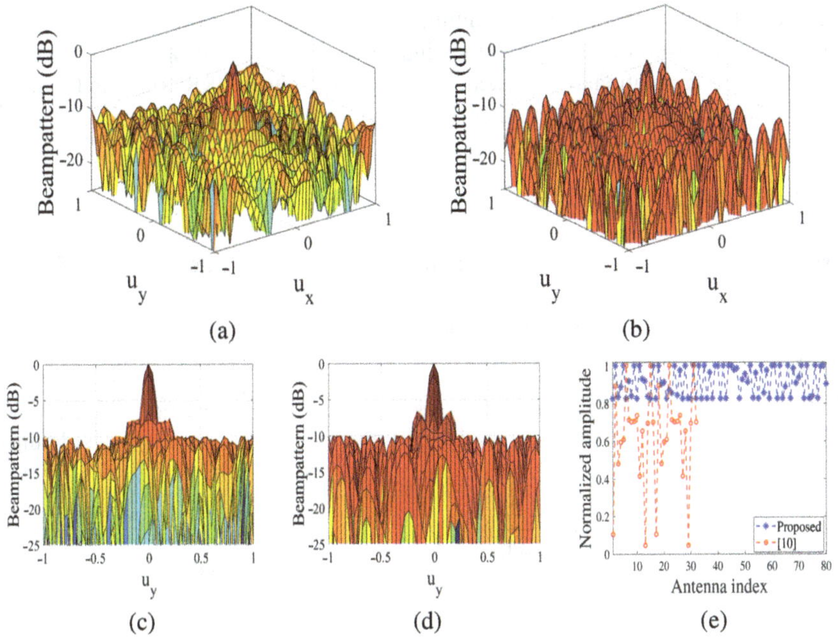

Figure 4.9 (a) Synthesized 3D power pattern by the proposed algorithm; (b) synthesized 3D power pattern by [10]; (c) view of power pattern in u_y-directions of (a); (d) view of power pattern in u_y-directions of (b); and (e) normalized excitation magnitudes for UCA.

Table 4.6 Performance of different schemes for UCA

	DRR	SLL	RMLL (dB)	CPU time (s)
Proposed	1.21	−10	0	4015.21
[10]	22.29	−10	0	1470.56

Hence, the suggested algorithm not only combines the desired power pattern but also achieves a significantly reduced DRR compared to [10] for UCA and URA.

4.5 Conclusion

Introducing a novel optimization problem resembling a fraction is introduced to create an array beampattern while minimizing the DRR of the complex excitations. This problem is found to be complex and non-linear. To address this, we introduce additional variables and establish an equivalent optimization model that transforms the fractional objective function into a linear one. This model is then addressed using the alternating direction method of multiplier. A comparison between our proposed approach and several existing methods indicates that while our approach may not be the most computationally efficient, it does yield a lower DRR compared to other algorithms while ensuring that the desired radiation patterns are achieved.

Appendix A: Focused beam synthesis

Assume $f(x) = ax^2 + bx + c$ where $a, b, c \in \mathbb{R}, a \neq 0$. when $a > 0$, $f(x)$ is a convex function. Therefore, $f(x)$ achieves the minimal value at $x = -\frac{b}{2a}$ by virtue of the decrease and increase in $(-\infty, -\frac{b}{2a}]$ and $[-\frac{b}{2a}, \infty)$, respectively. Denote $x = \sqrt{\beta}$, $a = A_k = \sum_{m=1}^{M} (Q_{1m}^{(t)} U + Q_{2m}^{(t)} L) + \sum_{s=1}^{S} Q_{3s}^{(t)} \eta > 0$, $b = B_k = -2\sum_{s=1}^{S} Q_{3s}^{(t)} \sqrt{\eta} |\bar{v}_s^{(t)}| < 0$, and $c = C_k = \sum_{m=1}^{M} (Q_{1m}^{(t)} + Q_{2m}^{(t)}) |\bar{z}_m^{(t)}|^2 + \sum_{s=1}^{S} Q_{3s}^{(t)} |\bar{v}_s^{(t)}|^2 > 0$. Then if $\beta \in [\beta_{k-1}^{(t)}, \beta_k^{(t)}]$, the minimal point (x^*, f^*) is obtained as follows:

$$x^* = \begin{cases} \sqrt{\beta_k^{(t)}}, & \text{if } (-\frac{B_k}{2A_k}) \geq \sqrt{\beta_k^{(t)}} \\ \sqrt{\beta_{k-1}^{(t)}}, & \text{if } (-\frac{B_k}{2A_k}) \leq \sqrt{\beta_{k-1}^{(t)}} \\ -\frac{B_k}{2A_k}, & \text{otherwise} \end{cases} \tag{A.1}$$

Furthermore,

$$x^{*2} = \begin{cases} \beta_k^{(t)}, & \text{if } (-\frac{B_k}{2A_k})^2 \geq \beta_k^{(t)} \\ \beta_{k-1}^{(t)}, & \text{if } (-\frac{B_k}{2A_k})^2 \leq \beta_{k-1}^{(t)} \\ \frac{B_k^2}{4A_k^2}, & \text{otherwise} \end{cases} \tag{A.2}$$

that is, (A.2) is the solution to (4.21).

Appendix B: Solution to (4.13)

Similarly, for the second subproblem shown in (4.13), inserting (4.16) into (4.13) yields

$$\min_p \; p + \frac{\rho}{2} \sum_{n=1}^{N} (Q_{4n}^{(t)} (p - |\bar{y}_n^{(t)}|)^2 + Q_{5n}^{(t)} (1 - |\bar{y}_n^{(t)}|)^2), \tag{B.1}$$

where unit-step functions $Q_{4n}^{(t)}$ and $Q_{5n}^{(t)}$ [23] are:

$$Q_{4n}^{(t)} = \begin{cases} 1, & \text{if } |\bar{y}_n^{(t)}| \geq p \\ 0, & \text{otherwise} \end{cases} \tag{B.2}$$

and

$$Q_{5n}^{(t)} = \begin{cases} 1, & \text{if } |\bar{y}_n^{(t)}| \leq 1 \\ 0, & \text{otherwise} \end{cases} \tag{B.3}$$

Since $p \geq 1$, we remove the elements which are less than 1 in the selected turning points $|\bar{y}_1^{(t)}|, \cdots, |\bar{y}_N^{(t)}|$, reorder the remaining elements in ascending order as $p_1^{(t)}, \cdots, p_{\bar{K}}^{(t)}$, and obtain $(\bar{K}+1)$ subfunctions in $[p_0^{(t)}, p_1^{(t)}], \cdots, [p_{\bar{K}}^{(t)}, p_{\bar{K}+1}^{(t)}]$, where $p_0^{(t)} = p_L$ and $p_{\bar{K}+1}^{(t)} = p_U$. For the $\bar{k}$th segment $[p_{\bar{k}-1}^{(t)}, p_{\bar{k}}^{(t)}]$, ignoring the irrelevant term, (B.1) is expressed as a quadratic function of variable p:

$$\min_p F_p(p), \tag{B.4}$$

where

$$F_p(p) = A_{\bar{k}}p^2 + B_{\bar{k}}p + C_{\bar{k}}, \quad A_{\bar{k}} = \frac{\rho}{2} \sum_{n=1}^{N} Q_{4n}^{(t)}, \tag{B.5}$$

$$B_{\bar{k}} = 1 - \rho \sum_{n=1}^{N} Q_{4n}^{(t)} |\bar{y}_n^{(t)}|, \quad C_{\bar{k}} = \frac{\rho}{2} \sum_{n=1}^{N} Q_{4n}^{(t)} |\bar{y}_n^{(t)}|^2. \tag{B.6}$$

Note that (B.4) is also convex due to $A_{\bar{k}} > 0$ and thus we attain the local minimum $C_{\bar{k}} - \frac{B_{\bar{k}}^2}{4A_{\bar{k}}}$ at $p_{\bar{k}} = -\frac{B_{\bar{k}}}{2A_{\bar{k}}}$ when $-\frac{B_{\bar{k}}}{2A_{\bar{k}}}$ belongs to the region $[p_{\bar{k}-1}^{(t)}, p_{\bar{k}}^{(t)}]$. In addition, when $-\frac{B_{\bar{k}}}{2A_{\bar{k}}}$ lies in the region $[p_{\bar{k}}^{(t)}, +\infty]$ or $[-\infty, p_{\bar{k}-1}^{(t)}]$, the local minimum is located at $p_{\bar{k}} = p_{\bar{k}}^{(t)}$ or $p_{\bar{k}} = p_{\bar{k}-1}^{(t)}$, respectively. After obtaining the $(\bar{K}+1)$ locally minimal values of the $(\bar{K}+1)$ subregions, we select the smallest one and its corresponding optimal value is employed to determine $p^{(t+1)}$, namely,

$$p^{(t+1)} = \arg\min_{p_{\bar{k}}} \left\{ F_p(p_{\bar{k}}), \bar{k} = 1, \cdots, \bar{K}+1 \right\}. \tag{B.7}$$

Therefore, (B.7) is the solution to (B.1). Then substituting $p^{(t+1)}$ into (4.16) results in $y_n^{(t+1)}$.

References

[1] W. Croswell. Antenna theory, analysis, and design. *IEEE Antennas and Propagation Society Newsletter*. 1982;24(6):28–29.

[2] R. Vescovo. Consistency of constraints on nulls and on dynamic range ratio in pattern synthesis for antenna arrays. *IEEE Transactions on Antennas and Propagation*. 2007;55(10):2662–2670.

[3] B. Fuchs and S. Rondineau. Array pattern synthesis with excitation control via norm minimization. *IEEE Transactions on Antennas and Propagation*. 2016;64(10):4228–4234.

[4] D. J. Shpak. A method for the optimal pattern synthesis of linear arrays with prescribed nulls. *IEEE Transactions on Antennas and Propagation*. 2002;44(3):286–294.

[5] H. Steyskal, R. Shore, and R. Haupt. Methods for null control and their effects on the radiation pattern. *IEEE Transactions on Antennas and Propagation*. 2003;34(3):404–409.

[6] W. P. M. N. Keizer. Low-sidelobe pattern synthesis using iterative Fourier techniques coded in MATLAB [EM programmer's notebook]. *IEEE Transactions on Antennas and Propagation*. 2009;51(2):137–150.

[7] O. M. Bucci, G. Mazzarella, and G. Panariello. Reconfigurable arrays by phase-only control. *IEEE Transactions on Antennas and Propagation*. 1991;39(7):919–925.

[8] J. Liang, X. Fan, D. Zhou, and J. Li. Phase-only pattern synthesis for linear antenna arrays. *IEEE Antennas and Wireless Propagation Letters*. 2017;16:3232–3235.

[9] G. K. Mahanti, A. Chakraborty, and S. Das. Design of fully digital controlled reconfigurable array antennas with fixed dynamic range ratio. *Journal of Electromagnetic Waves and Applications*. 2007;21(1):97–106.

[10] S. E. Nai, W. Ser, Z. L. Yu, and H. Chen. Beampattern synthesis for linear and planar arrays with antenna selection by convex optimization. *IEEE Transactions on Antennas and Propagation*. 2010;58(12):3923–3930.

[11] G. Buttazzoni and R. Vescovo. Power synthesis for reconfigurable arrays by phase-only control with simultaneous dynamic range ratio and near-field reduction. *IEEE Transactions on Antennas and Propagation*. 2012;60(2):1161–1165.

[12] R. Vescovo. Null formation with excitation constraints in the pattern synthesis for circular arrays of antennas. *Electromagnetics*. 2001;21(3):213–230.

[13] J. Bai, Y. Liu, J. Cheng, P. You, and Q. Liu. Shaped power pattern antenna array synthesis with reduction of dynamic range ratio. *2016 Progress in Electromagnetic Research Symposium (PIERS)*. Shanghai, 2016: 2444–2447.

[14] G. K. Mahanti, S. Das, and A. Chakraborty. Design of phase-differentiated reconfigurable array antennas with minimum dynamic range ratio. *IEEE Antennas and Wireless Propagation Letters*. 2006;5(1):262–264.

[15] M. Comisso and R. Vescovo. Fast iterative method of power synthesis for antenna arrays. *IEEE Transactions on Antennas and Propagation*. 2009;57(7):1952–1962.

[16] X. Fan, J. Liang, and H. C. So. Beampattern synthesis with minimal dynamic range ratio. *Signal Processing*. 2018;152(11):411–416.

[17] Z.-Q. Luo, W.-K. Ma, A. M.-C. So, Y. Ye, and S. Zhang. Semidefinite relaxation of quadratic optimization problems. *IEEE Signal Processing Magazine*. 2010;27(3):20–34.

[18] CVX Toolbox: Version 2.1. Mar. 2017. [Online]. Available: http://cvxr.com/cvx/.

[19] F. E. S. Santos and J. A. R. Azevedo. Adapted raised cosine window function for array factor control with dynamic range ratio limitation. *2017 11th European Conference on Antennas and Propagation (EUCAP)*. Paris, 2017: 2020–2024

[20] J. Eckstein, and D. P. Bertsekas. On the Douglas-Rachford splitting method and the proximal point algorithm for maximal monotone operators. *Mathematical Programming*. 1992;55:293–318.

[21] D. Gabay. Applications of the method of multipliers to variational inequalities. *Augmented Lagrangian Methods: Applications to the Solution of Boundary-Value Problems*. North-Holland, Amsterdam, 1983.

[22] S. Boyd, N. Parikh, E. Chu, B. Peleato, and J. Eckstein. Distributed optimization and statistical learning via the alternating direction method of multipliers. *Foundations and Trends in Machine Learning*. 2011;3(1):1–122.

[23] J. Liang, H. C. So, J. Li, and A. Farina. Unimodular sequence design based on alternating direction method of multipliers. *IEEE Transactions on Signal Processing*. 2016;64(20):5367–5381.

[24] J. Liang, G. Yu, B. Chen, and M. Zhao. Decentralized dimensionality reduction for distributed tensor data across sensor networks. *IEEE Transactions on Neural Networks and Learning Systems*. 2016;27(11):2174–2186.

[25] J. Liang, X. Zhang, H. C. So, and D. Zhou. Sparse array beampattern synthesis via alternating direction method of multipliers. *IEEE Transactions on Antennas and Propagation*. 2018;66(5):2333–2345.

[26] L. Vandenberghe and S. Boyd. Semidefinite programming. *SIAM Review*. 2002;137(3):461–482.

[27] C. G. Tsinos, and B. Ottersten. An efficient algorithm for unit-modulus quadratic programs with application in beamforming for wireless sensor networks. *IEEE Signal Processing Letters*. 2018;25(2):169–173.

[28] Z. Wen, C. Yang, X. Liu, and S. Marchesini. Alternating direction methods for classical and ptychographic phase retrieval. *Inverse Problems*. 2012;28(11):461–482.

[29] M. Comisso and R. Vescovo. 3D power synthesis with reduction of near-field and dynamic range ratio for conformal antenna arrays. *IEEE Transactions on Antennas and Propagation*. 2011;59(4):1164–1174.

[30] T. Svantesson. Modeling and estimation of mutual coupling in a uniform linear array of dipoles. *Proceeding of the IEEE International Conference on Acoustics*. 1999.

[31] C. Liu, Z. Ye Z, and Y. Zhang. Autocalibration algorithm for mutual coupling of planar array. *Signal Processing*. 2010;90(3):784–794.

[32] Z. Ye, and C. Liu. 2-D DOA estimation in the presence of mutual coupling. *IEEE Transactions on Antennas and Propagation*. 2008;56(10):3150–3158.

[33] J. Liang, X. Zeng, W. Wang, and H. Chen. L-shaped array-based elevation and azimuth direction finding in the presence of mutual coupling. *Signal Processing*. 2010;91(10):1319–1328.

Chapter 5

Array beampattern synthesis with shape constraints and excitation range control

Xuan Zhang[1], Xiangrong Wang[1], Jiayi Huang[1] and Hing Cheung So[2]

This chapter considers the receive array beampattern synthesis problem, which aims at minimizing the peak sidelobe level under the constraints of controlled mainlobe ripples and Dynamic Range Ratio (DRR) of current excitations. Especially, when DRR = 1, the phase-only beampattern synthesis problem is solved, which targets at improving the power efficiency of the radar transmitter.

5.1 Background and introduction

Array beampattern synthesis is an important problem in the field of array signal processing. The goal is to make the array synthesize enhanced response in the desired direction while suppressing the interferences in other directions by properly adjusting the amplitudes and phases of the excitation weight vector. Besides, in order to maximize the utilization of transmitter power and accommodate the hardware implementation, it is desired to control the dynamic range ratio (DRR) on the complex excitations [1]. Conventional deterministic array synthesis methods, such as window function weighting schemes [2–4], cannot effectively control the DRR of the beamforming excitation vector, which may lead to a large dynamic range and decrease the transmit power efficiency. Therefore, it is well promoted that the optimization-based computational approach is applied in the field of array synthesis.

In this chapter, we propose a method based on primary-dual iteration to address the array beampattern synthesis problem of minimizing the peak sidelobe level (PSL) under the constraints of controlled mainlobe shape and DRR of current excitations. The proposed method can be utilized to synthesize a variety of desired-shaped beampatterns, such as focused beam, flat-top beam, and cosecant square beam as well as null beampattern. Additionally, our method can be applied in arbitrary array configurations, such as uniform linear array, planar array, and circular rings array, even in generic array with non-uniform spacing between array elements.

[1]School of Electronic and Information Engineering, Beihang University, China
[2]Department of Electrical Engineering, City University of Hong Kong, China

In particular, when DRR=1, the phase-only beampattern synthesis problem is resolved by our method, where the beams are uniformly excited by constant amplitudes of complex excitations. This situation is conducive to power amplifiers with maximum transmit energy available in the antenna system.

The introduction and background are discussed in Section 5.1. The review of previous work is introduced in Section 5.2, and the problem formulation is presented in Section 5.3. The proposed array beampattern synthesis method with shape constraints and excitation range control is described in Section 5.4. A series of representative numerical examples are shown in Section 5.5 to verify the effectiveness of the proposed method in terms of lowering the PSL and controlling the DRR. Finally, conclusions are drawn in Section 5.6.

The beampattern synthesis of antenna arrays has received continuously increasing attention since it finds tangible applications in radar, remote sensing, and millimeter-wave communications, to list a few [5–11]. Designing array beampatterns that satisfy different application criteria requires a trade-off between beam width, sidelobe level (SLL), shape constraints, excitation range, and many other factors. A typical one is the synthesis of sidelobes as low as possible to concentrate energy on the mainlobe. In [12], a weighted analytical expression was obtained to achieve a compromise between beam width and SLL. In [13], the relationship between the weighting coefficients of the weighted least-squares (WLS) cost function and the desired array response was established theoretically and an analytical solution for the WLS beampattern synthesis was proposed to ensure the desired beampattern without fine-tuning any user parameters. In [14], a novel artificial neural network (ANN)-based array synthesis method was proposed, which successfully synthesizes mask-constrained beampattern of focused or shaped beams for linear arrays with arbitrary given array geometries.

Besides lowering the sidelobes of the radiation pattern, it is usually necessary to place deep nulls on the pattern in the presence of strong interference. In [15], two methods for pattern nulling in an antenna array were presented by phase-only control. The methods iteratively minimized a cost function defined as a weighted sum of the squared amplitudes of the beampatterns in null directions and achieved the real-time array control. The literature [16] proposed a curved array split into two subarrays to improve the pattern null depth and width for reception pattern antennas, and a microstrip patch antenna in conjunction with a constraint least-mean-square (LMS) algorithm was used to calculate array patterns. In [17], the author proposed a phase-only control method that iteratively minimizes the array output power and forms nulls in the unknown interference direction while approximating the desired beampattern, where the task of phase-only excitation range control is to optimize only the phases to match an expected power pattern for an antenna array while fixing the excitation amplitudes.

Generally speaking, designing antenna arrays with equal amplitude excitation can make active arrays more attractive in terms of cost, reliability, and power efficiency. Especially for non-periodic arrays, it is useful for multibeam satellite applications to reduce the cost and complexity of direct radiating arrays while optimizing the level of sidelobes and grating lobes [18]. In [5], a method based on semidefinite relaxation (SDR) technique was applied in phase-only array synthesis. However, the synthesized beampattern by the sought solution cannot guarantee to meet the constant modulus constraint (CMC) as this technique drops a rank-one constraint deliberately for relaxation. In [19], an iterative method of array beampattern synthesis of arbitrary geometry was presented. This method is based on successive projections and allows to synthesize multiple desired patterns, each of which can be converted to any of the others by phase-only control. In [20], an improved least-squares optimization method for large conformal antennas was proposed and it performed phase synthesis with predetermined excitation amplitude of elements. Another relaxation scheme was adopted in [21] to transform the non-convex CMC to convex constraints. Unfortunately, this relaxation scheme cannot always warrant each excitation to satisfy CMC and the desired mainlobe responses need to be carefully specified as well. Besides, there are losses of degrees-of-freedom (DoFs) in [5] and [21], as they assume the fixed excitation magnitude to match the desired pattern shape, which may cause the pattern distortion due to the ill-suited magnitudes. To avoid this problem, [22] introduced a scalar variable in the pattern shaping constraints to increase the DoFs although the magnitudes are also preset, while [23] optimized an unspecified uniform magnitude for phase-only array synthesis. Nevertheless, all these works considered beampattern synthesis with a given pattern mask, especially with a pre-specified PSL and in the general sense, the available DoFs may not be fully utilized to produce a lower SLL. In addition, for the fixed amplitude of the phase-only problem, some DoFs of the excitation vector are lost, making it difficult to ensure that the generated beampattern matches the expected beam shape. In such scenarios, both amplitudes and phases control can produce better performance. Furthermore, from the viewpoints of hardware realization and utmost utilization of the transmitters' power, one of the important issues is to control the DRR of the current excitations, which is defined as [1]

$$\mathrm{DRR}(w) = \frac{\max_n\{|w_n|\}}{\min_n\{|w_n|\}}, \tag{5.1}$$

with w being the complex excitation vector, which facilitates a flexible control of array systems in terms of power efficiency, cost, and mutual coupling between antennas. Specifically, if we specify a upper threshold T_r on DRR and introduce an auxiliary variable μ, then the DRR constraint (DRRC), i.e., $\mathrm{DRR}(w) \leq T_r$ can be expressed as follows:

$$\mu \leq |w_n| \leq \mu T_r, n = 1, \cdots, N. \tag{5.2}$$

Further, when $T_r = 1$, (5.2) degenerates to the common constant modulus case, that is,

$$|w_n| = \mu, n = 1, \cdots, N. \tag{5.3}$$

This essentially corresponds to a phase-only beampattern synthesis, where only phase shifters are used for beamforming to gain maximum power efficiency [5,9, 21–26]. However, the constraints of uniform magnitude in (5.3) are non-convex, in turn causing the problem challenging to solve.

Excitation dynamic range control is a promising technique used to control the excitation level between different antennas in an array to reduce the dynamic range difference. By reducing the dynamic range ratio, the processing efficiency and performance of the array signal can be improved. Therefore, many scholars have carried out a lot of research on this. In [27], the author presented a fast method for conformal arrays that enables the synthesis of 3D co-polar and cross-polar beampatterns and simultaneously reduces the DRR of the array excitation amplitudes. The literature [28] proposed an iterative algorithm based on the alternating projection method for the power of beam scanning arrays of arbitrary geometry synthesis. This method reduces the mutual interference and cross-coupling phenomena between antenna elements by optimizing the layout and phase control strategy of the antenna array, thus improving the system performance. In [29], the authors proposed a method for designing a dual beam array antenna with minimum dynamic range ratio through discrete phase control. Note that the design of reconfigurable dual-beam linear antenna arrays with a discrete phase converter using a real-coded genetic algorithm (GA) was described to achieve control of the antenna beam direction and shape to meet the requirements of different scenarios. In [30], Xu Z *et al.* discussed how to achieve power pattern synthesis for line polarization in arbitrary antenna arrays with control of the dynamic range ratio, and they proposed an optimization method to adjust the amplitude and phase weights of each element to realize the power pattern synthesis of line polarization while trying to ensure a reasonable range of dynamic range ratio. The work in [31] proposed a new method to synthesize a pencil beam with constrained SLL and DRR in order to achieve flexible control and optimal design of the pencil beam, which is a very narrow primary flap beam with important applications in radar and communication systems. Besides, a simple method was proposed in [32] for designing symmetric flat-top Gaussian arrays with controllable DRR by optimizing the weight distribution. It can be seen that reducing the DRR of the excitations in the array synthesis has been a hot issue in the field of array signal processing.

5.2 Previous methods based on convex optimization

In this section, we first review several current advanced array synthesis methods based on convex optimization techniques and then introduce the proposed scheme based on non-convex optimization.

In previous works, a state-of-the-art approach to array synthesis problem is to employ the convex optimization theory based on semi-definite relaxation (SDR) technique. For example, in [5], the authors considered N elements placed at position $\vec{r_n}$ with $n = 1, \ldots, N$ and the far-field response $f(\theta)$ radiated by the array is given by

$$f(\theta) = a(\theta)^H \omega, \tag{5.4}$$

where ω is the complex excitation vector and $a(\theta)$ is the steering vector in the direction θ. By introducing the notation $f_i = f(\theta_i)$ and $a_i = a(\theta_i)$, the real-value form of (5.4) can be represented as

$$[\Re(f_i)\ \Im(f_i)]^T = A_i x \quad with\ A_i = \begin{bmatrix} \Re(a_i{}^T) & -\Im(a_i{}^T) \\ \Im(a_i{}^T) & \Re(a_i{}^T) \end{bmatrix}\ x = \begin{bmatrix} \Re(\omega) \\ \Im(\omega) \end{bmatrix} \tag{5.5}$$

where $\Re(\cdot)$ and $\Im(\cdot)$ represent the real and imaginary parts of a complex number, respectively, and $A_i \in \mathbb{R}^{2\times 2N}$, $x \in \mathbb{R}^{2N\times 1}$. Then, it follows from (5.5) that the power radiated by the array is given by

$$\begin{aligned} |f_i|^2 &= x^T Q_i x, \ \textit{with}\ Q_i = {A_i}^T A_i \\ &= \mathrm{Tr}(Q_i X), \ \textit{with}\ X = xx^T \in \mathbb{R}^{2N\times 2N}, \end{aligned} \tag{5.6}$$

where Tr() denotes the trace operation of a matrix.

Most of array synthesis problems correspond to finding array excitation x or matrix X such that the radiated power $|f_i|^2$ satisfies certain constraints for desired beampattern shapes, i.e., $\mathscr{C}_i, i = 1, \cdots, I$, which is formulated as

$$\textit{find } x \textit{ such that } x^T Q_i x \in \mathscr{C}_i, \ \textit{for } i = 1, \cdots, I, \tag{5.7}$$

or equivalently as

$$\textit{find } X \textit{ such that } \begin{cases} \mathrm{Tr}(Q_i X) \in \mathscr{C}_i, \ \textit{for } i = 1, \cdots, I \\ X \succeq 0 \\ \mathrm{rank}(X) = 1 \end{cases}. \tag{5.8}$$

Due to the rank-one constraint in (5.8), the problem (5.8) is not convex. However, by removing this constraint, one can obtain the following SDR problem:

$$\textit{find } X \textit{ such that } \begin{cases} \mathrm{Tr}(Q_i X) \in \mathscr{C}_i, \ \textit{for } i = 1, \cdots, I \\ X \succeq 0 \end{cases} \tag{5.9}$$

Note that the problem (5.9) is convex in general and can be easily solved through software such as CVX [33]. Once the solution $X^\star$ of (5.9) is determined, the remaining thing is to transform the global optimal solution $X^\star$ into a feasible point $\tilde{x}$ of the original synthesis problem (5.7). If $\mathrm{rank}(X^\star) = 1$, then $X^\star$ is not only a feasible point, but also the optimal solution of (5.7). If $\mathrm{rank}(X^\star) > 1$, a well-known convex heuristic algorithm to encourage low-rank solutions is to minimize the trace of X which equal to the sum of the eigenvalues of X, thereby minimizing the rank of X.

The above SDR technology can be applied to shaped beampattern synthesis problems. Specifically, with (5.6) and (5.9), the SDR of the shaped beam synthesis problem can be given by

$$\textit{find } X \succeq 0 \textit{ such that } X \in \mathscr{S}: \begin{cases} \mathrm{Tr}(Q_m X) \geq l_m, \ \textit{for } m = 1, \cdots, M \\ \mathrm{Tr}(Q_m X) \leq u_m, \ \textit{for } m = 1, \cdots, M \\ \mathrm{Tr}(Q_q X) \leq \rho_q, \ \ \textit{for } q = 1, \cdots, Q \end{cases} \tag{5.10}$$

where u_m and l_m represent the upper and lower boundaries of the mainlobe region with M sampled angles, respectively, and ρ_q represents the envelope upper boundary of the sidelobe region with Q sampled angles.

For the synthesis of arrays with phase-only control, the corresponding SDR form is given by

$$\textit{find } X \textit{ such that } \begin{cases} \mathrm{Tr}(Q_n X) = \alpha_n, \ n = 1, \cdots, N \\ X \in \mathscr{C} \\ X \succeq 0 \textit{ and rank}(X) = 1 \end{cases} \tag{5.11}$$

where α_n is the excitation amplitude defined as

$$\alpha_n = |\omega_n|^2 = x^T Q_n x = \mathrm{Tr}(Q_n X), \quad n = 1, \cdots, N \tag{5.12}$$

and Q_n is a $2N \times 2N$ dimensional diagonal matrix defined as

$$Q_n(i,i) = \begin{cases} 1, \ \textit{if} \ i = n \\ 1, \ \textit{if} \ i = n + N \\ 0, \ \textit{elsewhere} \end{cases} . \tag{5.13}$$

In summary, although the above relaxation technique makes the original problem convex and convenient to solve, it cannot accurately control the beampattern shape and usually violates the design requirements due to the relaxation error resulting from the ignored rank-one constraint.

Another representative work in [10] employed the convex optimization technique to synthesize array beampatterns while controlling the excitation range. They used a mixture of ℓ_1/ℓ_∞ norm to control the excitation vector associated with the constraints on the radiation pattern. Specifically, for a focused beampattern, they proposed the following optimization problem to realize this goal:

$$\min_{\hat{X}} \mathrm{Tr}(\hat{X}) \quad \text{s.t. } FB(\hat{X}) \begin{cases} \mathrm{Tr}\left(\hat{A}_0 \hat{X}\right) \geq 1 \\ \mathrm{Tr}\left(\hat{A}_s \hat{X}\right) \leq \rho_s^2, \quad \forall s \\ \hat{X} \succeq 0 \end{cases} \tag{5.14}$$

where $\hat{A}_0 = a(\theta_0)a(\theta_0)^H$ with θ_0 being the direction in the main beam, $\hat{X} = \omega\omega^H$, and $\hat{A}_s = a(\theta_s)a(\theta_s)^H$ with θ_s being the direction in the sidelobe region below a given upper ρ_s. In addition to this, if given a desired pattern mask $y^d(\theta)$, the synthesis of a desired shaped beampattern can be formulated as

$$||f(\theta)|^2 - |y^d(\theta)|^2| \leq \varepsilon, \quad \forall \theta \in \Theta_{sb}, \tag{5.15}$$

where the parameter ε denotes the error approximate to desired beam and Θ_{sb} denotes the sidelobe angular region. Similar to (5.14), the SDR form of designing a desired shaped beampattern is given by

$$\min_{\hat{X}} \mathrm{Tr}(\hat{X}) \quad \text{s.t. } SB(\hat{X}) \begin{cases} \left|\mathrm{Tr}\left(\hat{A}_s \hat{X}\right) - \left|y_s^d\right|^2 \right| \leq \varepsilon \quad \forall s \\ \hat{X} \succeq 0 \end{cases} \tag{5.16}$$

To control the excitation, the authors imposed different norm constraints on $\hat{X}$ or ω. In particular, the sparse arrays synthesis with uniform amplitude and phase excitations is formulated by solving the following minimization of ℓ_1 and ℓ_∞ norm:

$$\begin{aligned} &\min_{0 \leq \omega \leq 1} \|\omega\|_1 + \gamma\|\omega - \alpha\|_\infty \\ &\text{s.t. } FB(\omega) \begin{cases} \Re(a(\theta_0)^H \omega) \geq 1 \\ |a(\theta_s)^H \omega| \leq \rho_s, \quad \forall s \end{cases} \\ &\quad \text{or } SB(\omega) : |a(\theta_s)^H \omega - y_s^d| \leq \varepsilon, \quad \forall s \end{aligned} \tag{5.17}$$

where γ is a tradeoff parameter controlling the sparsity and the binary feature of the solution, and α is a scalar. When $\alpha = \frac{1}{2}$, the solution to (5.17) trends to be binary, which produces a sparse array and achieves the selection of antenna elements.

Furthermore, when the excitations are real numbers, the array synthesis with a reduced DRR boils down to the following convex problem:

$$\begin{aligned}&\min_{\boldsymbol{\omega}\in\mathbb{R}^N} \max(\boldsymbol{\omega}) - \min(\boldsymbol{\omega})\\ &\text{s.t. } FB(\boldsymbol{\omega}) \text{ or } SB(\boldsymbol{\omega}) \text{ with } \omega_n \geq 0 \quad \forall n.\end{aligned} \tag{5.18}$$

On the other hand, if $\boldsymbol{\omega} \in \mathbb{C}^N$ is a complex vector, a reduction in the DRR can be achieved by minimizing $|\max_n |\omega_n|^2 - \min_n |\omega_n|^2|$ with the help of the matrix $\hat{X} = \boldsymbol{\omega}\boldsymbol{\omega}^H$. In this case, the array synthesis with a minimized DRR can be expressed as follows:

$$\begin{aligned}&\min_{\hat{X}} \left[\max_n \left(\hat{X}_{n,n}\right) - \min_n \left(\hat{X}_{n,n}\right)\right] + \gamma\mathrm{Tr}(\hat{X})\\ &\text{s.t. } FB(\hat{X}) \text{ or } SB(\hat{X}) \text{ and } \hat{X} \succeq 0.\end{aligned} \tag{5.19}$$

If the excitations are desired to have a smooth variation, it is considered to minimize the total TV-norm to achieve the synthesis of the array beampattern, where TV-norm is a smoothing function defined as the sum of the absolute values of the differences between neighboring components. In this situation, the problem is formulated as follows:

$$\begin{aligned}&\min_{\hat{X}} \sum_{n=1}^{N} \left|\hat{X}_{n+1,n+1} - \hat{X}_{n,n}\right| + \gamma\mathrm{Tr}(\hat{X})\\ &\text{s.t. } FB(\hat{X}) \text{ or } SB(\hat{X}) \text{ and } \hat{X} \succeq 0.\end{aligned} \tag{5.20}$$

It is worth noting that although the above methods enable the control of antenna beampattern and excitations, they just minimize a difference between the maximum power and minimum power instead of directly controlling the ratio of the maximum one and minimum one. Therefore, the existing methods cannot precisely and sufficiently control the DRR of excitations.

Unlike the above-mentioned efforts, this chapter explores array synthesis with an SLL as low as possible and directly controls the mainlobe ripples and the DRR of excitations. Specially, when $T_r = 1$, the phase-only array synthesis is considered with an unspecified magnitude μ in (5.3). Moreover, in order to obtain a low SLL, the PSL is minimized subject to two non-convex constraints of controlled mainlobe ripples and the DRR of the excitations, which results in a non-convex and non-smooth optimization problem. To solve the resultant problem, a variable splitting technique is adopted to decouple the excitation vector w from the mainlobe and sidelobe constraints and an efficient primary-dual iterative method is subsequently proposed, which handles the two non-convex constraints in the primary-dual iteration. Such a splitting technique helps to decouple the excitation vector from the mainlobe and sidelobe constraints. Note that the variable splitting technique is an advanced mathematical method for solving optimization problems involving multiple variables or constraints. It works by decomposing a complex optimization problem into a series of simpler subproblems, each involving a subset of variables or constraints.

In this work, we skillfully utilize the variable splitting technique for non-convex array synthesis optimization problems and decompose the original problem into a series of subproblems that can be easily solved independently. Evidently, such a separable structure is suitable to be handled by the primary-dual iteration framework and greatly reduces the computational complexity. This is the mainly technical contribution of our method in the field of array beampattern synthesis.

5.3 Problem formulation

Consider a linear array consisting of N elements placed at locations $p_n = n\tilde{d}$ with $\tilde{d}$ being the inter-element spacing in a unit of wavelength λ for $n = 0, \ldots, N-1$, as seen in Figure 5.1. Without loss of generality, the problem of the one-dimension (1-D) array beampattern synthesis is addressed in this chapter, while the extension to the two-dimension (2-D) case is straightforward. Assume there is a plane wave impinging the array from the looking direction θ, and then the far-field radiation power, i.e., so-called array beampattern is given by

$$P(\theta) = |w^H a(\theta)|^2, \tag{5.21}$$

where $(\cdot)^H$ represents conjugate transpose, w is the complex excitation vector as defined above and $a(\theta)$ is the array steering vector for linear arrays defined by

$$a(\theta) = [1, e^{-j\frac{2\pi}{\lambda}\sin(\theta)} \cdots e^{-j\frac{2\pi}{\lambda}(N-1)\tilde{d}\sin(\theta)}]^H. \tag{5.22}$$

To adapt the array beampattern to different operation tasks, it is desired to synthesize various array beampatterns, such as focused beam, flat-top beam, and cosecant square beam. To this end, we denote the mainlobe region by Θ_m which is uniformly divided into M angular grid points of interest, i.e., $[\theta_1, \cdots, \theta_M]$, and introduce the following mainlobe shape constraint with prescribed ripples:

$$\mathbb{SB} : d_m - \varepsilon \leq |w^H a(\theta_m)|^2 \leq d_m + \varepsilon, m = 1, \cdots, M, \tag{5.23}$$

where d_m is the mth desired mainlobe response and ε denotes the maximum ripple tolerance. Particularly, when we only set $d_i = 1$ at certain single angle θ_i and $\varepsilon = 0$, the shaped constraint $\mathbb{SB}$ reduces to the following focused mainlobe constraint:

$$\mathbb{FB} : |w^H a(\theta_i)|^2 = 1, \tag{5.24}$$

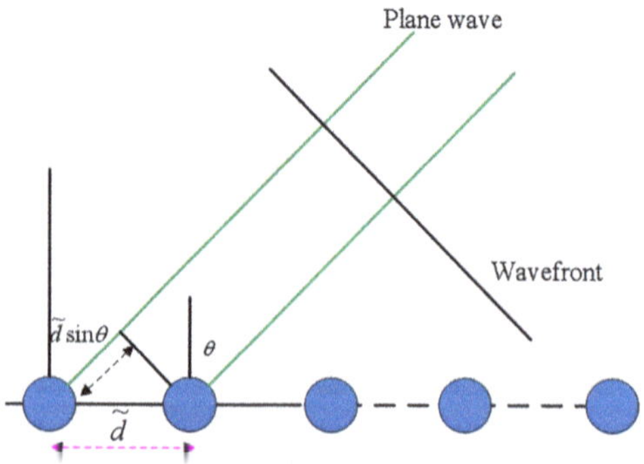

Figure 5.1 Uniform linear array diagram

where θ_i denotes the direction targeted by the focused beam.

Besides, to prevent the clutters and jammer into the array, it is expected to obtain as low SLL as possible in the sidelobe region which is denoted by Θ_s and uniformly discretized into S grid angles, i.e., $[\bar{\theta}_1, \cdots, \bar{\theta}_S]$. For this aim, we intend to suppress the following PSL among the sidelobe angular regions, i.e.,

$$\mathrm{PSL} = \max_{\bar{\theta}_s}\{|w^H a(\bar{\theta}_s)|^2\}, \bar{\theta}_s \in \Theta_s. \tag{5.25}$$

Clearly, a smaller PSL in (5.25) leads to a lower SLL. With the above consideration in mind, we propose to minimize the PSL, i.e., $\max\{|w^H a(\bar{\theta}_s)|^2\}_{s=1}^S$, while integrating the DRRC into the array synthesis, giving the following array optimization problem:

$$\mathscr{P}: \begin{cases} \min_{w} \ \max_{s}\{|w^H a(\bar{\theta}_s)|^2\}_{s=1}^S \\ \text{s.t. } d_m - \varepsilon \le |w^H a(\theta_m)|^2 \le d_m + \varepsilon, m = 1, \cdots, M, \\ \quad \mathrm{DRR}(w) \le T_r. \end{cases} \tag{5.26}$$

Note that $\mathscr{P}$ is a min–max optimization problem with a non-smooth objective subject to multiple non-constraints, which seems difficult to solve. In addition, when we set $T_r = 1$, the problem $\mathscr{P}$ reduces to the following phase-only one with an unspecified uniform amplitude variable μ:

$$\mathscr{P}_{\mathrm{u}}: \begin{cases} \min_{w,\mu} \ \max_{s}\{|w^H a(\bar{\theta}_s)|^2\}_{s=1}^S \\ \text{s.t. } d_m - \varepsilon \le |w^H a(\theta_m)|^2 \le d_m + \varepsilon, m = 1, \cdots, M, \\ \quad |w_n| = \mu, n = 1, \cdots, N. \end{cases} \tag{5.27}$$

It is worth noting that $\mathscr{P}_u$ optimizes weighted excitation amplitudes instead of the fixed one in the common phase-only constraint. This feature obviously offers more DoFs than the one with fixed amplitudes to further lower the SLL. However, $\mathscr{P}$ or $\mathscr{P}_u$ is challenging to solve due to the non-convex fractional DRRC and double-sided mainlobe shape constraints. Despite this, in the next section, we will derive a primal-dual iteration-based to solve $\mathscr{P}$ and correspondingly, $\mathscr{P}_{\mathrm{u}}$ is ready to be resolved.

5.4 Beampattern shaping with excitation range control

This section proposes a method to solve the problem $\mathscr{P}$. More precisely, to surmount the non-smoothness of the cost function in $\mathscr{P}$, we introduce an upper bound auxiliary variable η and rewrite $\mathscr{P}$ as follows:

$$\begin{aligned} &\min_{w,\eta} \ \eta^2 \\ &\text{s.t. } d_m - \varepsilon \le |w^H a(\theta_m)|^2 \le d_m + \varepsilon, m = 1, \cdots, M, \\ &\quad |w^H a(\bar{\theta}_s)| \le \eta, s = 1, \cdots, S, \\ &\quad \mathrm{DRR}(w) \le T_r. \end{aligned} \tag{5.28}$$

Note that (5.28) is a non-convex optimization problem due to two non-convex constraints, namely, the DRRC and the quadratic mainlobe ripple constraints, which make the problem difficult to solve. To facilitate solving (5.28) and splitting w from these non-convex constraints, we again introduce two sets of auxiliary variables $v_m, m = 1, \cdots, M$ for the mainlobe shape and $r_s, s = 1, \cdots, S$ for the sidelobe shape, and impose the following two linear equality constraints (LECs):

$$w^H a(\theta_m) = v_m, m = 1, \cdots, M, \tag{5.29}$$

$$w^H a(\bar{\theta}_s) = r_s, s = 1, \cdots, S, \tag{5.30}$$

on (5.28), leading to,

$$\begin{aligned} &\min_{w,\eta,v_m,r_s} \ \eta^2 \\ &\text{s.t. } d_m - \varepsilon \leq |v_m|^2 \leq d_m + \varepsilon, m = 1, \cdots, M, \\ &\quad |r_s| \leq \eta, s = 1, \cdots, S, \\ &\quad \text{DRR}(w) \leq T_r, \\ &\quad w^H a(\theta_m) = v_m, m = 1, \cdots, M, \\ &\quad w^H a(\bar{\theta}_s) = r_s, s = 1, \cdots, S. \end{aligned} \tag{5.31}$$

Apparently, such a variable splitting technique enables w decoupled from the quadratic constraints (5.28) attribute to these two LECs. Then, a primal-dual type algorithm mimicking the alternating direction method of multipliers (ADMM) [34] with a series of iterations is proposed to solve (5.31), which iteratively switches between minimizing the primal problem and maximizing the dual problem. The primal problem is decomposable and easy to solve while the dual problem is always convex and can be solved via a dual ascent method. Note that the new constraints, i.e., $d_m - \varepsilon \leq |v_m|^2 \leq d_m + \varepsilon$ and $|r_s| \leq \eta$, and the DRRC will play a separate role in their respective subproblems. For this purpose, we build the augmented Lagrangian function for (5.31) with respect to the LECs as follows:

$$\begin{aligned} L(w, \eta, r_s, v_m, \lambda_m, \kappa_s) \triangleq{}& \eta^2 + \Re[\sum_{m=1}^{M} \lambda_m^*(w^H a(\theta_m) - v_m)] \\ &+ \sum_{m=1}^{M} \frac{\rho_1}{2} |w^H a(\theta_m) - v_m|^2 + \Re[\sum_{s=1}^{S} \kappa_s^*(w^H a(\bar{\theta}_s) - r_s)] \\ &+ \sum_{s=1}^{S} \frac{\rho_2}{2} |w^H a(\bar{\theta}_s) - r_s|^2, \end{aligned} \tag{5.32}$$

where $(\cdot)^*$ take the conjugate part of a complex number, and λ_m, κ_s are Lagrange multipliers while ρ_1 and ρ_2 are the penalty parameters. Let $w^{(k)}$ denote the value of w at the kth iteration and the same notations are used for the other variables. Then, the iterative updating procedure of the proposed primal-dual algorithm is provided in Table 5.1, and the solutions to the respective subproblems from Steps 1–3 are derived as follows.

Table 5.1 Primal-dual method for minimum peak sidelobe beampattern synthesis with shape constraints and excitation range control

Input: $k=0$, initialize $w^{(0)}, v_m^{(0)}, \lambda_m^{(0)}, \kappa_s^{(0)}, I$.

Repeat:

Step 1: update: $\{r_s^{(k+1)}, \eta^{(k+1)}\}$ by solving:

$\min_{r_s,\eta} L(w^{(k)}, \eta, r_s, v_m^{(k)}, \lambda_m^{(k)}, \kappa_s^{(k)})$

s.t. $|r_s| \le \eta, s = 1, \cdots, S.$

Step 2: update: $v_m^{(k+1)}$ by solving:

$\min_{v_m} L(w^{(k)}, \eta^{(k+1)}, r_s^{(k+1)}, v_m, \lambda_m^{(k)}, \kappa_s^{(k)})$

s.t. $d_m - \varepsilon \le |v_m|^2 \le d_m + \varepsilon, m = 1, \cdots, M.$

Step 3: update: $w^{(k+1)}$ by solving:

$\min_w L(w, \eta^{(k+1)}, r_s^{(k+1)}, v_m^{(k+1)}, \lambda_m^{(k)}, \kappa_s^{(k)})$

s.t. $\mathrm{DRR}(w) \le T_r$.

Step 4: update dual variables:

$\lambda_m^{(k+1)} = \lambda_m^{(k)} + \rho_1(w^{(k+1)H}a(\theta_m) - v_m^{(k+1)}), m = 1, \cdots, M,$

$\kappa_s^{(k+1)} = \kappa_s^{(k)} + \rho_2(w^{(k+1)H}a(\bar{\theta}_s) - r_s^{(k+1)}), s = 1, \cdots, S.$

$k = k + 1$;

Until: $k = I$ or the maximal errors, e.g., $\max_m |w^H a(\theta_m) - v_m|$, and $\max_s |w^H a(\bar{\theta}_s) - r_s|$ are all less than 10^{-7}.

5.4.1 *Step 1: update* $\{r_s^{(k+1)}, \eta^{(k+1)}\}$

The subproblem in Step 1 with irrelevant items removed is equivalent to

$$\begin{aligned} &\min_{\eta, r_s} \ \eta^2 + \frac{\rho_2}{2}\sum_{s=1}^{S}|r_s - h_s|^2 \\ &\text{s.t. } |r_s| \le \eta, s = 1, \cdots, S, \end{aligned} \tag{5.33}$$

where $h_s = w^{(k)H}a(\bar{\theta}_s) + \frac{\kappa_s^{(k)}}{\rho}$. Clearly, (5.33) is separable for each s and for a fixed η, it admits a closed-form solution $r_s^\star$ by minimizing $f_s = \frac{\rho_2}{2}|r_s - h_s|^2$:

$$r_s^\star(\eta) = \begin{cases} h_s, & |h_s| \le \eta \\ \eta e^{j\angle h_s}, & \text{otherwise} \end{cases} \tag{5.34}$$

and inserting $r_s^\star(\eta)$ into f_s yields

$$\hat{f}_s(\eta) = \begin{cases} 0, & |h_s| \le \eta \\ \frac{\rho_2}{2}\eta^2 - \rho_2|h_s|\eta + \frac{\rho_2}{2}|h_s|^2, & \text{otherwise} \end{cases} \tag{5.35}$$

Table 5.2 Solving (5.36) *using bisection search*

Initialization: Set $\eta_1 = 0$ and $\eta_2 = \eta_0$ is a sufficient large number containing the optimal $\eta^\star$.
repeat: 1: update:$\eta = \frac{\eta_1+\eta_2}{2}$; 2: if $\nabla f(\eta) < 0$, then $\eta_1 = \eta$; else $\eta_2 = \eta$; **until:** $\|\eta_1 - \eta_2\| \leq 10^{-6}$; **return**: $\eta^\star = \frac{\eta_1+\eta_2}{2}$

We find that $\hat{f}_s(\eta)$ is a convex non-increasing differentiable function in terms of η, and by substituting it into the objective function in (5.33), we obtain the following problem of finding the optimal $\eta^\star$:

$$\min_{\eta \geq 0} f(\eta) = \eta^2 + \sum_{s=1}^{S} \hat{f}_s(\eta), \tag{5.36}$$

and since its derivative $\nabla f(\eta) = \sum_{s=1}^{S} \min\{\rho_2 \eta - \rho_2 |h_s|, 0\} + 2\eta$ increases with η, the optimal $\eta^\star$ can be determined by solving the equation $\nabla f(\eta) = 0$ by a simple bisection method, as shown in Table 5.2. Hence, once $\eta^\star$ is found, $r_s^{(k+1)}$ is updated by inserting $\eta^\star$ into (5.34).

In particular, when $\bar{\theta}_{s'}$ belongs to a null region Θ_n which is below a specified notch level $\hat{\eta}$, then $r_{s'}^{k+1}$ is updated by

$$r_{s'}^{k+1} = \begin{cases} h_{s'}, & |h_{s'}| \leq \hat{\eta} \\ \hat{\eta} e^{j\angle h_{s'}}, & \text{otherwise} \end{cases} \tag{5.37}$$

where $h_{s'} = w^{(k)H} a(\bar{\theta}_{s'}) + \frac{\kappa_{s'}^{(k)}}{\rho}$.

5.4.2 Step 2: update $v_m^{(k+1)}$

The subproblem in Step 2 with irrelevant items ignored is equivalent to

$$\begin{aligned} &\min_{v_m} \sum_{m=1}^{M} |v_m - z_m|^2 \\ &\text{s.t. } d_m - \varepsilon \leq |v_m|^2 \leq d_m + \varepsilon, m = 1, \cdots, M, \end{aligned} \tag{5.38}$$

with $z_m = w^{(k)H}a(\theta_m) + \frac{\lambda_m^{(k)}}{\rho_1}$. Clearly, (5.38) is also separable for each m, and it has a closed-form solution $v_m^{(k+1)}$:

$$v_m^{(k+1)} = \begin{cases} \sqrt{d_m+\varepsilon}e^{j\angle z_m}, & \text{if } |z_m| \geq \sqrt{d_m+\varepsilon} \\ \sqrt{d_m-\varepsilon}e^{j\angle z_m}, & \text{if } |z_m| \leq \sqrt{d_m-\varepsilon} \\ z_m, & \text{otherwise} \end{cases} \tag{5.39}$$

for $m = 1, \cdots, M$.

5.4.3 Step 3: update $\mathbf{w}^{(k+1)}$

The subproblem in Step 3 with irrelevant items ignored is equivalent to

$$\begin{aligned} &\min_{w} \ \|Aw - b\|^2 \\ &\text{s.t. } \mathrm{DRR}(w) \leq T_r, \end{aligned} \tag{5.40}$$

where

$$A = [\sqrt{\rho_1}a(\theta_1), \cdots, \sqrt{\rho_1}a(\theta_M), \sqrt{\rho_2}a(\bar{\theta}_1), \cdots, \sqrt{\rho_2}a(\bar{\theta}_S)]^H \tag{5.41}$$

$$b = [\sqrt{\rho_1}\delta_1, \cdots, \sqrt{\rho_1}\delta_M, \sqrt{\rho_2}\xi_1, \cdots, \sqrt{\rho_2}\xi_S]^H, \tag{5.42}$$

with $\delta_m = v_m^{(k+1)} - \frac{\lambda_m^{(k)}}{\rho}$ and $\xi_s = r_s^{(k+1)} - \frac{\kappa_s^{(k)}}{\rho}$.

To facilitate restraining w to satisfy the DRRC in (5.40), we linearize the cost function $\|Aw - b\|_2^2$ in the kth iteration $w^{(k)}$ with $\beta \geq \hat{\lambda}_{\max}(A^H A)$ where $\hat{\lambda}_{\max}(A^H A)$ denotes the maximum eigenvalue of the matrix $A^H A$, and rewrite (5.40) as follows:

$$\begin{aligned} &\min_{w} \ \Re\{[A^H(Aw^{(k)} - b)]^T(w - w^{(k)})\} + \frac{\beta}{2}\|w - w^{(k)}\|_2^2 \\ &\text{s.t. } \frac{\max_n\{|w_n|\}}{\min_n\{|w_n|\}} \leq T_r. \end{aligned} \tag{5.43}$$

After manipulations, (5.43) is equivalent to

$$\begin{aligned} &\min_{w} \ \|w - \tilde{w}\|_2^2 \\ &\text{s.t. } \frac{\max_n\{|w_n|\}}{\min_n\{|w_n|\}} \leq T_r, \end{aligned} \tag{5.44}$$

with $\tilde{w} = w^{(k)} - \frac{1}{\beta}A^H(Aw^{(k)} - b)$. It can be seen that (5.44) strives to seek the closest projection onto the DRRC set. To readily solve (5.43), we introduce an auxiliary variable μ to rewrite (5.44) as

$$\begin{aligned} &\min_{w,\mu} \ \sum_{n=1}^{N} |w_n - \tilde{w}_n|^2 \\ &\text{s.t. } \mu \leq |w_n| \leq \mu T_r, n = 1, \cdots, N, \end{aligned} \tag{5.45}$$

where $\tilde{w}_n$ is the n-th element of the vector $\tilde{w}$. Since (5.45) is similar to the peak-valley-power ratio (PVR) projection operation in [35], here we apply the interval splitting idea to solve (5.45) and the details can be seen in Appendix B of [35].

5.4.4 Convergence and computational complexity

Regarding the convergence of the proposed method, the primal-dual iterative nature of the proposed algorithm, just like ADMM, is guaranteed to have superior convergence properties as explained in [34]. Furthermore, we provide some mild assumptions to further prove that the proposed algorithm can find a locally optimal solution to the problem $\mathscr{P}$, as explained in Theorem 5.1.

Theorem 1. *Assume that the scaled dual variables* $\{\lambda_m^{(k)}\}_{m=1}^M$ *and* $\{\kappa_s^{(k)}\}_{s=1}^S$ *are bounded and have* $\lim_{k\to\infty} \lambda_m^{(k+1)} - \lambda_m^{(k)} = 0$ *and* $\lim_{k\to\infty} \kappa_s^{(k+1)} - \kappa_s^{(k)} = 0$ *as the iteration number increases, then the iteration sequence* $\{w^{(k)}, \eta^{(k)}, r_s^{(k)}, v_m^{(k)}, \lambda_m^{(k)}, \kappa_s^{(k)}\}$ *will converge to a limit point set* $\{w^{(\star)}, \eta^{(\star)}, r_s^{(\star)}, v_m^{(\star)}, \lambda_m^{(\star)}, \kappa_s^{(\star)}\}$ *which satisfies the KKT conditions of the problem* $\mathscr{P}$.

The proof is reported in Appendix A.1.

According to Theorem 1, we examine the maximal consensus errors between the auxiliary variables and beampattern responses, i.e., $\max_m |w^H a(\theta_m) - v_m|$ and $\max_s |w^H a(\bar{\theta}_s) - r_s|$ if they are all less than a tolerance value, e.g., 10^{-7}. If so, it implies that the assumptions of $\lim_{k\to\infty} \lambda_m^{(k+1)} - \lambda_m^{(k)} = 0$ and $\lim_{k\to\infty} \kappa_s^{(k+1)} - \kappa_s^{(k)} = 0$ hold and the proposed method has found a locally optimal solution.

As to the computational complexity of the proposed method, we consider the major computation steps, i.e., the multiplications in one iteration. In Table 5.1, Steps 1, 2, and 3 require $\mathscr{O}\{2S(N+1) + S\log S\}$, $\mathscr{O}\{2M(N+1)\}$, and $\mathscr{O}\{3N^2(M+S)\}$, respectively, and Step 4 needs $\mathscr{O}\{NM + NS\}$. Thus, the proposed algorithm has an approximate complexity of $\mathscr{O}\{3(N^2+N)(M+S) + S\log S\}$.

5.5 Numerical examples

In this section, a series of computer simulations are provided to evaluate the proposed method and there are different beampattern shapes synthesized. In all experiments, we initialize $w^{(0)}, v_m^{(0)}, \lambda_m^{(0)}, \kappa_s^{(0)}$ as zero, and the maximum iteration number I is set as 120000, $\eta_0 = 100$ in Table 5.2.

5.5.1 Experiment 1: focused beampatterns

First, we aim to synthesize a focused beam pointing toward the direction of $\theta_0 = 0°$ surrounded in the region $[-6°, 6°]$ with $\varepsilon = 0$ and $d_i = 1, i = 0$ in $\mathbb{FB}$, and the angle sampling interval is 1° via using a uniform linear array (ULA) of $N = 21$ antennas with half-wavelength uniformly spaced. The reconfigurable phase-only method (RPM) in [9] is compared to the proposed method with parameters of DRR = 1, $\rho_1 = 0.3$ and $\rho_2 = 0.6$, and the results are shown in Figure 5.2. We observe that our algorithm can achieve a lower PSL of −16.5 dB, compared to −13.02 dB in [9]. Clearly, a larger DRR threshold T_r creates more degrees of optimization freedom to synthesize a beampattern with a lower PSL. When T_r increases to 2, a lower PSL

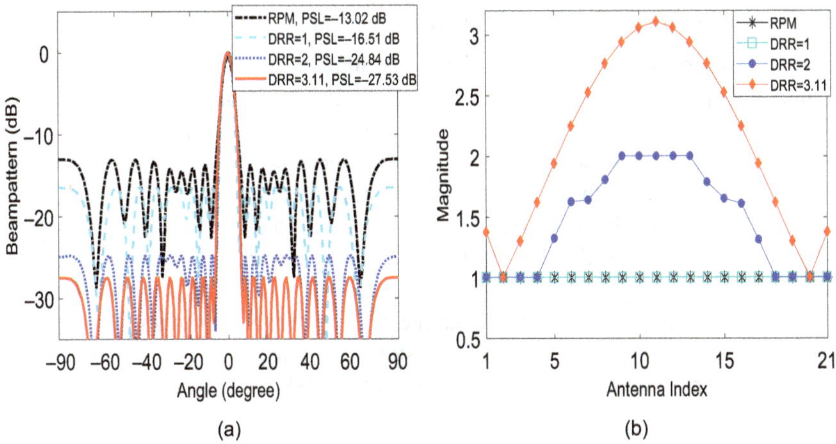

Figure 5.2 Synthesized focused beampatterns and excitation magnitudes in Experiment 1. (a) Synthesized beampatterns and (b) normalized magnitudes.

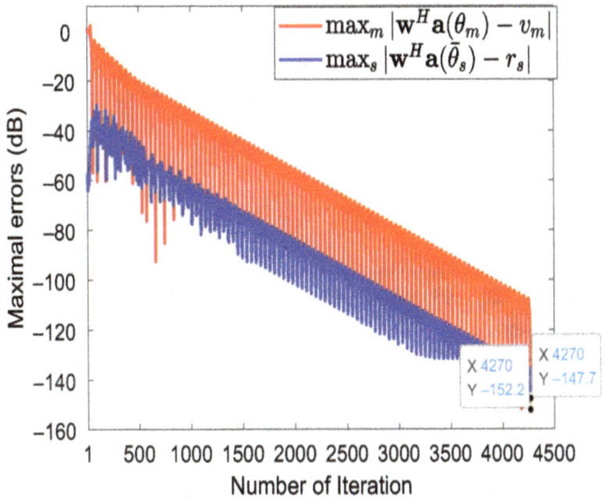

Figure 5.3 Maximum consensus errors versus the iteration number

of -24.84 dB is achieved. When T_r keeps increasing to 3.11, the PSL has stabilized at -27.53 dB. This implies the lowest SLL has been achieved regardless of the excitation range restriction of DRR.

For the test of convergence of our method, we plot the maximal consensus errors between the auxiliary variables and beampattern responses, i.e., $\max_m |w^H a(\theta_m) - v_m|$ and $\max_s |w^H a(\bar{\theta}_s) - r_s|$ versus the iteration number in Figure 5.3. It is seen that the proposed method converges at the 4270-th iteration, much smaller than the maximum iteration number, and the maximum errors of two LECs are -147.7 dB and -152.2 dB, respectively, which indicates that the scaled dual variables $\lambda_m^{(k)}, \kappa_s^{(k)}$ are unchanged as the iteration goes on and have found the optimal solutions at the convergence.

5.5.2 Experiment 2: flat-top beampatterns

Second, we consider synthesizing a flat-top beampattern using a ULA consisting of 32 antennas spaced at half wavelength and the mainlobe region is set as $[-14°, 14°]$ with the maximum ripple being ± 0.3 dB, whereas the sidelobe region is set as $[-90°, -20°] \cup [20°, 90°]$. First, we gradually increase T_r from 2 to 10 and the synthesized patterns by the proposed method with $\rho_1 = 1$ and $\rho_2 = 1.5$ are plotted in Figure 5.4(a) and (b). It is seen that all the generated beampatterns satisfy the design requirements and the PSL starts from -24.61 dB when DRR=2 and drops to -28.06 dB when DRR $= 5$, and finally comes to -37.49 dB with DRR $= 10$, which accounts for the more DoFs gained from the relaxed excitation ranges.

Furthermore, for the special case, we study the phase-only case with DRR $= 1$ and the constant modulus-shaped beam method (CMSB) in [21] and the phase-only

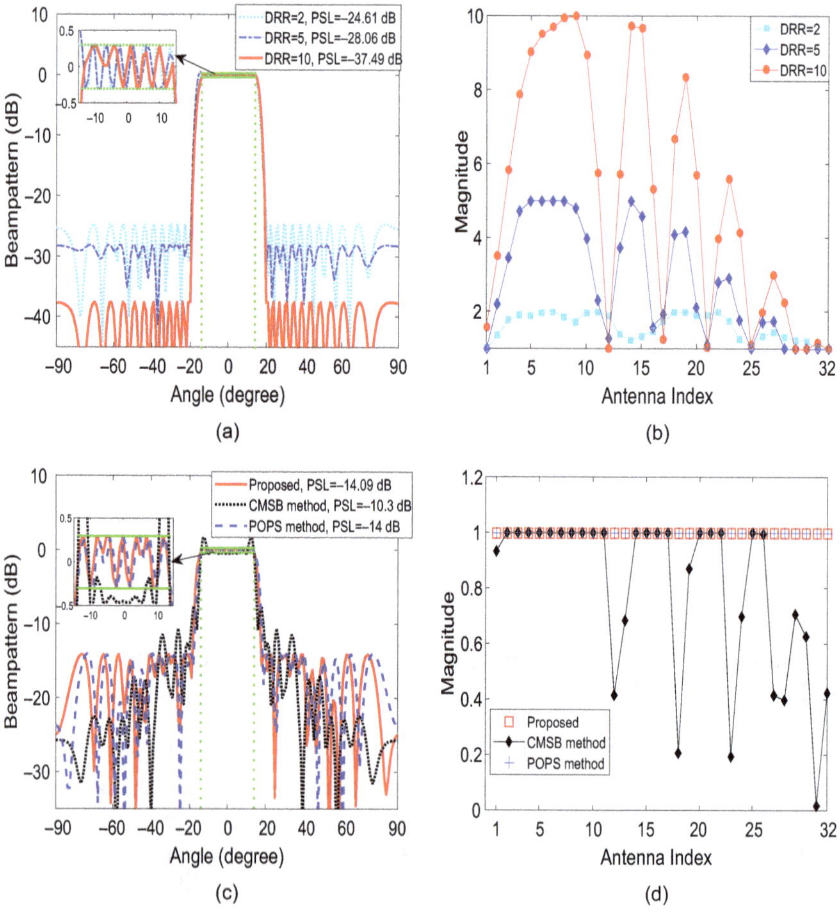

Figure 5.4 Synthesized flat-top beampatterns and excitation magnitudes in Experiment 2. (a) Synthesized beampatterns, (b) normalized magnitudes, (c) uniformly excited beampatterns and (d) normalized magnitudes.

pattern synthesis (POPS) method in [22] are used for comparison. The obtained beampatterns and the normalized magnitudes are depicted in Figure 5.4(c) and (d), respectively. Notably, the CMSB method not only violates the mainlobe ripple requirement but also fails to meet the CMC for each excitation owing to the relaxation behavior in the optimization. However, both our method and POPS method work well on the beampatterns with uniform magnitudes while the lowest PSL is achieved by ours, i.e., -14.09 dB slightly lower than that of the POPS method, i.e., -14 dB.

5.5.3 Experiment 3: cosecant square beampatterns

Next, we set with $\rho_1 = \rho_2 = 0.25$ in the proposed method to synthesize a cosecant square beampattern with the same array configuration requirements as [23], where a ULA consists of 32 antennas with inter-element spacing being one-third wavelength. The mainlobe region is specified in $0.1 \leq \sin\theta \leq 0.5$ with the maximum ripple less

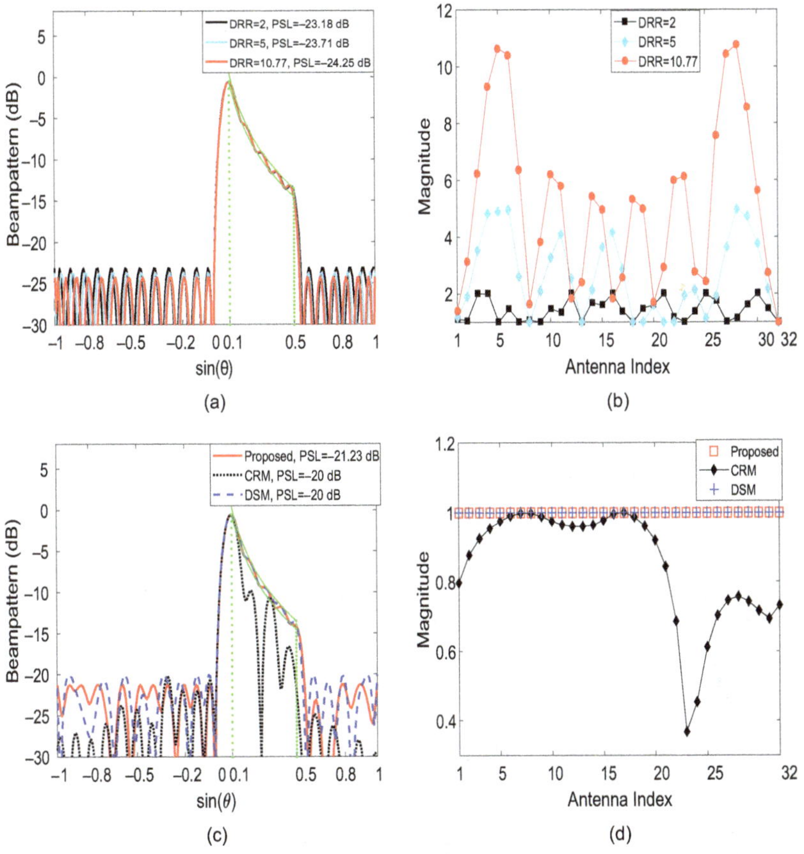

Figure 5.5 Synthesized cosecant square beampatterns and excitation magnitudes in Experiment 3. (a) Synthesized beampatterns, (b) normalized magnitudes, (c) uniformly excited beampatterns and (d) normalized magnitudes.

than 0.5 dB, and the sidelobe region is set as $-1 \leq \sin\theta \leq 0$ and $0.55 \leq \sin\theta \leq 1$. The results of the designed beampatterns and the excitation magnitudes under different DRRs are shown in Figure 5.5(a) and (b), respectively, where all the mainlobes exhibit controlled ripples and the changes of PSLs are small with the increase of DRR. For example, when DRR = 5, the PSL equals −23.71 dB slightly lower than −23.18 dB in the case of DRR=2, and the lowest PSL is attained at −24.25 dB with DRR=10.77. Evidently, a larger DRR on the excitation range leads to a lower PSL.

Besides, for the phase-only case, Figure 5.5(c) shows the synthesized uniformly excited beampatterns where the Convex Relaxation Method (CRM) in [5] and the Distributed beam Synthesis Method (DSM) in [23] are included for comparison. The optimized normalized magnitudes are plotted in Figure 5.5(d) and it is observed

Table 5.3 The antenna positions of the 10-element non-uniform linear array

Antenna Index	Position: unit of wavelength λ
1	0
2	0.52
3	1.23
4	1.84
5	2.37
6	2.66
7	3.39
8	3.75
9	4.39
10	5.02

Source: Part of Table II in [36].

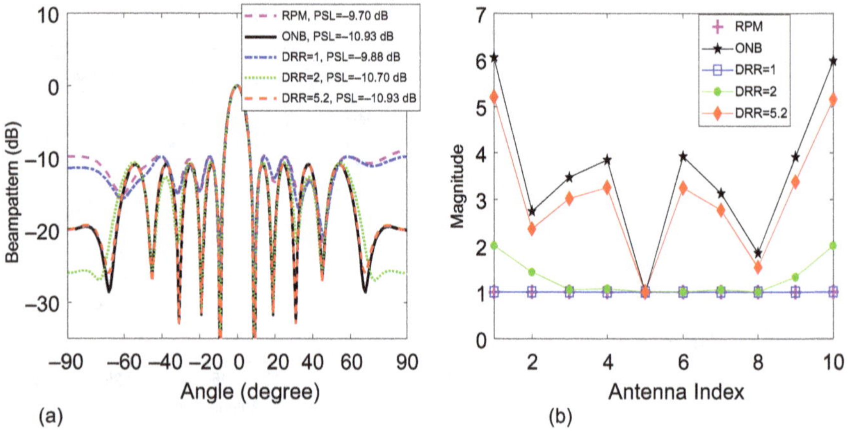

Figure 5.6 Synthesized focused beampatterns and excitation magnitudes in Experiment 4. (a) Synthesized beampatterns and (b) normalized magnitudes.

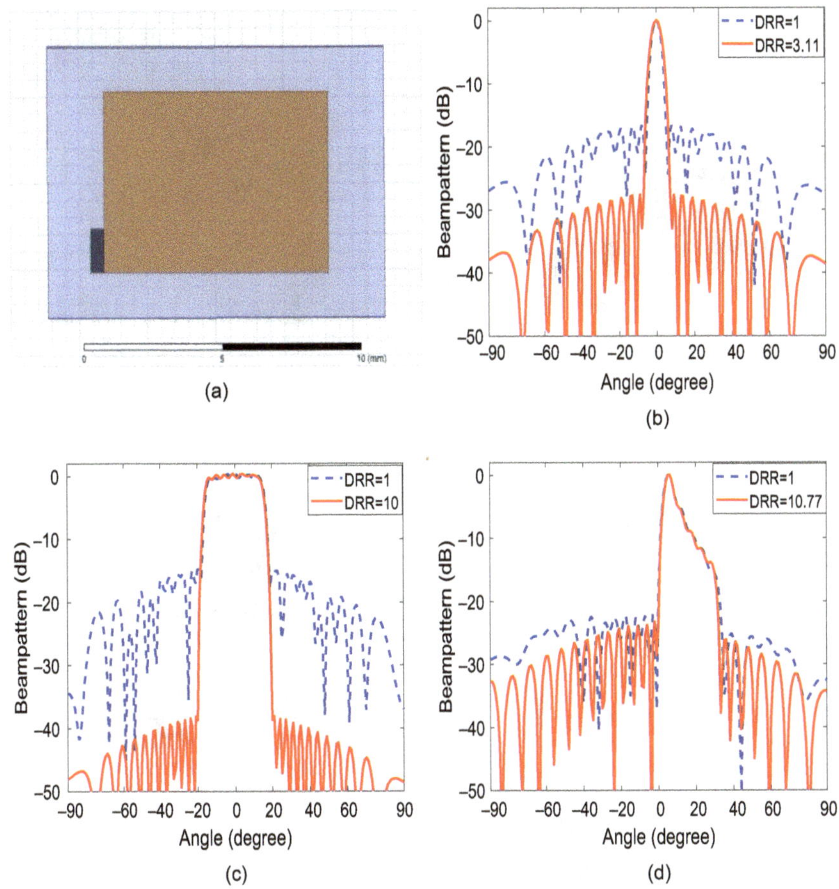

Figure 5.7 Synthesized beampatterns by using Ansys HFSS simulation software in Experiment 5. (a) Microstrip patch of unit element, (b) focused beam synthesis, (c) flat–top beam synthesis and (d) cosecant square beam synthesis.

that the proposed method and DSM algorithm completely satisfy the CMC and the mainlobe ripple constraints, yet CRM algorithm fails due to the rank-one constraint relaxed in SDR optimization. Besides, the proposed method achieves a lower PSL of −21.23 dB than that of DSM, i.e., −20 dB, which illustrates the superiority of the proposed method over other methods.

5.5.4 Experiment 4: extension to linear arbitrary arrays

In this example, we verify the applicability of the proposed method to linear arbitrary arrays. Specifically, we use an aperiodic and non-uniform array consisting of 10 elements to synthesize a focused beam whose beampattern shape requirements are the same as that in Experiment 1. The detailed array element positions are listed in Table 5.3. The RPM method in [9] and the Optimal Narrow Beam synthesis (ONB)

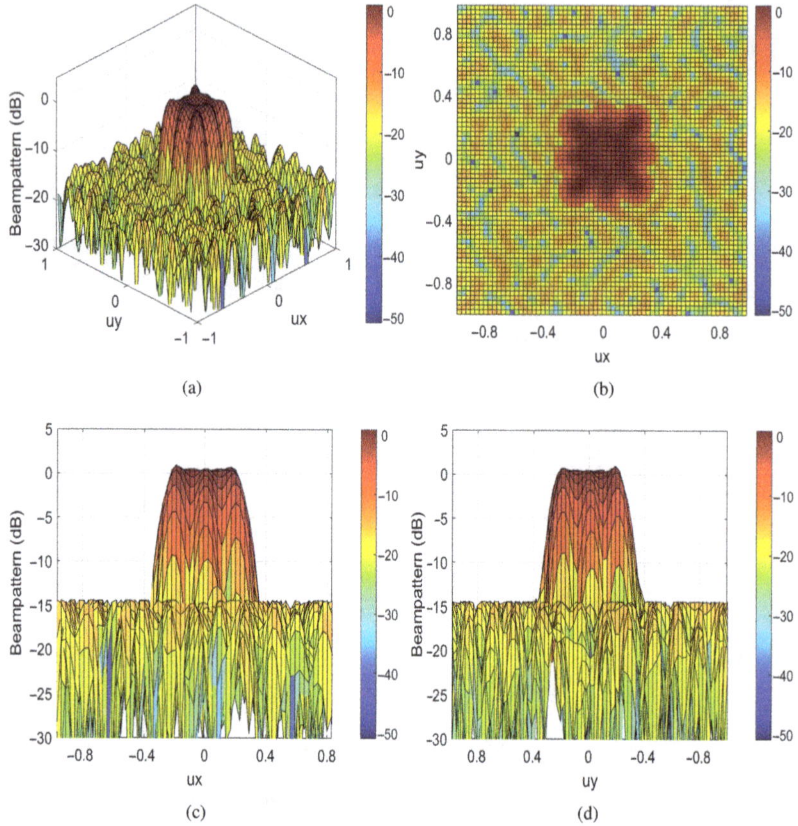

Figure 5.8 2-D Planar array phase-only beampattern synthesis with DRR = 1 in Experiment 6. (a) 3-D view of synthesized beampattern, (b) top view of synthesized beampattern, (c) view of beampattern in ux direction and (d) view of beampattern in uy direction.

method in [36] are compared with the proposed method in Figure 5.6. The results of the synthesized beampatterns show that the proposed method achieves a lower PSL of −9.88 dB than −9.70 dB of the RPM method in the phase-only case. Meanwhile, the minimum PSL is achieved at −10.93 dB same as that in the ONB method with the DRR constraint relaxed. Nevertheless, the excitation range obtained by the proposed method shows a lower DRR of 5.2 than 6.05 from the ONB method. This experiment proves that the proposed method has a wider adaptability in array configurations, especially for a non-uniform spaced array.

5.5.5 *Experiment 5: performance evaluation of the proposed method via using Ansys HFSS software*

In this example, we examine the performance of the proposed method in the actual array antenna unit by using the full-wave 3D electromagnetic Ansys high-frequency

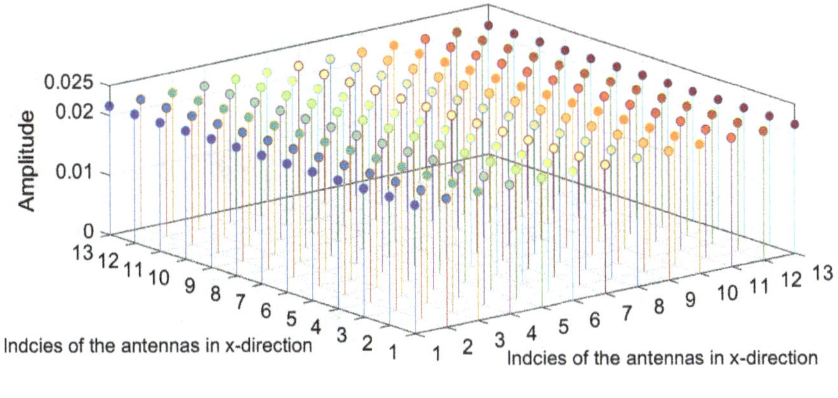

(a)

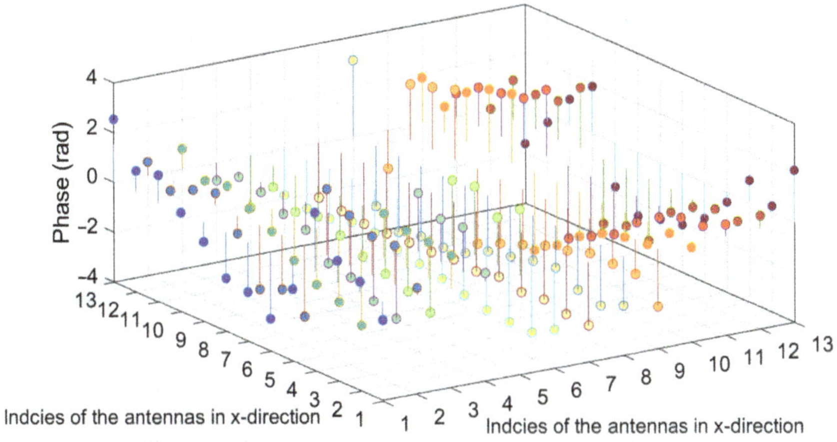

(b)

Figure 5.9 Optimized amplitudes and phases in the phase-only control. (a) Uniformly excited amplitudes and (b) synthesized phases.

simulation software (HFSS). Each array element is considered to be a frequently used microstrip patch antenna with a length of 5.25 mm and width of 8.2 mm, as shown in Figure 5.7(a), and it works at 16 GHz and is fed from the edge of the 50 ohm microstrip line with a width of 1.26 mm. The thickness of the dielectric layer is 0.5 mm.

Figure 5.7(b)–(d) shows the synthesized focused beam and flat-top beam as well as cosecant square beam under the requirements of Experiments 1–3, respectively. It is seen that the excitation weight vector optimized by the proposed method can still synthesize the desired beampattern shape in the actual microstrip patch antenna array. This verifies the practicability and effectiveness of the proposed method which enables obtaining a lower SLL in the practical beampattern.

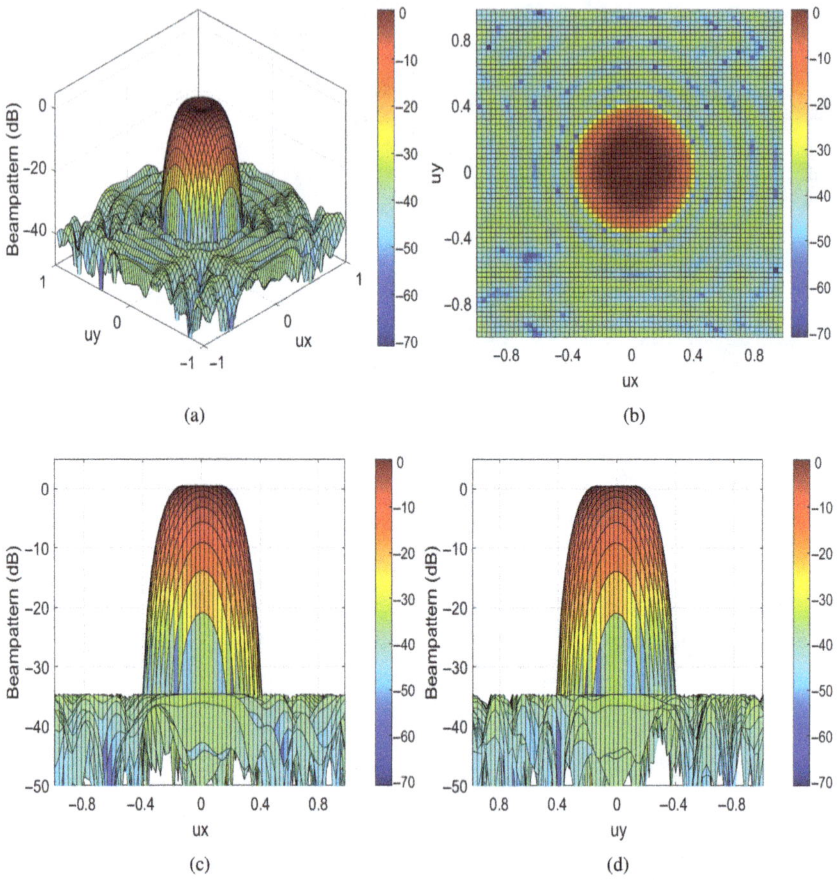

Figure 5.10 2-D planar array minimum sidelobe beampattern synthesis with DRR = 100. (a) 3-D view of synthesized beampattern, (b) top view of synthesized beampattern, (c) view of beampattern in ux direction and (d) view of beampattern in uy direction.

5.5.6 *Experiment 6: 2-D planar array beampattern synthesis*

In this experiment, we extend the proposed method to synthesize 2-D beampatterns in planar arrays. First of all, for rectangular planar arrays, we consider a 13×13 uniform rectangular array with half-wavelength spaced both in *x*- and *y*-directions, The desired beampattern is assumed to have a circular-shaped mainlobe in the region of $ux^2 + uy^2 \leq 0.2^2$ with controlled maximum ripples less than ± 0.5 dB, where $ux = \sin\phi\cos\theta$ and $uy = \sin\phi\sin\theta$ with θ and ϕ being the azimuth and polar angles, respectively, and the sidelobe region is set as $ux^2 + uy^2 > 0.4^2$. We apply the proposed method to control the excitation dynamic range for synthesizing such a desired beampattern in $ux - uy$ axes.

Figure 5.8 shows the synthesized beampattern under DRR = 1, which implies that the excitation vector has a uniform amplitude and is only controlled by different phases as plotted in Figure 5.9. It is seen that the resulting beampattern has a desired mainlobe shape with controlled ripples and the minimal sidelobe level is arrived at −14.4 dB. Next, we increase the DRR of the excitation vector to 100 for exploring a lower sidelobe. Figure 5.10 shows the designed beampattern under DRR = 100. Clearly, a more regular shape like circle is formed in the mainlobe and the peak sidelobe level is achieved at −34.62 dB. From the comparison in Figures 5.8 and 5.10, it is evident that a larger DRR on the excitation vector results in a more DoF utilized to further suppress the sidelobe level.

Besides, we apply the proposed method to a special planar array synthesis, i.e., concentric ring arrays, where array elements lie on different circles and share a common center. Figure 5.11(a) shows the considered concentric ring array with 6 rings and 130 elements in total. The elements in each ring are equally spaced by

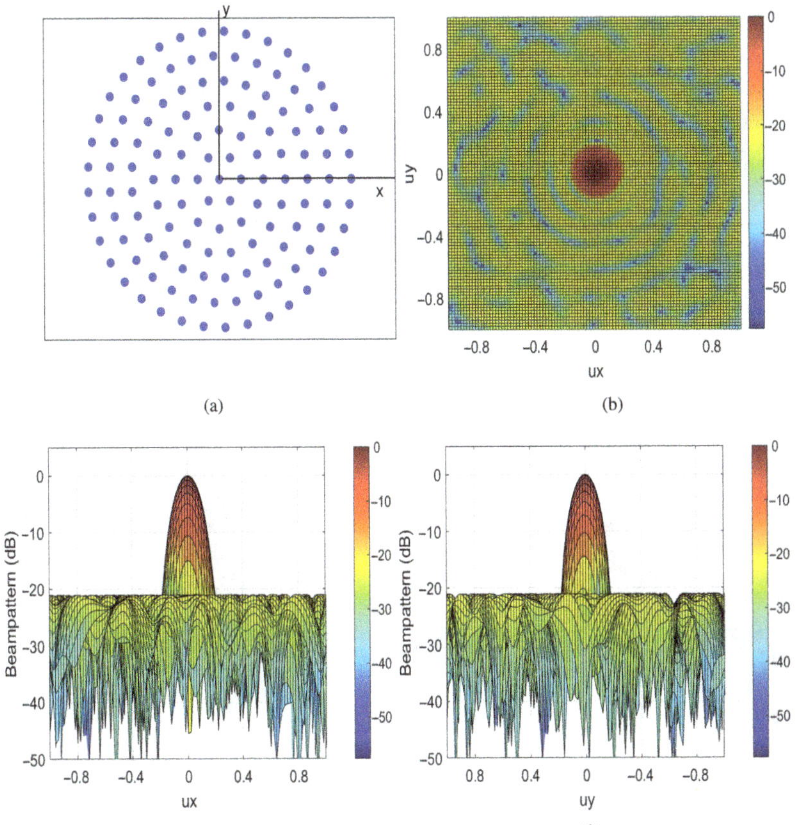

Figure 5.11 2-D circular ring array phase-only beampattern synthesis with DRR = 1. (a) Uniform circular ring array with six rings, (b) top view of synthesized beampattern, (c) view of beampattern in ux direction and (d) view of beampattern in uy direction.

half-wavelength, i.e., $d_l = \frac{\lambda}{2}$, and the l-th ring has the radius of $r_l = \frac{l\lambda}{2}$. The number of elements in the l-th ring is given by

$$N_l = \left\lfloor \frac{2\pi r_l}{d_l} \right\rfloor = \lfloor 2\pi l \rfloor, \tag{5.46}$$

where $\lfloor \cdot \rfloor$ denotes the largest integer less than or equal to the argument.

Figure 5.11(b)–(d) shows the synthesized focused beampattern with phase-only control when DRR=1 and it is observed that the main beam is targeted at the spatial position of $ux = 0, uy = 0$ and the peak sidelobe level attains -21.04 dB. Then, by relaxing the DRRC with $T_r = 4$, we obtain the lowest sidelobe level equal to -34.78 dB in Figure 5.12. Evidently, the case of more relaxed DRR on the range of excitation amplitudes leads to a lower sidelobe but also generates a wider

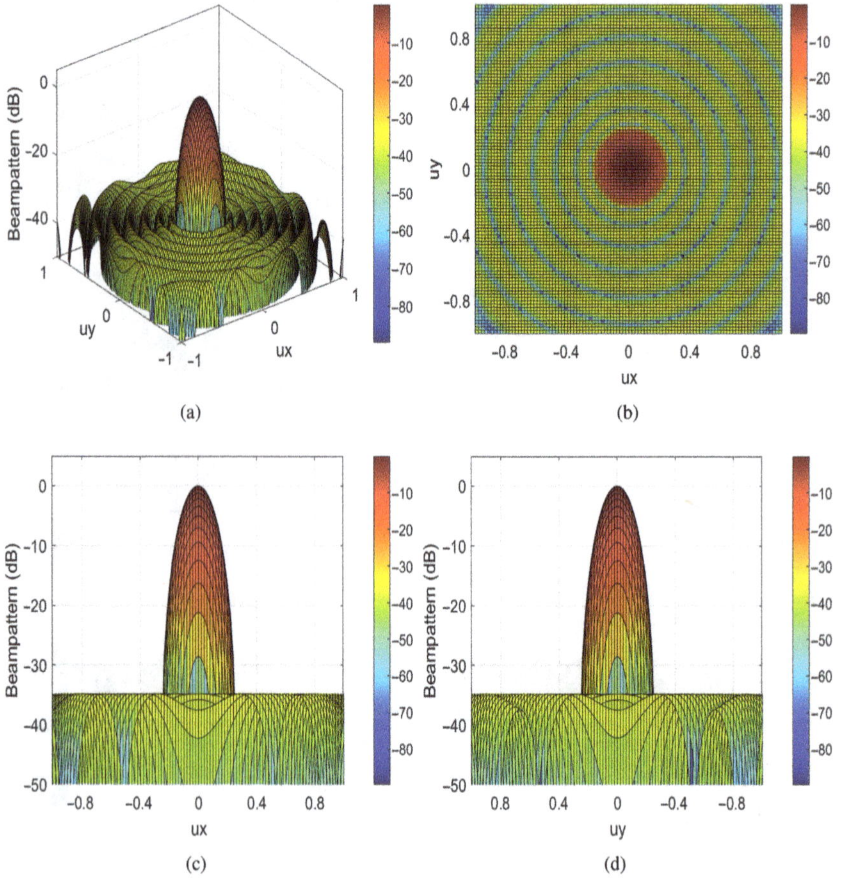

Figure 5.12 2-D Circular ring array minimum sidelobe beampattern synthesis with DRR = 4. (a) 3-D view of synthesized beampattern, (b) top view of synthesized beampattern, (c) view of beampattern in ux direction and (d) view of beampattern in uy direction.

main beam by comparing Figure 5.12(b) with Figure 5.11(b). Therefore, in practical applications, it usually needs to make a satisfying compromise for engineers among the sidelobe level and the width of mainlobe as well as the DRR of the excitation vector.

5.5.7 Experiment 7: array beampattern synthesis with null formation

In the last example, we utilize the proposed method to synthesize array beampatterns with null formation. The results of designed beampatterns are shown in Figure 5.13. Figure 5.13(a) and (b) plots the obtained focused beam and shaped beam by separately using the consistent array configuration in Experiments 1 and 2, where the

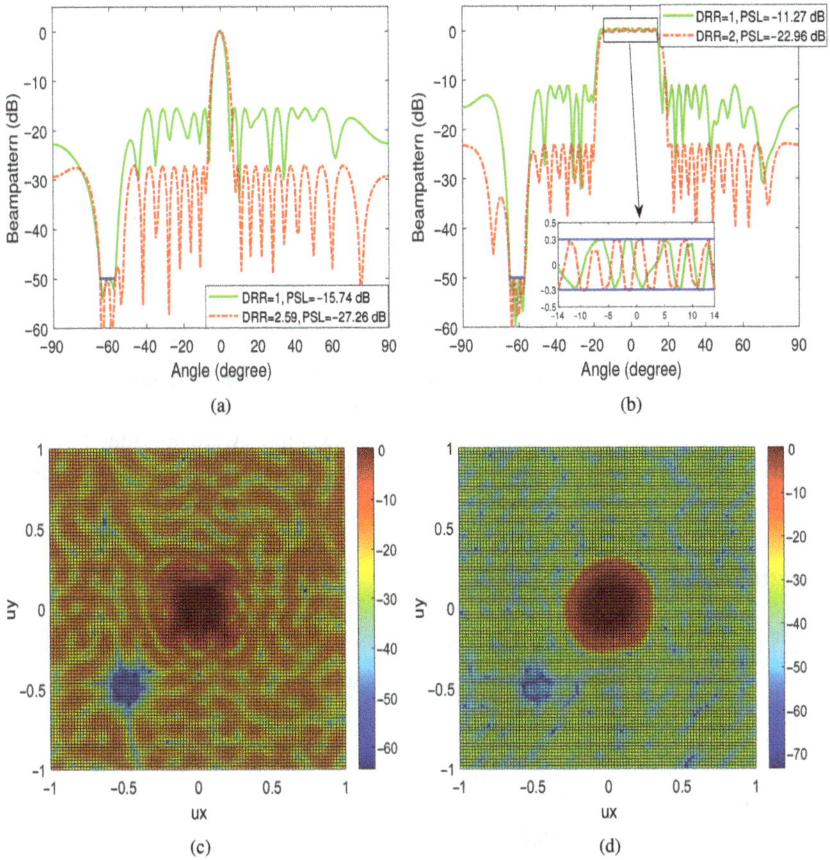

Figure 5.13 Null pattern synthesis with minimum peak sidelobe. (a) Focused beam synthesis with null, (b) shaped beam synthesis with null, (c) top view of phase-only beampattern with null and (d) top view of synthesized null pattern with DRR = 10.

null region is set as $\Theta_n = [-65°, -57°]$ with a preset null depth equal to -50 dB. From that, we can see all the optimized beampatterns meet the shape requirements with an accurate control on the notch depth and the mainlobe ripple range. Expectedly, the design of a larger DRR leads to a lower PSL. Next, we employ the proposed method to synthesize 2-D beampatterns in a planar array, where a 14×14 uniform rectangular array with inter-element spacing being half-wavelength is considered. The mainlobe region is set as $ux^2 + uy^2 \leq 0.13^2$ with the largest ripples less than ± 0.5 dB, and the sidelobe region is set as $ux^2 + uy^2 \geq 0.33^2$. The null region is set as $(ux + 0.5)^2 + (uy + 0.5)^2 \leq 0.1^2$ and the null depth is assigned as -50 dB. Figure 5.13(c) and (d) shows the top views of the obtained beampatterns with notches corresponding to phase-only (DRR=1) and DRR=10, respectively. Notably, a desired notch of -50 dB is formed in the prescribed null region in their respective beampattern. Besides, a larger DRR=10 results in a more regular circular shape in the mainlobe region and a lower SLL, i.e., -31.66 dB < -15.34 dB with DRR $= 1$, which thanks to increased DoFs.

5.6 Conclusions

In this chapter, we presented a primal-dual iteration method of synthesizing beampatterns with shape constraints and DRR control on the excitation range. The proposed method enables to synthesize a variety of shaped-desired patterns such as focused beampattern and flat-top beampattern as well as cosecant square beampattern. In particular, our method is able to obtain the lowest SLL under uniform excitation amplitudes when DRR $= 1$, which is in favor of the maximization of the antenna power efficiency. Also, a desired null pattern can be synthesized by our method with a preset notch depth. In addition, the proposed method is not only applied in linear uniform arrays but also can be extended to arbitrary arrays with non-uniform spacing and arbitrary planar arrays. A series of numerical examples were provided to demonstrate the efficiency of the proposed method and the superiority over several existing schemes.

Appendix A.1

Since $\lim_{k\to\infty} \lambda_m^{(k+1)} - \lambda_m^{(k)} = 0$ and $\lim_{k\to\infty} \kappa_s^{(k+1)} - \kappa_s^{(k)} = 0$, it then follows from Step 4 that,

$$\lim_{k\to\infty} w^{(k+1)H} a(\theta_m) - v_m^{(k+1)} = 0. \tag{A.1}$$

$$\lim_{k\to\infty} w^{(k+1)H} a(\bar{\theta}_s) - r_s^{(k+1)} = 0. \tag{A.2}$$

Moreover, from the bounded sets imposed on v_m and r_s, we have that the sequences $\{v_m^{(k)}\}_{m=1}^M$ and $\{r_s^{(k)}\}_{s=1}^S$ updated by Step 1 and Step 2 are also bounded, respectively. In addition, from the following inequalities:

$$|w^{(k+1)H} a(\theta_m) - v_m^{(k+1)}| - |w^{(k+1)H} a(\theta_m)|^2 \leq |v_m^{(k+1)}|^2 \tag{A.3}$$

$$|w^{(k+1)H} a(\bar{\theta}_s) - r_s^{(k+1)}| - |w^{(k+1)H} a(\bar{\theta}_s)|^2 \leq |r_s^{(k+1)}|^2 \tag{A.4}$$

and the assumptions (A.1) and (A.2), we have that the sequence $\{w^{(k)}\}$ are also bounded. Therefore, there exists a subsequence of $\{w^{(k)}, \eta^{(k)}, r_s^{(k)}, v_m^{(k)}, \lambda_m^{(k)}, \kappa_s^{(k)}\}$ which converges to $\{w^{(\star)}, \eta^{(\star)}, r_s^{(\star)}, v_m^{(\star)}, \lambda_m^{(\star)}, \kappa_s^{(\star)}\}$ and has the following equalities:

$$0 = \lim_{k\to\infty} w^{(k+1)H} a(\theta_m) - v_m^{(k+1)} = w^{(\star)H} a(\theta_m) - v_m^{(\star)}. \quad \text{(A.5)}$$

$$0 = \lim_{k\to\infty} w^{(k+1)H} a(\bar{\theta}_s) - r_s^{(k+1)} = w^{(\star)H} a(\bar{\theta}_s) - r_s^{(\star)}. \quad \text{(A.6)}$$

Furthermore, since $\{r_s^{(\star)}, \eta^{(\star)}\}$ and $v_m^{(\star)}, w^{(\star)}$ are determined by their respective subproblems, all of them meet own optimality conditions and the Lagrangian $L(w, \eta, r_s, v_m, \lambda_m, \kappa_s)$ degrades into the Lagrangian of the original problem $\mathscr{P}$ without quadratic augmented terms at the convergence, which implies that the sequence $\{w^{(\star)}, \eta^{(\star)}, r_s^{(\star)}, v_m^{(\star)}, \lambda_m^{(\star)}, \kappa_s^{(\star)}\}$ also satisfies the KKT conditions of $\mathscr{P}$. The proof is completed.

References

[1] Vescovo R. Consistency of constraints on nulls and on dynamic range ratio in pattern synthesis for antenna arrays. IEEE Transactions on Antennas and Propagation. 2007;55(10):2662–2670.

[2] Li D, Tan X, and Feng Z. Design of far field beamforming problem by window function. In: *2018 IEEE 3rd International Conference on Cloud Computing and Internet of Things (CCIOT)*; 2018. p. 534–538.

[3] Dolph CL. A current distribution for broadside arrays which optimizes the relationship between beam width and side-lobe level. *Proceedings of the IRE*. 1946;34(6):335–348.

[4] Taylor TT. Design of line-source antennas for narrow beamwidth and low side lobes. *Transactions of the IRE Professional Group on Antennas and Propagation*. 1955;3(1):16–28.

[5] Fuchs B. Application of convex relaxation to array synthesis problems. *IEEE Transactions on Antennas and Propagation*. 2014;62(2):634–640.

[6] Toso G, Mangenot C, and Roederer A. Sparse and thinned arrays for multiple beam satellite applications. In: *Proc. 2nd Eur. Conf. Antennas Propag.*; 2007. p. 1–4.

[7] Hur S, Kim T, Love DJ, *et al.* Millimeter wave beamforming for wireless backhaul and access in small cell networks. *IEEE Transactions on Communications*. 2013;61(10):4391–4403.

[8] Wang X, Aboutanios E, and Amin MG. Thinned array beampattern synthesis by iterative soft-thresholding-based optimization algorithms. *IEEE Transactions on Antennas and Propagation*. 2014;62(12):6102–6113.

[9] Vescovo R. Reconfigurability and beam scanning with phase-only control for antenna arrays. *IEEE Transactions on Antennas and Propagation*. 2008;56(6):1555–1565.

[10] Fuchs B and Rondineau S. Array pattern synthesis with excitation control via norm minimization. *IEEE Transactions on Antennas and Propagation*. 2016;64(10):4228–4234.

[11] Fan X, Liang J, Zhang Y, *et al.* Shaped power pattern synthesis with minimization of dynamic range ratio. *IEEE Transactions on Antennas and Propagation*. 2019;67(5):3067–3078.

[12] Dolph CL. A current distribution for broadside arrays which optimizes the relationship between beam width and side-lobe level. *Proc IRE*. 1946;35(6):335–348.

[13] Cheng G and Chen H. An analytical solution for weighted least-squares beampattern synthesis using adaptive array theory. *IEEE Transactions on Antennas and Propagation*. 2021;69(9):6034–6039.

[14] Cui C, Li WT, Ye XT, *et al.* An effective artificial neural network-based method for linear array beampattern synthesis. *IEEE Transactions on Antennas and Propagation*. 2021;69(10):6431–6443.

[15] Khzmalyan AD and Kondrat'yev AS. Fast iterative methods for phase-only synthesis of antenna array pattern nulls. *Electronics Letters*. 1995;31(8): 601–602.

[16] Byun G, Choo H, and Kim S. Improvement of pattern null depth and width using a curved array with two subarrays for CRPA systems. *IEEE Transactions on Antennas and Propagation*. 2015;63(6):2824–2827.

[17] Khzmalyan AD and Kondratiev AS. The phase-only shaping and adaptive nulling of an amplitude pattern. *IEEE Transactions on Antennas and Propagation*. 2003;51(2):264–272.

[18] Toso G, Mangenot C, and Roederer AG. Sparse and thinned arrays for multiple beam satellite applications. In: *The Second European Conference on Antennas and Propagation, EuCAP 2007*; 2007. p. 1–4.

[19] Buttazzoni G and Vescovo R. Power synthesis for reconfigurable arrays by phase-only control with simultaneous dynamic range ratio and near-field reduction. *IEEE Transactions on Antennas and Propagation*. 2012;60(2): 1161–1165.

[20] Vaskelainen LI. Constrained least-squares optimization in conformal array antenna synthesis. *IEEE Transactions on Antennas and Propagation*. 2007;55(3):859–867.

[21] Cao P, Thompson JS, and Haas H. Constant modulus shaped beam synthesis via convex relaxation. *IEEE Antennas and Wireless Propagation Letters*. 2017;16:617–620.

[22] Liang J, Fan X, Fan W, *et al.* Phase-only pattern synthesis for linear antenna arrays. *IEEE Antennas and Wireless Propagation Letters*. 2017;16: 3232–3235.

[23] Zhang X, Liang J, Fan X, *et al.* Reconfigurable array beampattern synthesis via conceptual sensor network modeling and computation. *IEEE Transactions on Antennas and Propagation*. 2020;68(6):4512–4525.

[24] Subramanian R, Alphones A, and Suganthan P. Genetic algorithm based design of a reconfigurable antenna array with discrete phase shifter. *Microwave and Optical Technology Letters*. 2005;45:461–465.

[25] Gies D and Rahmat-Samii Y. Particle swarm optimization for reconfigurable phase-differentiated array design. *Microwave and Optical Technology Letters*. 2003;38:168–175.

[26] Xiong J, Wang W, Shao H, *et al.* Frequency diverse array transmit beampattern optimization with genetic algorithm. *IEEE Antennas and Wireless Propagation Letters*. 2017;16:469–472.

[27] Comisso M and Vescovo R. Fast co-polar and cross-polar 3D pattern synthesis with dynamic range ratio reduction for conformal antenna arrays. *IEEE Transactions on Antennas and Propagation*. 2013;61(2):614–626.

[28] Buttazzoni G and Vescovo R. Phase-controlled beam-scanning with near-field and DRR reduction for arbitrary antenna arrays. In: *2010 IEEE Antennas and Propagation Society International Symposium*; 2010. p. 1–4.

[29] Sinhamahapatra TK, Ahmed A, Mahanti GK, *et al.* Design of discrete phase-only dual-beam array antennas with minimum dynamic range ratio. In: *2007 IEEE Applied Electromagnetics Conference (AEMC)*; 2007. p. 1–4.

[30] Xu Z, Liu Y, Li M, *et al.* Linearly polarized shaped power pattern synthesis with dynamic range ratio control for arbitrary antenna arrays. *IEEE Access*. 2019;7:53621–53628.

[31] Vodvarka K, Bellotti MJ, and Vucic M. Synthesis of L_1 pencil beams with constrained sidelobe level and dynamic range ratio. In: *2022 16th European Conference on Antennas and Propagation (EuCAP)*; 2022. p. 1–5.

[32] Molnar G, Ljubenko D, and Sakic M. Flat-top Gaussian arrays with dynamic range ratio control. In: *2022 16th European Conference on Antennas and Propagation (EuCAP)*; 2022. p. 1–5.

[33] Grant M and Boyd S. *CVX: Matlab Software for Disciplined Convex Programming*, version 2.1; 2014. http://cvxr.com/cvx.

[34] Boyd S, Parikh N, Chu E, *et al.* Distributed optimization and statistical learning via the alternating direction method of multipliers. *Foundations and Trends® in Machine Learning*. 2011;3(1):1–122.

[35] Zhang X, Wang X, So HC, *et al.* Min–Max optimization for MIMO radar waveform design with improved power efficiency. *IEEE Transactions on Signal Processing*. 2022;70:6112–6127.

[36] Fuchs B and Fuchs JJ. Optimal narrow beam low sidelobe synthesis for arbitrary arrays. *IEEE Transactions on Antennas and Propagation*. 2010;58(6):2130–2135.

Chapter 6
Pattern synthesis via array response control

Xuejing Zhang[1], *Zishu He*[1] *and Bin Liao*[2]

This chapter delves into the techniques of pattern synthesis through array response control, introducing several distinct methods: the accurate array response control (A2RC) algorithm, the multipoint accurate array response control (MA2RC) algorithm, optimal and precise array response control (OPARC) algorithm, the weight vector orthogonal decomposition (WORD) algorithm, and flexible array response control via oblique projection (FARCOP) algorithm. These algorithms are rooted in the adaptive array theory, aiming to provide precise, optimal, flexible, and robust control over array response levels. Collectively, these methods present a comprehensive and effective framework for pattern synthesis through precise array response control. They offer promising solutions for wireless communications, radar systems, and beyond, enabling fine-tuned control of array responses and pattern synthesis in a wide range of applications.

6.1 Introduction

Array pattern synthesis is (also called beamforming) an important research direction in array processing. It is a signal processing technology that uses sensor arrays to send and receive signals directionally. Pattern design and synthesis play an important role in the high performance of the array system [1–7]. For example, in a radar system, it is usually necessary to form a deep null in the interference direction to achieve interference suppression. In some communication systems, it is necessary to design a multi-beam pattern to realize data transmission to multiple users. In satellite remote sensing applications, it is necessary to design a wide main lobe pattern to expand the detection area. In pattern synthesis, antenna weights need to be designed so that the resulting pattern meets specific requirements. Pattern synthesis can be used on both the signal transmitting end and the signal receiving end. With the complexity of the electromagnetic environment, how to design a directional pattern that is flexible, robust, fast, and meets specific hardware requirements has important theoretical and application values.

[1]School of Information and Communication Engineering, University of Electronic Science and Technology of China, China
[2]Guangdong Key Laboratory of Intelligent Information Processing, College of Electronics and Information Engineering, Shenzhen University, China

In the past few decades, beampattern synthesis has received extensive attention from scholars at home and abroad, and many direction beampattern synthesis technologies have emerged one after another. The classical algorithms [8–10] have given the closed-form expressions; however, they are limited to some specific array geometries or array patterns. To realize a pattern synthesis method suitable for arbitrary arrays, scholars propose a global optimization solution based on random methods such as genetic algorithm (GA) [11], particle swarm optimization (PSO) method [12] and simulated annealing (SA) method [13]. However, since the above method requires a global search, it is usually computationally expensive. Especially when large-scale arrays are used, the global search method needs to take a long time to obtain satisfactory pattern results. Another type of pattern synthesis method [14–17] is based on the principle of adaptive array [18–20]. In this type of method, virtual interference is added iteratively to minimize the deviation between the synthesized pattern and the desired pattern. It should be noted that the existing directional patterns based on adaptive array theory comprehensively use empirical methods [15–17] to select interference power and cannot accurately control the level of a given direction.

More recently, a scheme of zeroing the pattern based on the selection of antenna elements has been proposed in [21,22]. This type of method uses the state switching of the radio frequency (RF) switches to control the beampattern. In [23], the concept of the almost difference sets (ADSs) has been proposed. It is used to design the pattern of sparse planar array. The peak sidelobe level (PSLs) can be predicted by using the almost difference set and the array spectrum, so that low sidelobe pattern synthesis can be realized. In [24], A new sparse regularization method is used to realize the pattern synthesis of continuous cluster arrays. A maximum efficiency pattern synthesis method utilizing generalized eigenvalue decomposition is proposed in [25]. It should be noted that the above methods in [21–25] use empirical methods to select interference power, it is impossible to accurately control the level of a given direction.

Another type of pattern synthesis algorithm is implemented based on convex optimization theory [26]. For instance, in [27], the pattern synthesis problem is modeled as a convex optimization problem, and on this basis, the interior point method is used to solve the problem. However, this method is only suitable for some specific pattern synthesis problems. When the desired pattern contains the lower limit constraint requirements, the problem obtained is usually non-convex, so that convex optimization theory cannot be used to model and solve the problem. To realize the synthesis of the directional pattern when the lower limit of the level is included, in [28], they consider symmetric linear and planar arrays, and convert the non-convex problem into a convex problem by introducing a conjugate symmetric weight vector, thus realizing the lower limit of the level under the special array. In [29], the idea of semi-definite relaxation (SDR) [30] is introduced into the pattern synthesis. The non-convex pattern is integrated for relaxation modeling. This scheme uses an iterative method to solve the weight vector, thereby reducing the relaxation error caused by the relaxation operation. However, since the relaxed problem is different from the original problem, the semi-definite relaxation can only get an approximate solution.

There are also some convex optimization methods [31–33] that use the toolkit [34] to solve. In addition, some scholars respectively proposed the use of least squares method [35,36], fast Fourier transformation (FFT) [37], and excitation matching method [38] to solve the problem. It should be pointed out that none of the above methods can achieve precise directional control. Therefore, even if there is a slight change in the desired pattern, the above methods need to redesign the pattern. In this chapter, we give an overview of the accurate array response control-based pattern synthesis algorithms.

6.2 Array response control

In this section, we review the work of pattern synthesis using array response control first. Then, we will introduce several algorithms of accurate array response control.

6.2.1 *Adaptive array theory*

We consider an N-element array, the steering vector in direction θ can be written as

$$\boldsymbol{a}(\theta) = [g_1(\theta)e^{-j\omega\tau_1(\theta)}, \cdots, g_N(\theta)e^{-j\omega\tau_N(\theta)}]^{\mathrm{T}} \tag{6.1}$$

where $g_n(\theta)$ $(n = 1, \cdots, N)$ denotes the element pattern of the nth sensor, $\tau_n(\theta)$ is the time-delay between the nth element and the reference one, ω represents the operating frequency. We can then express the array response as

$$f(\theta) = \boldsymbol{w}^{\mathrm{H}}\boldsymbol{a}(\theta) \tag{6.2}$$

where $\boldsymbol{w}$ is the weight vector. In pattern synthesis, the problem is how to design an appropriate weight vector to make the amplitude response $|f(\theta)|$ satisfy specific requirements.

According to the theory above, the weight vector $\boldsymbol{w}$ can be optimally obtained by maximizing the output signal-to-interference-plus-noise ratio (SINR). In this case, the optimal weight vector $\boldsymbol{w}_{opt}$ can be expressed as

$$\boldsymbol{w}_{opt} = \alpha\boldsymbol{R}_{n+i}^{-1}\boldsymbol{a}(\theta_0) \tag{6.3}$$

where α is a normalization factor, $\boldsymbol{R}_{n+i}$ denotes the $N \times N$ noise-plus-interference covariance matrix, and $\boldsymbol{a}(\theta_0)$ is the signal steering vector. In the case of single interference, assuming the noise is spatially white and independent of the interference signal, the covariance matrix $\boldsymbol{R}_{n+i}$ can be written as

$$\boldsymbol{R}_{n+i} = \sigma_n^2\boldsymbol{I} + \sigma_i^2\boldsymbol{a}(\theta_i)\boldsymbol{a}^{\mathrm{H}}(\theta_i) \tag{6.4}$$

where σ_n^2 and σ_i^2 stands for the noise power and interference power, $\boldsymbol{a}(\theta_i)$ is the steering vector of interference.

6.2.2 A^2RC [1]

If we apply the matrix inversion principle, the weight vector in (6.3) can be rewritten as

$$\begin{aligned} \boldsymbol{w}_{opt} &= \alpha \boldsymbol{R}_{n+i}^{-1}\boldsymbol{a}(\theta_0) \\ &= \frac{\alpha}{\sigma_n^2}\left(\boldsymbol{a}(\theta_0) - \frac{\frac{\sigma_i^2}{\sigma_n^2}\boldsymbol{a}(\theta_i)\boldsymbol{a}^{\mathrm{H}}(\theta_i)\boldsymbol{a}(\theta_0)}{1+\frac{\sigma_i^2}{\sigma_n^2}\|\boldsymbol{a}(\theta_i)\|_2^2}\right) \end{aligned} \tag{6.5}$$

where the factor α/σ_n^2 will not affect the performance. Therefore, it will be omitted in the following analysis.

We express the optimal weight vector (denoted as $\boldsymbol{w}_\star$) as

$$\boldsymbol{w}_\star = \boldsymbol{w}_0 + \mu\boldsymbol{a}(\theta_i) \tag{6.6}$$

where $\boldsymbol{w}_0 \triangleq \boldsymbol{a}(\theta_0)$ is the initial weight, μ is a complex number which is determined by the interference-to-noise ratio (INR).

According to the result in (6.6), we can adjust the array response at a given direction θ_i by designing the complex constant μ. More precisely, if we need to adjust the normalized array power response at θ_1 to ρ_1, then there will exist a μ_1 and the weight vector $\boldsymbol{w}_1$ as below to satisfy the level requirement at the direction θ_i.

$$\boldsymbol{w}_1 = \boldsymbol{w}_0 + \mu_1\boldsymbol{a}(\theta_1). \tag{6.7}$$

Moreover, since $\boldsymbol{w}_0$ and $\boldsymbol{a}(\theta_1)$ are known, the remaining task is to find an appropriate μ_1 to fulfill the response requirement. Assuming μ_1 has been obtained, if we continue to control the response at another direction θ_2 to ρ_2, the weight vector will be adjusted to $\boldsymbol{w}_2 = \boldsymbol{w}_1 + \mu_2\boldsymbol{a}(\theta_2)$ with a constant μ_2. In the same way, at the kth step, we will get

$$\boldsymbol{w}_k = \boldsymbol{w}_{k-1} + \mu_k\boldsymbol{a}(\theta_k). \tag{6.8}$$

Obviously, the array response at a given direction can be controlled by adjusting the weight vector available instead of redesigning the weight vector completely.

In the analysis above, it is necessary to determine the μ_k in every step. Therefore, we define

$$L_\star^{(k)}(\theta_k, \theta_0) = \frac{|\boldsymbol{w}_k^{\mathrm{H}}\boldsymbol{a}(\theta_k)|^2}{|\boldsymbol{w}_k^{\mathrm{H}}\boldsymbol{a}(\theta_0)|^2}. \tag{6.9}$$

To make the normalized power response at θ_k to ρ_k, i.e.,

$$L_\star^{(k)}(\theta_k, \theta_0) = \rho_k \tag{6.10}$$

we can derive that

$$\mu_k = -\frac{\boldsymbol{Q}_k^*(1,2)}{\boldsymbol{Q}_k(2,2)} + \frac{\sqrt{-\det(\boldsymbol{Q}_k)}}{|\boldsymbol{Q}_k(2,2)|}e^{j\phi} \tag{6.11}$$

where ϕ can be any value in $[0, 2\pi)$ $\boldsymbol{Q}_k = \boldsymbol{C}_{k-1}^{\mathrm{H}}\left(\boldsymbol{a}(\theta_k)\boldsymbol{a}^{\mathrm{H}}(\theta_k) - \rho_k\boldsymbol{a}(\theta_0)\boldsymbol{a}^{\mathrm{H}}(\theta_0)\right)\boldsymbol{C}_{k-1}$, $\boldsymbol{C}_{k-1} = \left[\boldsymbol{w}_{k-1} \quad \boldsymbol{a}(\theta_k)\right]$, the result above describes the distribution of μ_k more intuitively from a geometric point of view, which is also more conducive to analyze the implicit properties and results.

6.2.3 MA^{2}RC and M^{2}A^{2}RC [2]

6.2.3.1 MA^{2}RC

In the A^{2}RC algorithm, we have already known that the array weight vector is initialized with $\boldsymbol{a}(\theta_0)$ and updated as

$$\boldsymbol{w}_{k+1} = \boldsymbol{w}_{k,\star} + \mu_{k+1}\boldsymbol{a}(\theta_{k+1}) \tag{6.12}$$

where $\boldsymbol{w}_{k,\star}$ denotes the weight vector at the kth step and θ_{k+1} denotes the direction which needs to be adjusted in the $k+1$th step. Moreover, given the desired response level at θ_{k+1} (denoted by ρ_{k+1}), then the corresponding μ_{k+1} which satisfies

$$L^{(k+1)}(\theta_{k+1}, \theta_0) = \frac{|\boldsymbol{w}_{k+1}^{\mathrm{H}}\boldsymbol{a}(\theta_{k+1})|^2}{|\boldsymbol{w}_{k+1}^{\mathrm{H}}\boldsymbol{a}(\theta_0)|^2} = \rho_{k+1} \tag{6.13}$$

can be any value on a circle.

At the $(k+1)$th step, the array weight vector can be written as

$$\boldsymbol{w}_{k+1,\star} = \boldsymbol{w}_{k,\star} + \mu_{k+1,\star}\boldsymbol{a}(\theta_{k+1}). \tag{6.14}$$

From (6.13), it is noticed that, if we multiply a non-zero factor c to the weight vector $\boldsymbol{w}_{k+1,\star}$, the new weight vector $c\boldsymbol{w}_{k+1,\star}$ will not change the response level at θ_{k+1}. Therefore, the response will not change even if we add a vector Δ_{k+1}, which is orthogonal to both $\boldsymbol{a}(\theta_{k+1})$ and $\boldsymbol{a}(\theta_0)$ to $c\boldsymbol{w}_{k+1,\star}$. This can be described as

$$\frac{|(c\boldsymbol{w}_{k+1,\star} + \Delta_{k+1})^{\mathrm{H}}\boldsymbol{a}(\theta_{k+1})|^2}{|(c\boldsymbol{w}_{k+1,\star} + \Delta_{k+1})^{\mathrm{H}}\boldsymbol{a}(\theta_0)|^2} = \rho_{k+1} \tag{6.15}$$

for $\forall c \neq 0$, $\Delta_{k+1} \perp \boldsymbol{a}(\theta_0)$ and $\Delta_{k+1} \perp \boldsymbol{a}(\theta_{k+1})$.

Suppose that in the kth step the weight vector is $\boldsymbol{w}_{k,\star}$ and we have already obtained M weight vectors by using the A^{2}RC algorithm, i.e.,

$$\boldsymbol{w}_{k+1,m} = \boldsymbol{w}_{k,\star} + \mu_{k+1,m}\boldsymbol{a}(\theta_{k+1,m}) \tag{6.16}$$

where $\mu_{k+1,m}$ and $\boldsymbol{w}_{k+1,m}$ $(m = 1, \cdots, M)$ are computed by using the formula below

$$L(\theta_{k+1,m}, \theta_0) = \frac{|\boldsymbol{w}_{k+1,m}^{\mathrm{H}}\boldsymbol{a}(\theta_{k+1,m})|^2}{|\boldsymbol{w}_{k+1,m}^{\mathrm{H}}\boldsymbol{a}(\theta_0)|^2} = \rho_{k+1,m}. \tag{6.17}$$

For any $m = 1, \cdots, M$, we were able to adjust the response at $\theta_{k+1,m}$ direction to $\rho_{k+1,m}$ by utilizing the corresponding weight vector $w_{k+1,m}$ right now. Then we will think about that how to obtain a unique weight vector which is able to simultaneously control the responses at M directions. In other words, we should find an appropriate weight vector $\overline{\boldsymbol{w}}_{k+1}$ which satisfies

$$L(\theta_{k+1,m}, \theta_0) = \frac{|\overline{\boldsymbol{w}}_{k+1}^{\mathrm{H}}\boldsymbol{a}(\theta_{k+1,m})|^2}{|\overline{\boldsymbol{w}}_{k+1}^{\mathrm{H}}\boldsymbol{a}(\theta_0)|^2} = \rho_{k+1,m} \tag{6.18}$$

In fact, the following $\overline{w}_{k+1}$ satisfies (6.18):

$$\overline{w}_{k+1} = \bar{c}_1 \boldsymbol{H}_1 \begin{bmatrix} -\boldsymbol{F}^{\dagger}\boldsymbol{q} + \boldsymbol{f}_n \\ 1 \end{bmatrix}, \quad \bar{c}_1 \neq 0,\ \forall \boldsymbol{f}_n \in \mathcal{N}(\boldsymbol{F}) \tag{6.19}$$

where $\boldsymbol{F}$ and $\boldsymbol{q}$ are given by

$$\boldsymbol{F} = \begin{bmatrix} (\boldsymbol{I}_N - \boldsymbol{H}_2\boldsymbol{H}_2^{\dagger})\boldsymbol{U}_{12} \\ (\boldsymbol{I}_N - \boldsymbol{H}_3\boldsymbol{H}_3^{\dagger})\boldsymbol{U}_{12} \\ \vdots \\ (\boldsymbol{I}_N - \boldsymbol{H}_M\boldsymbol{H}_M^{\dagger})\boldsymbol{U}_{12} \end{bmatrix} \in \mathbb{C}^{N(M-1)\times(N-2)} \tag{6.20}$$

$$\boldsymbol{q} = \begin{bmatrix} (\boldsymbol{I}_N - \boldsymbol{H}_2\boldsymbol{H}_2^{\dagger})\boldsymbol{w}_{k+1,1} \\ (\boldsymbol{I}_N - \boldsymbol{H}_3\boldsymbol{H}_3^{\dagger})\boldsymbol{w}_{k+1,1} \\ \vdots \\ (\boldsymbol{I}_N - \boldsymbol{H}_M\boldsymbol{H}_M^{\dagger})\boldsymbol{w}_{k+1,1} \end{bmatrix} \in \mathbb{C}^{N(M-1)} \tag{6.21}$$

where $\dagger$ denotes the pseudo-inverse of a matrix, and more details about the definitions of $\boldsymbol{F}$ and $\boldsymbol{q}$ are given in [2].

6.2.3.2 *M^2A^2RC*

Although the MA^2RC method has been able to simultaneously and accurately control the array responses at multiple directions, it still may lead to the shift of the beam axis. Supposing θ_0 is the direction of desired beam axis, the beam axis corresponds to the weight vector $\overline{w}_{k+1}$ in (6.19) may not be equal to θ_0. To deal with this problem, a modified multi-point accurate array response control (M^2A^2RC) algorithm is introduced in [2].

More precisely, we consider adding a derivative constraint on the basis of MA^2RC which is expressed as

$$\frac{\partial P(\theta)}{\partial \theta}\Big|_{\theta=\theta_0} = 0 \tag{6.22}$$

where $P(\theta) = \mathbf{w}^{\mathrm{H}}\boldsymbol{a}(\theta)\boldsymbol{a}^{\mathrm{H}}(\theta)\boldsymbol{w}$ denotes the array power response, θ_0 is the direction of desired beam axis.

$$\widetilde{w}_{k+1} = c_1 \begin{bmatrix} \Xi & w_{k+1,1} \end{bmatrix} \begin{bmatrix} (\boldsymbol{C}^{\dagger}\boldsymbol{k} + z_n)^{\mathrm{T}} & 1 \end{bmatrix}^{\mathrm{T}}, c \neq 0,\ \forall z_n \in \mathcal{N}(\boldsymbol{C}) \tag{6.23}$$

More details about the above $\widetilde{w}_{k+1}$ can be found in [2].

6.2.4 *OPARC [5,6]*

It is known that the optimal weight vector $\boldsymbol{w}_{\mathrm{opt}}$, which maximizes the SINR, is given by [18]

$$\boldsymbol{w}_{\mathrm{opt}} = \boldsymbol{R}_{n+i}^{-1}\boldsymbol{a}(\theta_0) \tag{6.24}$$

Note that the above SINR can be expressed as $G \cdot \sigma_s^2/\sigma_n^2$, where G is defined as

$$G = \frac{|\boldsymbol{w}^{\mathrm{H}}\boldsymbol{a}(\theta_0)|^2}{\boldsymbol{w}^{\mathrm{H}}\boldsymbol{T}_{n+i}\boldsymbol{w}} \tag{6.25}$$

with $\boldsymbol{T}_{n+i} \triangleq \boldsymbol{R}_{n+i}/\sigma_n^2$ standing for the normalized noise-plus-interference covariance matrix, i.e.,

$$\boldsymbol{T}_{n+i} = \frac{\boldsymbol{R}_{n+i}}{\sigma_n^2} = \boldsymbol{I} + \sum_{\ell=1}^{k} \beta_\ell \boldsymbol{a}(\theta_\ell)\boldsymbol{a}^{\mathrm{H}}(\theta_\ell) \tag{6.26}$$

where $\beta_\ell \triangleq \sigma_\ell^2/\sigma_n^2$ denotes the interference-to-noise ratio (INR). Actually, G stands for the amplification multiple of the input signal-to-noise ratio (SNR), which is called as the array gain. The optimal weight vector can achieve the maximization of array gain.

From (6.24)–(6.39), we can notice that the optimal weight vector $\boldsymbol{w}_{\mathrm{opt}}$ depends on $\boldsymbol{R}_{n+i}$ or $\boldsymbol{T}_{n+i}$. Considering the data-independent array response control, we need to design the weight vector w to make the normalized beam pattern $L(\theta, \theta_0) \triangleq |\boldsymbol{w}^{\mathrm{H}}\boldsymbol{a}(\theta)|^2/|\boldsymbol{w}^{\mathrm{H}}\boldsymbol{a}(\theta_0)|^2$ meet some specific requirement, where $\boldsymbol{a}(\theta)$ and θ_0 are given. In this chapter, what we consider is to control the response accurately at a single direction. Since $\boldsymbol{R}_{n+i}$ and $\boldsymbol{T}_{n+i}$ are unavailable, we can achieve that by designing a virtual normalized noise-plus-interference covariance matrix (VCM), denoted as $\boldsymbol{T}_k$. Note that the $\boldsymbol{T}_k$ is not produced by real data, thus it may not have real physical meaning. Besides, we do not assume $\boldsymbol{T}_k$ is positive definite nor Hermitian. Utilizing the concept of VCM, the data-dependent adaptive array theory can be applied to the data-independent beamforming. What we will discuss below is how to interfere virtually (e.g., designing the INR of the virtual interference) to update the weight vector $\boldsymbol{w}_{k-1,\mathrm{opt}} = \boldsymbol{T}_{k-1}^{-1}\boldsymbol{a}(\theta_0)$ to $\boldsymbol{w}_{k,\mathrm{opt}}$ optimally with the response at θ_k being the desired level ρ_k.

Specifically, suppose that the response levels of the $k-1$ directions have been controlled by adding $k-1$ virtual interference, denote the corresponding VCM as $\boldsymbol{T}_{k-1}$. For the given direction θ_k and desired response level ρ_k, we add the kth interference at θ_k and design the interference–noise ratio β_k optimally. From (6.39), we notice that the VCM can be updated as

$$\boldsymbol{T}_k = \boldsymbol{T}_{k-1} + \beta_k \boldsymbol{a}(\theta_k)\boldsymbol{a}^{\mathrm{H}}(\theta_k). \tag{6.27}$$

Using the Woodbury Lemma, we have

$$\boldsymbol{T}_k^{-1} = \boldsymbol{T}_{k-1}^{-1} - \frac{\beta_k \boldsymbol{T}_{k-1}^{-1}\boldsymbol{a}(\theta_k)\boldsymbol{a}^{\mathrm{H}}(\theta_k)\boldsymbol{T}_{k-1}^{-1}}{1 + \beta_k \boldsymbol{a}^{\mathrm{H}}(\theta_k)\boldsymbol{T}_{k-1}^{-1}\boldsymbol{a}(\theta_k)}. \tag{6.28}$$

Correspondingly, the optimal weight vector can be expressed as $\boldsymbol{w}_{k,\mathrm{opt}} = \boldsymbol{T}_k^{-1}\boldsymbol{a}(\theta_0)$. According to (6.24) and (6.28), we can express $\boldsymbol{w}_{k,\mathrm{opt}}$ as

$$\boldsymbol{w}_{k,\mathrm{opt}} = \boldsymbol{w}_{k-1,\mathrm{opt}} + \gamma_k \boldsymbol{T}_{k-1}^{-1}\boldsymbol{a}(\theta_k) \tag{6.29}$$

where $\boldsymbol{w}_{k-1,\mathrm{opt}} = \boldsymbol{T}_{k-1}^{-1}\boldsymbol{a}(\theta_0)$ denotes the previous optimal weight vector and γ_k is given by

$$\gamma_k = -\frac{\beta_k \boldsymbol{a}^{\mathrm{H}}(\theta_k)\boldsymbol{T}_{k-1}^{-1}\boldsymbol{a}(\theta_0)}{1+\beta_k \boldsymbol{a}^{\mathrm{H}}(\theta_k)\boldsymbol{T}_{k-1}^{-1}\boldsymbol{a}(\theta_k)} \triangleq \Psi_k(\beta_k) \tag{6.30}$$

where $\Psi_k(\cdot)$ stands for the mapping from β_k to γ_k.

The formulas (6.29)–(6.30) give the update form of the weight vector when optimizing the SINR. However, it still can not adjust the response level at θ_k to the desired value ρ_k. To meet the requirement of the level control, we next consider the following question. For the given $\boldsymbol{w}_{k-1,\mathrm{opt}} = \boldsymbol{T}_{k-1}^{-1}\boldsymbol{a}(\theta_0)$, does there exist γ_k (or equivalently β_k) that are able to adjust the response at θ_k to ρ_k? If it does, what the value it should be?

The solution in (6.29)–(6.30) gives the optimal solution for maximizing the SINR, which may not meet the response level ρ_k at θ_k. To meet this response level requirement, we next consider the following questions first. Given the previous weight vector $\boldsymbol{w}_{k-1,\mathrm{opt}} = \boldsymbol{T}_{k-1}^{-1}\boldsymbol{a}(\theta_0)$, does there exist γ_k (or equivalently β_k) such that the response level at θ_k is precisely ρ_k? and what value it should be if it exists? To answer the above questions, we denote the weight vector as

$$\boldsymbol{w}_k = \boldsymbol{w}_{k-1} + \gamma_k \boldsymbol{v}_k \tag{6.31}$$

where the subscript $(\cdot)_{\mathrm{opt}}$ is omitted for notational simplicity and $\boldsymbol{v}_k$ is defined as

$$\boldsymbol{v}_k \triangleq \boldsymbol{T}_{k-1}^{-1}\boldsymbol{a}(\theta_k). \tag{6.32}$$

6.2.5 WORD [3]

Now, considering covariance matrix $\boldsymbol{R}_{n+i}$ in (6.4), we can obtain $\boldsymbol{R}_{n+i}^{-1}$ by applying the Woodbury lemma as follows:

$$\begin{aligned}
\boldsymbol{R}_{n+i}^{-1} &= \frac{1}{\sigma_n^2}\left(\boldsymbol{I} - \frac{\sigma_i^2\|\boldsymbol{a}(\theta_i)\|_2^2}{\sigma_n^2+\sigma_i^2\|\boldsymbol{a}(\theta_i)\|_2^2}\cdot\frac{\boldsymbol{a}(\theta_i)\boldsymbol{a}^{\mathrm{H}}(\theta_i)}{\|\boldsymbol{a}(\theta_i)\|_2^2}\right)\\
&= \frac{1}{\sigma_n^2}\left(\boldsymbol{I} - \boldsymbol{P}_{[\boldsymbol{a}(\theta_i)]} + \frac{\sigma_n^2}{\sigma_n^2+\sigma_i^2\|\boldsymbol{a}(\theta_i)\|_2^2}\boldsymbol{P}_{[\boldsymbol{a}(\theta_i)]}\right)\\
&= \frac{1}{\sigma_n^2}\left(\boldsymbol{P}_{[\boldsymbol{a}(\theta_i)]}^{\perp} + \beta\boldsymbol{P}_{[\boldsymbol{a}(\theta_i)]}\right)
\end{aligned} \tag{6.33}$$

where

$$\boldsymbol{P}_{[\boldsymbol{a}(\theta_i)]} = \boldsymbol{a}(\theta_i)(\boldsymbol{a}^{\mathrm{H}}(\theta_i)\boldsymbol{a}(\theta_i))^{-1}\boldsymbol{a}^{\mathrm{H}}(\theta_i) \tag{6.34}$$

and

$$\boldsymbol{P}_{[\boldsymbol{a}(\theta_i)]}^{\perp} = \boldsymbol{I} - \boldsymbol{P}_{[\boldsymbol{a}(\theta_i)]}. \tag{6.35}$$

β is a real number associated with σ_n^2, σ_i^2, and $\boldsymbol{a}(\theta_i)$, which can be expressed as

$$\beta = \frac{\sigma_n^2}{\sigma_n^2 + \sigma_i^2 \|\boldsymbol{a}(\theta_i)\|_2^2}. \tag{6.36}$$

From (6.33), we can notice that $\boldsymbol{R}_{n+i}^{-1}$ is a linear combination of the projection matrix $\boldsymbol{P}_{[\boldsymbol{a}(\theta_i)]}^{\perp}$ and $\boldsymbol{P}_{[\boldsymbol{a}(\theta_i)]}$. Therefore, the optimal weight vector in (6.24) can be rewritten as

$$\begin{aligned} \boldsymbol{w}_\star &= \left(\boldsymbol{P}_{[\boldsymbol{a}(\theta_i)]}^{\perp} + \beta \boldsymbol{P}_{[\boldsymbol{a}(\theta_i)]}\right) \boldsymbol{a}(\theta_0) \\ &= \boldsymbol{w}_{(0)\perp} + \beta \boldsymbol{w}_{(0)\|} \\ &= \left[\boldsymbol{w}_{(0)\perp} \quad \boldsymbol{w}_{(0)\|}\right] \left[1 \ \beta\right]^{\mathrm{T}} \end{aligned} \tag{6.37}$$

where the common factor $1/\sigma_n^2$ is omitted, since it does not affect the performance of the beamforming. In (6.37), $\boldsymbol{w}_{(0)\perp}$ and $\boldsymbol{w}_{(0)\|}$ are given by

$$\boldsymbol{w}_{(0)\perp} = \boldsymbol{P}_{[\boldsymbol{a}(\theta_i)]}^{\perp} \boldsymbol{w}_{(0)} \tag{6.38a}$$

$$\boldsymbol{w}_{(0)\|} = \boldsymbol{P}_{[\boldsymbol{a}(\theta_i)]} \boldsymbol{w}_{(0)} \tag{6.38b}$$

where $\boldsymbol{w}_{(0)}$ denotes the quiescent weight vector as below:

$$\boldsymbol{w}_{(0)} = \boldsymbol{a}(\theta_0) \tag{6.39}$$

Actually, the $\boldsymbol{w}_{(0)}$ above only corresponds to the optimal weight vector in the presence of white noise only, when $\sigma_i^2 = 0$ or $\beta = 1$.

According to (6.37), it can be seen that the $\boldsymbol{w}_\star$ is updated from $\boldsymbol{w}_0$ (by using the parameter β). Hence, we can regard $\boldsymbol{w}_\star$ and $\boldsymbol{w}_0$ as the current weight vector and the previous weight vector separately. Based on this idea, modulating the response at the direction θ_k to a certain level (e.g., ρ_k) by a given weight vector $\boldsymbol{w}_{(k+1)}$ can be updated as

$$\boldsymbol{w}_{(k)} = \left[\boldsymbol{w}_{(k-1)\perp} \quad \boldsymbol{w}_{(k-1)\|}\right] \left[1 \quad \beta\right]^{\mathrm{T}} \tag{6.40}$$

where $\boldsymbol{w}_{(k-1)\perp}$ and $\boldsymbol{w}_{(k-1)\|}$ are defined according to (6.38) as

$$\boldsymbol{w}_{(k-1)\perp} = \boldsymbol{P}_{[\boldsymbol{a}(\theta_k)]}^{\perp} \boldsymbol{w}_{(k-1)} \tag{6.41a}$$

$$\boldsymbol{w}_{(k-1)\|} = \boldsymbol{P}_{[\boldsymbol{a}(\theta_k)]} \boldsymbol{w}_{(k-1)} \tag{6.41b}$$

and β can be determined from the following formula:

$$L_{(k)}(\theta_k, \theta_0) = \frac{|\boldsymbol{w}_{(k)}^{\mathrm{H}} \boldsymbol{a}(\theta_k)|^2}{|\boldsymbol{w}_{(k)}^{\mathrm{H}} \boldsymbol{a}(\theta_0)|^2} = \rho_k. \tag{6.42}$$

Actually, if $0 \leq \rho_k \leq 1$ and $\boldsymbol{a}^{\mathrm{H}}(\theta_k)\boldsymbol{a}(\theta_k) > |\boldsymbol{a}^{\mathrm{H}}(\theta_k)\boldsymbol{a}(\theta_0)|$, then problem can be analytically solved with two solutions given by

$$\beta_a = \frac{-\Re(\boldsymbol{B}(1,2)) + d}{\boldsymbol{B}(2,2)} \tag{6.43a}$$

$$\beta_b = \frac{-\Re(\boldsymbol{B}(1,2)) - d}{\boldsymbol{B}(2,2)} \tag{6.43b}$$

where $d = \sqrt{\Re^2(\boldsymbol{B}(1,2)) - \boldsymbol{B}(1,1)\boldsymbol{B}(2,2)}$, and $\Re(\cdot)$ represents the real part of a complex number. The new WORD scheme has been established [3].

6.2.6 C²−WORD [4]

In the above WORD algorithm, only two candidates (i.e., β_a and β_b) are available for the parameter β_k, and both of them are real-valued. Actually, there must exist a complex-valued β_k that can obtain the same response at the direction θ_k with the real-valued β_k. Apparently, it is more reasonable to consider the β_k in the WORD algorithm. This will lead to the complex-coefficient weight vector orthogonal decomposition algorithm below.

Specifically, for the given weight vector $\boldsymbol{w}_{k-1}$, to adjust the array response level of θ_k to the desired level ρ_k, we propose an updated formula of the weight vector as below:

$$\boldsymbol{w}_k = \begin{bmatrix} \boldsymbol{w}_\perp & \boldsymbol{w}_\parallel \end{bmatrix} \begin{bmatrix} 1 & \beta_k \end{bmatrix}^{\mathrm{T}}, \ \beta_k \in \mathbb{C}. \tag{6.44}$$

Different from the weight vector update of WORD in (6.40), the parameter β_k in (6.44) can be complex-valued.

6.2.7 Robust C²−WORD [4]

The C²-WORD algorithm developed in the previous subsection can control the array response level of a given direction in the absence of steering vector uncertainties. To realize array response control in the case where the steering vector is present, the next problem is the robust sidelobe control. For the convenience of later derivations, we first define the normalized amplitude response as below

$$V_a(\theta) = |\boldsymbol{w}^{\mathrm{H}}\boldsymbol{a}(\theta)| / |\boldsymbol{w}^{\mathrm{H}}\boldsymbol{a}(\theta_0)|. \tag{6.45}$$

Note that the above $V_a(\theta)$ is different from the normalized power response $L(\theta, \theta_0)$ defined in the previous subsection. The relationship between the two responses satisfies $V_a^2(\theta) = L(\theta, \theta_0)$.

Clearly, $V_a(\theta)$ describes the array magnitude response in the absence of array uncertainties. In practice, the steering vector may usually be influenced by the antenna, for instance, gain-phase error, element position error, mutual coupling error, and so on. After considering the error, the actual steering vector $\boldsymbol{b}(\theta)$ can be expressed as

$$\boldsymbol{b}(\theta) = \boldsymbol{a}(\theta) + \Delta(\theta) \tag{6.46}$$

where $\Delta(\theta)$ is the unknown steering vector uncertainty which may depend on θ. The actual normalized magnitude response (denoted by $V_b(\theta)$) can be expressed as

$$V_b(\theta) = |\boldsymbol{w}^{\mathrm{H}}\boldsymbol{b}(\theta)| / |\boldsymbol{w}^{\mathrm{H}}\boldsymbol{b}(\theta_0)| \tag{6.47}$$

Obviously, $V_b(\theta)$ is different from $V_a(\theta)$ in the usual cases. Besides, note that we still use the output at the direction θ_0 as our normalization factor, though the actual beam axis may deviate from the ideal direction θ_0. Therefore, we consider how to keep the actual magnitude response $V_b(\theta)$ lower than some specific level in the robust sidelobe control.

To proceed, we first present a boundary analysis on the actual magnitude response $V_b(\theta)$. Next, we reasonably assume that the norm of the uncertainty vector $\Delta(\theta)$ is

$$\|\Delta(\theta)\|_2 \leq \varepsilon(\theta) \tag{6.48}$$

where $\varepsilon(\theta)$ is the upper limit of $\Delta(\theta)$'s norm. Then, according to the triangle inequality property, we have

$$\begin{aligned} V_b(\theta) &= \frac{|\boldsymbol{w}^{\mathrm{H}}(\boldsymbol{a}(\theta) + \Delta(\theta))|}{|\boldsymbol{w}^{\mathrm{H}}(\boldsymbol{a}(\theta_0) + \Delta(\theta_0))|} \\ &\leq \frac{|\boldsymbol{w}^{\mathrm{H}}\boldsymbol{a}(\theta)| + \varepsilon(\theta)\|\boldsymbol{w}\|_2}{|\boldsymbol{w}^{\mathrm{H}}\boldsymbol{a}(\theta_0)| - \varepsilon(\theta_0)\|\boldsymbol{w}\|_2} \\ &= \frac{V_a(\theta) + \varepsilon(\theta) \cdot \|\boldsymbol{w}\|_2 / |\boldsymbol{w}^{\mathrm{H}}\boldsymbol{a}(\theta_0)|}{1 - \varepsilon(\theta_0) \cdot \|\boldsymbol{w}\|_2 / |\boldsymbol{w}^{\mathrm{H}}\boldsymbol{a}(\theta_0)|} \\ &\triangleq V_u(\theta) \end{aligned} \tag{6.49}$$

and

$$\begin{aligned} V_b(\theta) &\geq \frac{|\boldsymbol{w}^{\mathrm{H}}\boldsymbol{a}(\theta)| - \varepsilon(\theta)\|\boldsymbol{w}\|_2}{|\boldsymbol{w}^{\mathrm{H}}\boldsymbol{a}(\theta_0)| + \varepsilon(\theta_0)\|\boldsymbol{w}\|_2} \\ &= \frac{V_a(\theta) - \varepsilon(\theta) \cdot \|\boldsymbol{w}\|_2 / |\boldsymbol{w}^{\mathrm{H}}\boldsymbol{a}(\theta_0)|}{1 + \varepsilon(\theta_0) \cdot \|\boldsymbol{w}\|_2 / |\boldsymbol{w}^{\mathrm{H}}\boldsymbol{a}(\theta_0)|} \\ &\triangleq V_l(\theta). \end{aligned} \tag{6.50}$$

Then, we have

$$0 \leq V_l(\theta) \leq V_b(\theta) \leq V_u(\theta) \tag{6.51}$$

where $V_u(\theta)$ and $V_l(\theta)$ stand for the worst upper and lower boundaries of magnitude response, respectively. According to (6.51), the actual response $V_b(\theta)$ can vary in the range $[V_l(\theta), V_u(\theta)]$. Additionally, we have already assumed in the formula (6.49) that

$$|\boldsymbol{w}^{\mathrm{H}}\boldsymbol{a}(\theta_0)| - \varepsilon(\theta_0)\|\boldsymbol{w}\|_2 > 0. \tag{6.52}$$

Otherwise, $V_u(\theta) < 0$, and (6.49) does not hold true.

In the previous subsection, a boundary analysis of the array response is presented. In this subsection, we focus on the robust one-point sidelobe control, i.e., making the response level of an actual sidelobe level lower than a specific value in the presence of steering vector uncertainties.

More specifically, we denote the desired magnitude response as $V_d(\theta)$. Give the previous weight vector $\boldsymbol{w}_{k-1}$ and a sidelobe angle θ_k to be controlled, where

k stands for the kth step. We need to find a new weight vector $\boldsymbol{w}_k$ that makes the actual magnitude response level of θ_k lower than $V_d(\theta_k)$.

To simplify notations, we still use the symbol and meaning of $V_a(\theta)$, $V_b(\theta)$, $V_u(\theta)$ $V_l(\theta)$ in the two preceding subsections, to describe the amplitude pattern corresponding to $\boldsymbol{w}_k$. Then, the problem of one-point robust sidelobe control can be formulated as

$$\text{find} \quad \boldsymbol{w}_k \tag{6.53a}$$

$$\text{subject to} \quad V_b(\theta_k) \leq V_d(\theta_k). \tag{6.53b}$$

Note that $V_b(\theta_k)$ contains the unknown disturbance vector. As a result, it is not easy to adjust $V_b(\theta_k)$ as desired.

6.2.8 FARCOP [7]

Suppose that the noise is spatially white and Q independent interference. Then, the normalized noise-plus-interference covariance matrix is given by

$$\Xi_{n+i} \triangleq \frac{\boldsymbol{R}_{n+i}}{\sigma_n^2} = \boldsymbol{I} + \sum_{q=1}^{Q} \beta_q \boldsymbol{a}(\theta_q)\boldsymbol{a}^{\mathrm{H}}(\theta_q) \tag{6.54}$$

where σ_n^2 is the noise power, $\beta_q \triangleq \sigma_q^2/\sigma_n^2$, σ_q^2, and θ_q represent the interference-to-noise ratio (INR), power, and direction of the qth interference, respectively. Then, the optimal weight vector can be equivalently expressed as

$$\boldsymbol{w}_\star = \Xi_{n+i}^{-1}\boldsymbol{a}(\theta_0) \tag{6.55}$$

which leads to the same output SINR as $\boldsymbol{w}_{\mathrm{opt}}$ in (6.24).

It can be seen that the optimal weight $\boldsymbol{w}_\star$ depends on the matrix Ξ_{n+i} (or $\boldsymbol{R}_{n+i}$). However, the matrix is unavailable when designing the data-independent array response pattern. Hence, we need to design a weight vector $\boldsymbol{w}$ satisfying

$$L(\theta, \theta_0) \triangleq \frac{|\boldsymbol{w}^{\mathrm{H}}\boldsymbol{a}(\theta)|^2}{|\boldsymbol{w}^{\mathrm{H}}\boldsymbol{a}(\theta_0)|^2} \tag{6.56}$$

To obtain the corresponding data matrix, the concept of the virtual normalized noise-plus-interference covariance matrix (VCM) was proposed in [5]. The analysis shows that the pattern level in some directions can be adjusted by imposing virtual interference. However, it is tough to calculate the power of corresponding virtual interference, which is mainly because of the non-correspondence (interactive influence) between the array pattern response and the INR. To avoid the disadvantages above, we propose a flexible algorithm of pattern control.

To begin with, we rewrite the normalized covariance matrix in (6.54) as

$$\Xi_{n+i} = \boldsymbol{I} + \boldsymbol{A}(\theta_1, \cdots, \theta_Q)\Sigma\boldsymbol{A}^{\mathrm{H}}(\theta_1, \cdots, \theta_Q) \tag{6.57}$$

where $A(\theta_i, \cdots, \theta_j)$ and Σ are defined, respectively, as

$$A(\theta_i, \cdots, \theta_j) \triangleq [a(\theta_i), \cdots, a(\theta_j)] \tag{6.58a}$$

$$\Sigma = \text{Diag}\left([\beta_1, \beta_2, \cdots, \beta_Q]\right). \tag{6.58b}$$

Substituting (6.57) into (6.55) and applying the Woodbury matrix lemma to Ξ_{n+i}, we can get

$$\begin{aligned} w_\star &= \left(I + A(\theta_1, \cdots, \theta_Q)\Sigma A^{\mathrm{H}}(\theta_1, \cdots, \theta_Q)\right)^{-1} a(\theta_0) \\ &= \underbrace{\left[a(\theta_0) \quad A(\theta_1, \cdots, \theta_Q)\right]}_{\triangleq \breve{A}} \left[1 \quad u^{\mathrm{T}}\right]^{\mathrm{T}} \end{aligned} \tag{6.59}$$

where $\breve{A} \triangleq A(\theta_0, \theta_1, \cdots, \theta_Q) \in \mathbb{C}^{N\times(Q+1)}$, and $u \in \mathbb{C}^Q$ is given by

$$u = -\left(I + \Sigma A^{\mathrm{H}}(\theta_1, \cdots, \theta_Q)A(\theta_1, \cdots, \theta_Q)\right)^{-1} \Sigma A^{\mathrm{H}}(\theta_1, \cdots, \theta_Q)a(\theta_0). \tag{6.60}$$

We can find from (6.59) that the coefficient of the linear combination of optimal weight vectors $a(\theta_0), a(\theta_1), \cdots, a(\theta_Q)$ are determined by u. Obviously, we cannot adjust the response at θ_q by adjusting the INR simply, and then keep the responses at other $Q-1$ directions unchanging. This is because that θ_q not only depends on β_q, but also is related to all INRs $\beta_1, \cdots, \beta_Q$. Therefore, it is hard to adjust responses of several patterns accurately by adjusting INR directly.

However, we are able to control the response in multiple directions independently. Let us first define

$$v(i,j) \triangleq a^{\mathrm{H}}(\theta_i)a(\theta_j) \tag{6.61}$$

Assuming the $a(\theta_0), a(\theta_1), \cdots, a(\theta_Q)$ are linearly independent and $v(q,0) \neq 0$ for $\forall q \in \{1, \cdots, Q\}$. Then we can obtain that the weight vector $w_\star$ in (6.59) is re-represented as w_{OP}, which satisfies

$$\begin{aligned} w_{\mathrm{OP}} &= \left(\left(I - E^{\mathrm{H}}_{\breve{A}_{0-}|0}\right) + \sum_{q=1}^{Q} \eta_q E^{\mathrm{H}}_{q|\breve{A}_{q-}}\right) a(\theta_0) \\ &= c w_\star \end{aligned} \tag{6.62}$$

where c is a constant, $E_{\breve{A}_{0-}|0}$ and $E_{q|\breve{A}_{q-}}$ denote the project matrices as

$$E_{\breve{A}_{0-}|0} \triangleq E_{\breve{A}_{0-}|a(\theta_0)} \tag{6.63a}$$

$$E_{q|\breve{A}_{q-}} \triangleq E_{a(\theta_q)|\breve{A}_{q-}},\ q = 1, \cdots, Q \tag{6.63b}$$

where $\breve{A}_{i-}$ is the matrix obtained after removing $a(\theta_i)$ from the matrix $\breve{A}$, $i = 0, 1, \cdots, Q$, i.e.,

$$\breve{A}_{i-} \triangleq A(\theta_0, \theta_1, \cdots, \theta_{i-1}, \theta_{i+1}, \cdots, \theta_Q) \in \mathbb{C}^{N\times Q}. \tag{6.64}$$

In the previous subsection, we obtain the equivalent representation of the optimal weight. Next, we will parameterize the weight vector and propose the FARCOP algorithm.

For the sake of notational simplicity, we first define

$$\boldsymbol{\Psi}(\eta) \triangleq \left(\boldsymbol{I} - \boldsymbol{E}^{\mathrm{H}}_{\breve{\boldsymbol{A}}_{0-}|0}\right) + \sum_{q=1}^{Q} \eta_q \boldsymbol{E}^{\mathrm{H}}_{q|\breve{\boldsymbol{A}}_{q-}} \tag{6.65}$$

where η is determined as

$$\eta \triangleq [1, \eta_1, \eta_2, \cdots, \eta_Q]^{\mathrm{T}} \in \mathbb{C}^{Q+1}. \tag{6.66}$$

Hence, we have

$$\boldsymbol{w}_{\mathrm{OP}} = \boldsymbol{\Psi}(\eta)\boldsymbol{a}(\theta_0) \tag{6.67}$$

From the formula above, it is known that we can obtain $\boldsymbol{w}_{\mathrm{OP}}$ and achieve the level control at $\theta_1, \cdots, \theta_Q$ by changing the quiescent weight $\boldsymbol{a}(\theta_0)$ linearly. Then, we consider two questions below. Whether the linear transformation above can be expanded to a general transformation which can be used for any given weight vectors or not? Whether it can keep the responses unchanged when adjusting levels at some directions?

Formula (6.67) offers a new perspective on controlling the array response at multiple points. It shows that we can control the response level at some specific angles by the linear transformation of the given weight vectors. On the basis of (6.67), we introduce the FARCOP algorithm next, to achieve the array response control based on the given weight vector $\boldsymbol{w}_{\mathrm{pre}}$.

More specifically, for a given $\boldsymbol{w}_{\mathrm{pre}}$, we consider looking for new weight vector $\boldsymbol{w}_{\mathrm{new}}$ to adjust the response levels at $\theta_1, \theta_2, \cdots, \theta_Q$ to $\rho_1, \rho_2, \cdots, \rho_Q$. In the FARCOP algorithm, we transform the $\boldsymbol{w}_{\mathrm{new}}$ linearly and construct a new weight vector $\boldsymbol{w}_{\mathrm{new}}$ by

$$\boldsymbol{w}_{\mathrm{new}} = \boldsymbol{\Psi}(\eta)\boldsymbol{w}_{\mathrm{pre}} \tag{6.68}$$

where the transformation matrix $\boldsymbol{\Psi}(\eta)$ is given in (6.65).

6.3 Simulations

In this section, representative simulations are presented to illustrate the application of the proposed FARCOP algorithm to array pattern synthesis. Various approaches, including the convex programming (CP) method, the $\mathrm{A^2RC}$ method, the $\mathrm{MA^2RC}$ method, and the multiple-point OPARC method, are compared.

6.3.1 Nonuniform sidelobe synthesis for a large ULA

In this subsection, consider a uniform linear array with $N = 100$ elements, whose element spacing is half of wavelength. The desired pattern steers to $\theta_0 = 60°$ with nonuniform sidelobe levels. Specifically, in the region $[-20°, 30°]$, the upper level needs to be -45 dB; meanwhile, the level needs to be -35 dB in the rest of the region. It can be seen that the desired pattern is similar to a Chebshev pattern with a -35 dB uniform sidelobe to some extent. Therefore, we utilize the initial weight of the FARCOP algorithm as the Chebshev weight with a -35 dB sidelobe attenuation, to simplify the synthesis procedure.

Under this assumption, we choose $Q_k = 41$ sidelobe peak angles in each step, then use the FARCOP algorithm to adjust their corresponding level to the desired response level. Since $Q \ll N$, the calculation in each step is reduced greatly. In addition, the array is centro-symmetric and the initial weight vector is conjugate centro-symmetric. We take the optimal parameter vector $\eta_\star$ by (6.51) and obtain a weight vector which has a closed-form expression in each step of response control. Hence, the computation complexity of the proposed algorithm is further reduced. Results are shown in Figure 6.1, we can see that only $k = 9$ steps are required to synthesize a satisfactory beampattern.

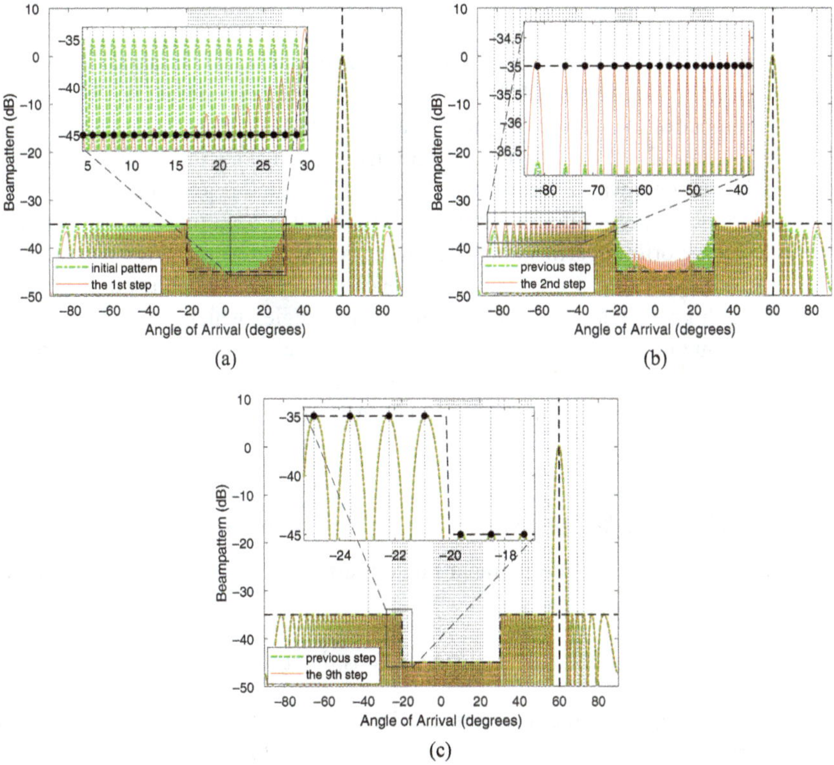

Figure 6.1 Resultant patterns at different steps when carrying out a nonuniform sidelobe synthesis for a nonuniform linear array: (a) synthesized pattern at the first step; (b) synthesized pattern at the second step; and (c) synthesized pattern at the ninth step.

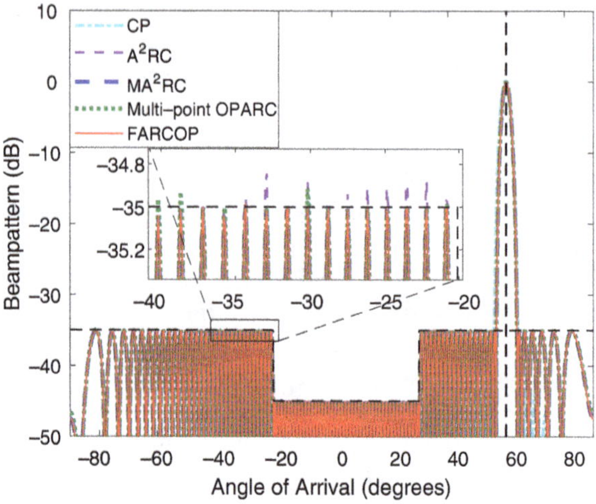

Figure 6.2 Synthesized patterns for a large ULA

To have fairer comparisons, we take the same number of iterations steps, i.e., $k = 9$ steps when we use the MA^{2}RC method [2] and the multi-point OPARC method [6]. The number of the selected angles in each step is also set as $Q_k = 41$. From the results in Figure 6.2, we can see that the pattern envelope of CP method is not aligned with the desired one, where we can tell that the CP method is not able to control the beampattern according to the requirement. Besides, it is seen that the MA^{2}RC and the FARCOP algorithm perform better then the A^{2}RC algorithm (after 300 iteration steps). Additionally, we notice that the sidelobe level of the multi-point OPARC algorithm is higher than our desired level at some angles which is mainly because the multi-point OPARC algorithm is initialized by a quiescent pattern instead of the Chebshev pattern that is closer to the desired level. Hence, more iteration steps are needed in the application of the multi-point OPARC algorithm to achieve the desired pattern synthesis.

6.3.2 Uniform sidelobe synthesis for a nonisotropic random array

In this subsection, we consider a linear random array with 21 elements. The pattern of the nth element is given by

$$g_n(\theta) = [\cos(\pi l_n \sin(\theta + \zeta_n)) + \cos(\pi l_n)] / \cos(\theta + \zeta_n)$$

where ζ_n and l_n represent the direction and length of the element, respectively. The beam steers to $\theta_0 = 20°$ and the desired pattern has a -25dB uniform sidelobe. We take the quiescent weight vector $\boldsymbol{a}(\theta_0)$ as the initial weight vector. Then we can obtain a satisfactory pattern by FARCOP method (with $k = 30$ steps). The result is shown in Figure 6.3. Notice the region $[-5°, 10°]$, we can see that both the A^{2}RC algorithm (with 200 iteration steps) and the multi-point OPARC algorithm (with the same step number as FARCOP) produce responses that are higher than the desired levels at

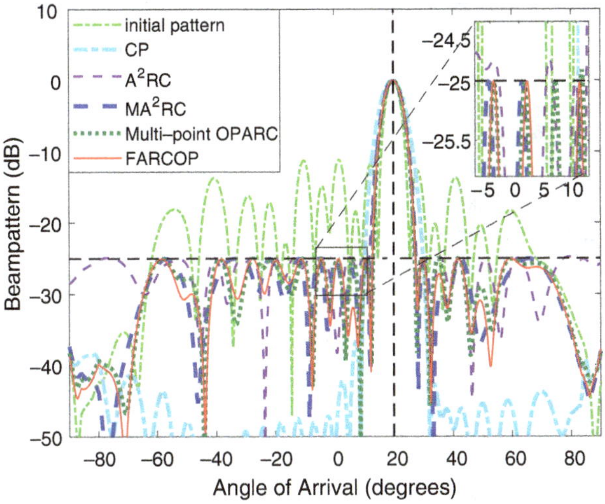

Figure 6.3 Synthesized patterns with uniform sidelobe for a nonisotropic array

certain angles. Naturally, it may require more synthesis steps for these two methods to achieve satisfactory beampatterns. The CP method leads to a pattern with sidelobe much lower than the prescribed level. For the proposed FARCOP algorithm, Figure 6.3 shows that the envelope of the synthesized pattern is aligned with the desired pattern. Moreover, it takes shorter time than MA^2RC (with the same step number as FARCOP) to synthesize a satisfactory beampattern.

6.4 Conclusion

In summary, this chapter demonstrates the diversity and richness of array response control techniques available for pattern synthesis. Each algorithm introduced offers unique advantages and capabilities, tailored to specific needs and challenges in the field. Taken together, these techniques represent a significant advancement in the state of the art of array response control and pattern synthesis, paving the way for future innovations and applications in this exciting area of research.

References

[1] X. Zhang, Z. He, B. Liao, X. Zhang, Z. Cheng, and Y. Lu, "A^2RC: An accurate array response control algorithm for pattern synthesis," *IEEE Trans. Signal Process.*, vol. 65, pp. 1810–1824, 2017.

[2] X. Zhang, Z. He, B. Liao, X. Zhang, and W. Peng, "Pattern synthesis with multipoint accurate array response control," *IEEE Trans. Antennas Propag.*, vol. 65, pp. 4075-4088, 2017.

[3] X. Zhang, Z. He, B. Liao, X. Zhang, and W. Peng, 'Pattern synthesis for arbitrary arrays via weight vector orthogonal decomposition," *IEEE Trans. Signal Process.*, vol. 66, pp. 1286–1299, 2018.

[4] X. Zhang, Z. He, and B. Liao, "Robust sidelobe control via complex-coefficient weight vector orthogonal decomposition," *IEEE Trans. Antennas Propag.*, vol. 67, pp. 5411–5425, 2019.

[5] X. Zhang, Z. He, X.-G. Xia, B. Liao, X. Zhang, and Y. Yang, "OPARC: optimal and precise array response control algorithm – Part I: Fundamentals," *IEEE Trans. Signal Process.*, vol. 67, pp. 652–667, 2019.

[6] X. Zhang, Z. He, X.-G. Xia, B. Liao, X. Zhang, and Y. Yang, "OPARC: optimal and precise array response control algorithm – Part II: Multi-points and applications," *IEEE Trans. Signal Process.*, vol. 67, pp. 668–683, 2019.

[7] X. Zhang, Z. He, and B. Liao, "Flexible array response control via oblique projection," *IEEE Trans. Signal Process.*, vol. 67, pp. 3126–3139, 2019.

[8] C. L. Dolph, "A current distribution for broadside arrays which optimizes the relationship between beam width and side-lobe level," *Proc. IRE*, vol. 34, pp. 335–348, 1946.

[9] H. Unz, "Linear arrays with arbitrarily distributed elements," *IRE Trans. Antennas Propag.*, vol. 8, pp. 222–223, 1960.

[10] A. Koretz and B. Rafaely, "Dolph–Chebyshev beampattern design for spherical arrays," *IEEE Trans. Signal Process.*, vol. 57, pp. 2417–2420, 2009.

[11] K. K. Yan and Y. Lu, "Sidelobe reduction in array-pattern synthesis using genetic algorithm," *IEEE Trans. Antennas Propag.*, vol. 45, pp. 1117–1122, 1997.

[12] D. W. Boeringer and D. H. Werner, "Particle swarm optimization versus genetic algorithms for phased array synthesis," *IEEE Trans. Antennas Propag.*, vol. 52, pp. 771–779, 2004.

[13] V. Murino, A. Trucco, and C. S. Regazzoni, "Synthesis of unequally spaced arrays by simulated annealing," *IEEE Trans. Signal Process.*, vol. 44, pp. 119–122, 1996.

[14] C. Y. Tseng and L. J. Griffiths, "A simple algorithm to achieve desired patterns for arbitrary arrays," *IEEE Trans. Signal Process.*, vol. 40, pp. 2737–2746, 1992.

[15] C. A. Olen and R. T. Compton, "A numerical pattern synthesis algorithm for arrays," *IEEE Trans. Antennas Propag.*, vol. 38, pp. 1666–1676, 1990.

[16] W. A. Swart and J. C. Olivier, "Numerical synthesis of arbitrary discrete arrays," *IEEE Trans. Antennas Propag.*, vol. 41, pp. 1171–1174, 1993.

[17] P. Y. Zhou and M. A. Ingram, "Pattern synthesis for arbitrary arrays using an adaptive array method," *IEEE Trans. Antennas Propag.*, vol. 47, pp. 862–869, 1999.

[18] H. L. Van Trees, *Optimum Array Processing*. New York: Wiley, 2002.

[19] I. S. Reed, J. D. Mallett, and L. E. Brennan, "Rapid convergence rate in adaptive arrays," *IEEE Trans. Aerosp. Electron. Syst.*, vol. AES-10, pp. 853–863, 1974.

[20] S. T. Smith, "Optimum phase-only adaptive nulling," *IEEE Trans. Signal Process.*, vol. 47, pp. 1835–1843, 1999.

[21] P. Rocca, R. L. Haupt, and A. Massa, "Interference suppression in uniform linear arrays through a dynamic thinning strategy," *IEEE Trans. Antennas Propag.*, vol. 59, pp. 4525–4533, 2011.

[22] L. Poli, P. Rocca, M. Salucci, and A. Massa, "Reconfigurable thinning for the adaptive control of linear arrays," *IEEE Trans. Antennas Propag.*, vol. 61, pp. 5068–5077, 2013.

[23] G. Oliveri, L. Manica, and A. Massa, "ADS-based guidelines for thinned planar arrays," *IEEE Trans. Antennas Propag.*, vol. 58, pp. 1935–1948, 2010.

[24] G. Oliveri, M. Salucci, and A. Massa, "Synthesis of modular contiguously clustered linear arrays through a sparseness-regularized solver," *IEEE Trans. Antennas Propag.*, vol. 64, pp. 4277–4287, 2016.

[25] G. Oliveri, L. Poli, and A. Massa, "Maximum efficiency beam synthesis of radiating planar arrays for wireless power transmission," *IEEE Trans. Antennas Propag.*, vol. 61, pp. 2490–2499, 2013.

[26] S. Boyd and L. Vandenberghe, *Convex Optimization*. Cambridge: Cambridge University Press, 2004.

[27] H. Lebret and S. Boyd, "Antenna array pattern synthesis via convex optimization," *IEEE Trans. Signal Process.*, vol. 45, pp. 526–532, 1997.

[28] S. E. Nai, W. Ser, Z. L. Yu, and H. Chen, "Beampattern synthesis for linear and planar arrays with antenna selection by convex optimization," *IEEE Trans. Antennas Propag.*, vol. 58, pp. 3923–3930, 2010.

[29] B. Fuchs, "Application of convex relaxation to array synthesis problems," *IEEE Trans. Antennas Propag.*, vol. 62, pp. 634–640, 2014.

[30] Z.-Q. Luo, W.-K. Ma, A. M.-C. So, Y. Ye, and S. Zhang, "Semidefinite relaxation of quadratic optimization problems," *IEEE Signal Process. Mag.*, vol. 27, pp. 20–34, 2010.

[31] H. G. Hoang, H. D. Tuan, and B. N. Vo, "Low-dimensional SDP formulation for large antenna array synthesis," *IEEE Trans. Antennas Propag.*, vol. 55, pp. 1716–1725, 2007.

[32] P. J. Kajenski, "Phase only antenna pattern notching via a semidefinite programming relaxation," *IEEE Trans. Antennas Propag.*, vol. 60, pp. 2562–2565, 2012.

[33] F. Wang, V. Balakrishnan, P. Y. Zhou, J. J. Chen, R. Yang, and C. Frank, "Optimal array pattern synthesis using semidefinite programming," *IEEE Trans. Signal Process.*, vol. 51, pp. 1172–1183, 2003.

[34] CVX Research, Inc., *CVX: Matlab software for disciplined convex programming*, San Ramon, CA, USA, September 2012.

[35] F. Wang, R. Yang, and C. Frank, "A new algorithm for array pattern synthesis using the recursive least squares method," *IEEE Signal Process. Lett.*, vol. 10, pp. 235–238, 2003.

[36] S. Zhan and Z. Feng, "A new array pattern synthesis algorithm using the two-step least-squares method," *IEEE Signal Process. Lett.*, vol. 12, pp. 250–253, 2005.

[37] K. Yang, Z. Zhao, Z. Nie, J. Ouyang, and Q. H. Liu, "Synthesis of conformal phased arrays with embedded element pattern decomposition," *IEEE Trans. Antennas Propag.*, vol. 59, pp. 2882–2888, 2011.
[38] L. Manica, P. Rocca, and A. Massa, "Design of subarrayed linear and planar array antennas with SLL control based on an excitation matching approach," *IEEE Trans. Antennas Propag.*, vol. 57, pp. 1684–1691, 2009.

Chapter 7

Wideband beampattern synthesis using single digital beamformer with integer time delay filters

Guolong Cui[1], Qinghui Lu[2], Xianxiang Yu[1], Xiangrong Wang[3] and Lingjiang Kong[1]

7.1 Introduction

Wideband digital beamforming (WDB) has been paid more attention across various applications, including radar, sonar, and communications [1–3]. And three main strategies are employed to address the unwanted aperture effect [4] generated by directly applying phase shift techniques in narrowband beamforming to WDB. The first category involves employing discrete Fourier transforms (DFTs) [5–7] based on the subband stitching concept, where the data received by each sensor is buffered and converted to the frequency domain through DFT. The second strategy utilizes tapped delay-lines (TDLs) or structure of FIR/IIR filters in time-domain [8–11], each with a group of coefficients. The last strategy is to apply the fractional delay filters (FDFs) or integer time delay filters (ITDFs) in time-domain [12–14] to construct the wideband beamformer, as the time delay remains consistent across different frequency bins. As evident from the aforementioned description, the wideband beamformers in the first two strategies require multiple beamformers, each corresponding to one frequency bin, so from the perspective of engineering practice, more multipliers need to be used in comparison to the third strategy. Additionally, although a higher delay accuracy can be achieved by the FDFs in the last strategy, the ITDF consumes less resources and is easier to implement through the delay flip-flop in the digital processor, making it more suitable for practical engineering. Therefore, we propose a novel wideband digital beamformer (DBF) to reduce unnecessary resource wastage, requiring only a group of weighting coefficients and ITDFs.

Moreover, current works related to the last category mainly emphasize the aperture effect, ignoring the beampattern shape. For example, the authors in [12] proposed a wideband transmit beamforming method using integer-time-delayed and phase-shifted waveforms. However, it only ensures that the transmitting beam has no

[1]School of Information and Communication Engineering, University of Electronic Science and Technology of China, China
[2]School of Electronic Information and Electrical Engineering, Shanghai Jiao Tong University, China
[3]School of Electronic and Information Engineering, Beihang University, China

significant distortion without considering the constraint on sidelobe interference suppression. Since the received wideband signals are inevitably disrupted by spectrum overlap interferences, especially in spectrum-congested environments [15,16], it is necessary to perform beamforming to eliminate interference in the spatial-spectral region of interest (SSRI). Therefore, we are committed to studying the wideband beampattern synthesis using ITDFs to mitigate interference [17,18].

To this end, a novel wideband beamforming approach is introduced in this chapter. Given certain spectral features of the wideband system and prior knowledge of potential interference locations, this method is designed to create a deep notch in the SSRI to mitigate interference. Specifically, we investigate a wideband beamforming design aimed at minimizing the peak sidelobe level (PSL) of the beampattern within the SSRI. Concurrently, the white noise gain and mainlobe level constraints are imposed to ensure an acceptable output signal-to-noise ratio (SNR) gain. To tackle the complex optimization problem with non-convex constraints [19], we present both the alternating optimization (AO) algorithm and the convex approximation (CA) algorithm and evaluate their performance in terms of convergence, complexity, and null depth.

In summary, the contributions of this chapter are outlined as follows:

- We present a novel time-domain wideband DBF that can control the beam shape over the whole bandwidth by using only one set of amplitude-phase weighting coefficients and ITDFs. Compared with the wideband beamformers using multiple groups of weighting coefficients, it effectively reduces resource consumption.
- To reduce the interference effect in the SSRI, we minimize the received beampattern PSL within the SSRI involving additional constraints on white noise gain and minimum mainlobe level. In particular, we only need three specially chosen frequency bins in the SSRI to control the interference suppression over the whole bandwidth. By doing so, the number of null constraints is greatly reduced, and thus the computational load is alleviated.
- We introduce two competitive algorithms to address the non-convex multi-constraint optimization problem. The AO algorithm is well-suited for scenarios with massive wideband arrays or high real-time requirements. Meanwhile, the CA algorithm is suitable for small wideband arrays or offline scenarios.
- We evaluate the effectiveness of the proposed wideband DBF through extensive simulations employing linear frequency modulated (LFM) waveforms.

The remainder of this chapter is structured as follows. In Section 7.2, we introduce a novel wideband DBF. Section 7.3 proposes a model centered around minimizing beampattern PSL in the SSRI, while also considering white noise gain and minimum mainlobe level constraints. Section 7.4 addresses the solution of the non-convex problem using the AO algorithm, while Section 7.5 introduces the CA algorithm for the same purpose. Section 7.6 presents the validation of the proposed method through numerical simulations. Finally, Section 7.7 provides conclusions and suggests potential tracks for future research.

7.2 System model

Consider a uniform linear array (ULA) composed of M sensors positioned at the location of d_m. Suppose that the transmit wideband signal $\tilde{s}(t)$ is represented as

$$\tilde{s}(t) = \text{rect}(t/T)\, s(t)\, e^{j2\pi f_c t}, \tag{7.1}$$

where $\text{rect}(t)$ is a rectangular pulse, f_c is the center frequency, T denotes the pulse duration, $s(t) = A(t)\, e^{j\varphi(t)}$ represents the baseband signal, $A(t)$ and $\varphi(t)$ stand for the envelope and phase of the baseband signal, respectively.

Then, the wideband signal $x_m(t)$ propagating from the direction of θ received by the mth channel is given by

$$\begin{aligned} x_m(t) &= \tilde{s}(t - \tau_m) + n(t) \\ &= \text{rect}\left(\frac{t-\tau_m}{T}\right) A(t-\tau_m)\, e^{j\varphi(t-\tau_m)} e^{j2\pi f_c (t-\tau_m)} + n(t), \end{aligned} \tag{7.2}$$

where $n(t)$ denotes the Gaussian white noise with zero mean, and $\tau_m = d_m \sin\theta / c$, $c = 3 \times 10^8 m/s$ represents the speed of light. The baseband signal after down-converting $x_m(t)$ can be represented as

$$h_m(t) = \text{rect}\left(\frac{t-\tau_m}{T}\right) A(t-\tau_m)\, e^{j\varphi(t-\tau_m)} e^{-j2\pi f_c \tau_m} + n(t), \tag{7.3}$$

We consider a novel wideband DBF with ITDF, as shown in Figure 7.1, which possesses the following characteristics:

- This method introduces complex weights w_m and integer delay D_m for each channel in the direction of θ, where $\mathbf{w} = [w_1, w_2, \cdots, w_M]^T$, $D_m = \langle \tau_m \cdot F_s \rangle / F_s$, with F_s being the sampling rate of baseband signal and $\langle \cdot \rangle$ denoting the rounding operation [12].
- The ITDF structure can eliminate the aperture effect of the wideband array antenna [4], that is, the phenomenon that the antenna beam direction shifts due to changes in signal frequency, thereby reducing the aperture fill time.

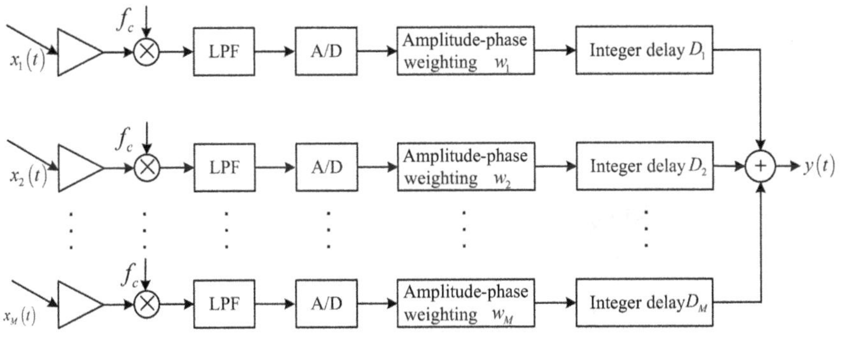

Figure 7.1 A receiver structure for wideband DBF based on ITDF [20]

- Only one DBF is required in this structure to control the beam shape across the entire bandwidth, effectively suppressing interference signals and greatly reducing the computational complexity of digital signal processing (DSP).

Hence, the output signal $y(t)$ can be represented as

$$\begin{aligned} y(t) &= \sum_{m=1}^{M} w_m^* h_m(t) \otimes \delta(t + D_m) \\ &= \mathbf{w}^\dagger \mathbf{n}(t) + \mathrm{rect}\left([t - (\tau_m - D_m)]/T\right) \\ &\times \sum_{m=1}^{M} w_m^* A\left[t - (\tau_m - D_m)\right] e^{j\varphi[t-(\tau_m - D_m)]} e^{-j2\pi f_c \tau_m}. \end{aligned} \tag{7.4}$$

In the special case where the received signal is the LFM waveform, the envelope and phase of the baseband signal can be expressed as

$$\begin{cases} A(t) = A, \\ \varphi(t) = \pi\mu_0 t^2 + \varphi_0, \end{cases} \tag{7.5}$$

where A is the constant amplitude, $\mu_0 = B/T$ with B being the bandwidth, and φ_0 is the initial phase. Then, the output signal is represented as

$$\begin{aligned} y(t) = {} & \mathbf{w}^\dagger \mathbf{n}(t) + \mathrm{rect}\left(\frac{t - (\tau_m - D_m)}{T}\right) A e^{j\left(\pi\mu_0 t^2 + \varphi_0\right)} \\ & \times \sum_{m=1}^{M} w_m^* \left[e^{-j2\pi f_c \tau_m} e^{-j2\pi(\tau_m - D_m)\mu_0 t} e^{j\pi\mu_0(\tau_m - D_m)^2}\right]. \end{aligned} \tag{7.6}$$

We define that

$$a_m(\theta, \Delta f) = e^{-j2\pi f_c \frac{d_m \sin\theta}{c}} e^{-j2\pi\Delta f\left(\frac{d_m \sin\theta}{c} - D_m\right)} e^{j\pi\mu_0\left(\frac{d_m \sin\theta}{c} - D_m\right)^2}, \tag{7.7}$$

where $\mathbf{a}(\theta, \Delta f) = [a_1(\theta, \Delta f), a_2(\theta, \Delta f), \cdots, a_M(\theta, \Delta f)]^T$ can be regarded as the steering vector of the wideband signal, $\Delta f = \mu_0 t$, $\Delta f \in [-B/2, B/2]$. For other waveforms, the form of $\mathbf{a}(\theta, \Delta f)$ varies. For example, when considering the non-LFM (NLFM) signal with the frequency modulation function $f(t) = f_c + \mu(t)t$, where $\mu(t)$ is the frequency modulation slope, Δf in (7.7) can be expressed as $\Delta f = \mu(t - (\tau_m - D_m))t$.

Hence, the radiated beampattern of the wideband array is written as

$$F(\theta, \Delta f) = \sum_{m=1}^{M} w_m^* a_m(\theta, \Delta f). \tag{7.8}$$

For computational convenience, discrete angles and frequency bins are required. Consequently, the discrete angles corresponding to nulling in the SSRI can be denoted as $\Omega_Q = \left\{\hat{\theta}_q\right\}_{q=1}^{Q}$. The frequency bins across the entire bandwidth ($[-B/2, B/2]$) can be represented as $\Omega_P = \{\Delta f_p\}_{p=1}^{P}$.

In the following, we introduce the design criteria, the associated constraints, and the formulated optimization problem.

7.3 Problem formulation

7.3.1 Design criteria of beamforming

We aim to minimize the beampattern PSL in the SSRI while satisfying additional constraints. Hence, the objective function is defined as the peak power of the notch region over all frequency bins,

$$\max_{\hat{\theta}_q, \Delta f} \left| \mathbf{w}^{\dagger} \mathbf{a} \left(\hat{\theta}_q, \Delta f_p \right) \right|^2, \hat{\theta}_q \in \Omega_Q, \Delta f_p \in \Omega_P, \tag{7.9}$$

By introducing an upper bound variable η [21], we can formulate the problem (7.9) as,

$$\begin{cases} \min\limits_{\mathbf{w},\eta} \ \eta \\ \text{s.t.} \ \ \left| \mathbf{w}^{\dagger} \mathbf{a} \left(\hat{\theta}_q, \Delta f_p \right) \right|^2 \leq \eta, \hat{\theta}_q \in \Omega_Q, \Delta f_p \in \Omega_P. \end{cases} \tag{7.10}$$

It can be observed from (7.10) that there are QP constraints, leading to prohibitive computational complexity and resource consumption. In practice, a satisfactory nulling depth can be achieved by considering the nulling in the corresponding directional intervals of three specific frequency bins $\Delta f_p \in \Omega_3, \Omega_3 = \{-B/2, 0, B/2\}$ [20]. Therefore, the optimization problem with reduced constraints can be formulated as

$$\begin{cases} \min\limits_{\mathbf{w},\eta} \ \eta \\ \text{s.t.} \ \ \left| \mathbf{w}^{\dagger} \mathbf{a} \left(\hat{\theta}_q, \Delta f_p \right) \right|^2 \leq \eta, \hat{\theta}_q \in \Omega_Q, \Delta f_p \in \Omega_3. \end{cases} \tag{7.11}$$

7.3.2 White noise gain constraint

Assume that $P_{\mathbf{n},in}$ is defined as the input Gaussian white noise power, and $\mathbf{S_n}\{p\} = S_{\mathbf{n}}\{p\}\mathbf{I}$ is the spatial spectral matrix of the pth frequency bin, the output noise power $P_{\mathbf{n},o}$ can be obtained:

$$\begin{aligned} P_{\mathbf{n},o} &= \frac{1}{P^2} \sum_{p=0}^{P-1} \mathbf{W}^{\dagger}\{p\} \mathbf{S_n}\{p\} \mathbf{W}\{p\} \\ &= \frac{1}{P^2} \sum_{p=0}^{P-1} S_{\mathbf{n}}\{p\} \mathbf{w}^{\dagger} \mathbf{I} \mathbf{w} \\ &= \mathbf{w}^{\dagger} \mathbf{w} P_{\mathbf{n},in}, \end{aligned} \tag{7.12}$$

where $\mathbf{W}\{p\}$ denotes the weight vector corresponding to the pth frequency bin, and $\mathbf{w} = \mathbf{W}\{p\}, p = 1, 2, \cdots, P$ represents the proposed single time-domain wideband beamformer. Therefore, to accumulate noises incoherently [5], we introduce the following constraint related to the weight vector $\mathbf{w}$,

$$\mathbf{w}^{\dagger} \mathbf{w} = 1. \tag{7.13}$$

7.3.3 Mainlobe level constraint

For a focused beam, considering the constraint in (7.13), the output SNR gain is maximum when $\mathbf{w} = \mathbf{a}(\theta_0, \Delta f_p)/\sqrt{M}$, where θ_0 denotes the direction of the beam axis. However, when considering multi-constraint beamforming, it is inevitable that the output SNR gain will decrease. Hence, we impose the following constraint on the lower bound of the mainlobe level to control gain loss [22],

$$\left|\mathbf{w}^\dagger \mathbf{a}(\theta_0, 0)\right|^2 \geq P_l, \tag{7.14}$$

where P_l represents the lower bounds of the beam axis direction. In addition, this constraint serves as a reference for the measurement of the null depth.

7.3.4 Optimization problem

Combining the objective function and constraints discussed in (7.9)–(7.14), the optimization problem can be written as

$$\mathscr{P} \begin{cases} \min\limits_{\mathbf{w},\eta} \ \eta \\ \text{s.t.} \ \left|\mathbf{w}^\dagger \mathbf{a}\left(\hat{\theta}_q, \Delta f_p\right)\right|^2 \leq \eta, \hat{\theta}_q \in \Omega_Q, \Delta f_p \in \Omega_3 \\ \quad\ \left|\mathbf{w}^\dagger \mathbf{a}(\theta_0, 0)\right|^2 \geq P_l, \\ \quad\ \mathbf{w}^\dagger \mathbf{w} = 1. \end{cases} \tag{7.15}$$

It is worth noting that there is no closed-form solution to the problem $\mathscr{P}$, since both the minimum mainlobe level constraint $\left|\mathbf{w}^\dagger \mathbf{a}(\theta_0, 0)\right|^2 \geq P_l$ and white noise gain constraint $\mathbf{w}^\dagger \mathbf{w} = 1$ are non-convex [19]. Hence, we present two iterative optimization algorithms for different scenarios to address this non-convex problem $\mathscr{P}$.

7.4 AO algorithm

In this section, we introduce the AO algorithm based on the ADMM structure to tackle the non-convex problem $\mathscr{P}$. The principle of this algorithm is to decompose the original problem into multiple subproblems and solve them iteratively. It has the advantages of low computational complexity and fast convergence.

By introducing auxiliary variables y_0 and $y_{q,p}$, the optimization problem $\mathscr{P}$ can be transformed into

$$\begin{cases} \min\limits_{\mathbf{w},\eta,y_0,y_{q,p}} \ \eta \\ \text{s.t.} \ y_{q,p} = \mathbf{w}^\dagger \mathbf{a}\left(\hat{\theta}_q, \Delta f_p\right), \hat{\theta}_q \in \Omega_Q, \Delta f_p \in \Omega_3, \\ \quad\ \left|y_{q,p}\right|^2 \leq \eta, q = 1, 2, \cdots, Q, q = 1, 2, 3, \\ \quad\ y_0 = \mathbf{w}^\dagger \mathbf{a}(\theta_0, 0), \\ \quad\ \left|y_0\right|^2 \geq P_l, \\ \quad\ \mathbf{w}^\dagger \mathbf{w} = 1. \end{cases} \tag{7.16}$$

Firstly, the augmented Lagrangian function is shown as

$$\begin{aligned}&\mathscr{L}\left(\mathbf{w},\eta,y_0,y_{q,p},\alpha,\beta_{q,p},\rho_0,\rho_y\right)\\&=\eta+\sum_{q=1}^{Q}\sum_{p=1}^{3}\left(\operatorname{Re}\left\{\beta_{q,p}^*\left(y_{q,p}-\mathbf{w}^\dagger\mathbf{a}\left(\hat{\theta}_q,\Delta f_p\right)\right)\right\}\right.\\&\left.+\frac{\rho_y}{2}\left\|\left(y_{q,p}-\mathbf{w}^\dagger\mathbf{a}\left(\hat{\theta}_q,\Delta f_p\right)\right)\right\|_2^2\right)\\&+\operatorname{Re}\left\{\alpha^*\left(y_0-\mathbf{w}^\dagger\mathbf{a}\left(\theta_0,0\right)\right)\right\}+\frac{\rho_0}{2}\left\|\left(y_0-\mathbf{w}^\dagger\mathbf{a}\left(\theta_0,0\right)\right)\right\|_2^2,\end{aligned}\tag{7.17}$$

where α and $\beta_{q,p}$ $(q=1,2,\cdots,Q,p=1,2,3)$ are Lagrange multipliers of the constraints $y_0-\mathbf{w}^\dagger\mathbf{a}(\theta_0,0)=0$ and $y_{q,p}-\mathbf{w}^\dagger\mathbf{a}\left(\hat{\theta}_q,\Delta f_p\right)=0$, $\rho_0>0$ and $\rho_y>0$ represent the penalty parameters.

Then, following the ADMM framework [23], we combine the augmented Lagrangian function with the remaining constraints to decompose the problem into multiple subproblems that can be solved iteratively. Assuming that $\mathbf{w}^{(k)}$, $\eta^{(k)}$, $y_0^{(k)}$, $y_{q,p}^{(k)}$, $\alpha^{(k)}$, $\beta_{q,p}^{(k)}$ are obtained at kth iteration, optimization variables at the $(k+1)$th iteration can be updated via the following iterative steps.

- **Step 1**: update $\mathbf{w}^{(k+1)}$ by solving

$$\begin{aligned}\min_{\mathbf{w}}\ &\mathscr{L}\left(\mathbf{w},\eta^{(k)},y_0^{(k)},y_{q,p}^{(k)},\alpha^{(k)},\beta_{q,p}^{(k)}\right)\\\text{s.t.}\ \ &\mathbf{w}^\dagger\mathbf{w}=1.\end{aligned}\tag{7.18}$$

- **Step 2**: update $y_{q,p}^{(k+1)}$ and $\eta^{(k+1)}$ by solving

$$\begin{aligned}\min_{\eta,y_{q,p}}\ &\mathscr{L}\left(\mathbf{w}^{(k+1)},\eta,y_0^{(k)},y_{q,p},\alpha^{(k)},\beta_{q,p}^{(k)}\right)\\\text{s.t.}\ \ &\left|y_{q,p}\right|^2\leq\eta,q=1,2,\cdots,Q,p=1,2,3.\end{aligned}\tag{7.19}$$

- **Step 3**: update $y_0^{(k+1)}$ by solving

$$\begin{aligned}\min_{y_0}\ &\mathscr{L}\left(\mathbf{w}^{(k+1)},\eta^{(k+1)},y_0,y_{q,p}^{(k+1)},\alpha^{(k)},\beta_{q,p}^{(k)}\right)\\\text{s.t.}\ \ &\left|y_0\right|^2\geq P_l.\end{aligned}\tag{7.20}$$

- **Step 4**: update $\alpha^{(k+1)}$ and $\beta_{q,p}^{(k+1)}$ as following

$$\alpha^{(k+1)}=\alpha^{(k)}+\rho_0\left(y_0^{(k+1)}-\mathbf{w}^{(k+1)^\dagger}\mathbf{a}\left(\theta_0,0\right)\right),\tag{7.21}$$

$$\begin{aligned}\beta_{q,p}^{(k+1)}=\beta_{q,p}^{(k)}+\rho_y\left(y_{q,p}^{(k+1)}-\mathbf{w}^{(k+1)^\dagger}\mathbf{a}\left(\hat{\theta}_q,\Delta f_p\right)\right),\\q=1,2,\cdots,Q,p=1,2,3.\end{aligned}\tag{7.22}$$

The solution methods for the above subproblems in (7.18)–(7.20) will be introduced sequentially in the following subsections.

7.4.1 Step 1: update $\mathbf{w}^{(k+1)}$

Simplifying the equation related to **w** and ignoring the irrelevant terms, the optimization problem concerning in (7.18) can be rewritten as

$$\begin{aligned} \min_{\mathbf{w}} \quad & \mathbf{w}^{\dagger}\mathbf{X}\mathbf{w} + \mathbf{d}^{\dagger}\mathbf{w} + \mathbf{w}^{\dagger}\mathbf{d} \\ \text{s.t.} \quad & \mathbf{w}^{\dagger}\mathbf{w} = 1, \end{aligned} \tag{7.23}$$

where

$$\mathbf{X} = \rho_0 \mathbf{a}\left(\theta_0, 0\right) \mathbf{a}^{\dagger}\left(\theta_0, 0\right) + \rho_y \sum_{q=1}^{Q} \sum_{p=1}^{3} \mathbf{a}\left(\hat{\theta}_q, \Delta f_p\right) \mathbf{a}^{\dagger}\left(\hat{\theta}_q, \Delta f_p\right), \tag{7.24}$$

$$\begin{aligned} \mathbf{d} &= -\left(\rho_0 y_0^{(k)} + \alpha^{(k)}\right)^{*} \mathbf{a}\left(\theta_0, 0\right) \\ &\quad - \sum_{q=1}^{Q} \sum_{p=1}^{3} \left(\rho_y y_{q,p}^{(k)} + \beta_{q,p}^{(k)}\right)^{*} \mathbf{a}\left(\hat{\theta}_q, \Delta f_p\right). \end{aligned} \tag{7.25}$$

Then, we get the Lagrangian $\mathscr{H}\left(\mathbf{w}, \gamma\right)$ for problem (7.23) [24],

$$\mathscr{H}\left(\mathbf{w}, \gamma\right) = \mathbf{w}^{\dagger}\mathbf{X}\mathbf{w} + \mathbf{d}^{\dagger}\mathbf{w} + \mathbf{w}^{\dagger}\mathbf{d} + \gamma\left(\mathbf{w}^{\dagger}\mathbf{w} - 1\right). \tag{7.26}$$

Taking the partial derivatives of $\mathscr{H}\left(\mathbf{w}, \gamma\right)$ with respect to **w** and γ, respectively, and equating the results to zero yield the two Lagrangian equations,

$$\mathbf{X}\mathbf{w} + \mathbf{d} + \gamma\mathbf{w} = 0, \tag{7.27}$$

$$\mathbf{w}^{\dagger}\mathbf{w} - 1 = 0. \tag{7.28}$$

From (7.27), we can obtain the analytical solution of **w**, that is,

$$\mathbf{w} = -\left(\mathbf{X} + \gamma\mathbf{I}_N\right)^{-1}\mathbf{d}. \tag{7.29}$$

Substitute (7.29) into (7.28), then we will get $h\left(\gamma\right)$:

$$h\left(\gamma\right) = \mathbf{d}^{\dagger}\left(\mathbf{X} + \gamma\mathbf{I}_N\right)^{-2}\mathbf{d} - 1, \tag{7.30}$$

which shows that the solution to $h(\gamma) = 0$ gives the optimal value γ.

Considering that **X** is a positive definite matrix, we perform a singular value decomposition for it,

$$\mathbf{X} = \mathbf{U}\mathbf{S}\mathbf{U}^{\dagger}, \tag{7.31}$$

where the diagonal elements of singular value matrix $\mathbf{S} = diag\left\{s_1, s_2, \cdots, s_N\right\}$ are singular values of positive definite matrix **X**, and are arranged in descending order, i.e.: $s_1 \geq s_2 \geq \cdots \geq s_N \geq 0$. In addition, $\mathbf{U} = \left[\mathbf{u}_1, \mathbf{u}_2, \cdots, \mathbf{u}_N\right]$ is the eigenvector matrix of **X**.

Then, substituting (7.31) into (7.30) rewrites $h\left(\gamma\right)$ as

$$\begin{aligned} h\left(\gamma\right) &= \mathbf{d}^{\dagger}\mathbf{U}\left(\mathbf{S} + \gamma\mathbf{I}_N\right)^{-2}\mathbf{U}^{\dagger}\mathbf{d} - 1 \\ &= \sum_{n=1}^{N} \frac{\left|\mathbf{d}^{\dagger}\mathbf{u}_n\right|^2}{\left(s_n + \gamma\right)^2} - 1. \end{aligned} \tag{7.32}$$

As can be seen from (7.32),

$$\lim_{\gamma\to -s_N} h(\gamma) = +\infty, \tag{7.33}$$

$$\lim_{\gamma\to +\infty} h(\gamma) = -1. \tag{7.34}$$

And the derivative of $h(\gamma)$ is given by,

$$\frac{\partial h(\gamma)}{\partial \gamma} = -2\sum_{n=1}^{N} \frac{\left|\mathbf{d}^{\dagger}\mathbf{u}_n\right|^2}{(s_n+\gamma)^3}, \tag{7.35}$$

where $\partial h(\gamma)/\partial\gamma < 0, \gamma \in (-s_N, +\infty)$. Hence, $h(\gamma)$ is a monotonically decreasing function. According to (7.33) and (7.34), there must be an optimal value $\tilde{\gamma}$ that satisfies $h(\tilde{\gamma}) = 0$. Therefore, we can obtain the optimal value $\tilde{\gamma}$ by the bisection method. Then, the optimal value $\tilde{\gamma}$ is substituted into (7.29) to get $\mathbf{w}^{(k+1)}$.

7.4.2 Step 2: update $y_{q,p}^{(k+1)}$, $\eta^{(k+1)}$

Ignoring the terms unrelated to η and $y_{q,p}$, the optimization problem in (7.19) can be expressed as

$$\begin{aligned} &\min_{y_{q,p},\eta} \ \eta + \sum_{q=1}^{Q}\sum_{p=1}^{3}\left|y_{q,p} - \mathbf{w}^{(k+1)\dagger}\mathbf{a}\left(\hat{\theta}_q, \Delta f_p\right) + \frac{\beta_{q,p}^{(k)}}{\rho_y}\right|^2 \\ &\text{s.t.} \quad \left|y_{q,p}\right|^2 \le \eta, q = 1, 2, \cdots, Q, p = 1, 2, 3. \end{aligned} \tag{7.36}$$

Assuming that the solution $\eta^{(k+1)}$ at $(k+1)$th iteration has been obtained [21], the problem of $y_{q,p}$ can be represented by the following equation:

$$\begin{aligned} &\min_{y_{q,p}} \ \sum_{q=1}^{Q}\sum_{p=1}^{3}\left|y_{q,p} - \mathbf{w}^{(k+1)\dagger}\mathbf{a}\left(\hat{\theta}_q, \Delta f_p\right) + \frac{\beta_{q,p}^{(k)}}{\rho_y}\right|^2 \\ &\text{s.t.} \quad \left|y_{q,p}\right|^2 \le \eta^{(k+1)}, q = 1, 2, \cdots, Q, p = 1, 2, 3. \end{aligned} \tag{7.37}$$

Obviously, the solution $y_{q,p}^{(k+1)}$ is as follows:

$$y_{q,p}^{(k+1)} = \begin{cases} \sqrt{\eta^{(k+1)}}\frac{\tilde{y}_{q,p}}{|\tilde{y}_{q,p}|}, |\tilde{y}_{q,p}| > \sqrt{\eta^{(k+1)}}, \\ \tilde{y}_{q,p}, |\tilde{y}_{q,p}| \le \sqrt{\eta^{(k+1)}}. \end{cases} \tag{7.38}$$

where $\tilde{y}_{q,p} = \mathbf{w}^{(k+1)\dagger}\mathbf{a}\left(\hat{\theta}_q, \Delta f_p\right) - \beta_{q,p}^{(k)}\big/\rho_y$.

By substituting (7.38) into (7.36), we can obtain an optimization problem that only depends on η:

$$\min_{\eta} \ \eta + \frac{\rho_y}{2}\sum_{q=1}^{Q}\sum_{p=1}^{3} v_{q,p}(\sqrt{\eta} - |\tilde{y}_{q,p}|)^2 \tag{7.39}$$

where $v_{q,p} = \begin{cases} 1, |\tilde{y}_{q,p}| > \sqrt{\eta}, \\ 0, |\tilde{y}_{q,p}| \le \sqrt{\eta}. \end{cases}$

To solve (7.39), we sort $|\tilde{y}_{q,p}|$ in ascending order and remove duplicate items. Then, a new sequence can be obtained: $z_1, \cdots, z_N$ $(N \leq (3Q-1))$, and they satisfy $z_1 < \cdots < z_N$. Combining the above discussion, we can define an index set $\Omega_n = \{(\hat{q},\hat{p}) \,||\tilde{y}_{\hat{q},\hat{p}}| \geq z_n, \hat{q} = 1,2,\cdots,Q, \hat{p} = 1,2,3\}$.

Letting $l = \sqrt{\eta}$, (7.39) can be rewritten as follows:

$$G_n(l) = \begin{cases} \tilde{a}_n \left(l - \tilde{b}_n\right)^2 + \tilde{c}_n, l \in [z_{n-1}, z_n], n \in [2, N], \\ \tilde{a}_1 \left(l - \tilde{b}_1\right)^2 + \tilde{c}_1, l < z_1, n = 1, \end{cases} \tag{7.40}$$

where $\chi = \sum\limits_{(q,p)\in\Omega_n} 1$, $\gamma = \sum\limits_{(q,p)\in\Omega_n} |\tilde{y}_{q,p}|$, $\tilde{b}_n = \rho_y\gamma/(2+\rho_y\chi)$, $\tilde{a}_n = 1 + \frac{\rho_y}{2}\chi$, $\tilde{c}_n = -\frac{\rho_y^2\gamma^2}{4+2\rho_y\chi} + \frac{\rho_y}{2}\sum\limits_{(q,p)\in\Omega_n} |\tilde{y}_{q,p}|^2$.

It is evident from (7.40) that $G_n(l)$ is a piecewise function with one variable that can be minimized to obtain its minimum value. Then, we can get $\eta^{(k+1)}$ according to $l = \sqrt{\eta}$.

7.4.3 Step 3: update $y_0^{(k+1)}$

Simplifying the problem in (7.20) and getting rid of the irrelevant terms [25,26], we can obtain $y_0^{(k+1)}$ by solving the following problem:

$$\begin{aligned} \min_{y_0} \quad & \left| y_0 - \mathbf{w}^{(k+1)\dagger}\mathbf{a}(\theta_0, 0) + \frac{\alpha^{(k)}}{\rho_0} \right|^2 \\ \text{s.t.} \quad & |y_0|^2 \geq P_l. \end{aligned} \tag{7.41}$$

And the solution to the objective function is

$$\tilde{y}_0 = \mathbf{w}^{(k+1)\dagger}\mathbf{a}(\theta_0, 0) - \alpha^{(k)} \Big/ \rho_0. \tag{7.42}$$

Hence, the solution $y_0^{(k+1)}$ is

$$y_0^{(k+1)} = \begin{cases} \tilde{y}_0, |\tilde{y}_0| \geq \sqrt{P_l}, \\ \sqrt{P_l}\tilde{y}_0/|\tilde{y}_0|, |\tilde{y}_0| < \sqrt{P_l}. \end{cases} \tag{7.43}$$

Iterative computation steps 1–4 continue until the exit conditions are met or the maximum number of iteration K_{max} is reached. The exit conditions are set as $\Delta r_y^{(k+1)} \leq \varsigma_y$ and $\Delta r_0^{(k+1)} \leq \varsigma_0$, where ς_y and ς_0 are the maximum tolerance errors of nulling and mainlobe regions, respectively, and the residuals $\Delta r_y^{(k+1)}$ and $\Delta r_0^{(k+1)}$ are defined as,

$$\Delta r_0^{(k+1)} = \left| y_0^{(k+1)} - \mathbf{w}^{(k+1)\dagger}\mathbf{a}(\theta_0, 0) \right|, \tag{7.44}$$

$$\Delta r_y^{(k+1)} = \left\| y_{q,p}^{(k+1)} - \mathbf{w}^{(k+1)\dagger}\mathbf{a}(\theta_q, \Delta f_p) \right\|_\infty. \tag{7.45}$$

Finally, the optimal solution to the problem $\mathscr{P}$ can be obtained, and the related algorithm is summarized in Algorithm 7.1.

Algorithm 7.1 AO algorithm for solving problem $\mathscr{P}$

Input: $f_c, B, T, F_s, P_l, M, d_m, \theta_0, \hat{\theta}_q, \rho_0, \rho_y, K_{max}, \varsigma_0, \varsigma_y$;
Output: $\mathbf{w}$;

```
Initialize w^(0), y_{q,p}^(0), y_0^(0), α^(0), β_{q,p}^(0);
For k = 1, ··· , K_max
    Step 1: update w^(k+1) per (7.29), (7.35);
    Step 2: update y_{q,p}^(k+1), η^(k+1) per (7.38), (7.40);
    Step 3: update y_0^(k+1) per (7.43);
    Step 4: update α^(k+1), β_{q,p}^(k+1) per (7.21), (7.22);
    If Δr_y^(k+1) ≤ ς_y and Δr_0^(k+1) ≤ ς_0
        Break;
    End If;
End For;
Obtain w = w^(k+1);
```

7.4.4 Performance analysis

We analyze the computational complexity of the AO algorithm in one iteration. In the process of updating $\mathbf{w}^{(k+1)}$, considering that $\mathbf{X}$ is a matrix determined before optimization, the computational complexity can be greatly reduced and is mainly reflected in solving (7.32), so the computational complexity of this step is $\mathcal{O}\left(M^{2.373}\right)$. And the computational complexity of of $\eta^{(k+1)}$ and $y_{q,p}^{(k+1)}$ is mainly related to solving (7.38) and (7.40), with a specific value of $\mathcal{O}\left(3QM+3Q\right)$. As to updating $y_0^{(k+1)}$, (7.42) plays a decisive role in determining the computational complexity, with the computational complexity of $\mathcal{O}\left(M\right)$. In addition, the computational complexity of updating $\alpha^{(k+1)}$ and $\beta_{q,p}^{(k+1)}$ is $\mathcal{O}\left(3QM+M\right)$.

In the ADMM framework, ensuring that non-convex problem $\mathscr{P}$ converges to a stationary point [27,28] relies on setting appropriate penalty parameters ρ_0 and ρ_y. To emphasize this issue, we analyze the convergence of the AO algorithm under different penalty parameters in the following simulations.

7.5 CA algorithm

In this section, a robust iterative algorithm CA is employed to tackle the non-convex problem $\mathscr{P}$, which approximates the non-convex constraints using their convex upper bound functions.

Firstly, problem $\mathscr{P}$ can be rewritten as

$$\left\{\begin{array}{ll}\min\limits_{\mathbf{w},\eta} & \eta \\ \text{s.t.} & \mathbf{w}^{\dagger}\mathbf{R}_{q,p}\mathbf{w} \leq \eta, q=1,2,\cdots,Q, p=1,2,3, \\ & \mathbf{w}^{\dagger}\mathbf{R}_0\mathbf{w} \geq P_l, \\ & \mathbf{w}^{\dagger}\mathbf{w} \leq 1, \\ & \mathbf{w}^{\dagger}\mathbf{w} \geq 1,\end{array}\right. \tag{7.46}$$

where $\mathbf{R}_{q,p} = \mathbf{a}\left(\hat{\theta}_q, \Delta f_p\right)\mathbf{a}^\dagger\left(\hat{\theta}_q, \Delta f_p\right), \hat{\theta}_q \in \Omega_Q, \Delta f_p \in \Omega_3, \mathbf{R}_0 = \mathbf{a}(\theta_0, 0)\,\mathbf{a}^\dagger(\theta_0, 0)$.

Assuming we have obtained $\mathbf{w}^{(k-1)}$ in the $(k-1)$th iteration, the next step is to approximate the non-convex constraints with their corresponding convex upper bound functions in the kth iteration.

7.5.1 Approximate mainlobe level constraint

As can be seen from the optimization problem $\mathscr{P}$, the mainlobe level constraint is applied only to a desired direction θ_0. Hence, we express the mainlobe level constraint in the following form without losing degrees of freedom (DoFs) [29–31],

$$\begin{aligned}&\mathrm{Re}\left\{\mathbf{w}^\dagger \mathbf{a}(\theta_0, 0)\right\} \geq \sqrt{P_l},\\ &\mathrm{Im}\left\{\mathbf{w}^\dagger \mathbf{a}(\theta_0, 0)\right\} = 0,\end{aligned} \tag{7.47}$$

both of which are linear functions. However, it is worth noting that when considering the wide mainlobe [31], some DoFs are lost in the imaginary part constraint.

7.5.2 Approximate white noise gain constraint

The concave white noise gain constraint $\mathbf{w}^\dagger\mathbf{w} \geq 1$ can be deformed into the form $1 - S(\mathbf{w}) \leq 0$, where $S(\mathbf{w}) = \mathbf{w}^\dagger\mathbf{w}$ is a convex function with multiple variables in quadratic form. According to the Taylor expansion formula [32,33], local linearization of $S(\mathbf{w})$ is given by

$$S(\mathbf{w}) = S\left(\mathbf{w}^{(k-1)}\right) + \mathrm{Re}\left\{\nabla^\dagger S\left(\mathbf{w}^{(k-1)}\right)\left(\mathbf{w} - \mathbf{w}^{(k-1)}\right)\right\} + o\left(\mathbf{w} - \mathbf{w}^{(k-1)}\right), \tag{7.48}$$

where $\nabla S(\mathbf{w}) = 2\mathbf{w}$, $o(\cdot)$ is defined as a positive infinitely small quantity.

Hence, the convex upper bound function of the concave white noise gain constraint can be represented as

$$1 - \mathbf{w}^{(k-1)\dagger}\mathbf{w}^{(k-1)} - \mathrm{Re}\left\{2\mathbf{w}^{(k-1)\dagger}\left(\mathbf{w} - \mathbf{w}^{(k-1)}\right)\right\} \leq 0, \tag{7.49}$$

And (7.49) can be deformed as

$$1 + \mathbf{w}^{(k-1)\dagger}\mathbf{w}^{(k-1)} - 2\mathrm{Re}\left\{\mathbf{w}^{(k-1)\dagger}\mathbf{w}\right\} \leq 0. \tag{7.50}$$

Combining with the constraints $\mathbf{w}^\dagger\mathbf{w} \leq 1$ and $\mathbf{w}^{(k-1)\dagger}\mathbf{w}^{(k-1)} = 1$, we can obtain the following inequality,

$$\begin{aligned}\left\|\mathbf{w}^{(k-1)} - \mathbf{w}\right\|^2 &= \mathbf{w}^\dagger\mathbf{w} + \mathbf{w}^{(k-1)\dagger}\mathbf{w}^{(k-1)} - 2\mathrm{Re}\left\{\mathbf{w}^{(k-1)\dagger}\mathbf{w}\right\}\\ &\leq 1 + \mathbf{w}^{(k-1)\dagger}\mathbf{w}^{(k-1)} - 2\mathrm{Re}\left\{\mathbf{w}^{(k-1)\dagger}\mathbf{w}\right\}\\ &\leq 0.\end{aligned} \tag{7.51}$$

It is obvious from (7.51) that the only solution is $\mathbf{w} = \mathbf{w}^{(k-1)}$. Specifically, the white noise gain constraint of the problem $\mathscr{P}$ in (7.15) restricts the solution space to a complex spherical space. Hence, its solution remains the initial feasible solution $\mathbf{w}^{(0)}$

after the first approximation, which indicates that the objective function does not exhibit a decreasing trend with increasing iterations.

To address this issue, we introduce a non-negative slack variable u in the fourth constraint of problem $\mathscr{P}$, minimizing the norm of u alongside the original cost [34]. Thus, the optimization problem for the kth iteration can be formulated as

$$\mathscr{P}_1 \begin{cases} \min\limits_{\mathbf{w},\eta,u} \ \eta + \rho_1^{(k)} u + \rho_2 \left\| \mathbf{w} - \mathbf{w}^{(k-1)} \right\|^2 \\ \text{s.t.} \quad \mathbf{w}^{\dagger}\mathbf{R}_{q,p}\mathbf{w} \le \eta, q \in [1, Q], p = 1, 2, 3, \\ \qquad \operatorname{Re}\left\{ \mathbf{w}^{\dagger}\mathbf{a}(\theta_0, 0) \right\} \ge \sqrt{P_l}, \\ \qquad \operatorname{Im}\left\{ \mathbf{w}^{\dagger}\mathbf{a}(\theta_0, 0) \right\} = 0, \\ \qquad \mathbf{w}^{\dagger}\mathbf{w} \le 1, \\ \qquad 1 - u - \mathbf{w}^{(k-1)\dagger}\mathbf{w}^{(k-1)} \\ \qquad - \operatorname{Re}\left\{ 2\mathbf{w}^{(k-1)\dagger}\left(\mathbf{w} - \mathbf{w}^{(k-1)} \right) \right\} \le 0, \\ \qquad u \ge 0, \\ \qquad u \le u^{(k-1)}, \end{cases} \tag{7.52}$$

where the proximal term $\left\| \mathbf{w} - \mathbf{w}^{(k-1)} \right\|^2$ is involved in the objective function to ensure that a unique solution is obtained when CA converges. The last constraint guarantees that u is a non-increasing sequence,

$$\rho_1^{(k)} = \begin{cases} \rho_1^{(k-1)}, u^{(k-1)} \le \xi_1 u^{(k-2)}, \\ \xi_2 \rho_1^{(k-1)}, else, \end{cases} \tag{7.53}$$

$\xi_1 < 1$ and $\xi_2 > 1$ are close to 1, ρ_1 and ρ_2 denote predetermined parameters, and $\rho_1^{(0)}$ can be initialized using a small positive constant. As can be seen from (7.53) that if the inequality $u^{(k-1)} \le \xi_1 u^{(k-2)}$ is not true, ρ_1 will increase, causing u in the problem $\mathscr{P}$ to gradually approach 0 over the iterative process [35]. Then, we can obtain the optimization solution.

The residual in each iteration of the algorithm is defined as

$$\Delta r^{(k)} = \left| \left(\eta'^{(k)} - \eta'^{(k-1)} \right) / \eta'^{(k-1)} \right|, \tag{7.54}$$

where $\eta'^{(k)} = \eta^{(k)} + \rho_1^{(k)} u^{(k)}$ represents the improved objective function that omits the proximal term. The exit condition of the CA algorithm is $\Delta r \le \varsigma$.

Finally, the interior point method (IPM) [19] can be used to solve this convex problem $\mathscr{P}_1$ with a series of SOCP constraints. The convex optimization toolbox CVX in MATLAB [36,37] can be applied, with the total computational complexity of $\mathscr{O}\left(M^3\right)$. The CA method of solving problem $\mathscr{P}$ is summarized in Algorithm 7.2.

7.5.3 Performance analysis

The convergence of the problem in (7.52) is explained in detail in this subsection. $\left\{ \rho_1^{(k)} \right\}_{k=1}^{\infty}$ is a non-decreasing sequence due to (7.53). Therefore, we prove the convergence in two cases, one in which $\rho_1^{(k)}$ is bounded (i.e., there exists $k \ge \tilde{k}$ such

Algorithm 7.2 CA algorithm for solving problem $\mathscr{P}$

Input: $f_c, B, T, F_s, P_l, M, d_m, \theta_0, \hat{\theta}_q, \mathbf{w}^{(0)}, \rho_1^{(0)}, \rho_2, \varsigma, \xi_1, \xi_2$;
Output: $\mathbf{w}$;

1: $k = 0$
2: While $\Delta r > \varsigma$
3: $k = k + 1$;
4: Get the solution $\rho_1^{(k)}$ per (7.53);
5: Get the solution $\mathbf{w}^{(k)}$ and $u^{(k)}$ of the convex problem per (7.52);
6: Get the residual $\Delta r^{(k)}$;
7: If $\Delta r^{(k)} \leq \varsigma$
8: Break;
9: Else
10: Go to step (3);
11: End If;
12: End While;
13: Obtain $\mathbf{w} = \mathbf{w}^{(k)}$;

that $\rho_1^{(k)}$ is a constant), and the other in which $\rho_1^{(k)}$ tends to infinity as the number of iterations k increases. The specific proof process is as follows.

1) *Bounded penalty parameter case:* Let $\left(\hat{\mathbf{w}}^T, \hat{\eta}, \hat{u}\right)$ be the optimal solution to problem in (7.52), then

a) The sequence of the improved objective function $\left\{\eta'^{(k)}\right\}_{k=1}^{\infty}$ is non-increasing.

b) The sequence $\left\{\eta'^{(k)}\right\}_{k=1}^{\infty}$ converges to a finite value $\hat{\eta}'$.

c) Assuming that Slater's constraint qualification holds at each iteration and $u \to 0$ as $k \to \infty$, the limit point $\left(\hat{\mathbf{w}}^T, \hat{\eta}\right)$ to $\left(\mathbf{w}^{(k)T}, \eta^{(k)}\right)$ obtained by using Algorithm 7.2 is a Karush-Kuhn-Tucker (KKT) point for the problem in (7.15) according to [34].

The proof is as follows. Since the penalty parameter $\rho_1^{(k)}$ is bounded, we have

$$\lim_{k\to\infty} \rho_1^{(k)} = \rho_1^{(k-1)} = \hat{\rho}_1, \tag{7.55}$$

Considering that $\left(\mathbf{w}^{(k-1)T}, \eta^{(k-1)}, u^{(k-1)}\right)$ is a feasible solution of the kth iteration of the convex optimization problem in (7.52), and its optimal solution $\left(\mathbf{w}^{(k)T}, \eta^{(k)}, u^{(k)}\right)$ can be found in polynomial time. Then, we can obtain

$$\begin{aligned}
\eta'^{(k)} &= \eta^{(k)} + \rho_1^{(k)} u^{(k)} \\
&\leq \eta^{(k)} + \rho_1^{(k)} u^{(k)} + \rho_2 \left\| \mathbf{w}^{(k)} - \mathbf{w}^{(k-1)} \right\|^2 \\
&\leq \eta^{(k-1)} + \rho_1^{(k)} u^{(k-1)} + \rho_2 \left\| \mathbf{w}^{(k-1)} - \mathbf{w}^{(k-1)} \right\|^2 \\
&= \eta^{(k-1)} + \rho_1^{(k)} u^{(k-1)} \\
&= \eta^{(k-1)} + \rho_1^{(k-1)} u^{(k-1)} \\
&= \eta'^{(k-1)}
\end{aligned} \tag{7.56}$$

where $\left\|\mathbf{w}^{(k)}-\mathbf{w}^{(k-1)}\right\|^2 \geq 0$,

Since $\rho_1^{(k)}u^{(k)} \geq 0$, we can easily get $\eta^{(k)}+\rho_1^{(k)}u^{(k)} \geq \eta^{(k)}$. According to the constraints $\mathbf{w}^{\dagger}\mathbf{R}_{q,p}\mathbf{w} \leq \eta, q=1,2,\cdots,Q, p=1,2=3$ and $\mathbf{w}^{\dagger}\mathbf{w}=1$ in problem (7.52), we have,

$$\begin{aligned}\eta &\geq \max_{q,p}\ \mathbf{w}^{\dagger}\mathbf{R}_{q,p}\mathbf{w}\\ &= \mathbf{w}^{\dagger}\mathbf{R}_{\hat{q},\hat{p}}\mathbf{w}\\ &\geq \lambda_{\max}\left(\mathbf{R}_{\hat{q},\hat{p}}\right)\end{aligned} \tag{7.57}$$

where $(\hat{q},\hat{p})$ denotes the index corresponding to the maximum null level value, $\lambda_{\max}\left(\mathbf{R}_{\hat{q},\hat{p}}\right)$ is the maximum eigenvalue of $\mathbf{R}_{\hat{q},\hat{p}}$. The theory of eigenvalue decomposition and the constraint $\mathbf{w}^{\dagger}\mathbf{w}=1$ are the basis of the last inequality in (7.57).

From (7.55) to (7.57), we can conclude that the improved objective function is a monotonically decreasing sequence with a lower bound, which converges to a finite value.

2) *Unbounded penalty parameter case:* Since $\lim_{k\to\infty}\rho_1^{(k)}=\infty$, we can get the following conclusions:

a) The objective function depends on u, and is minimized when $u=0$.

b) The optimization problem in this case is convergent without null depth control.

The proof is as follows. Since $\lim_{k\to\infty}\rho_1^{(k)}=\infty$ occurs in this case, the optimization problem (7.52) at kth iteration can be simplified as

$$\begin{cases}\min\limits_{\mathbf{w},\eta,u}\ \rho_1^{(k)}u\\ \text{s.t.}\quad \mathbf{w}^{\dagger}\mathbf{R}_{q,p}\mathbf{w}\leq\eta, q\in[1,Q], p=1,2,3,\\ \qquad \operatorname{Re}\left\{\mathbf{w}^{\dagger}\mathbf{a}(\theta_0,0)\right\}\geq\sqrt{P_l},\\ \qquad \operatorname{Im}\left\{\mathbf{w}^{\dagger}\mathbf{a}(\theta_0,0)\right\}=0,\\ \qquad \mathbf{w}^{\dagger}\mathbf{w}\leq 1,\\ \qquad 1-u-\mathbf{w}^{(k-1)\dagger}\mathbf{w}^{(k-1)}\\ \qquad -\operatorname{Re}\left\{2\mathbf{w}^{(k-1)\dagger}\left(\mathbf{w}-\mathbf{w}^{(k-1)}\right)\right\}\leq 0,\\ \qquad u\geq 0,\\ \qquad u\leq u^{(k-1)},\end{cases} \tag{7.58}$$

Obviously, the minimum value of the objective function in the above optimization problem is 0, in which case $u=0$ and it satisfies the relevant constraints. The solution to $\left(\mathbf{w}^T,\eta\right)$ only needs to satisfy the following multi-constraint problem:

$$\begin{cases}\text{find}\quad \mathbf{w},\eta\\ \text{s.t.}\quad \mathbf{w}^{\dagger}\mathbf{R}_{q,p}\mathbf{w}\leq\eta, q\in[1,Q], p=1,2,3,\\ \qquad \operatorname{Re}\left\{\mathbf{w}^{\dagger}\mathbf{a}(\theta_0,0)\right\}\geq\sqrt{P_l},\\ \qquad \operatorname{Im}\left\{\mathbf{w}^{\dagger}\mathbf{a}(\theta_0,0)\right\}=0,\\ \qquad \mathbf{w}^{\dagger}\mathbf{w}\leq 1,\\ \qquad 1-u-\mathbf{w}^{(k-1)\dagger}\mathbf{w}^{(k-1)}\\ \qquad -\operatorname{Re}\left\{2\mathbf{w}^{(k-1)\dagger}\left(\mathbf{w}-\mathbf{w}^{(k-1)}\right)\right\}\leq 0.\end{cases} \tag{7.59}$$

The feasible solution to the above problem is readily available (i.e., $\mathbf{w} = \mathbf{a}(\theta_0, 0)/\sqrt{M}$), because it is a multi-constrained convex optimization problem without objective function. However, in this unbounded case, we can only guarantee the convergence of the problem (7.52), but the null depth may not necessarily be low. This is because the objective function only includes u, without optimizing the null level parameter η.

7.6 Numerical results

This section verifies the proposed time-domain wideband beamforming strategy by numerical simulations. The first part of this section describes the rationality of the wideband DBF in comparison with traditional narrowband beamformers. The second part illustrates the validity of the proposed design model through the comparison of different design criteria. Finally, the performance of the AO algorithm and CA algorithm is compared in convergence, computational efficiency, and beampattern synthesis. In all simulations, a ULA with $M = 16$ isotropic antennas is considered. We set $f_c = 1$ GHz and $T = 40$ μs.

7.6.1 Structure rationality analysis

To validate the proposed wideband DBF based on ITDF effectively reduces the aperture fill time, the wideband beampattern results of the traditional narrowband beamformer, the wideband DBF based on FDF (refer to [13,38]), and the proposed wideband DBF are compared.

Let the bandwidth of the received signal be $B = 100$ MHz, which means the highest frequency $f_{\max} = 1.05$ GHz and the lowest frequency $f_{\min} = 0.95$ GHz. Other parameters are set as follows: $\theta_0 = 45°$, $d = c/2f_{\max}$, $F_s = 4B$, $\mathbf{w} = \mathbf{a}(\theta_0, 0)/\sqrt{M}$. Considering the device performance and cost factors, mainstream analog-to-digital (A/D) converters typically have sampling rates below 2 GHz [40], which generally meet the sampling rate requirements of the proposed method. The spatial angular range shown in this part is set to $[0°, 90°]$, with an interval of 0.02°, to facilitate the viewing of the aperture effect.

From Figure 7.2(a) and (b), it is evident that traditional methods using phase shifters to change the beam direction cannot meet the requirements of wideband phased array radar applications due to the angle error of $f_{\max}$ and $f_{\min}$ which cannot be neglected. However, as illustrated in Figure 7.2(c)–(f), both the wideband DBF based on FDF and the proposed wideband DBF based on ITDF can effectively reduce the aperture fill time and improve the beam pointing accuracy. Specifically, based on the aperture effect of phased arrays and the maximum beam offset angle [39], the proposed method can tolerate the generated angular errors within different frequency bins when the sampling rate F_s meets

$$F_s \geq BM/0.886\,(M - 1). \tag{7.60}$$

Therefore, this work proposes a wideband DBF based on ITDF at an appropriate sampling rate, thereby reducing unnecessary computations introduced by the FDF.

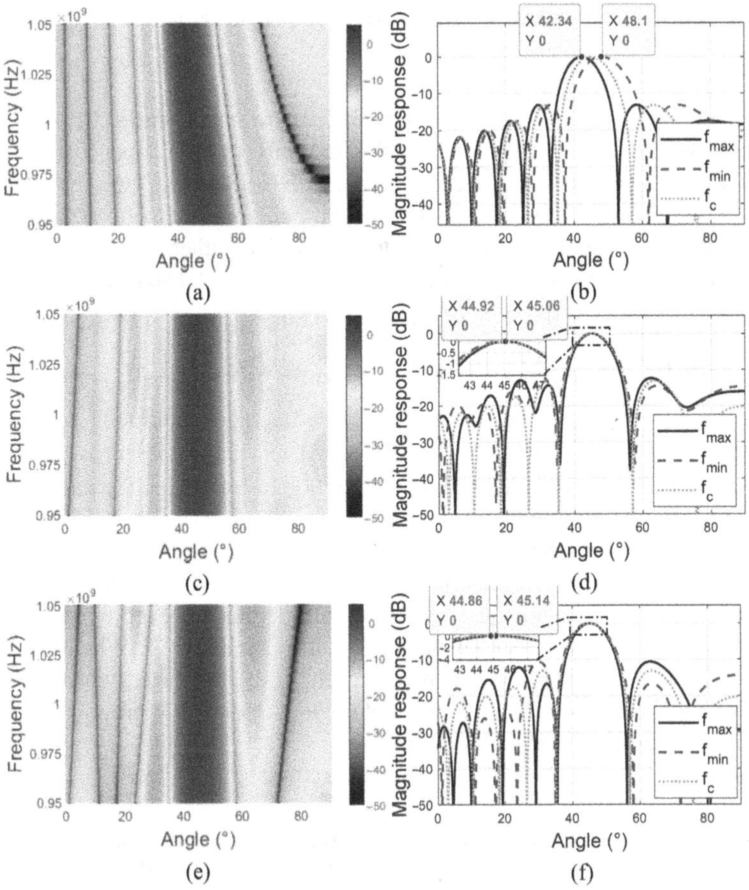

Figure 7.2 Comparison of structural rationality: (a) normalized 2-D beampattern of the traditional narrowband beamformer, (b) the normalized beampattern slices at $f_{\max}$, $f_{\min}$, and f_c of (a), (c) normalized 2-D beampattern of the wideband DBF based on FDF, (d) the normalized beampattern slices at $f_{\max}$, $f_{\min}$, and f_c of (c), (e) normalized 2-D beampattern of the proposed wideband DBF with low sampling rate, (f) the normalized beampattern slices at $f_{\max}$, $f_{\min}$, and f_c of (e).

7.6.2 Model validity analysis

In this part, comparing the null depth at different bandwidths is used to illustrate the validity of the proposed design model. In addition to the case of wideband beamforming without optimization (WBBO, $\mathbf{w} = \mathbf{a}(\theta_0, 0)/\sqrt{M}$), we also compare the narrowband beamforming design criteria (i.e., the case of the proposed design model considering only the null level constraint at the center frequency f_c (NLCCF)).

The fractional bandwidth (FB) mentioned in [41] is defined as

$$B_F = B/f_c. \tag{7.61}$$

And theoretically, $0 < B_F \leq 0.01$ stands for narrowband radar, $0.01 < B_F \leq 0.25$ means wideband radar, and $B_F > 0.25$ represents ultra-wideband radar. Since this work focuses primarily on wideband beamforming, we consider different FBs between 0.02 and 0.30, with an interval of 0.02. In addition to the parameter settings mentioned earlier, other parameters are set as follows: $\theta_0 = 20°$, $\hat{\theta}_q = [-90°, -70°]$, $d = c/(2 \times 1.15 f_c)$, $F_s = 10B$, $P_l = 10\log_{10}\left(20\log_{10}\left(\mathbf{a}(\theta_0, 0)^\dagger \mathbf{a}(\theta_0, 0)\big/\sqrt{M}\right) - 0.6\right)$, $K_{\max} = 2 \times 10^4$, $\varsigma_0 = \varsigma_y = 10^{-4}$, $\mathbf{w}^{(0)} = \mathbf{a}(\theta_0, 0)/\sqrt{M}$, $y_0^{(0)} = \mathbf{w}^{(0)\dagger}\mathbf{a}(\theta_0, 0)$, $y_{q,p}^{(0)} = \mathbf{w}^{(0)\dagger}\mathbf{a}\left(\hat{\theta}_q, \Delta f_p\right)$, ρ_0 and ρ_y are set to appropriate values between 0.1 and 50, the Lagrange multipliers $\alpha^{(0)}$ and $\beta_{q,p}^{(0)}$ are initialized randomly between 0 and 1, $\varsigma = 10^{-4}$, $\xi_1 = 0.999$, $\xi_2 = 1.001$, and $\rho_1^{(0)} = 10^{-4}$. The angular range from $-90°$ to $90°$ is considered for this simulation and the angular sampling interval is 0.5°.

Figure 7.3 reflects the null depths, obtained by WBBO, NLCCF, and the proposed design model, with the increase of FBs. As can be seen from the shaded area in Figure 7.3 (the wideband case), the null depths achieved by the proposed design model are significantly deeper than that of the NLCCF. When the FB fulfills $B_F \in (0.01, 0.14]$, the null depth can reach 50 dB. However, when $B_F \geq 0.26$ (the ultra-wideband case), the nulling is not obvious. Hence, our proposed method is more suitable for wideband beamforming.

Next, the comparison of the beampattern corresponding to $B_F = 0.10$ will be analyzed with emphasis.

Figure 7.4 illustrates the beampattern performance corresponding to the wideband signal bandwidth range of $f_{\max} = 1.05$ GHz, $f_{\min} = 0.95$ GHz, and $f_c = 1$ GHz

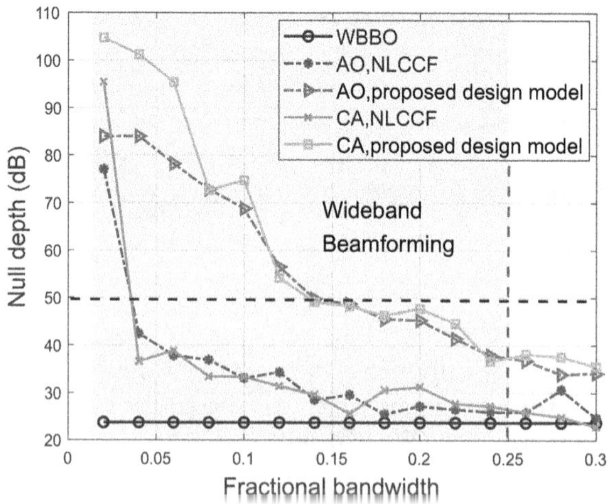

Figure 7.3 Comparison of model validity: the null depth versus FB

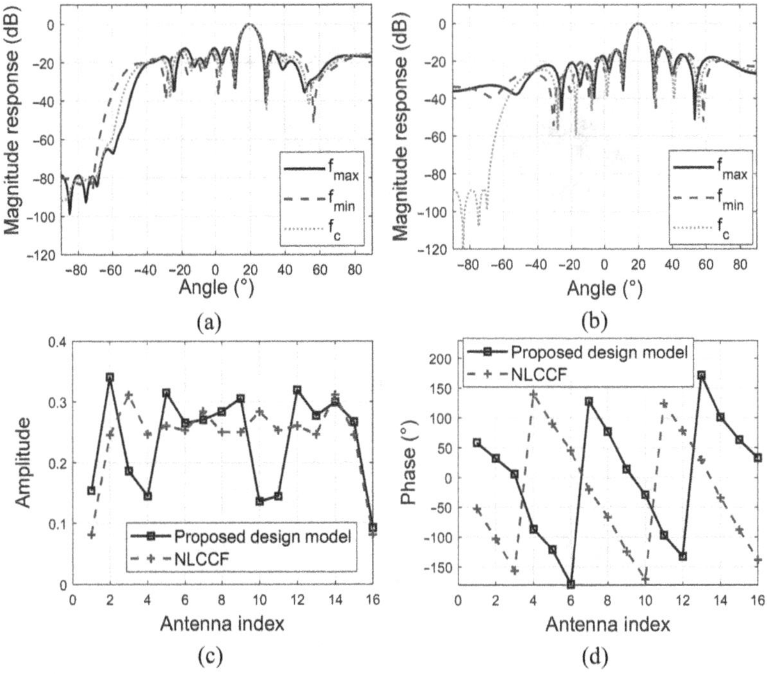

Figure 7.4 (a) The normalized beampattern sectional drawing at $f_{\max}$, $f_{\min}$, and f_c via AO algorithm of the proposed design model, (b) the normalized beampattern sectional drawing at $f_{\max}$, $f_{\min}$, and f_c via AO algorithm of the NLCCF model, (c) amplitudes of the excitations, (d) phases of the excitations.

under two different design models. From Figure 7.4(a), it can be observed that employing the proposed design model generates 80 dB of the null depth at these three frequency bins, whereas Figure 7.4(b) shows that the NLCCF model can only produce effective nulling in the beampattern corresponding to the center frequency f_c. Additionally, the amplitude and phase of the array element excitations are depicted in Figure 7.4(c) and (d).

Furthermore, the normalized beampatterns of wideband beamforming corresponding to these three models are displayed in Figure 7.5. Figure 7.5(c) and (d) illustrates the angle-frequency beampattern of wideband beamforming with the proposed design model via the AO algorithm. It can be clearly observed that a relatively consistent null depth throughout the whole bandwidth can be generated by the proposed design model. In fact, the null depth up to 70 dB with respect to the mainlobe level can be observed. When the null constraints are considered only at the null interval at the center frequency f_c, the wideband beampattern obtained through the AO algorithm is as presented in Figure 7.5(e) and (f). The wideband beamforming produces low null depth only around the central frequency bandwidth (marked with black dashed rectangle in Figure 7.5(f)), with nulling in higher or lower frequency parts of the region $\hat{\theta}_q = [-90^\circ, -70^\circ]$ not prominent. In particular, the null

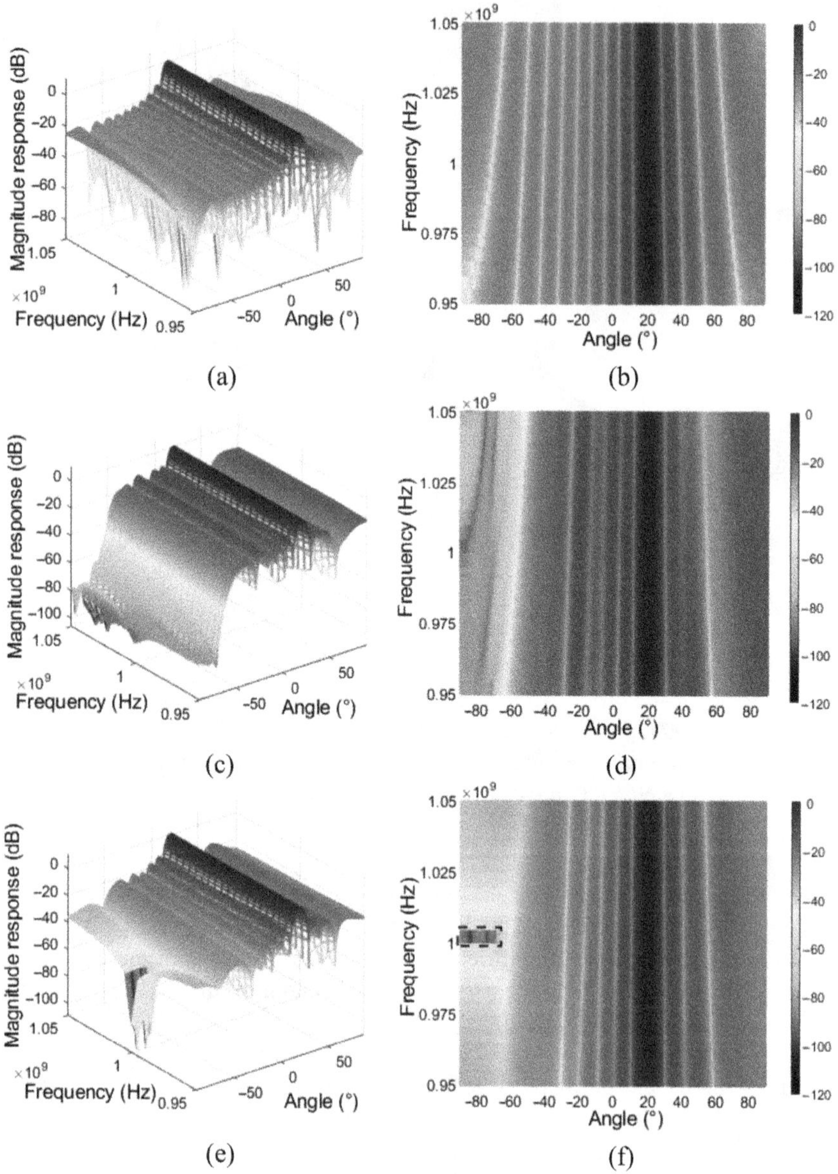

Figure 7.5 The normalized beampattern of wideband beamforming: (a) 3-D exhibitions of synthesized beampattern with WBBO, (b) 2-D exhibitions of (a), (c) 3-D exhibitions of synthesized beampattern with the proposed design model via AO, (d) 2-D exhibitions of (c), (e) 3-D exhibitions of synthesized beampattern with NLCCF via AO, and (f) 2-D exhibitions of (e).

depth is 33.08 dB. Therefore, these performance behaviors highlight that the proposed design model with null level constraints at three special frequency bins can effectively achieve a deep nulling along with the whole frequency band.

7.6.3 Algorithm performance comparison

In this part, different penalty parameters are set to compare the performance of the AO and CA algorithms. To illustrate the problem more fully, we consider both isotropic uniform linear array (IULA) and nonisotropic linear random array (NLRA).

7.6.3.1 IULA case

In the simulation of this case, the parameter settings are as follows: $\theta_0 = 20°$, $\hat{\theta}_q = [-50°, -40°]$, $P_l = 10\log_{10}\left(20\log_{10}\left(\mathbf{a}(\theta_0,0)^\dagger \mathbf{a}(\theta_0,0)\big/\sqrt{M}\right) - 0.5\right)$, $B = 0.06f_c$, $F_s = 10B$, $d = \lambda_{\max}/2$ ($\lambda_{\max} = c/f_{\max}$ is the wavelength of the highest frequency $f_{\max} = 1.03\,\text{GHz}$), $\mathbf{w}^{(0)} = \mathbf{a}(\theta_0,0)/\sqrt{M}$. The angular space from $-90°$ to $90°$ are considered for this simulation with an angular sampling interval is $0.5°$. Additionally, the parameter settings introduced in the AO algorithm are as follows: $\varsigma_0 = \varsigma_y = 10^{-4}$, $K_{\max} = 2\times10^4$, $y_m^{(0)} = \mathbf{w}^{(0)^\dagger}\mathbf{a}(\theta_m,0)$, $y_{q,\Delta f}^{(0)} = \mathbf{w}^{(0)^\dagger}\mathbf{a}\left(\hat{\theta}_q,\Delta f\right)$, the Lagrange multipliers $\alpha^{(0)}$ and $\beta_{q,\Delta f}^{(0)}$ are initialized randomly between 0 and 1. In different simulations, the penalty parameters are set as: $\rho_0 = 1$ and $\rho_y = 10$, $\rho_0 = 0.5$ and $\rho_y = 5$, $\rho_0 = \rho_y = 0.01$, $\rho_0 = \rho_y = 10^{-5}$, respectively. The parameter settings introduced in the CA algorithm are as follows: $\varsigma = 10^{-4}$, $\xi_1 = 0.999$, and $\xi_2 = 1.001$. And the penalty parameters are set as: $\rho_1^{(0)} = 10^{-1}$, $\rho_1^{(0)} = 10^{-2}$, $\rho_1^{(0)} = 10^{-3}$, and $\rho_1^{(0)} = 10^{-4}$, respectively.

Figure 7.6(a)–(d) illustrates the convergence performance of the AO algorithm. It can be observed from these figures that the convergence rates of residuals vary with different penalty parameter settings. Specifically, the AO algorithm converges fastest when $\rho_0 = \rho_y = 0.01$, while it fails to converge when $\rho_0 = \rho_y = 10^{-5}$, indicating the inability to satisfy the constraints of this non-convex problem. Furthermore, the convergence of the CA algorithm is depicted in Figure 7.7. The results demonstrate that different penalty parameters affect the convergence rate of the CA algorithm, but it still exhibits convergent.

Table 7.2 summarizes the performance of these two algorithms, from which we can draw the following conclusions:

(a) CA algorithm achieves nearly identical null depths under different penalty parameters, whereas the null depths obtained by the AO algorithm vary with different penalty parameters, indicating its robustness. This is because the CA algorithm has a theoretically proven convergence, making it easier to find good solutions compared with AO algorithm.

(b) Under different penalty parameter settings, each iteration time (in seconds) of the AO algorithm is much faster than that of CA algorithm. This is because the CA algorithm solves the convex problem via the IPM in each iteration, which

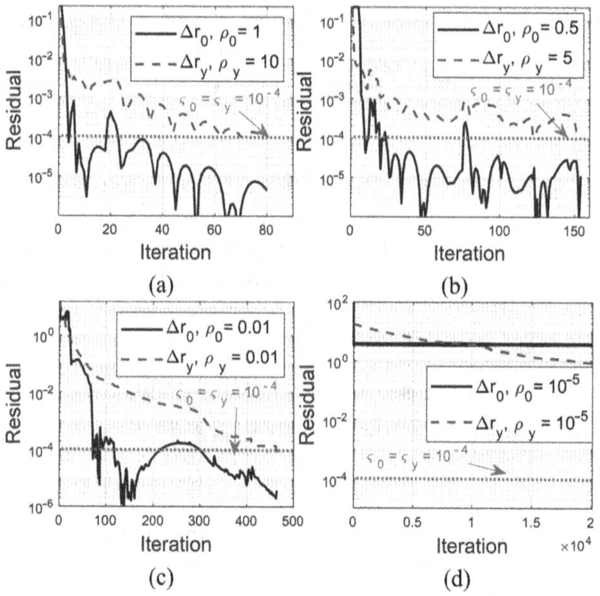

Figure 7.6 The convergence of AO algorithm: (a) residual versus iteration in the case of $\rho_0 = 1$ and $\rho_y = 10$, (b) residual versus iteration in the case of $\rho_0 = 0.5$ and $\rho_y = 5$, (c) residual versus iteration in the case of $\rho_0 = \rho_y = 0.01$, and (d) residual versus iteration in the case of $\rho_0 = \rho_y = 10^{-5}$.

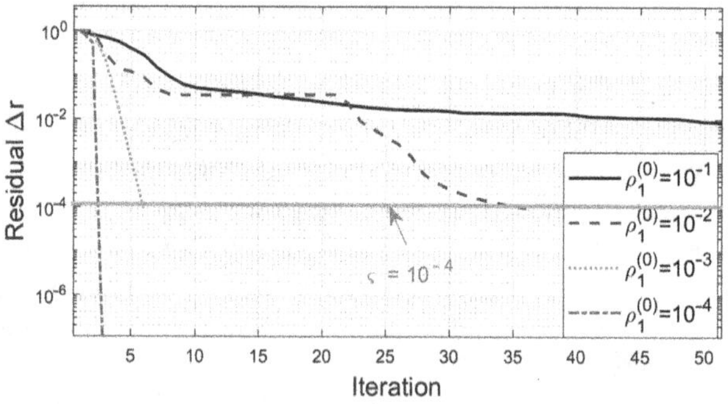

Figure 7.7 The convergence of CA algorithm: residual Δr versus iteration

has high computational complexity. In contrast, the AO algorithm iteratively solves subproblems with relatively lower computational complexity.

(c) The convergence rates of these algorithms are influenced by the penalty parameter values. In the case of convergence, as the penalty parameters increase for both AO and CA algorithms, the weight of the original objective function η diminishes. As a result, the AO algorithm converges faster while the CA algorithm converges slower.

Table 7.1 The performance comparison of AO algorithm and CA algorithm

Algorithm	Penalty parameters	Null depth(dB)	Iteration	Time(s)
AO	$\rho_0 = 1,\ \rho_y = 10$	27.53	80	1.23
	$\rho_0 = 0.5,\ \rho_y = 5$	40.48	154	2.31
	$\rho_0 = \rho_y = 0.01$	61.23	461	7.36
	$\rho_0 = \rho_y = 10^{-5}$	15.63	20000	326.70
CA	$\rho_1^{(0)} = 10^{-1}$	61.31	251	496.03
	$\rho_1^{(0)} = 10^{-2}$	61.32	36	76.62
	$\rho_1^{(0)} = 10^{-3}$	61.33	6	14.74
	$\rho_1^{(0)} = 10^{-4}$	61.33	3	8.26

Table 7.2 Parameters of the nonisotropic linear random array

m	$d(\lambda_1)$	$l\,(\lambda)$	$\xi\,(^\circ)$	m	$d\,(\lambda_1)$	$l\,(\lambda)$	$\xi\,(^\circ)$
1	0.00	0.23	−2.76	9	4.09	0.30	−1.13
2	0.50	0.24	1.68	10	4.55	0.22	4.16
3	1.02	0.25	3.44	11	5.06	0.27	−4.99
4	1.51	0.25	−1.56	12	5.59	0.23	−0.38
5	2.04	0.29	2.81	13	6.08	0.27	−0.76
6	2.56	0.25	1.75	14	6.54	0.27	−0.39
7	3.11	0.29	−4.93	15	7.02	0.21	2.70
8	3.61	0.26	1.02	16	7.49	0.23	−1.78

Hence, we can adapt the AO algorithm for scenarios with high real-time requirements or massive wideband arrays, and CA algorithm for offline scenarios or small wideband arrays.

7.6.3.2 NLRA case

In this case, we consider an NLRA [42] in phased array radar systems, with specific parameters as shown in Table 7.2. The individual pattern for the mth element is given by:

$$g_m(\theta) = \frac{\cos\left[\pi l_m \sin(\theta + \xi_m)\right] - \cos(\pi l_m)}{\cos(\theta + \xi_m)} \tag{7.62}$$

where l_m and ξ_m represent the orientation and length of the radiation element, respectively. In addition, $\theta_0 = 20^\circ$, $\theta_q = [-65^\circ, -55^\circ]$, $B = 0.08f_c$, $F_s = 10B$, $\lambda_1 = c/f_{\max}$ is the wavelength of the highest frequency $f_{\max} = 1.04$ GHz, $\mathbf{w}^{(0)} = \mathbf{a}(\theta_0, 0)/\sqrt{M}$, $P_l = 10\log_{10}\left(20\log_{10}\left(\mathbf{a}(\theta_0,0)^\dagger\,(\mathbf{g}(\theta_0)\,\mathbf{a}(\theta_0,0))\big/\sqrt{M}\right) - 0.6\right)$. This simulation considers the angular space from -90° to 90° with an angular sampling interval of 0.5°. In addition to the parameter settings mentioned earlier, the parameter settings introduced in the AO algorithm are as follows: $\varsigma_0 = \varsigma_y = 10^{-4}$, $K_{\max} = 2 \times 10^4$, $y_m^{(0)} = \mathbf{w}^{(0)^\dagger}\mathbf{a}(\theta_m, 0)$, $y_{q,\Delta f}^{(0)} = \mathbf{w}^{(0)^\dagger}\mathbf{a}\left(\hat{\theta}_q, \Delta f\right)$, $\rho_0 = \rho_y = 0.01$, the Lagrange

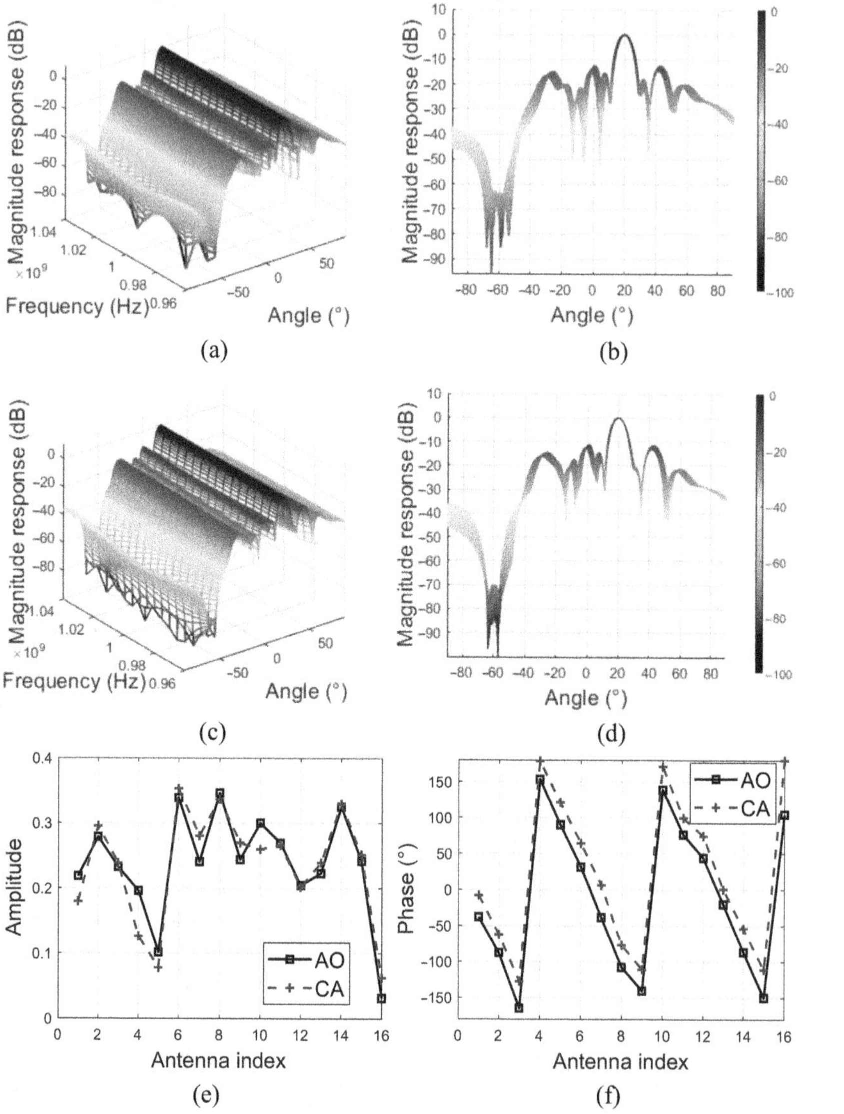

Figure 7.8 The wideband beampattern synthesis comparison in NLRA case: (a) normalized 3-D exhibitions of synthesized beampattern via AO, (b) 2-D exhibitions of (a), (c) normalized 3-D exhibitions of synthesized beampattern via CA, (d) 2-D exhibitions of (c), (e) amplitudes of the excitations, and (f) phases of the excitations.

multipliers $\alpha^{(0)}$ and $\beta^{(0)}_{q,\Delta f}$ are initialized randomly between 0 and 1. And the parameter settings introduced in the CA algorithm are as follows: $\varsigma = 10^{-4}$, $\xi_1 = 0.999$ and $\xi_2 = 1.001$, $\rho_1^{(0)} = 10^{-10}$.

The wideband beampattern synthesis results optimized via AO algorithm and CA algorithm are described in Figure 7.8(a) and (b) and Figure 7.8(c) and (d), respectively. It is evident that both the AO and CA algorithms can achieve a relatively consistent null depth up to 60 dB across the entire bandwidth. The corresponding array element excitation magnitudes and phases are depicted in Figure 7.8(e) and(f). The effectiveness of the proposed methods can be clearly observed when applied to an NLRA.

7.7 Conclusion

In this chapter, we proposed a novel time-domain wideband DBF method that utilizes one set of amplitude-phase weighting and ITDF to control the beam shape over the whole bandwidth. To suppress sidelobe interferences in a spectrally crowded environment, we synthesized low nulls over the whole bandwidth via minimizing the maximal null level at three specially chosen frequency bins. Two competitive algorithms were proposed to address the resulting non-convex problem, suitable for different scenarios. Finally, the effectiveness of the proposed WDB was validated through numerical simulations. As the focus of this work is on the theoretical development of digital wideband beamforming, future research will involve the practical implementation of the proposed design using DSP chips or field programmable gate array (FPGA). Other potential researches derived from this work include extending the proposed framework to consider joint design of weighting coefficients and antenna locations to achieve the desired wideband or ultra-wideband beampattern [7].

References

[1] Liu W. and Weiss S. *Wideband Beamforming: Concepts and Techniques*. Chichester: Wiley, 2010.

[2] Cheung C., Shah R., and Parker M. 'Time delay digital beamforming for wideband pulsed radar implementation'. *2013 IEEE International Symposium on Phased Array Systems and Technology*; Waltham, MA, 2013, pp. 448–455.

[3] Chen X., Shu T., Yu K., and Yu W. 'Wideband jamming cancellation for wideband phased array radar: digital beamforming architecture and preliminary results'. *2019 IEEE Radar Conference*; Boston, MA, USA, 2019, pp. 1–6.

[4] Li T. and Wang X. 'Wideband digital beamforming by implementing digital fractional filter at baseband'. *2013 International Conference on Communications, Circuits and Systems*; Chengdu, 2013, pp. 182–185.

[5] Trees H.K.V. *Optimum Array Processing*. New York, NY, USA: Wiley, 2002.

[6] Liu W., Koh C.L., and Weiss S. 'Constrained adaptive broadband beamforming algorithm in frequency domain'. *Processing Workshop Proceedings, 2004 Sensor Array and Multichannel Signal*; Barcelona, Spain, 2004, pp. 94–98.

[7] Hamza S.A. and Amin M.G. 'Sparse array beamforming design for wideband signal models'. *IEEE Transactions on Aerospace and Electronic Systems*. 2021;57(2):1211–1226.

[8] Feng L., Cui G., Yu X., Liu R., and Lu Q. 'Wideband frequency-invariant beamforming with dynamic range ratio constraints'. *Signal Processing*. 2021;181.

[9] Yang X., Li S., Sun Y., Long T., and Sarkar T.K. 'Robust wideband adaptive beamforming with null broadening and constant beamwidth'. *IEEE Transactions on Antennas and Propagation*. 2019;67(8):5380–5389.

[10] Liu Y., Cheng J., Xu K.D., Yang S., Liu Q.H., and Guo Y.J. 'Reducing the number of elements in the synthesis of a broadband linear array with multiple simultaneous frequency-invariant beam patterns'. *IEEE Transactions on Antennas and Propagation*. 2018;66(11):5838–5848.

[11] Liu Y., Zhang L., Zhu C., and Liu Q.H. 'Synthesis of nonuniformly spaced linear arrays with frequency-invariant patterns by the generalized matrix pencil methods'. *IEEE Transactions on Antennas and Propagation*. 2015;63(4):1614–1625.

[12] Gao Y., Jiang D., and Liu M. 'Wideband transmit beamforming using integer-time-delayed and phase-shifted waveforms'. *Electronics Letters*. 2017;53(6):376–378.

[13] Guo Q. and Sun C. 'Time-domain nearfield wideband beamforming based on fractional delay filters'. *2011 IEEE 3rd International Conference on Communication Software and Networks*; Xian, China, 2011, pp. 421–425.

[14] Wang M. and Wu S. 'A time domain beamforming method of UWB pulse array'. *IEEE International Radar Conference*; Arlington, VA, 2005, pp. 697–702.

[15] Aubry A., Maio A.De, Piezzo M., and Farina A. 'Radar waveform design in a spectrally crowded environment via nonconvex quadratic optimization'. *IEEE Transactions on Aerospace and Electronic Systems*. 2014;50(2): 1138–1152.

[16] Tang B., Li J., and Liang J. 'Alternating direction method of multipliers for radar waveform design in spectrally crowded environments'. *Signal Processing*. 2018;142:398–402.

[17] Amin M.G., Wang X., Zhang Y.D., Ahmad F., and Aboutanios E. 'Sparse arrays and sampling for interference mitigation and DOA estimation in GNSS'. *Proceedings of the IEEE*. 2016;104(6):1302–1317.

[18] Zhang X., He Z., Liao B., Yang Y., Zhang J., and Zhang X. 'Flexible array response control via oblique projection'. *IEEE Transactions on Antennas and Propagation*. 2019;67(12):3126–3139.

[19] Boyd S. and Vandenberghe L. *Convex Optimization*. Cambridge, U.K.: Cambridge Univ. Press, 2004.

[20] Lu Q., Liu R., Feng L., Cui G., Zhang Z., and Zhou L. 'Cognitive wideband beamforming for sparse array'. *2020 IEEE Radar Conference*; Florence, Italy, 2020, pp. 1–6.

[21] Gemechu A.Y., Cui G., Yu X., and Kong L. 'Beampattern synthesis with sidelobe control and applications'. *IEEE Transactions on Antennas and Propagation*. 2020;68(1):297–310.

[22] Zhang X., He Z., Zhang X., and Peng W. 'High-performance beampattern synthesis via linear fractional semidefinite relaxation and quasi-convex optimization'. *IEEE Transactions on Antennas and Propagation*. 2018;66(7):3421–3431.

[23] Boyd S., Parikh N., Chu E., Peleato B., and Eckstein J. 'Distributed optimization and statistical learning via the alternating direction method of multipliers'. *Foundations and Trends in Machine Learning*. 2011;3(1):1–122.

[24] Liang J., Fan X., So H.C., and Zhou D. 'Array beampattern synthesis without specifying lobe level masks'. *IEEE Transactions on Antennas and Propagation*. 2020;68(6):4526–4539.

[25] Feng L., Cui G., Yu X., and Kong L. 'Beampattern synthesis via the constrained subarray layout optimization'. *IEEE Transactions on Antennas and Propagation*. 2021;69(1):182–194.

[26] Fan X., Liang J., Zhang Y., So H.C., and Zhao X. 'Shaped power pattern synthesis with minimization of dynamic range ratio'. *IEEE Transactions on Antennas and Propagation*. 2019;67(5):3067–3078.

[27] Hong M., Luo Z., and Razaviyayn M. 'Convergence analysis of alternating direction method of multipliers for a family of nonconvex problems'. *2015 IEEE International Conference on Acoustics, Speech and Signal Processing*; Brisbane, QLD. 2015, pp. 3836–3840.

[28] Wang Y., Yin W., and Zeng J. 'Global convergence of ADMM in nonconvex nonsmooth optimization'. *Journal of Scientific Computing*. 2019;78:29–63.

[29] Isernia T., Iorio P.Di, and Soldovieri F. 'An effective approach for the optimal focusing of array fields subject to arbitrary upper bounds'. *IEEE Transactions on Antennas and Propagation*. 2000;48(12):1837–1847.

[30] Vorobyov S.A., Gershman A.B., and Luo Z. Q. 'Robust adaptive beamforming using worst-case performance optimization: a solution to the signal mismatch problem'. *IEEE Transactions on Signal Processing*. 2003;51(2): 313–324.

[31] Zhang T. and Ser W. 'Robust beampattern synthesis for antenna arrays with mutual coupling effect'. *IEEE Transactions on Antennas and Propagation*. 2011;59(8):2889–2895.

[32] Ding S. 'Study on the order of Taylor expansion with applications'. *2010 International Conference on Educational and Network Technology*, Qinhuangdao, China, 2010, pp. 54–57.

[33] Wang X., Greco M.S., and Gini F. 'Adaptive sparse array beamformer design by regularized complementary antenna switching'. *IEEE Transactions on Signal Processing*. 2021;69:2302–2315.

[34] Yu X., Alhujaili K., Cui G., and Monga V. 'MIMO radar waveform design in the presence of multiple targets and practical constraints'. *IEEE Transactions on Signal Processing*. 2020;68:1974–1989.

[35] Yu X., Cui G., Yang J., Li J., and Kong L. 'Quadratic optimization for unimodular sequence design via an ADPM framework'. *IEEE Transactions on Signal Processing*. 2020;68:3619–3634.

[36] Grant M. and Boyd S. *CVX package[Online]*. 2012. Available from http://www.cvxr.com/cvx.r.
[37] *CVX: MATLAB Software for Disciplined Convex Programming*. San Ramon, CA, USA: CVX Res., Inc., 2012.
[38] Valimaki V. and Laakso T. I. 'Principles of fractional delay filters'. *2000 IEEE International Conference on Acoustics, Speech, and Signal Processing. Proceedings*; Istanbul, Turkey, 2000, pp. 3870–3873.
[39] Gu J., Hu J., and Li C. 'A method saluting the instantaneous bandwidth of broadband phased array radar'. *Shipboard Electronic Countermeasure*. 2019;42(5):95–97.
[40] Analog Devices. *Standard High Speed A/D Converters >20 MSPS[Online]*. 2021. Available from https://www.analog.com/en/parametricsearch/11814#/.
[41] Wicks M., Mokole E., Blunt S., Schneible R., and Amuso V. *Principles of Waveform Diversity and Design*. Raleigh, USA: SciTech Publishing Inc., 2011.
[42] Fuchs B. and Fuchs J.J. 'Optimal narrow beam low sidelobe synthesis for arbitrary arrays'. *IEEE Transactions on Antennas and Propagation*. 2010;58(6):2130–2135.

Chapter 8
Hybrid beamforming design for dual-function radar-communication system

Ziyang Cheng[1], Bowen Wang[1], Zishu He[1] and Bin Liao[2]

Dual-function radar-communication (DFRC), as an integration of radar and communication functionality, has been envisioned as a promising technology for B5G and 6G. Nowadays, by integrating millimeter wave (mmWave) DFRC and massive multiple-input multiple-output (MIMO), the DFRC can achieve high-precision sensing while guaranteeing high-throughput communications. However, implementing mmWave massive MIMO DFRC systems with fully digital beamforming architecture is impractical due to the substantial power consumption and prohibitive hardware cost. As a cost-effective alternative, the DFRC equipped with hybrid beamforming (HBF) architecture is envisioned as a good trade-off between system performance and hardware complexity.

In this chapter, two cases of HBF design for mmWave DFRC systems are studied. Specifically, in the first case, we consider a scenario in which a single-carrier DFRC system communicates with a single multi-antenna user while detecting an extended target in the presence of extended clutters. In the second case, a multi-user and multi-carrier DFRC system is investigated, where this system leverages multi-carrier signals to achieve high-accuracy direction of arrival estimation and ensure high-quality communication services. Two state-of-the-art HBF design algorithms are proposed for the two cases under consideration. Numerical results are provided to validate the effectiveness of the proposed algorithms and superiority of mmWave DFRC with HBF architecture.

8.1 Introduction

The number of wireless devices has been growing dramatically to meet the demand of the Internet of Everything (IoE). Particularly, the upcoming 5G technology seeks to enable various connected services, due to many new applications beyond personal

[1]School of Information and Communication Engineering, University of Electronic Science and Technology of China, China
[2]Guangdong Key Laboratory of Intelligent Information Processing, College of Electronics and Information Engineering, Shenzhen University, China

communications [1]. As a consequence, the radio frequency (RF) spectrum is becoming increasingly congested. As two key applications of RF technology, radar and communication occupy a large portion of the frequency spectrum. Conventionally, they were operated at different frequency bands to avoid mutual interference. In the case of a congested spectrum, nevertheless, it is difficult or even impossible to allocate independent bands for them. Actually, the bands of radar systems have overlaid with those of communication. For example, the S-band, which ranges from 2GHz to 4GHz, is utilized by airborne early warning radars [2], and communication devices including long-term evolution (LTE) and WiMax systems [3]. Besides, weather radars generally operate at C-band from 4GHz to 8GHz, which is also used for 802.11 wireless local area network (WLAN) [4]. In addition, the millimeter wave (mmWave) band has been utilized by automotive radar, high-resolution imaging radar, and communication as well. Therefore, it is critical to develop joint radar and communication (JRC) techniques to achieve communication and radar spectrum sharing (CRSS) [2].

Generally speaking, there exist two popular ways for CRSS. The first one is called radar-communication co-existence (RCC) [5]. In this scenario, both communication devices and radar systems could work independently. However, mutual interference, as a key design metric, should be taken into account for both radar and communication. The second way for CRSS is dual-function radar-communication (DFRC) [6], which provides a unique platform for both radar and communication transmissions. This allows us to realize integrated sensing and communication.

For RCC implementation, a number of spectrally compatible waveform (SCW) methods have been developed [7–10]. In these methods, radar waveforms are designed with desired spectrum nulls, thereby limiting the interfering energy produced by radar systems to a level tolerable for communications. However, RCC do not make efficient use of the spectrum, due to the fact that the communication devices only operate at the frequency bands which the radar does not occupy. Even worse, for narrowband radars, such as search radars, there are not enough degrees of freedoms (DoFs) to form nulls at the frequency bands of communications. Therefore, more efficient co-designs have been investigated in [11–17].

In the early work [11] for RCC design, it is assumed the prior knowledge of channel state information (CSI) on the interference channels between the communication is known to radar. Based on the known CSI, the RCC is achieved by projecting the radar waveform onto the null space of the interference channels. This method is shown to be competitive to the traditional radar waveforms in terms of the target identification capability while guaranteeing the RCC in the same band. However, it may fail once the response of the target falls into the row space of the CSI matrix. To improve the performance for both radar and communication, a joint design of the radar sub-sampling matrix and communication covariance matrix is proposed in [12]. The problem is formulated by minimizing the interference caused to radar subject to power and capacity constraints for co-existence of multiple-input multiple-output (MIMO) communication system and a matrix completion MIMO radar. This method was then extended to the case of signal-dependent clutter in [13]. For the co-existence of pulsed radar and communication systems in the presence of signal-dependent interference, the co-design of radar pulse codes and communication encoding matrix is considered in [16]. The problem is formulated by maximizing the compound rate, i.e., the rate with and without radar interference, subject to radar signal-to-interference-plus-noise ratio (SINR) constraint.

It is worth noting that the aforementioned co-designs for RCC are detrimental to reduced size, hardware cost, and power consumption. To overcome these disadvantages, DFRC systems [18–25] have recently received extensive attention of researchers. For example, similar to the concept of space division multiple access (SDMA) in wireless communication, one DFRC strategy is to perform target detection within the mainlobe of the beampattern and communication within the sidelobe. In [21], the communication function is implemented by controlling the sidelobe level, and different levels signify different symbols. This can be viewed as amplitude shift keying (ASK). To increase the data rate of DFRC, a phase-shift keying (PSK) is achieved by utilizing frequency hopping (FH) codes in [23], where communication symbols are embedded in each FH waveform. Note that these methods assume line-of-sight (LOS) channel. For non-LOS (NLOS) channel, a beamforming scheme is proposed to implement the MIMO radar and downlink multi-user MIMO (MU-MIMO) communication simultaneously in [24].

When a fully digital architecture is deployed in the DFRC system, each antenna needs one radio frequency (RF) chain including a mixer and a digital-to-analog converter (DAC). This architecture is preferred in terms of the system performance. However, it could result in prohibitive cost and power consumption, especially for millimeter wave (mmWave) systems, in which the number of antennas is usually very large. A possible way to deal with this problem is adopting the hybrid beamforming (HBF) architecture, which uses phase shifters (PSs) to realize analog beamforming and demands a much smaller number of RF chains (than the number of antennas) to implement digital beamforming. In fact, the HBF architecture has recently attracted much research interest and been proved to be promising for communication-only systems [26–31]. However, for DFRC systems, the HBF architecture has not been adequately investigated yet [32–38].

To address this gap, this chapter focuses on the HBF design for mmWave DFRC systems. The main contents of this chapter are summarized as follows:

***First**, HBF design for single-carrier DFRC system:* An HBF design based on the subarray connection architecture is proposed for a single-carrier DFRC system in the presence of extended target and clutters. We derive the communication spectral efficiency (SE) and radar SINR with respect to the transmit HBF and radar receiver, and formulate the HBF design problem as the SE maximization subject to the radar SINR and power constraints. To solve the formulated nonconvex problem, the joinT Hybrid bEamforming and Radar rEceiver OptimizatioN (THEREON) is proposed, where the radar receiver is optimized via the generalized eigenvalue decomposition, and the transmit HBF is updated with a low complexity in a parallel manner using the consensus alternating direction method of multipliers (consensus-ADMM). Numerical simulations demonstrate the efficacy of the proposed algorithm and show that the solution provides a good trade-off between the number of phase shifters and the performance gain of the DPS HBF.

***Second**, HBF design for multi-carrier DFRC system:* The issues of transmit HBF design and direction-of-arrival (DOA) estimation in multi-carrier DFRC systems. In the designed system, communication symbols are embedded into radar pulse intervals with multiple orthogonal waveforms, and the HBF is optimized to focus the transmitted energy within the spatial sectors of interest by taking the communication quality of service (QoS) requirement for MUs into account. A consensus-ADMM framework based on weighted mean-square error minimization (WMMSE) is devised to tackle the resultant nonconvex problem. Additionally, the MUltiple SIgnal Classification (MUSIC)-based DOA estimation with the designed HBF architecture is presented, and the corresponding Cramér–Rao bound (CRB) is derived. Numerical simulations are presented to demonstrate the effectiveness of the proposed designs.

The rest of this chapter is organized as follows. First, the system model and HBF design method for the single-carrier mmWave DFRC system are introduced in Section 8.2. Then, we present the consensus-ADMM-based HBF design method for multi-carrier mmWave DFRC system in Section 8.3. Finally, we draw a conclusion in Section 8.4.

8.2 HBF design for single-carrier DFRC system

8.2.1 System model and problem formulation

In this section, we formulate the system model and optimization problem for the single-carrier DFRC system with HBF architecture. We consider a scenario as shown in Figure 8.1(a), where a DFRC vehicle sends communication symbols to a recipient vehicle receiver while detecting a target vehicle of interest in the presence of stationary clutters (such as trees, ground, buildings, etc.) simultaneously. The system architecture is depicted in Figure 8.1(b), where we assume a time-division duplex (TDD) DFRC system with N_{Tx} antennas and N_{RF} RF chains adopting a non-overlapping subarray architecture. Each subarray has $M = N_{\text{Tx}}/N_{\text{RF}}$ antennas connected to an RF chain. The recipient vehicle receiver with N_{Rx} antennas employs the fully digital beamforming structure.

8.2.1.1 Transmit model

At the transmitter, the symbol block $\mathbf{s}_l$ in l-th subpulse is precoded by a digital precoding matrix $\mathbf{F}_{\text{D},l} \in \mathbb{C}^{N_{\text{RF}} \times N_s}$, at first, where N_s is the number of data streams. Subsequently, the baseband signal is up-converted to the RF domain via N_{RF} RF chains and processed by analog PSs. Different from the conventional subarray architecture where each antenna is connected to a single PS, we consider exploiting double PSs to provide additional amplitude control for the HBF, the diagram of which is sketched in Figure 8.1(b).

Without loss of generality, each PS has a constant magnitude $1/\sqrt{N_{\text{Tx}}}$, and the synthesized value of each DPS module meets $Ae^{\jmath\varphi}$ with $A \in [0, 2/\sqrt{N_{\text{Tx}}}]$ and $\varphi \in [0, 2\pi]$. Thus, the proposed analog precoder can be expressed as $\mathbf{F}_{\text{RF}} = \mathbf{F}_{\text{set}}\mathbf{P}$, where $\mathbf{F}_{\text{set}} = \text{diag}\{f_1, \cdots, f_{N_{\text{Tx}}}\}$ with $f_m = A_m e^{\jmath\varphi_m}$,

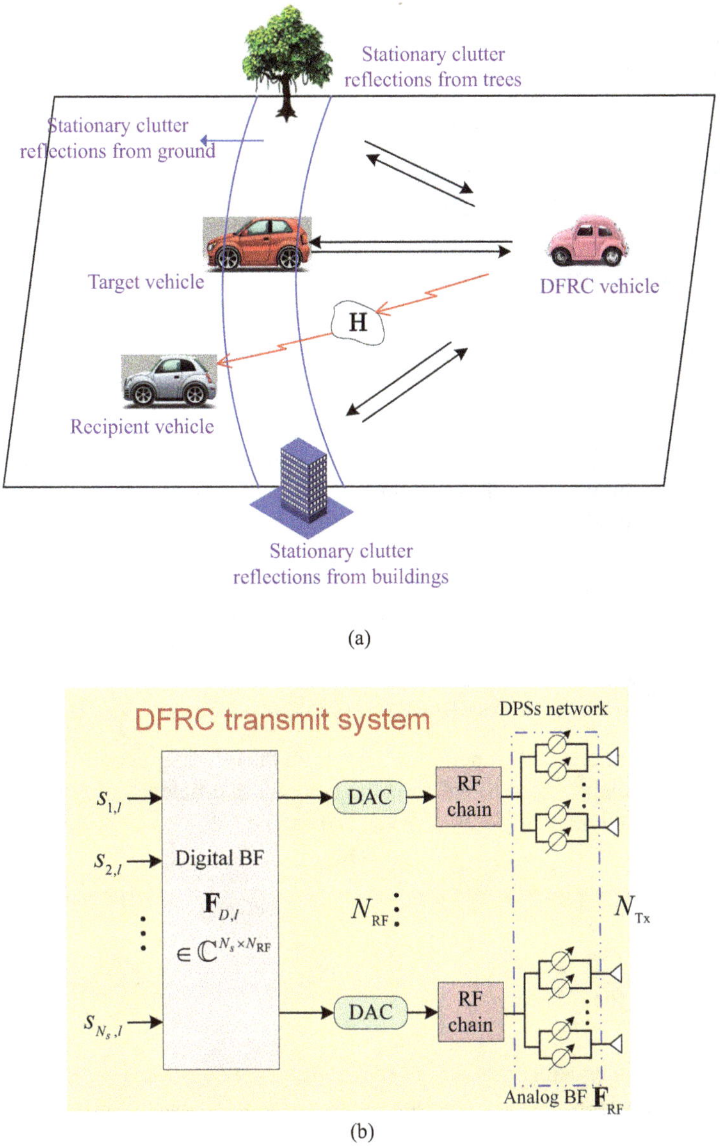

Figure 8.1 (a) Illustration of the considered scenario for the DFRC system. (b) Overview of the DPS-based HBF DFRC system.

$A_m \in [0, 2/\sqrt{N_{\mathrm{Tx}}}]$,$\varphi_m \in [0, 2\pi]$, $\forall m = 1, \cdots, N_{\mathrm{Tx}}$, $\mathbf{P} = \mathrm{Bdiag}\{\mathbf{1}_M, \cdots, \mathbf{1}_M\} \in \mathbb{C}^{N_{\mathrm{Tx}} \times N_{\mathrm{RF}}}$ is a binary matrix indicating the antenna selection in a subarray. Thus, the complex baseband discrete-time signal at the transmitter can be written as

$$\mathbf{x}[l] = \mathbf{F}_{\mathrm{RF}}\mathbf{F}_{\mathrm{D},l}\mathbf{s}_l, \tag{8.1}$$

where $\mathbf{s}_l$ is the normalized symbol sequence corresponding to the l-th subpulse with $\mathbb{E}\{\mathbf{s}_l\mathbf{s}_l^H\} = \mathbf{I}_{N_s}$. Assuming L subpulses are contained in one pulse duration, and collecting all L transmit vectors into a matrix $\mathbf{X} \in \mathbb{C}^{N_{\text{Tx}} \times L}$, we have

$$\mathbf{X} = \mathbf{F}_{\text{RF}}[\mathbf{F}_{\text{D},1}\mathbf{s}_1, \cdots, \mathbf{F}_{\text{D},L}\mathbf{s}_L]. \tag{8.2}$$

With the transmit model at hand, we will elaborate on the communication and radar models in the following sections:

8.2.1.2 Communication model

At the recipient vehicle receiver, the signal corresponding to the l-th subpulse is modeled as

$$\mathbf{c}[l] = \mathbf{H}\mathbf{F}_{\text{RF}}\mathbf{F}_{\text{D},l}\mathbf{s}_l + \mathbf{z}_c[l], \tag{8.3}$$

where $\mathbf{H} \in \mathbb{C}^{N_{\text{Rx}} \times N_{\text{Tx}}}$ is the CSI from the transmitter to the recipient vehicle and assumed to be known through some channel estimation techniques [12,39,40] such as the pilot method. $\mathbf{z}_c[l]$ is additive Gaussian noise vector with zero mean and variance σ_c^2.

The recipient vehicle adopts an $N_{\text{Rx}} \times N_s$ digital combiner $\mathbf{U}_l = [\mathbf{u}_{1,l}, \cdots, \mathbf{u}_{N_s,l}]$, to estimate the symbol block of the l-th subpulse, then the estimated $\hat{\mathbf{s}}_l$ can be modeled as

$$\hat{\mathbf{s}}_l = \mathbf{U}_l^H\mathbf{c}[l] = \mathbf{U}_l^H\mathbf{H}\mathbf{F}_{\text{RF}}\mathbf{F}_{\text{D},l}\mathbf{s}_l + \mathbf{U}_l^H\mathbf{z}_c[l], \tag{8.4}$$

For the communication function, we focus on the hybrid precoder design to maximize the SE, which is used to describe the bandwidth efficiency of communication systems. Concretely, the SE $R_l(\mathbf{F}_{\text{D},l}, \mathbf{F}_{\text{RF}}, \mathbf{U}_l)$ for the l-th subpulse is defined as [29]

$$\begin{aligned} &R_l(\mathbf{F}_{\text{D},l}, \mathbf{F}_{\text{RF}}, \mathbf{U}_l) \text{ [bits/s/Hz]} \\ &\quad = \log\left|\mathbf{I}_{N_{\text{Rx}}} + \mathbf{U}_l\mathbf{C}_l^{-1}\mathbf{U}_l^H\mathbf{H}\mathbf{F}_{\text{RF}}\mathbf{F}_{\text{D},l}\mathbf{F}_{\text{D},l}^H\mathbf{F}_{\text{RF}}^H\mathbf{H}^H\right|, \end{aligned} \tag{8.5}$$

where $\mathbf{C}_l = \sigma_c^2\mathbf{U}_l^H\mathbf{U}_l$.

8.2.1.3 Radar model

For the radar function, we assume that the radar receive array with N_{Rad} elements adopts a full-digital beamforming structure, and consider a scenario where the radar receiver needs to detect the target vehicle of interest in the presence of clutter. In the mmWave band, the scattering of the target is extended in distance due to the high range resolution. To be more specific, let θ_t be the angle of a generic extended target and $t(k), k = 0, \cdots, L_{\text{tar}} - 1$ be the finite impulse response (FIR) of the extended target with L_{tar} being the support length of the FIR [41,42]. Then, the received vector is modeled as

$$\mathbf{r}[n] = \mathbf{H}_t(\theta_t)e^{\jmath 2\pi n f_d/f_s}\sum_{l=1}^{L} t(n-l)\mathbf{x}[l] + \mathbf{j}[n] + \mathbf{z}_r[n], \tag{8.6}$$

where $\mathbf{H}_t(\theta_t) = \mathbf{a}_{Rr}(\theta_t)\mathbf{a}_t^H(\theta_t)$ is the spatial steering matrix, $f_d = \frac{2v_r}{\lambda}$ is the Doppler shifts of the target with v_r being the radial velocity of the target, f_s is the sampling frequency, $\mathbf{z}_r[n]$ is a zero-mean Gaussian noise vector with variance σ_r^2, and $\mathbf{j}[n]$ is interference term from the stationary clutters. Assuming that the clutter is divided into K clutter bins located at $\theta_i, \forall i = 1, \cdots, K$, then $\mathbf{j}[n]$ is expressed as $\mathbf{j}[n] = \sum_{i=1}^{K} \mathbf{H}_i(\theta_i) \sum_{l=1}^{L} j_i(n-l)\mathbf{x}[l]$, where $\mathbf{H}_i(\theta_i) = \mathbf{a}_{Rr}(\theta_i)\mathbf{a}_t^H(\theta_i)$ is the spatial steering matrix of the i-th clutter bin and $j_i[k], k = 0, \cdots, L_{c,i}-1$ denotes the FIR of the i-th clutter bin with $L_{c,i}$ being the support length.

We define $\mathbf{t} = [t(0), \cdots, t(L_{\text{tar}}-1)]^T$ and $\mathbf{j}_i = [j_i(0), \cdots, j_i(L_{c,i}-1)]^T$ and assume that both $\mathbf{t}$ and $\{\mathbf{j}_i\}$ are zero mean random vectors with covariance matrix being $\boldsymbol{\Sigma}_t = \mathbb{E}\{\mathbf{t}\mathbf{t}^H\}$ and $\boldsymbol{\Sigma}_{c,i} = \mathbb{E}\{\mathbf{j}_i\mathbf{j}_i^H\}$, respectively. Let $L_{\text{obs}} = L + \max\{L_{\text{tar}}, \{L_{c,i}\}\} - 1$ being the receiver observation length. After defining $\mathbf{R} = [\mathbf{r}[1], \cdots, \mathbf{r}[L_{\text{obs}}]] \in \mathbb{C}^{N_{\text{Rx}} \times L_{\text{obs}}}$ and $\mathbf{Z}_r = [\mathbf{z}_r[1], \cdots, \mathbf{r}_r[L_{\text{obs}}]] \in \mathbb{C}^{N_{\text{Rx}} \times L_{\text{obs}}}$, the model can be written in the matrix form as follows:

$$\mathbf{R} = \mathbf{H}_t(\theta_t)\mathbf{X}\mathbf{T}\mathscr{F}_d + \sum_{i=1}^{K} \mathbf{H}_i(\theta_i)\mathbf{X}\mathbf{J}_i + \mathbf{Z}_r, \tag{8.7}$$

where

$$\mathbf{T} = \begin{bmatrix} t(0) & \cdots & t(L_{tar}-1) & & 0 \\ & \ddots & & \ddots & \\ 0 & & t(0) & \cdots & t(L_{tar}-1) \end{bmatrix} \in \mathbb{C}^{L \times L_{\text{obs}}},$$

$$\mathbf{J}_i = \begin{bmatrix} j_i(0) & \cdots & j_i(L_{c,i}-1) & & 0 \\ & \ddots & & \ddots & \\ 0 & & j_i(0) & \cdots & j_i(L_{c,i}-1) \end{bmatrix} \in \mathbb{C}^{L \times L_{\text{obs}}}.$$

and $\mathscr{F}_d = \text{diag}\{e^{\jmath 2\pi f_d/f_s}, \cdots, e^{\jmath 2\pi L_{\text{obs}} f_d/f_s}\}$.

The received signal $\mathbf{R}$ is filtered via the receive beamformer $\mathbf{V} \in \mathbb{C}^{N_{\text{Rad}} \times L_{\text{obs}}}$, then the output SINR can be written as

$$\text{SINR}(\mathbf{F}_{\text{RF}}, \mathbf{F}_{\text{D}}, \mathbf{V}) = \frac{\mathbb{E}\left\{\left|\text{Tr}\left\{\mathbf{V}^H\mathbf{H}_t(\theta_t)\mathbf{X}\mathbf{T}\mathscr{F}_d\right\}\right|^2\right\}}{\sum_{i=1}^{K}\mathbb{E}\left\{\left|\text{Tr}\left\{\mathbf{V}^H\mathbf{H}_i(\theta_i)\mathbf{X}\mathbf{J}_i\right\}\right|^2\right\} + \mathbb{E}\left\{\left|\text{Tr}\left\{\mathbf{V}^H\mathbf{Z}_r\right\}\right|^2\right\}}, \tag{8.8}$$

where $\mathbf{F}_{\text{D}} = [\mathbf{F}_{\text{D},1}, \cdots, \mathbf{F}_{\text{D},L}]$. The following proposition will be used when designing the HBF and radar receive filter.

Proposition 1. *The SINR in* (8.8) *can be equivalently expressed as*

$$\text{SINR}(\mathbf{F}_{\text{RF}}, \mathbf{F}_{\text{D}}, \mathbf{V}) = \frac{\mathbf{v}^H \mathbf{\Theta}_t(\mathbf{F}_{\text{RF}}, \mathbf{F}_{\text{D}})\mathbf{v}}{\mathbf{v}^H \mathbf{\Theta}_c(\mathbf{F}_{\text{RF}}, \mathbf{F}_{\text{D}})\mathbf{v} + \sigma_r^2 \mathbf{v}^H \mathbf{v}} \tag{8.9a}$$

$$= \frac{\sum_{l=1}^{L} \text{Tr}\left\{\mathbf{F}_l \mathbf{F}_l^H \mathbf{\Phi}_t[l,l]\right\}}{\sum_{l=1}^{L} \text{Tr}\left\{\mathbf{F}_l \mathbf{F}_l^H \mathbf{\Phi}_c[l,l]\right\} + \sigma_r^2 \mathbf{v}^H \mathbf{v}}, \tag{8.9b}$$

where $\mathbf{F}_l = \mathbf{F}_{\text{RF}}\mathbf{F}_{\text{D},l}$, $\mathbf{v} = \text{vec}(\mathbf{V})$. $\mathbf{\Theta}_t(\mathbf{F}_{\text{RF}}, \mathbf{F}_{\text{D}}) \triangleq (\mathbf{I}_{L_{\text{obs}}} \otimes \mathbf{H}_t(\theta_t))(\sum_{i=1}^{L_{\text{tar}}} \sum_{j=1}^{L_{\text{tar}}} \mathbf{\Sigma}_t[i,j] \; \mathbf{\Gamma}_{ij}) \; (\mathbf{I}_{L_{\text{obs}}} \otimes \mathbf{H}_t(\theta_t))^H$, $\mathbf{\Theta}_c(\mathbf{F}_{\text{RF}}, \mathbf{F}_{\text{D}}) \triangleq \sum_{i=1}^{K} \mathbf{I}_{L_{\text{obs}}} \otimes \mathbf{H}_i(\theta_i))(\sum_{k=1}^{L_{c,i}} \sum_{l=1}^{L_{c,i}} \mathbf{\Sigma}_{c,i}[l,k]\mathbf{\Gamma}_{lk}^{c,i}) \; (\mathbf{I}_{L_{\text{obs}}} \otimes \; \mathbf{H}_i(\theta_i))^H$, $\mathbf{\Phi}_t \triangleq (\mathbf{I}_L \otimes \mathbf{H}_t^H(\theta_t))\widetilde{\mathbf{V}}\mathbf{\Sigma}_t\widetilde{\mathbf{V}}^H(\mathbf{I}_L \otimes \mathbf{H}_t(\theta_t))$, $\mathbf{\Phi}_c \triangleq \sum_{i=1}^{K} \; (\mathbf{I}_L \otimes \mathbf{H}_i^H(\theta_i))\widehat{\mathbf{V}}_i\mathbf{\Sigma}_{c,i}\widehat{\mathbf{V}}_i^H(\mathbf{I}_L \otimes \mathbf{H}_i(\theta_i))$. *The* $\mathbf{\Phi}_t[l,l]$ *and* $\mathbf{\Phi}_c[l,l]$ *represent the l-th block diagonal elements of* $\mathbf{\Phi}_t$ *and* $\mathbf{\Phi}_c$*, respectively.*

Predictably, (8.9a) will be useful in optimizing radar receive filter $\mathbf{V}$ with the fixed hybrid beamformer $(\mathbf{F}_{\text{RF}}, \mathbf{F}_{\text{D}})$. While the alternative (8.9b) is benefit to designing the $(\mathbf{F}_{\text{RF}}, \mathbf{F}_{\text{D}})$ with a given $\mathbf{V}$.

8.2.1.4 Problem formulation

According to above models, a meaningful criterion of jointly optimizing the hybrid digital/analog precoder $(\mathbf{F}_{\text{D}}, \mathbf{F}_{\text{set}})$, communication combiner $\{\mathbf{U}_l\}$ and radar receive filter $\mathbf{V}$ is to maximize the communication SE while keeping the SINR requirement for radar target. Mathematically, our problem of interest can be formulated as

$$\max_{\mathbf{F}_{\text{D}}, \mathbf{F}_{\text{set}}, \{\mathbf{U}_l\}, \mathbf{V}} \quad \sum_{l=1}^{L} R_l(\mathbf{F}_{\text{D},l}, \mathbf{F}_{\text{RF}}, \mathbf{U}_l) \tag{8.10a}$$

$$\text{s.t.} \quad \text{SINR}(\mathbf{F}_{\text{RF}}, \mathbf{F}_{\text{D}}, \mathbf{V}) \geq \gamma, \tag{8.10b}$$

$$\mathbf{F}_{\text{set}} = \text{diag}\{f_1, \cdots, f_{N_{\text{Tx}}}\}, f_m = A_m e^{\jmath \varphi_m}, \forall m \tag{8.10c}$$

$$0 \leq A_m \leq 2/\sqrt{N_{\text{Tx}}}, \; \varphi_m \in [0, 2\pi], \forall m, \tag{8.10d}$$

$$\text{Tr}\left(\mathbf{F}_{\text{RF}}\mathbf{F}_{\text{D}}\mathbf{F}_{\text{D}}^H\mathbf{F}_{\text{RF}}^H\right) \leq \mathscr{E}, \tag{8.10e}$$

where the constraint (8.10b) is the SINR requirement for the radar target with γ being the SINR threshold, the constraints (8.10c) and (8.10c) are feasible conditions for the DPS, and the constraint (8.10e) is the total energy requirement for the DFRC system with $\mathscr{E}$ being the total energy budget.

Note that this optimization problem involves a nonconvex objective function and nonconvex constraints (8.10b)-(8.10e), and hence, it is NP-hard [43] and challenging to solve.

8.2.2 *Proposed HBF algorithm for multi-carrier DFRC system*

In this section, we will solve the HBF design problem in an alternating optimization manner. The proposed algorithm is summarized in Algorithm 8.1. Specifically, the radar receiver $\mathbf{V}$ is optimized for a given $(\mathbf{F}_{\mathrm{D}}, \mathbf{F}_{\mathrm{set}}, \{\mathbf{U}_l\})$, and in turn, $(\mathbf{F}_{\mathrm{D}}, \mathbf{F}_{\mathrm{set}}, \{\mathbf{U}_l\})$ are jointly optimized for a given $\mathbf{V}$. In the following, we will illustrate the solutions to radar receiver $\mathbf{V}$ and HBF beamformer $(\mathbf{F}_{\mathrm{D}}, \mathbf{F}_{\mathrm{set}}, \{\mathbf{U}_l\})$, respectively.

8.2.2.1 Optimization of radar receiver

Note that the objective function in (8.10) is independent of $\mathbf{V}$, and thus, we only need to find a feasible solution $\mathbf{V}$ to meet the SINR requirement (8.10b). To this end, the radar filter $\mathbf{V}$ can be determined by maximizing the SINR value as

$$\max_{\mathbf{V}} \ \mathrm{SINR}(\mathbf{F}_{\mathrm{RF}}, \mathbf{F}_{D,l}, \mathbf{V}) = \frac{\mathbf{v}^H \boldsymbol{\Theta}_t(\mathbf{F}_{\mathrm{RF}}, \mathbf{F}_{\mathrm{D}})\mathbf{v}}{\mathbf{v}^H \boldsymbol{\Theta}_c(\mathbf{F}_{\mathrm{RF}}, \mathbf{F}_{\mathrm{D}})\mathbf{v} + \sigma_r^2 \mathbf{v}^H \mathbf{v}}, \tag{8.11}$$

of which the optimal solution $\mathbf{v}^\star$ can be achieved by taking the generalized eigenvalue decomposition (EVD) of $\left(\boldsymbol{\Theta}_t(\mathbf{F}_{\mathrm{RF}}, \mathbf{F}_{\mathrm{D}}), \boldsymbol{\Theta}_c(\mathbf{F}_{\mathrm{RF}}, \mathbf{F}_{\mathrm{D}}) + \sigma_r^2 \mathbf{I}_{N_{\mathrm{Rad}} N_s}\right)$, i.e.,

$$\mathbf{v}^\star = \mathscr{P}\left(\boldsymbol{\Theta}_t^{-1}(\mathbf{F}_{\mathrm{RF}}, \mathbf{F}_{\mathrm{D}}) \left(\boldsymbol{\Theta}_c(\mathbf{F}_{\mathrm{RF}}, \mathbf{F}_{\mathrm{D}}) + \sigma_r^2 \mathbf{I}_{N_{\mathrm{Rad}} N_s}\right)\right), \tag{8.12}$$

where the operator $\mathscr{P}(\,\cdot\,)$ denotes the principal eigenvector.

8.2.2.2 Optimization of hybrid beamformer and combiner

For a fixed $\mathbf{V}$, the subproblem with respect to $(\mathbf{F}_{\mathrm{D}}, \mathbf{F}_{\mathrm{set}}, \{\mathbf{U}_l\})$ is

$$\begin{aligned} \max_{\mathbf{F}_{\mathrm{D}}, \mathbf{F}_{\mathrm{set}}, \{\mathbf{U}_l\}} \quad & \sum_{l=1}^{L} R_l(\mathbf{F}_{\mathrm{D},l}, \mathbf{F}_{\mathrm{RF}}, \mathbf{U}_l) \\ \text{s.t.} \quad & (8.10\text{b}),\ (8.10\text{c}),\ (8.10\text{d}),\ \text{and}\ (8.10\text{e}) \end{aligned} \tag{8.13}$$

Algorithm 8.1 joinT Hybrid bEamforming and Radar rEceiver OptimizatioN (THEREON)

1: **Input:** Initial variables $\mathbf{F}_{\mathrm{set}}(0), \mathbf{F}_{D,l}(0)$ and iteration number $N_{\mathrm{THER}}^{\max}$
2: Set $t = 0$
3: **repeat**
4: Update $\mathbf{V}(t+1)$ according to (8.12)
5: Update $\mathbf{F}_{\mathrm{set}}(t+1), \mathbf{F}_{D,l}(t+1)$ by using Algorithm 8.2
6: $t = t + 1$
7: **until** $t = N_{\mathrm{THER}}^{\max}$
8: **Output:** $\mathbf{V}^\star = \mathbf{V}(t), \mathbf{F}_{\mathrm{set}}^\star = \mathbf{F}_{\mathrm{set}}(t), \mathbf{F}_{D,l}^\star = \mathbf{F}_{D,l}(t)$

Since $\mathbf{F}_{\text{set}}$ and $\mathbf{F}_{\text{D}}$ are coupled in constraints (8.10b) and (8.10e), this subproblem is difficult to solve. By introducing auxiliary variables $\mathbf{X}_l, \mathbf{Z}_l \in \mathbb{C}^{N_{\text{Tx}} \times N_s}, \forall l$, we decouple $\mathbf{F}_{\text{set}}$ and $\mathbf{F}_{\text{D}}$ and recast problem (8.13) into

$$\max_{\mathbf{F}_{\text{D}},\mathbf{F}_{\text{set}},\{\mathbf{U}_l\},\{\mathbf{X}_l\},\{\mathbf{Z}_l\}} \quad \sum_{l=1}^{L} R_l(\mathbf{X}_l, \mathbf{U}_l), \tag{8.14a}$$

$$\text{s.t.} \quad \frac{\sum_{l=1}^{L} \operatorname{Tr}\left\{\mathbf{Z}_l \mathbf{Z}_l^H \mathbf{\Phi}_t[l,l]\right\}}{\sum_{l=1}^{L} \operatorname{Tr}\left\{\mathbf{Z}_l \mathbf{Z}_l^H \mathbf{\Phi}_c[l,l]\right\} + \sigma_r^2 \mathbf{v}^H \mathbf{v}} \geq \gamma, \tag{8.14b}$$

$$\mathbf{X}_l = \mathbf{Z}_l = \mathbf{F}_{\text{set}} \mathbf{P} \mathbf{F}_{\text{D},l}, \forall l \tag{8.14c}$$

$$\sum_{l=1}^{L} \operatorname{Tr}\left(\mathbf{X}_l \mathbf{X}_l^H\right) \leq \mathscr{E}, \tag{8.14d}$$

$$\text{Constraints (8.10c), (8.10d),} \tag{8.14e}$$

where $R_l(\mathbf{X}_l, \mathbf{U}_l) = \log\left|\mathbf{I}_{N_{\text{Rx}}} + \mathbf{U}_l \mathbf{C}_l^{-1} \mathbf{U}_l^H \mathbf{H} \mathbf{X}_l \mathbf{X}_l^H \mathbf{H}^H\right|$.

It is observed that the introduction of auxiliary variables $\mathbf{X}_l, \mathbf{Z}_l, \forall l$ results in decoupling $\mathbf{F}_{\text{set}}$, $\mathbf{P}$ and $\mathbf{F}_{\text{D},l}$ in original objective function and imposing constraints (8.14b) and (8.14d) on $\mathbf{Z}_l, \forall l$ and $\mathbf{X}_l, \forall l$ respectively. This will enable us to construct the ADMM subproblems with respect to problem (8.14), each of which can be solved with a closed form solution. Concretely, placing the equality constraints $\mathbf{X}_l = \mathbf{Z}_l = \mathbf{F}_{\text{set}} \mathbf{P} \mathbf{F}_{\text{D},l}$ into the augmented Lagrangian function of (8.14) yields

$$\begin{aligned}
\mathscr{L} &= \sum_{l=1}^{L} \tilde{\mathscr{L}}_l(\mathbf{W}_l, \mathbf{X}_l, \mathbf{U}_l, \mathbf{Z}_l, \mathbf{F}_{\text{set}}, \mathbf{F}_{\text{D},l}, \mathbf{D}_{1,l}, \mathbf{D}_{2,l}), \\
&= \sum_{l=1}^{L} \Big\{ R_l(\mathbf{X}_l, \mathbf{U}_l) + \Re\left(\operatorname{Tr}\left\{\mathbf{D}_{1,l}^H (\mathbf{X}_l - \mathbf{F}_{\text{set}} \mathbf{P} \mathbf{F}_{\text{D},l})\right\}\right) + \frac{\rho_2}{2} \|\mathbf{X}_l - \mathbf{Z}_l\|_F^2 \\
&\quad + \frac{\rho_1}{2} \|\mathbf{X}_l - \mathbf{F}_{\text{set}} \mathbf{P} \mathbf{F}_{\text{D},l}\|_F^2 + \Re\left(\operatorname{Tr}\left\{\mathbf{D}_{2,l}^H (\mathbf{X}_l - \mathbf{Z}_l)\right\}\right) \Big\},
\end{aligned} \tag{8.15}$$

where $\mathbf{D}_{1,l}, \mathbf{D}_{2,l} \in \mathbb{C}^{N_{\text{Tx}} \times N_s}$ are dual variables corresponding to the equalities $\mathbf{X}_l = \mathbf{F}_{\text{set}} \mathbf{P} \mathbf{F}_{\text{D},l}$ and $\mathbf{X}_l = \mathbf{Z}_l$, respectively, and $\rho_1, \rho_2 > 0$ are the penalty parameters.

We split the optimized primal variables into two blocks $(\mathbf{W}_l, \mathbf{X}_l, \mathbf{U}_l)$ and $(\mathbf{Z}_l, \mathbf{F}_{\text{set}}, \mathbf{F}_{\text{D},l})$. In what follows, we shall present the update procedures of the two primal blocks and dual block $(\mathbf{D}_{1,l}, \mathbf{D}_{2,l})$.

Stage 1: *Optimization of* $(\mathbf{W}_l, \mathbf{X}_l, \mathbf{U}_l)$.

For fixed $(\mathbf{Z}_l, \mathbf{F}_{\text{set}}, \mathbf{F}_{\text{D},l})$ and $(\mathbf{D}_{1,l}, \mathbf{D}_{2,l})$, $(\mathbf{W}_l, \mathbf{X}_l, \mathbf{U}_l)$ is updated by solving

$$\begin{aligned}
&\min_{\{\mathbf{W}_l\},\{\mathbf{X}_l\},\{\mathbf{U}_l\}} \quad \sum_{l=1}^{L} \tilde{\mathscr{L}}_l(\mathbf{W}_l, \mathbf{X}_l, \mathbf{U}_l, \mathbf{Z}_l, \mathbf{F}_{\text{set}}, \mathbf{F}_{\text{D},l}, \mathbf{D}_{1,l}, \mathbf{D}_{2,l}) \\
&\text{s.t..} \quad \sum_{l=1}^{L} \operatorname{Tr}\left(\mathbf{X}_l \mathbf{X}_l^H\right) \leq \mathscr{E}.
\end{aligned} \tag{8.16}$$

Nevertheless, it is still difficult to find the solution of (8.16) due to the nonconvex function $R_l(\mathbf{X}_l, \mathbf{U}_l)$. To solve problem (8.16), by using the WMMSE approach and coordinate descent (CD) algorithm, problem (8.16) can be split into the following three subproblems:
(a): The update of $\mathbf{U}_l$ needs solving

$$\min_{\mathbf{U}_l} \ \mathrm{Tr}\left\{\mathbf{E}_l(\mathbf{X}_l, \mathbf{U}_l)\mathbf{W}_l\right\}. \tag{8.17}$$

where $\mathbf{E}_l(\mathbf{X}_l, \mathbf{U}_l) = \left(\mathbf{I}_{N_s} - \mathbf{U}_l^H\mathbf{H}\mathbf{X}_l\right)\left(\mathbf{I}_{N_s} - \mathbf{U}_l^H\mathbf{H}\mathbf{X}_l\right)^H + \sigma_c^2\mathbf{U}_l^H\mathbf{U}_l$. Its optimal solution of $\mathbf{U}_l$ can be attained via the first-order optimality condition given by

$$\mathbf{U}_l = \left(\mathbf{H}\mathbf{X}_l\mathbf{X}_l^H\mathbf{H}^H + \sigma_c^2\mathbf{I}_{\mathrm{Rx}}\right)^{-1}\mathbf{H}\mathbf{X}_l. \tag{8.18}$$

(b): The update of $\mathbf{W}_l$ needs solving

$$\min_{\mathbf{W}_l} \ \mathrm{Tr}\left\{\mathbf{E}_l(\mathbf{X}_l, \mathbf{U}_l)\mathbf{W}_l\right\} - \log\left|\mathbf{W}_l\right|, \tag{8.19}$$

which has the optimal solution given by

$$\mathbf{W}_l = \mathbf{E}_l^{-1}(\mathbf{X}_l, \mathbf{U}_l) = \left(\mathbf{I}_{N_s} - \mathbf{X}_l^H\mathbf{H}^H\mathbf{U}_l\right)^{-1}. \tag{8.20}$$

(c): The update of $\mathbf{X}_l$ needs solving

$$\begin{aligned}\min_{\mathbf{X}_l} \quad & \mathrm{Tr}\left\{\mathbf{E}_l(\mathbf{X}_l, \mathbf{U}_l)\mathbf{W}_l\right\} + \Re\left(\mathrm{Tr}\left\{\mathbf{D}_{1,l}^H\left(\mathbf{X}_l - \mathbf{F}_{\mathrm{set}}\mathbf{P}\mathbf{F}_{\mathrm{D},l}\right)\right\}\right)\\ & + \frac{\rho_1}{2}\left\|\mathbf{X}_l - \mathbf{F}_{\mathrm{set}}\mathbf{P}\mathbf{F}_{\mathrm{D},l}\right\|_F^2 + \Re\left(\mathrm{Tr}\left\{\mathbf{D}_{2,l}^H\left(\mathbf{X}_l - \mathbf{Z}_l\right)\right\}\right)\\ & + \frac{\rho_2}{2}\left\|\mathbf{X}_l - \mathbf{Z}_l\right\|_F^2\\ \text{s.t.} \quad & \sum_{l=1}^{L}\mathrm{Tr}\left(\mathbf{X}_l\mathbf{X}_l^H\right) \le \mathscr{E}.\end{aligned} \tag{8.21}$$

Subproblem (8.21) is convex, whose optimal solutions can be derived via the Karush-Kuhn-Tucker (KKT) conditions [43].

Stage 2: *Optimization of* $(\mathbf{Z}_l, \mathbf{F}_{\mathrm{set}}, \mathbf{F}_{\mathrm{D},l})$

For fixed $(\mathbf{W}_l, \mathbf{X}_l, \mathbf{U}_l)$ and $(\mathbf{D}_{1,l}, \mathbf{D}_{2,l})$, $(\mathbf{Z}_l, \mathbf{F}_{\mathrm{set}}, \mathbf{F}_{\mathrm{D},l})$ is updated by solving

$$\begin{aligned}\min_{\mathbf{Z}_l, \mathbf{F}_{\mathrm{set}}, \mathbf{F}_{\mathrm{D},l}} \quad & \textstyle\sum_{l=1}^{L}\tilde{\mathscr{L}}_l\left(\mathbf{W}_l, \mathbf{X}_l, \mathbf{U}_l, \mathbf{Z}_l, \mathbf{F}_{\mathrm{set}}, \mathbf{F}_{\mathrm{D},l}, \mathbf{D}_{1,l}, \mathbf{D}_{2,l}\right)\\ \text{s.t.} \quad & \sum_{l=1}^{L}\mathrm{Tr}\left(\mathbf{Z}_l\mathbf{Z}_l^H\mathbf{M}[l,l]\right) \ge \alpha, \text{(8.10c)}, \text{(8.10d)},\end{aligned} \tag{8.22}$$

where $\mathbf{M}[l,l] = \mathbf{\Phi}_t[l,l] - \gamma\mathbf{\Phi}_c[l,l]$, and $\alpha = \gamma\sigma_r^2\|\mathbf{v}\|^2$. We note that the CD method is able to solve the problem (8.22), which can be split into the following three subproblems:
(a): The update of $\mathbf{Z}_l$ needs solving

$$\begin{aligned}\min_{\mathbf{Z}_l} \quad & \sum_{l=1}^{L}\Re\left(\mathrm{Tr}\left\{\mathbf{D}_{2,l}^H\left(\mathbf{X}_l - \mathbf{Z}_l\right)\right\}\right) + \frac{\rho_2}{2}\left\|\mathbf{X}_l - \mathbf{Z}_l\right\|_F^2\\ \text{s.t.} \quad & \sum_{l=1}^{L}\mathrm{Tr}\left(\mathbf{Z}_l\mathbf{Z}_l^H\mathbf{M}[l,l]\right) \ge \alpha.\end{aligned} \tag{8.23}$$

Similar to the solution to problem (8.21), the optimal solution to problem (8.23) is obtained by analyzing the KKT conditions [43].
(b): The variables $\mathbf{F}_{\mathrm{D},l}, \forall l$ are updated in parallel by solving

$$\min_{\mathbf{F}_{\mathrm{D},l}} \sum_{l=1}^{L} \Re\left(\mathrm{Tr}\left\{\mathbf{D}_{1,l}^{H}\left(\mathbf{X}_l - \mathbf{F}_{\mathrm{set}}\mathbf{P}\mathbf{F}_{\mathrm{D},l}\right)\right\}\right) + \frac{\rho_1}{2}\left\|\mathbf{X}_l - \mathbf{F}_{\mathrm{set}}\mathbf{P}\mathbf{F}_{\mathrm{D},l}\right\|^2, \tag{8.24}$$

whose closed-form solution is

$$\begin{aligned} \mathbf{F}_{\mathrm{D},l} &= \left(\mathbf{P}^H\mathbf{F}_{\mathrm{set}}^H\mathbf{F}_{\mathrm{set}}\mathbf{P}\right)^{-1}\mathbf{P}^H\mathbf{F}_{\mathrm{set}}^H\left(\frac{1}{\rho_1}\mathbf{D}_{1,l} + \mathbf{X}_{1,l}\right) \\ &= \mathrm{diag}^{-1}\Big(\sum_{i=1}^{N_{\mathrm{Tx}}} A_i^2 p_{i,1}, \cdots, \sum_{i=1}^{N_{\mathrm{Tx}}} A_i^2 p_{i,N_{\mathrm{RF}}}\Big)\mathbf{P}^H\mathbf{F}_{\mathrm{set}}^H\left(\frac{1}{\rho_1}\mathbf{D}_{1,l} + \mathbf{X}_{1,l}\right). \end{aligned} \tag{8.25}$$

(c): The variable $\mathbf{F}_{\mathrm{set}}$ is updated by the following problem:

$$\begin{aligned} \min_{\mathbf{F}_{\mathrm{set}}} \quad & \sum_{l=1}^{L}\left\|\mathbf{X}_l - \mathbf{F}_{\mathrm{set}}\mathbf{P}\mathbf{F}_{\mathrm{D},l} + \frac{1}{\rho_1}\mathbf{D}_{1,l}\right\|^2 \\ \text{s.t.} \quad & \mathbf{F}_{\mathrm{set}} = \mathrm{diag}\left(f_1, \cdots, f_{N_{\mathrm{Tx}}}\right), f_m = A_m e^{\jmath\varphi_m}, \forall m \\ & 0 \le A_m \le 2/\sqrt{N_{\mathrm{Tx}}}, \varphi_m \in [0, 2\pi], \forall m. \end{aligned} \tag{8.26}$$

Let $\mathbf{\Pi}_l = \mathbf{X}_l + \frac{1}{\rho_1}\mathbf{D}_{1,l}$ and $\mathbf{Y}_l = \mathbf{P}\mathbf{F}_{\mathrm{D},l}$, problem (8.26) can be decomposed into

$$\begin{aligned} \min_{\{f_m\}} \quad & \sum_{l=1}^{L}\|\mathbf{\Pi}_l[m,:] - f_m\mathbf{Y}_l[m,:]\|_2^2 \\ \text{s.t.} \quad & f_m = A_m e^{\jmath\varphi_m},\ 0 \le A_m \le 2/\sqrt{N_{\mathrm{Tx}}}, \forall m, \end{aligned} \tag{8.27}$$

whose closed-form solution is

$$A_m = \begin{cases} \dfrac{\left|\sum_{l=1}^{L}\mathbf{\Pi}_l[m,:]\mathbf{Y}_l^H[m,:]\right|}{\sum_{l=1}^{L}\|\mathbf{Y}_l[m,:]\|_2^2}, & \dfrac{\left|\sum_{l=1}^{L}\mathbf{\Pi}_l[m,:]\mathbf{Y}_l^H[m,:]\right|}{\sum_{l=1}^{L}\|\mathbf{Y}_l[m,:]\|_2^2} \le \dfrac{2}{\sqrt{N_{\mathrm{Tx}}}} \\ \dfrac{2}{\sqrt{N_{\mathrm{Tx}}}}, & \dfrac{\left|\sum_{l=1}^{L}\mathbf{\Pi}_l[m,:]\mathbf{Y}_l^H[m,:]\right|}{\sum_{l=1}^{L}\|\mathbf{Y}_l[m,:]\|_2^2} > \dfrac{2}{\sqrt{N_{\mathrm{Tx}}}} \end{cases} \tag{8.28}$$

and

$$\varphi_m = \angle\left(\sum_{l=1}^{L}\mathbf{\Pi}_l[m,:]\mathbf{Y}_l^H[m,:]\right). \tag{8.29}$$

After obtaining A_m and φ_m, the phase values of phase shifters #1 and #2 in the DPS element are

$$\psi_{1,m} = \varphi_m + \arccos\left(A_m/2\right), \quad \psi_{2,m} = \varphi_m - \arccos\left(A_m/2\right). \tag{8.30}$$

Stage 3: *Optimization of* $(\mathbf{D}_{1,l}, \mathbf{D}_{2,l})$

For fixed $(\mathbf{W}_l, \mathbf{X}_l, \mathbf{U}_l)$ and $(\mathbf{Z}_l, \mathbf{F}_{\text{set}}, \mathbf{F}_{\text{D},l})$, $(\mathbf{D}_{1,l}, \mathbf{D}_{2,l})$ are updated by [44]:

$$\begin{aligned}\mathbf{D}_{1,l} &= \mathbf{D}_{1,l} + \rho_1 \left(\mathbf{X}_l - \mathbf{F}_{\text{set}}\mathbf{P}\mathbf{F}_{\text{D},l}\right), \\ \mathbf{D}_{2,l} &= \mathbf{D}_{2,l} + \rho_1 \left(\mathbf{X}_l - \mathbf{Z}_l\right).\end{aligned} \tag{8.31}$$

The consensus-ADMM for solving problem (8.13) is summarized in Algorithm 8.2.

8.2.3 Simulation results

This section provides various numerical simulations to examine the performance of the proposed hybrid beamforming design for the DFRC system.

Unless otherwise mentioned, in all simulations, we assume a DFRC system with $N_{\text{Tx}} = 32$ transmit antennas. The radar receive array with $N_{\text{Rad}} = 4$ is considered. The transmitter sends $N_s = 4$ data symbols per subpulse to a user equipped with 2 antennas. We assume a communication environment with $N_{path} = 16$ clusters, and the noises at users are modeled as additive White Gaussian with the covariances of $\sigma_c^2 = 0.1$.

For radar scenario, we assume that the Tx/Rx arrays boresight directions are used as the reference for the azimuth θ, and that an extended target located at $\theta_t = 0°$ (as illustrated in Figure 8.2). The number of subpulses in each pulse is $L = 16$, which is the case for $\mu = 4$ in the recently published 5G NR standard [45]. In this case, each subframe contains 16 slots, and the length of each slot is 62.5 us. For modeling the impulse response of the extended target, we use the exponentially shaped covariance to model the target second-order statistic matrix $\mathbf{\Sigma}_t = \mathbb{E}\{\mathbf{t}\mathbf{t}^H\}$, that is, $\mathbf{\Sigma}_t(m,n) = \sigma_\alpha^2 \eta_\alpha^{-|m-n|}$, $1 \leq m, n \leq L_{\text{tar}}$ with $L_{\text{tar}} = 6$, $\sigma_\alpha^2 = 10$ and $\eta_\alpha = 15$. For the signal-dependent interference (i.e., clutters), we consider a homogeneous clutter environment composed of $K = 31$ azimuth cells, the azimuth angle of the i-th

---

Algorithm 8.2 Consensus-ADMM for solving problem (8.13)

1: **Input:** Initial variables $\mathbf{F}_{\text{set}}(0), \mathbf{F}_{D,l}(0)$, $\mathbf{X}_l(0)$, $\mathbf{Z}_l(0)$, $\mathbf{D}_{1,l}(0)$, $\mathbf{D}_{2,l}(0)$ and $\rho_1, \rho_2 > 0$
2: Set $k = 0$
3: **repeat**
4: Update $\mathbf{U}_l(k+1)$ according to (8.18)
5: Update $\mathbf{W}_l(k+1)$ in parallel by (8.20)
6: Update $\mathbf{X}_l(k+1), \forall l$ in parallel by analyzing KKT conditions
7: Update $\mathbf{Z}_l(k+1), \forall l$ in parallel by analyzing KKT conditions
8: Update $\mathbf{F}_{\text{D},l}(k+1), \forall l$ according to (8.25)
9: Update $\mathbf{F}_{\text{set}}(k+1)$ according to (8.28)
10: Update $\mathbf{D}_{1,l}(k+1)$ and $\mathbf{D}_{2,l}(k+1)$ according to (8.31)
11: $k = k+1$
12: **until** $k = N_{\text{ADMM}}^{\max}$.
13: **Output:** $\mathbf{F}_{\text{set}}^\star = \mathbf{F}_{\text{set}}(k), \mathbf{F}_{D,l}^\star = \mathbf{F}_{D,l}(k)$

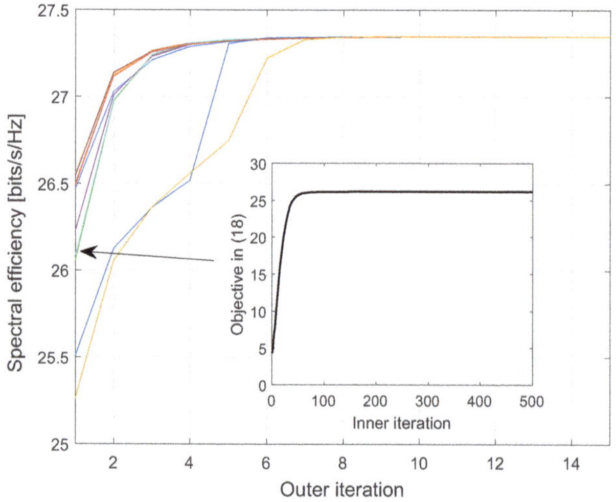

Figure 8.2 The convergence performance of the THEREON method for different initial points when considering the intended SINR requirement $\gamma = 12$, $N_{\text{Tx}} = 32$ and $N_{\text{RF}} = 4$

cell is $\theta_i = 2\pi(i-1)/K$. All clutter second-order statistic matrices $\mathbf{\Sigma}_{c,i} = \mathbb{E}\{\mathbf{j}_i\mathbf{j}_i^H\}$ are identically modeled as $\mathbf{\Sigma}_t$ with $\mathbf{\Sigma}_{c,i}(m,n) = \sigma_\beta^2 \eta_\beta^{-|m-n|}, \forall i, 1 \leq m, n \leq L_{c,i}$ with $L_{c,i} = 8$, $\sigma_\beta^2 = 1$ and $\eta_\beta = 1.2$. As for the radar receive noise, we assume corruption by a white noise with the variance $\sigma_r^2 = 0.1$.

Firstly, we examine the convergence performance of the proposed algorithm THEREON for solving problem (8.10). We consider the DFRC system with $N_{\text{RF}} = 4$ RF chains and the total energy of the system is $\mathscr{E} = 10$. The intended SINR requirement for the target is $\gamma = 12$ dB. We set the penalty parameters as $\rho_1 = \rho_2 = 20$. Figure 8.2 analyzes the effect of 10 different initial points on the convergence performance of the proposed THEREON framework for solving problem (8.10). For each initial point, we assume the entries of initial $\mathbf{F}_{\text{RF}}$ are $e^{\jmath\Phi}$, where Φ obeys the uniform distribution over $(0, 2\pi]$ and entries of initial $\mathbf{F}_{\text{D}}$ obey $\mathscr{CN}(0,1)$. As shown in the figure, the converged objective values are the same for different initial points can converge to the same value as the outer iteration (i.e., first update the radar filter $\mathbf{V}$ for given $(\mathbf{F}_{\text{D}}, \mathbf{F}_{\text{set}}, \{\mathbf{U}_l\})$, and then update $(\mathbf{F}_{\text{D}}, \mathbf{F}_{\text{set}}, \{\mathbf{U}_l\})$ with aid of the consensus-ADMM for given $\mathbf{V}$) goes on. In addition, for an instance of the THEREON framework, we also plot the convergence of the objective values of problem (8.14) versus the inner iteration number by using the consensus-ADMM algorithm. The result shows that the objective value obtained by the consensus-ADMM is able to converge to a sub-optimal value with the increasing iteration number.

Figure 8.3 plots the SE value under the proposed DPS (denoted by "Prop. DPS") architecture versus the intended radar SINR requirement γ. For comparison purposes, the fully digital beamformer (denoted by "fully digital") which provides the upper-bound SE, the two-stage method (denoted by "two-stage") in [27,46] and the conventional single-phase shifter (SPS) architecture (denoted by "Conv. SPS") are also considered. The results show that the obtained SE values decrease along with the γ, this is because when the intended γ is higher, the less DoFs can be used to maximize the communication SE. Thus there is a trade-off between the radar SINR behavior and communication performance. In addition, Figure 8.3 also shows that both the proposed DPS and conventional SPS structures with the consensus-ADMM achieve better SE values than the two-stage method in [27,46] which seeks to minimize the distance of the optimal fully digital beamformers. Moreover, the proposed DPS achieves a better performance consistently over different radar SINR requirements with an SE gap of about 3 bps/Hz in comparison with the conventional SPS.

Figure 8.4 displays the SE value versus the intended SINR requirement γ for different numbers of RF chains $N_{\mathrm{RF}} = 2, 4, 8, 16$ when considering $N_{\mathrm{Tx}} = 32$ and the

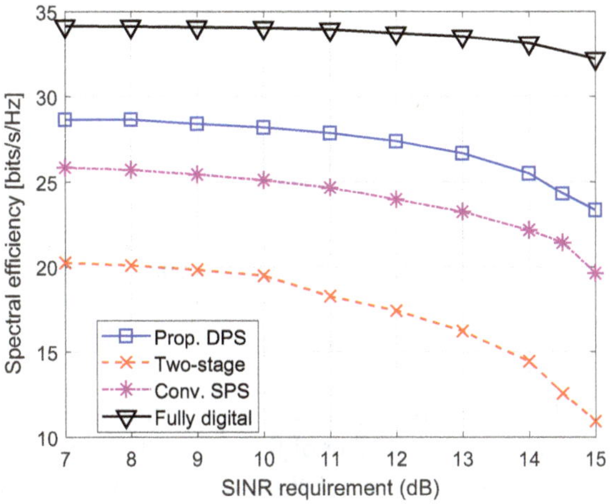

Figure 8.3 The achieved SE values of different methods versus the intended radar SINR requirement γ when considering $N_{\mathrm{Tx}} = 32$, $N_{\mathrm{RF}} = 4$ and $\mathscr{E} = 10$

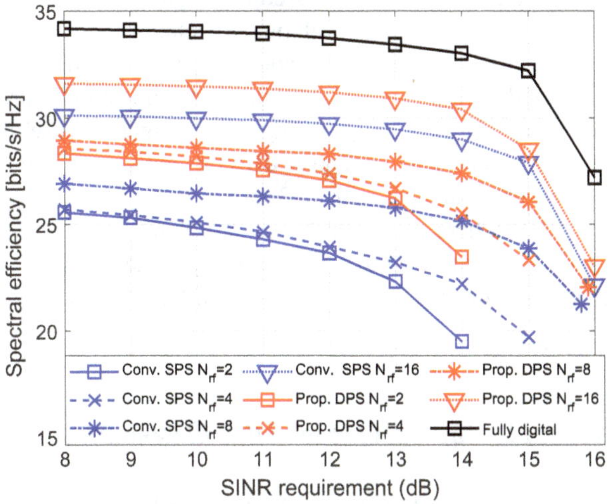

Figure 8.4 The achieved SE value versus the intended radar SINR requirement γ for different numbers of RF chains when considering $N_{\mathrm{Tx}} = 32$ and $\mathscr{E} = 10$

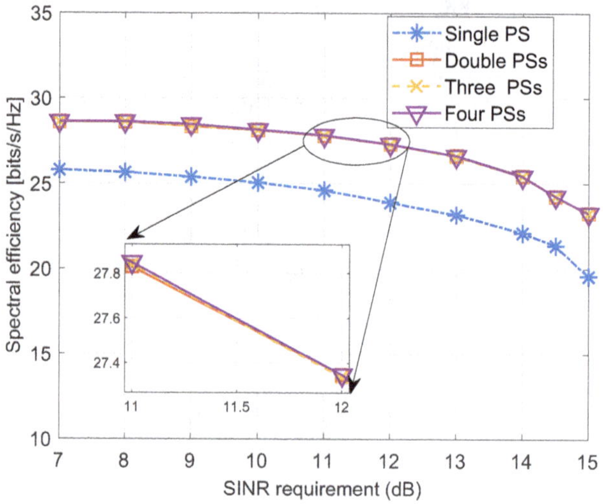

Figure 8.5 The achieved SE values based on different PS structures for different radar SINR requirements with a fixed HBF power budget

total energy $\mathscr{E} = 10$. For the case $N_{\mathrm{RF}} = 2$, the number of streams is considered to be 2. As expected, the larger the number of RF chains, the higher the achieved communication SE. Besides, we also note that as the N_{RF} increases, the gap between the proposed DPS and conventional SPS becomes smaller and smaller and that the degradation trend of the SE becomes larger and larger as the γ increases. Furthermore, Figure 8.4 shows that when the N_{RF} increases, achieving the radar SINR tends to be easier. This is because if the N_{RF} is larger, the larger the degrees of freedom (DoFs) in the optimization design can be used to suppress clutter, resulting in better radar SINR. This phenomenon agrees with our expectations.

Finally, we investigate the influence of adding more phase shifters into each DPS structure when the total HBF power budget is fixed. As shown in Figure 8.5, there are two points we can see clearly: (1) Comparing to the single PS case, the other cases have the extra performance gain around 4 dB constantly at different radar SINR level. (2) For all the PS structures with 2, 3, and 4 PS's, the corresponding curves are close to each other, where the differences are probably caused by the numerical accuracy. This indicates that, for a fixed power budget, the DPS structure is already capable to fully exploit the amplitude controlling in terms of improving the system performance.

8.3 HBF design for multi-carrier DFRC system

8.3.1 System model and problem formulation

In section 8.2, we consider a single-carrier DFRC system with a single user in the presence of an extended target. In this section, we will explore the HBF design for multi-carrier DFRC system with multiple users.

Specifically, we suppose that an M_t-antenna DFRC system estimates the DOAs of targets in the region of interest and sends communication symbols to U downlink users simultaneously. The transmit array is assumed to be uniform linear array (ULA) with half-wavelength inter-element spacing. The number of antennas at the u-th user is N_{cu}. To reduce the number of RF chains, the transmitter adopts a fully connected HBF architecture. In other words, each RF chain is connected to all antennas through analog phase shifters, as shown in Figure 8.6. Since each user is commonly configured with a small number of antennas, we assume each user employs the fully digital beamforming structure. The system operates in time division duplex (TDD) model [47].

8.3.1.1 Transmit model

Suppose that the concatenation of each user's data stream is MPSK signaling, which are embedded into U subcarrier signals $\phi_u(t) = \frac{1}{\sqrt{T_0}} e^{\jmath 2\pi u \Delta f t}, u = 1, \cdots, U$*, where Δf is the frequency step. It is assumed that $\phi_u(t), u = 1, \cdots, U$, are orthogonal with unit energy, i.e., $\int_{T_0} \phi_u(t)\phi_{\bar{u}}^*(t)\mathrm{d}t = \begin{cases} 1, & u = \bar{u} \\ 0, & u \neq \bar{u} \end{cases}$, where t is fast time index within one pulse, T_0 is transmit pulsewidth. Specifically, during each transmit pulse T_0, N symbols $\mathbf{s}_u = [s_{u,1}, \cdots, s_{u,N}]^T$ can be embedded, the symbol width is Δt, i.e., $T_0 = N\Delta t$. It is assumed that the symbols for different users are independent with zero-mean and covariance matrix $\mathbb{E}\left\{\mathbf{s}_u \mathbf{s}_u^H\right\} = \mathbf{I}_N$ [29].

During the transmit pulse T_0, the modulated set of the signals is defined as

$$\psi_u(t) = \sqrt{\frac{E}{UT_0}} \sum_{n=1}^{N} s_{u,n} e^{\jmath 2\pi u \Delta f t} \mathrm{rect}\left(t - (n-1)\Delta t\right) \tag{8.32}$$

where E represents the total transmit power, and $\mathrm{rect}\,(t) = \begin{cases} 1, & 0 < t < \Delta t \\ 0, & \text{otherwise.} \end{cases}$.

It can be verified that if $\Delta f = k/\Delta t, k = \pm 1, \pm 2, \cdots$, the modulated signals $\psi_u(t), u = 1, \cdots, U$, are orthogonal, i.e., we have

$$\begin{aligned} \int_{T_0} \psi_u(t)\psi_{\bar{u}}^*(t)\mathrm{d}t &= \frac{E}{UT_0} \sum_{n=1}^{N} s_{u,n} s_{\bar{u},n}^* \int_{\Delta t} e^{\jmath 2\pi (u-\bar{u})\Delta f t} \mathrm{rect}\left(t - (n-1)\Delta t\right)\mathrm{d}t \\ &= \begin{cases} \frac{E}{U}, & u = \bar{u} \\ 0, & u \neq \bar{u}. \end{cases} \end{aligned} \tag{8.33}$$

Here we note that the transmit power associated with the uth waveform is magnified by factor $\frac{E}{U}$ due to dividing the fixed total transmit power E over U. As a consequence, the smaller the U, the larger the power per waveform can be resulted in. The transmit system first performs digital precoding by using $\mathbf{V}_D = [\mathbf{V}_{D1}, \cdots, \mathbf{V}_{DU}] \in \mathbb{C}^{N_{\mathrm{RF}} \times UN}$, with the output given by $\mathbf{y}(t) = [y_1(t), \cdots, y_{N_{\mathrm{RF}}}(t)]^T$, where $y_{n_t}(t), n_t = 1, \cdots, N_{\mathrm{RF}}$, is

$$y_{n_t}(t) = \sqrt{\frac{E}{UT_0}} \sum_{u=1}^{U} \sum_{n=1}^{N} V_{Du}\left(n_t, n\right) s_{u,n} e^{\jmath 2\pi u \Delta f t} \mathrm{rect}\left(t - (n-1)\Delta t\right) \tag{8.34}$$

*In this chapter, we assume that the number of users is known at the transmitter in advance, and set the number of subcarriers equal to that of users.

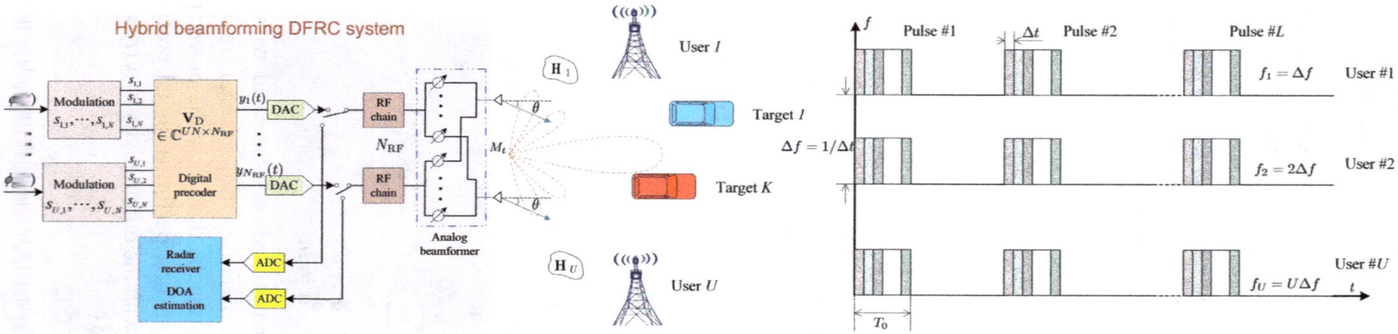

Figure 8.6 *(a) Overview of hybrid beamforming DFRC system, where the transmitter adopts the fully connected architecture, and the radar receiver shares the same platform with the transmitter. (b) The signal frame structure for the DFRC system in (a).*

which is up-converted to the carrier frequency through N_{RF} RF chains. Then, an RF beamformer $\mathbf{V}_{\mathrm{RF}} \in \mathbb{C}^{M_t \times N_{\mathrm{RF}}}$, realized by analog phase shifters, is utilized to form the final transmit signal. The entries of $\mathbf{V}_{\mathrm{RF}}$ have a constant modulus $\frac{1}{\sqrt{M_t}}$, i.e., $|\mathbf{V}_{\mathrm{RF}}(i,j)| = \frac{1}{\sqrt{M_t}}, i = 1, \cdots, M_t; j = 1, \cdots, N_{\mathrm{RF}}$. According to the above model, the transmitted signal at the transmit array can be written as

$$\mathbf{x}(t) = \mathbf{V}_{\mathrm{RF}}\mathbf{y}(t) \tag{8.35}$$

8.3.1.2 Communication model

At the u-th user, the received signal can be modeled as

$$\mathbf{c}_u(t) = \mathbf{H}_u\mathbf{V}_{\mathrm{RF}}\mathbf{y}(t) + \mathbf{z}_u(t) \tag{8.36}$$

where $\mathbf{H}_u \in \mathbb{C}^{M_r \times N_{cu}}$ is the complex channel matrix from the transmitter to the u-th user array. We assume that the CSIs are perfectly estimated by the channel estimation techniques [12,39,40], and are known both by the transmitter and users. $\mathbf{z}_u(t)$ is additive Gaussian noise vector with zero mean and covariance matrix $\sigma_z^2\mathbf{I}_{M_r}$. The user u first down-converts the received signal to baseband via the RF chains, and the baseband signal can be modeled as

$$\begin{aligned}\mathbf{q}_u &= \sum_{n=1}^{N}\int_{T_0}\mathbf{c}_u(t)e^{-\jmath 2\pi u\Delta f t}\mathrm{rect}\left(t-(n-1)\Delta t\right)\mathrm{d}t \\ &= \sqrt{\frac{E}{UT_0}}\Delta t\mathbf{H}_u\mathbf{V}_{\mathrm{RF}}\mathbf{V}_{Du}\mathbf{s}_u + \mathbf{n}_u\end{aligned} \tag{8.37}$$

where $\mathbf{n}_u = \sum_{n=1}^{N}\int_{T_0}\mathbf{z}_u(t)e^{-\jmath 2\pi u\Delta f t}\mathrm{rect}\left(t-(n-1)\Delta t\right)\mathrm{d}t$, which follows $\mathcal{CN}(\mathbf{0}, T_0\sigma_z^2\mathbf{I}_{M_r})$. Finally, adopting an $N_{cu} \times N$ digital combiner $\mathbf{W}_u = [\mathbf{w}_{u,1}, \cdots, \mathbf{w}_{u,N}]$, then the estimated symbol of the n-th transmitted stream of the u-th user can be obtained as

$$\begin{aligned}\hat{s}_{u,n} &= \mathbf{w}_{u,n}^H\mathbf{q}_u \\ &= \sqrt{\frac{E}{UT_0}}\Delta t\mathbf{w}_{u,n}^H\mathbf{H}_u\mathbf{v}_{u,n}s_{u,n} + \sqrt{\frac{E}{UT_0}}\Delta t\sum_{m\neq n}\mathbf{w}_{u,n}^H\mathbf{H}_u\mathbf{v}_{u,m}s_{u,m} + \mathbf{w}_{u,n}^H\mathbf{n}_u\end{aligned} \tag{8.38}$$

where $\mathbf{v}_{u,n}$ denotes the n-th column of the matrix $\mathbf{V}_{\mathrm{RF}}\mathbf{V}_{Du}$.

Accordingly, the sum-rate of the u-th user is given by [26]

$$R_u(\mathbf{V}_{\mathrm{D}}, \mathbf{V}_{\mathrm{RF}}, \mathbf{W}_u) = \sum_{n=1}^{N}\log\left(1 + \frac{\left|\mathbf{w}_{u,n}^H\mathbf{H}_u\mathbf{v}_{u,n}\right|^2}{\sum_{m\neq n}\left|\mathbf{w}_{u,n}^H\mathbf{H}_u\mathbf{v}_{u,m}\right|^2 + \beta\mathbf{w}_{u,n}^H\mathbf{w}_{u,n}}\right) \tag{8.39}$$

where $\beta = \frac{UN^2\sigma_z^2}{E}$.

8.3.1.3 Radar model

For the radar function, we assume that the receive array shares the same platform with the transmitter and that K targets of interest are present. The received signal can be written as

$$\mathbf{r}(t,l) = \sum_{k=1}^{K} \alpha_k(l)\big(\mathbf{a}^T(\theta_k)\mathbf{x}(t)\big)\mathbf{a}(\theta_k) + \mathbf{z}_r(t,l) \tag{8.40}$$

where l is the slow-time index, $\alpha_k(l)$ denotes the reflection coefficient within the lth pulse and follows the Swerling-II model [48], i.e., it is assumed to be constant during each pulse duration but varying from pulse to pulse. More concretely, $\{\alpha_k(l)\}$ are assumed to be independent zero mean random variables with variances $\sigma^2_{\alpha,k}$. θ_k is spatial angle associated with the k-th target. $\mathbf{z}_r(t,l)$ is the $M_t \times 1$ vector of zero-mean white Gaussian noise with variance σ^2_{zr}.

The received signals are first combined via the analog beamformer $\mathbf{V}_{\mathrm{RF}}$, then the received baseband signal vector $\mathbf{g}(t,l) \in \mathbb{C}^{N_{\mathrm{RF}}\times 1}$ becomes

$$\mathbf{g}(t,l) = \sum_{k=1}^{K} \alpha_k(l)\mathbf{V}^H_{\mathrm{RF}}\mathbf{a}(\theta_k)\big(\mathbf{a}^T(\theta_k)\mathbf{x}(t)\big) + \mathbf{V}^H_{\mathrm{RF}}\mathbf{z}_r(t,l) \tag{8.41}$$

By matched filtering $\mathbf{g}(t,l)$ to each of the orthogonal basis waveforms $\psi_u(t)$, we can obtain the virtual data vector corresponding to the u-th waveform, as†

$$\begin{aligned}\mathbf{h}_u(l) &= \int_{T_0} \mathbf{g}(t,l)\psi_u^*(t)\mathrm{d}t \\ &= \frac{E}{UN}\sum_{k=1}^{K} \alpha_k\,(l)\,\big(\mathbf{a}^T(\theta_k)\mathbf{V}_{\mathrm{RF}}\mathbf{V}_{Du}\mathbf{1}_N\big)\mathbf{V}^H_{\mathrm{RF}}\mathbf{a}(\theta_k) + \mathbf{V}^H_{\mathrm{RF}}\widetilde{\mathbf{z}}_{ru}(l)\end{aligned} \tag{8.42}$$

where $\mathbf{1}_N$ denotes an $N \times 1$ vector whose all elements are 1, and $\widetilde{\mathbf{z}}_{ru}(l) = \int_{T_0} \mathbf{z}_r(t,l)\psi_u^*(t)\mathrm{d}t$ has zero mean and covariance matrix $\frac{E\sigma^2_{zr}}{U}\mathbf{I}_{M_t}$. From (8.42), we also note that the smaller the U, the larger the power per waveform can be resulted in.

Stacking the individual components $\mathbf{h}_u(l), u = 1, \cdots, U$ in one column vector, we can obtain the $UN_{\mathrm{RF}} \times 1$ virtual data vector as

$$\begin{aligned}\mathbf{h}_{\mathrm{all}}(l) &= \big[\mathbf{h}_1^T(l), \cdots, \mathbf{h}_U^T(l)\big]^T \\ &= \frac{E}{UN}\sum_{k=1}^{K} \alpha_k\,(l)\,\big(\mathbf{D}^T\mathbf{a}(\theta_k)\big) \otimes \big(\mathbf{V}^H_{\mathrm{RF}}\mathbf{a}(\theta_k)\big) + \widetilde{\mathbf{z}}_{\mathrm{all}}\,(l)\end{aligned} \tag{8.43}$$

where $\mathbf{D} = \big[\mathbf{V}_{\mathrm{RF}}\mathbf{V}_{D1}\mathbf{1}_N, \cdots, \mathbf{V}_{\mathrm{RF}}\mathbf{V}_{DU}\mathbf{1}_N\big]^T$ and $\widetilde{\mathbf{z}}_{\mathrm{all}}\,(l) = \big[\widetilde{\mathbf{z}}_{r1}^T(l)\mathbf{V}^*_{\mathrm{RF}}, \cdots, \widetilde{\mathbf{z}}_{rU}^T(l)\mathbf{V}^*_{\mathrm{RF}}\big]^T$ whose covariance matrix is $\frac{E\sigma^2_{zr}}{U}\mathrm{Bdiag}\{\mathbf{V}^H_{\mathrm{RF}}\mathbf{V}_{\mathrm{RF}}, \cdots, \mathbf{V}^H_{\mathrm{RF}}\mathbf{V}_{\mathrm{RF}}\} \in \mathbb{C}^{UN_{\mathrm{RF}}\times UN_{\mathrm{RF}}}$. In this chapter, we assume that $\mathbf{V}^H_{\mathrm{RF}}\mathbf{V}_{\mathrm{RF}} = \mathbf{I}_{N_{\mathrm{RF}}}$ to keep the effective noise being white Gaussian noise, this will be beneficial to apply the subspace-based DOA estimator for the DFRC system with HBF architecture [49].

†Without loss of generality, zero range shift of the target is assumed in this model for convenience. Actually, the matched filtering step is performed for each range bin, namely, the time-delayed received signal is matched filtered to a time-delayed version of the waveforms $\psi_u^*(t), \forall u$.

Remark 1. *The virtual signal mode* (8.42) *provides the two important factors for radar application. (i) As compared to SIMO (single-input multiple-output) radar, which transmits a single waveform, the benefit of this model is to obtain the virtual aperture to increase the resolution of DOA estimation by adding the number of the orthogonal waveforms, and (ii) In this model, the transmit energy can be focused on a certain spatial sector to increase the SNR per virtual element by carefully designing the* $\mathbf{V}_D$ *and* $\mathbf{V}_{\mathrm{RF}}$*, resulting in improving the SNR per each virtual antenna. In addition, as discussed earlier, the larger the U will lead to less power per waveform. This scheme is similar to the transmit beamspace technique in [50].*

For DOA estimation, the amount of the transmit energy focused on the potential spatial region of interest should be maximized and the amount of energy out of the region should be minimized. Thus, we propose to optimize $\mathbf{V}_D$ and $\mathbf{V}_{\mathrm{RF}}$ to minimize the mean squared error (MSE) between the spatial spectrum $P(\theta)$ and a desired one $d(\theta)$ which is predefined to guarantee excellent directional transmission for radar applications [14,51]. More specifically, assuming that the spatial range $(-90°, 90°)$ is discredited as $\mathscr{K}$ points, the spatial spectrum MSE can be expressed as

$$\mathrm{MSE}(\mathbf{V}_{\mathrm{D}}, \mathbf{V}_{\mathrm{RF}}) = \sum_{q=1}^{\mathscr{K}} \left| P(\theta_q) - \tau^2 d(\theta_q) \right|^2 \tag{8.44}$$

where $P(\theta)$ is defined as $P(\theta) = \int_{T_0} \left| \mathbf{a}^T(\theta)\mathbf{x}(t) \right|^2 \mathrm{d}t$, which can be written explicitly as $P(\theta) = \mathbf{a}^T(\theta)\mathbf{V}_{\mathrm{RF}}\big(\sum_{u=1}^{U} \mathbf{V}_{Du}\mathbf{V}_{Du}^H \big)\mathbf{V}_{\mathrm{RF}}^H \mathbf{a}^*(\theta)$, and τ is a scaling parameter to be optimized.

8.3.1.4 Problem formulation

According to the above models for both radar and communication, a meaningful formulation of the joint optimization of the digital and analog beamformer is to minimize the spatial spectrum MSE while ensuring the QoS requirement for each user. Mathematically, the problem can be formulated as

$$\min_{\tau, \mathbf{V}_D, \mathbf{V}_{\mathrm{RF}}, \{\mathbf{W}_u\}} \quad \mathrm{MSE}(\mathbf{V}_{\mathrm{D}}, \mathbf{V}_{\mathrm{RF}}) \tag{8.45a}$$

$$\text{s.t.} \quad R_u(\mathbf{V}_{\mathrm{D},u}, \mathbf{V}_{\mathrm{RF}}, \mathbf{W}_u) \geq \gamma_u, \forall u \tag{8.45b}$$

$$\left| \mathbf{V}_{\mathrm{RF}}(i,j) \right| = 1/\sqrt{M_t}, \forall i,j, \tag{8.45c}$$

$$\mathbf{V}_{\mathrm{RF}}^H \mathbf{V}_{\mathrm{RF}} = \mathbf{I}_{N_{\mathrm{RF}}}, \tag{8.45d}$$

$$\sum_{u=1}^{U} \mathrm{Tr}\left(\mathbf{V}_{\mathrm{RF}}\mathbf{V}_{Du}\mathbf{V}_{Du}^H\mathbf{V}_{\mathrm{RF}}^H \right) = U, \tag{8.45e}$$

where (8.45b) is the QoS requirements for users with γ_u being the rate requirement for the u-th user, (8.45c) is the constant modulus constraint for analog beamformer, (8.45d) is the orthogonality constraint which guarantees the DOA estimation by using the subspace-based techniques, such as MUSIC [49], (8.45e) is the power constraint for the transmit beamformer. Note that the constraint (8.45e) enable the total transmit power of the transmitted signal $\mathbf{x}(t)$ in (8.35) to be E. In addition, according to the constraint (8.45d), the constraint (8.45e) is equivalent to $\sum_{u=1}^{U} \mathrm{Tr}\left(\mathbf{V}_{Du}\mathbf{V}_{Du}^H\right) = U$.

It is seen that the above optimization problem involves a fourth-order objective function and nonconvex constraints (8.45b)-(8.45e), and hence, it is NP-hard [43] and cannot be efficiently solved. To this end, in what follows, the consensus-ADMM algorithm [44] based on the WMMSE framework [52] is devised to tackle the problem (8.45).

8.3.2 *Proposed HBF algorithm for multi-carrier DFRC system*

8.3.2.1 Reformulation of problem (8.45)

Obviously, it is seen that problem (8.45) with QoS constraints is very complicated and is challenging to solve directly by using the ADMM framework. However, the problem can be recast as a relatively simple form based on the WMMSE framework. To apply the WMMSE, we need to decouple the relation between $\mathbf{V}_{\mathrm{D}}$ and $\mathbf{V}_{\mathrm{RF}}$ in constraints (8.45b). Thus, we introduce several auxiliary variables $\mathbf{T}_1, \cdots, \mathbf{T}_U$ to convert the problem (8.45) as

$$\min_{\tau,\mathbf{T},\mathbf{V}_D,\mathbf{V}_{\mathrm{RF}}} \quad \mathrm{MSE}(\mathbf{T}) \tag{8.46a}$$

$$\text{s.t.} \quad R_u(\mathbf{T}_u, \mathbf{W}_u) \geq \gamma_u, \tag{8.46b}$$

$$\mathbf{T}_u = \mathbf{V}_{\mathrm{RF}}\mathbf{V}_{Du}, \forall u \tag{8.46c}$$

$$\sum_{u=1}^{U} \mathrm{Tr}\left(\mathbf{V}_{Du}\mathbf{V}_{Du}^H\right) = U, \tag{8.46d}$$

$$\text{Constraints (8.45c), (8.45d).} \tag{8.46e}$$

where $\mathbf{T} = [\mathbf{T}_1, \cdots, \mathbf{T}_U] \in \mathbb{C}^{M_t \times UN}$, and $\mathrm{MSE}(\mathbf{T})$ and $R_u(\mathbf{T}_u, \mathbf{W}_u)$ are separately defined as

$$\mathrm{MSE}(\mathbf{T}) = \sum_{q=1}^{\mathcal{K}} \left|\mathbf{a}^T(\theta_q)\mathbf{T}\mathbf{T}^H\mathbf{a}^*(\theta) - \tau^2 d(\theta_q)\right|^2 \tag{8.47}$$

and

$$R_u(\mathbf{T}_u, \mathbf{W}_u) = \sum_{n=1}^{N} \log\left(1 + \frac{\left|\mathbf{w}_{u,n}^H\mathbf{H}_u\mathbf{t}_{u,n}\right|^2}{\sum_{m\neq n}\left|\mathbf{w}_{u,n}^H\mathbf{H}_u\mathbf{t}_{u,m}\right|^2 + \beta\mathbf{w}_{u,n}^H\mathbf{w}_{u,n}}\right) \tag{8.48}$$

where $\mathbf{t}_{u,n}$ denotes the beamforming vector for the n-th stream of the u-th user, corresponding to the n-th column of $\mathbf{T}_u$. By adopting the WMMSE approach and MMSE receiver $\mathbf{W}_u$ [52], the n-th column of $\mathbf{W}_u$ is given by

$$\mathbf{w}_{u,n} = \frac{1}{\zeta}\left(\mathbf{H}_u\mathbf{t}_{u,n}\mathbf{t}_{u,n}^H\mathbf{H}_u^H + \sum_{m\neq n}\mathbf{H}_u\mathbf{t}_{u,m}\mathbf{t}_{u,m}^H\mathbf{H}_u^H + \beta\mathbf{I}_{M_r}\right)^{-1}\mathbf{H}_u\mathbf{t}_{u,n} \tag{8.49}$$

where $\omega_{u,n} = 1/e_{u,n}$ is the weight with $e_{u,n}$ being the mean-square estimation error. After obtaining $\mathbf{W}_u$ and $\omega_{u,n}$, problem (8.46) can be reformulated as

$$\min_{\tau,\mathbf{T},\mathbf{V}_D,\mathbf{V}_{\mathrm{RF}}} \quad \mathrm{MSE}(\mathbf{T}) \tag{8.50a}$$

$$\text{s.t.} \quad \sum_{n=1}^{N} \omega_{u,n} e_{u,n} \leq \sum_{n=1}^{N} \log(\omega_{u,n}) + N - \gamma_u, \tag{8.50b}$$

$$\text{Constraints (8.46c), (8.46d), (8.46e).} \tag{8.50c}$$

To solve the problem (8.50) in a distributed manner, we introduce several auxiliary variables to further transform problem (8.50) as

$$\min_{\tau,\mathbf{T},\mathbf{V}_D,\mathbf{V}_{\mathrm{RF}},\{Q_{u,n}^{u,m}\}} \quad \mathrm{MSE}(\mathbf{T}) \tag{8.51a}$$

$$\text{s.t.} \quad \sum_{n=1}^{N} \omega_{u,n} \left(\left|Q_{u,n}^{u,n} - 1\right|^2 + \sum_{m \neq n} \left|Q_{u,n}^{u,m}\right|^2 \right) \leq \xi_u, \tag{8.51b}$$

$$Q_{u,n}^{u,m} = \zeta \mathbf{w}_{u,n}^H \mathbf{H}_u \mathbf{t}_{u,m}, \forall n, m, \tag{8.51c}$$

$$\text{Constraints (8.50c).} \tag{8.51d}$$

where $\xi_u = \sum_{n=1}^{N} \left(\log(\omega_{u,n}) - \omega_{u,n} T_0 \sigma_z^2 \left\|\mathbf{w}_{u,n}\right\|^2 \right) + N - \gamma_u$.

In addition, the fourth-order objective function in problem (8.51) makes the problem further difficult to straightforwardly solve. Inspiring by the concept of cyclic-algorithm-new (CAN) method in [53,54], we further modify the problem (8.51) by introducing a set of UN-dimensional auxiliary vectors $\{\varsigma_q\}_{q=1}^{\mathcal{K}}$ as

$$\min_{\tau,\mathbf{T},\{\varsigma_q\},\mathbf{V}_D,\mathbf{V}_{\mathrm{RF}},\{Q_{u,n}^{u,m}\}} \quad \sum_{q=1}^{\mathcal{K}} \left\|\mathbf{T}^T \mathbf{a}(\theta_q) - \tau \varsigma_q\right\|^2 \tag{8.52a}$$

$$\text{s.t.} \quad \left\|\varsigma_q\right\| = \sqrt{d(\theta_q)}, \forall q \tag{8.52b}$$

$$\text{Constraints (8.51b), (8.51c), (8.51d).} \tag{8.52c}$$

8.3.2.2 The proposed consensus-ADMM for solving problem (8.52)

Placing the equality constraints $\mathbf{T}_u = \mathbf{V}_{\mathrm{RF}} \mathbf{V}_{D,u}, \forall u$ and $Q_{u,n}^{u,m} = \zeta \mathbf{w}_{u,n}^H \mathbf{H}_u \mathbf{t}_{u,m}, \forall n, m$ of problem (8.52) into the augmented Lagrangian function (scaled form) of problem (8.52) yields

$$\begin{aligned}
&\mathcal{L}\left(\tau, \{\varsigma_q\}, \mathbf{T}, \mathbf{V}_D, \mathbf{V}_{\mathrm{RF}}, \left\{Q_{u,n}^{u,m}\right\}, \left\{\varpi_{u,n}\right\}, \left\{\lambda_{u,n}^{u,m}\right\}\right) \\
&= \sum_{q=1}^{\mathcal{K}} \left\|\mathbf{T}^T \mathbf{a}(\theta_q) - \tau \varsigma_q\right\|^2 + \frac{\rho_1}{2} \sum_{u=1}^{U} \sum_{n=1}^{N} \left\|\mathbf{t}_{u,n} - \mathbf{v}_{u,n} + \varpi_{u,n}\right\|^2 \\
&+ \frac{\rho_2}{2} \sum_{u=1}^{U} \sum_{n=1}^{N} \sum_{m=1}^{N} \left|Q_{u,n}^{u,m} - \zeta \mathbf{w}_{u,n}^H \mathbf{H}_u \mathbf{t}_{u,m} + \lambda_{u,n}^{u,m}\right|^2
\end{aligned} \tag{8.53}$$

where $\{\varpi_{u,n}\}$ and $\{\lambda_{u,n}^{u,m}\}$ are dual variables associated with the equality constraints $\mathbf{t}_{u,n} = \mathbf{v}_{u,n}$ with $\mathbf{v}_{u,n}$ being defined in (8.38) and $\mathcal{Q}_{u,n}^{u,m} = \zeta \mathbf{w}_{u,n}^H \mathbf{H}_u \mathbf{t}_{u,m}$, respectively. In addition, $\rho_1, \rho_2 > 0$ are the corresponding penalty parameters.

Based on the convergence requirements of the consensus-ADMM, we split the optimized primal variables into two blocks $(\tau, \{\varsigma_q\}, \mathbf{T})$ and $(\mathbf{V}_D, \mathbf{V}_{\mathrm{RF}}, \{\mathcal{Q}_{u,n}^{u,m}\})$. Therefore, at the $(k+1)$-th iteration, the main algorithmic steps of the proposed consensus-ADMM method are related to solving the following problems:

Stage 1: Optimize $(\tau(k+1), \{\varsigma_q(k+1)\}, \mathbf{T}(k+1))$

With the fixed $(\mathbf{V}_D(k), \mathbf{V}_{\mathrm{RF}}(k), \{\mathcal{Q}_{u,n}^{u,m}(k)\})$ and $(\{\varpi_{u,n}(k)\}, \{\lambda_{u,n}^{u,m}(k)\})$ by solving the following problem‡:

$$\begin{aligned} &\min_{\tau,\{\varsigma_q\},\mathbf{T}} \mathscr{L}\left(\tau, \{\varsigma_q\}, \mathbf{T}, \mathbf{V}_D, \mathbf{V}_{\mathrm{RF}}, \{\mathcal{Q}_{u,n}^{u,m}\}, \{\varpi_{u,n}\}, \{\lambda_{u,n}^{u,m}\}\right) \\ &\text{s.t.} \quad \|\varsigma_q\| = \sqrt{d(\theta_q)}, \forall q \end{aligned} \tag{8.54}$$

which can be decomposed into three problems:

(a): The update of ς_q needs solving

$$\min_{\{\varsigma_q\}} \sum_{q=1}^{\mathscr{K}} \left\|\mathbf{T}^T \mathbf{a}(\theta_q) - \tau\varsigma_q\right\|^2 \quad \text{s.t.} \quad \|\varsigma_q\| = \sqrt{d(\theta_q)}, \forall q \tag{8.55}$$

whose closed-form solution to problem (8.55) can be attained by using the Lagrangian multiplier method, as

$$\varsigma_q = \sqrt{d(\theta_q)}\mathbf{T}^T\mathbf{a}(\theta_q) / \left\|\mathbf{T}^T\mathbf{a}(\theta_q)\right\| \tag{8.56}$$

(b): The update of τ needs solving

$$\min_{\tau} \sum_{q=1}^{\mathscr{K}} \left\|\mathbf{T}^T \mathbf{a}(\theta_q) - \tau\varsigma_q\right\|^2 \tag{8.57}$$

whose closed-form solution can be readily obtained as

$$\tau = \sum_{q=1}^{\mathscr{K}} \Re\left\{\mathbf{a}^T(\theta_q)\mathbf{T}\varsigma_q^*\right\} / \sum_{q=1}^{\mathscr{K}} \|\varsigma_q\|^2 \tag{8.58}$$

(c) Solution to Problem (8.59)***:***

$$\begin{aligned} \min_{\mathbf{T}} &\sum_{q=1}^{\mathscr{K}} \left\|\mathbf{T}^T \mathbf{a}(\theta_q) - \tau\varsigma_q\right\|^2 + \frac{\rho_1}{2}\|\mathbf{T} - \mathbf{B}\|_F^2 \\ &+ \frac{\rho_2}{2} \sum_{u=1}^{U}\sum_{n=1}^{N}\sum_{m=1}^{N} \left|\mathcal{Q}_{u,n}^{u,m} - \zeta\mathbf{w}_{u,n}^H\mathbf{H}_u\mathbf{t}_{u,m} + \lambda_{u,n}^{u,m}\right|^2 \end{aligned} \tag{8.59}$$

‡For notational simplicity, the iteration index (k) are omitted in the following optimization problems.

where $\mathbf{B} = [\mathbf{B}_1, \cdots, \mathbf{B}_U] \in \mathbb{C}^{M_t \times UN}$ with $\mathbf{B}_u = [\mathbf{v}_{u,1} - \boldsymbol{\varpi}_{u,1}, \cdots, \mathbf{v}_{u,N} - \boldsymbol{\varpi}_{u,N}] \in \mathbb{C}^{M_t \times N}$. The closed-form solution of problem (8.59) is attained by taking its first-order optimality condition, as

$$\mathbf{T}_u = \left(\boldsymbol{\Omega} + \frac{\rho_1}{2}\mathbf{I}_{M_t} + \frac{\rho_2 \zeta^2}{2}\mathbf{H}_u^H \mathbf{W}_u \mathbf{W}_u^H \mathbf{H}_u\right)^{-1} \times \left(\mathbf{A}_u^{\varsigma} + \frac{\rho_1}{2}\mathbf{B}_u + \frac{\rho_2 \zeta}{2}\mathbf{H}_u^H \mathbf{W}_u \mathbf{Q}_u^{\lambda}\right) \tag{8.60}$$

where $\boldsymbol{\Omega} \triangleq \sum_{q=1}^{\mathcal{K}} \mathbf{a}^*(\theta_q)\mathbf{a}^T(\theta_q)$, $\mathbf{W}_u \triangleq [\mathbf{w}_{u,1}, \cdots, \mathbf{w}_{u,N}] \in \mathbb{C}^{N_{cu} \times N}$, $\mathbf{A}^{\varsigma} = [\mathbf{A}_1^{\varsigma}, \cdots, \mathbf{A}_U^{\varsigma}] \triangleq \sum_{q=1}^{\mathcal{K}} \tau \mathbf{a}^*(\theta_q)\boldsymbol{\varsigma}_q^T \in \mathbb{C}^{M_t \times UN}$ and the (m,n)-th entry of $\mathbf{Q}_u^{\lambda} \in \mathbb{C}^{N \times N}$ is $Q_{u,n}^{u,m} + \lambda_{u,n}^{u,m}$.

Stage 2: Optimize $(\mathbf{V}_D(k+1), \mathbf{V}_{\text{RF}}(k+1), \{Q_{u,n}^{u,m}(k+1)\})$

With the fixed $(\tau(k+1), \{\boldsymbol{\varsigma}_q(k+1)\}, \mathbf{T}(k+1))$ and $(\{\boldsymbol{\varpi}_{u,n}(k)\}, \{\lambda_{u,n}^{u,m}(k)\})$ by solving

$$\begin{aligned} \min_{\substack{\mathbf{V}_D, \mathbf{V}_{\text{RF}} \\ \{Q_{u,n}^{u,m}\}}} \; & \mathcal{L}\left(\tau, \{\boldsymbol{\varsigma}_q\}, \mathbf{T}, \mathbf{V}_D, \mathbf{V}_{\text{RF}}, \left\{Q_{u,n}^{u,m}\right\}, \left\{\boldsymbol{\varpi}_{u,n}\right\}, \left\{\lambda_{u,n}^{u,m}\right\}\right) \\ \text{s.t.} \; & \sum_{n=1}^{N} \omega_{u,n}\left(\left|Q_{u,n}^{u,n} - 1\right|^2 + \sum_{m \neq n}\left|Q_{u,n}^{u,m}\right|^2\right) \le \xi_u, \\ & |\mathbf{V}_{\text{RF}}(i,j)| = 1/\sqrt{M_t}, \forall i,j, \\ & \mathbf{V}_{\text{RF}}^H \mathbf{V}_{\text{RF}} = \mathbf{I}_{N_{\text{RF}}}, \; \sum_{u=1}^{U} \text{Tr}\left(\mathbf{V}_{Du}\mathbf{V}_{Du}^H\right) = U \end{aligned} \tag{8.61}$$

which can be decomposed into

(a): The update of $\left\{Q_{u,n}^{u,m}\right\}$ needs solving

$$\begin{aligned} \min_{\{Q_{u,n}^{u,m}\}} \; & \sum_{n=1}^{N}\sum_{m=1}^{N} \left|Q_{u,n}^{u,m} - \zeta \mathbf{w}_{u,n}^H \mathbf{H}_u \mathbf{t}_{u,m} + \lambda_{u,n}^{u,m}\right|^2 \\ \text{s.t.} \; & \sum_{n=1}^{N} \omega_{u,n}\left(\left|Q_{u,n}^{u,n} - 1\right|^2 + \sum_{m \neq n}\left|Q_{u,n}^{u,m}\right|^2\right) \le \xi_u, \end{aligned} \tag{8.62}$$

For the problem (8.62), we note that it is a quadratic optimization problem, and thus we can solve it by using the KKT conditions [43].

(b): The update of $\mathbf{V}_{\text{D}}$ needs solving

$$\min_{\mathbf{V}_{\text{D}}} \; \sum_{u=1}^{U} \|\mathbf{G}_u - \mathbf{V}_{\text{RF}}\mathbf{V}_{\text{D}u}\|_F^2 \qquad \text{s.t.} \; \sum_{u=1}^{U} \text{Tr}\left(\mathbf{V}_{Du}\mathbf{V}_{Du}^H\right) = U \tag{8.63}$$

For the problem (8.63), it is a conventional QP problem, whose closed-form solution can be achieved by the KKT conditions [43].

(c): The update of $\mathbf{V}_{\mathrm{RF}}$ needs solving

$$\begin{aligned} \min_{\mathbf{V}_{\mathrm{RF}}} \quad & \sum_{u=1}^{U} \|\mathbf{G}_u - \mathbf{V}_{\mathrm{RF}} \mathbf{V}_{\mathrm{D}u}\|_F^2 \\ \text{s.t.} \quad & |\mathbf{V}_{\mathrm{RF}}(i,j)| = 1/\sqrt{M_t}, \forall i,j, \quad \mathbf{V}_{\mathrm{RF}}^H \mathbf{V}_{\mathrm{RF}} = \mathbf{I}_{N_{\mathrm{RF}}} \end{aligned} \tag{8.64}$$

where $\mathbf{G}_u = [\mathbf{t}_{u,1} + \boldsymbol{\varpi}_{u,1}, \cdots, \mathbf{t}_{u,N} + \boldsymbol{\varpi}_{u,N}] \in \mathbb{C}^{M_t \times N}, \forall u$.

By defining $\boldsymbol{\Upsilon} \triangleq \sqrt{M_t}\mathbf{V}_{\mathrm{RF}}$, problem (8.64) can be further simplified as

$$\begin{aligned} \min_{\boldsymbol{\Upsilon}} \quad & \sum_{u=1}^{U} \|\mathbf{G}_u - \boldsymbol{\Upsilon} \mathbf{V}_{\mathrm{D}u}/\sqrt{M_t}\|_F^2 \\ \text{s.t.} \quad & |\boldsymbol{\Upsilon}(i,j)| = 1, \forall i,j, \\ & \boldsymbol{\Upsilon}^H \boldsymbol{\Upsilon} = M_t \mathbf{I}_{N_{\mathrm{RF}}} \end{aligned} \tag{8.65}$$

It can be observed that the feasible set of problem (8.65) can be interpreted as the intersection of the complex circle manifold $\mathcal{M}_c = \{\boldsymbol{\Upsilon} \in \mathbb{C}^{M_t \times N_{\mathrm{RF}}} : |\boldsymbol{\Upsilon}(i,j)| = 1, \forall i,j\}$ and Stiefel manifold $\mathcal{M}_s = \{\boldsymbol{\Upsilon} \in \mathbb{C}^{M_t \times N_{\mathrm{RF}}} : \boldsymbol{\Upsilon}^H \boldsymbol{\Upsilon} = M_t \mathbf{I}_{N_{\mathrm{RF}}}\}$ [55]. Therefore, we easily exploit the manifold optimization method [55–57] to solve problem (8.65).

Stage 3: Update $(\{\boldsymbol{\varpi}_{u,n}(t+1)\}, \{\lambda_{u,n}^{u,m}(t+1)\})$ using

$$\boldsymbol{\varpi}_{u,n}(t+1) = \boldsymbol{\varpi}_{u,n}(t) + (\mathbf{t}_{u,n} - \mathbf{v}_{u,n}), \forall u, n \tag{8.66a}$$

$$\lambda_{u,n}^{u,m}(t+1) = \lambda_{u,n}^{u,m}(t) + Q_{u,n}^{u,m} - \zeta \mathbf{w}_{u,n}^H \mathbf{H}_u \mathbf{t}_{u,m}, \forall n, m \tag{8.66b}$$

Stage 1 to ***Stage 3*** are repeated until a stopping condition is satisfied, such as, a maximum iteration number $N_{\mathrm{ADMM}}^{\max}$ is attained[§]. We summarize the above procedure of the consensus algorithm in Algorithm 8.3.

Algorithm 8.3 Consensus-ADMM for solving problem (8.52)

Input: Set the initial variables $\mathbf{T}(0)$, $\mathbf{V}_D(0)$, $\mathbf{V}_{\mathrm{RF}}(0)$, $\{Q_{u,n}^{u,m}(0)\}$, $\{\boldsymbol{\varpi}_{u,n}(0)\}$, $\{\lambda_{u,n}^{u,m}(0)\}$ and penalty parameters $\rho_1, \rho_2 > 0$.
Set $k = 0$.
repeat
  Update $\varsigma_q(k+1), \forall q$ in parallel according to (8.56).
  Update $\tau(k+1)$ according to (8.58).
  Update $\mathbf{T}_u(k+1), \forall u$ in parallel according to (8.60).
  Update $Q_{u,n}^{u,m}(t+1), \forall m, n$ in parallel by analyzing KKT conditions.
  Update $\mathbf{V}_{\mathrm{D}u}(k+1), \forall u$ in parallel according to KKT conditions.
  Update $\mathbf{V}_{\mathrm{RF}}(k+1), \forall u$ by utilizing manifold optimization method.
  Update $(\{\boldsymbol{\varpi}_{u,n}(k+1)\}, \{\lambda_{u,n}^{u,m}(k+1)\})$ according to (8.66).
  $k = k + 1$.
until $k = N_{\mathrm{ADMM}}^{\max}$.
Output: $\mathbf{T}^\star = \mathbf{T}(k)$, $\mathbf{V}_D^\star = \mathbf{V}_D(k)$, $\mathbf{V}_{\mathrm{RF}}^\star = \mathbf{V}_{\mathrm{RF}}(k)$.

[§] Note that the problem is nonconvex, and the general convergence of consensus-ADMM is still an open question. Nevertheless, it is clearly seen from the simulations that the objective function is convergent and the residuals of the proposed consensus-ADMM method approach zero with increasing iterations.

8.3.3 HBF-based DOA estimation

In this subsection, the DOA estimation method based on transmit HBF is devised for the QoS-constrained DFRC system.

8.3.3.1 MUSIC-based DOA estimation

Define $\boldsymbol{\theta} = [\theta_1, \cdots, \theta_K]$, the virtual data vector in (8.43) can be equivalently recast as

$$\mathbf{h}_{\text{all}}(l) = \frac{E}{UN}\widetilde{\mathbf{A}}(\boldsymbol{\theta})\boldsymbol{\alpha}(l) + \widetilde{\mathbf{z}}_{\text{all}}(l) \tag{8.67}$$

where $\boldsymbol{\alpha}(l) = [\alpha_1(l), \cdots, \alpha_K(l)]^T$ and $\widetilde{\mathbf{A}}(\boldsymbol{\theta}) = [\tilde{\mathbf{a}}(\theta_1), \cdots, \tilde{\mathbf{a}}(\theta_K)]$ with $\tilde{\mathbf{a}}(\theta_k)$ being defined as

$$\tilde{\mathbf{a}}(\theta_k) = \left(\mathbf{D}^T\mathbf{a}(\theta_k)\right) \otimes \left(\mathbf{V}_{\text{RF}}^H\mathbf{a}(\theta_k)\right) \tag{8.68}$$

Then, the covariance matrix of $\mathbf{h}_{\text{all}}(l)$ is given by

$$\mathbf{R}_{\text{vir}}(\boldsymbol{\theta}) \triangleq \mathbb{E}\left\{\mathbf{h}_{\text{all}}(l)\mathbf{h}_{\text{all}}^H(l)\right\} = \left(\frac{E}{UN}\right)^2 \widetilde{\mathbf{A}}(\boldsymbol{\theta})\mathbf{R}_\alpha\widetilde{\mathbf{A}}^H(\boldsymbol{\theta}) + \frac{E\sigma_{zr}^2}{U}\mathbf{I}_{UN_{\text{RF}}} \tag{8.69}$$

where $\mathbf{R}_\alpha \triangleq \mathbb{E}\left\{\boldsymbol{\alpha}(l)\boldsymbol{\alpha}^H(l)\right\} = \text{diag}\left([\sigma_{\alpha,1}^2, \cdots, \sigma_{\alpha,K}^2]\right)$. Owing to the finite observation time in practice, the exact knowledge of $\mathbf{R}_{\text{vir}}(\boldsymbol{\theta})$ is unavailable. Instead, the following sample covariance matrix is usually employed:

$$\widehat{\mathbf{R}}_{\text{vir}} = \frac{1}{L}\sum_{l=1}^{L}\mathbf{h}_{\text{all}}(l)\mathbf{h}_{\text{all}}^H(l) \tag{8.70}$$

where L denotes the number of sample sizes.

The traditional MUSIC algorithm needs the eigen decomposition of the covariance matrix $\widehat{\mathbf{R}}_{\text{vir}}$, defined as

$$\widehat{\mathbf{R}}_{\text{vir}} = \mathbf{E}_s\boldsymbol{\Lambda}_s\mathbf{E}_s^H + \mathbf{E}_n\boldsymbol{\Lambda}_n\mathbf{E}_n^H \tag{8.71}$$

where $\boldsymbol{\Lambda}_s$ and $\boldsymbol{\Lambda}_n$ contain the K largest (signal subspace) eigenvalues and the remaining $UN_{\text{RF}} - K$ (noise subspace) eigenvalues, respectively. the $UN_{\text{RF}} \times K$ matrix $\mathbf{E}_s$ and $UN_{\text{RF}} \times (UN_{\text{RF}} - K)$ matrix $\mathbf{E}_n$ are eigenvectors corresponding to $\boldsymbol{\Lambda}_s$ and $\boldsymbol{\Lambda}_n$, respectively.

Following the principle of the MUSIC estimator [49], the MUSIC spectrum can be thus formulated as

$$P_{\text{MUSIC}}(\theta) = \frac{\tilde{\mathbf{a}}^H(\theta)\tilde{\mathbf{a}}(\theta)}{\tilde{\mathbf{a}}^H(\theta)\mathbf{E}_n\mathbf{E}_n^H\tilde{\mathbf{a}}(\theta)} \tag{8.72}$$

Plugging (8.68) into (8.72) yields

$$P_{\text{MUSIC}}(\theta) = \frac{\mathbf{a}^H(\theta)\mathbf{D}^*\mathbf{D}^T\mathbf{a}(\theta) \cdot \mathbf{a}^H(\theta)\mathbf{V}_{\text{RF}}\mathbf{V}_{\text{RF}}^H\mathbf{a}(\theta)}{\left((\mathbf{D}^T\mathbf{a}(\theta)) \otimes (\mathbf{V}_{\text{RF}}^H\mathbf{a}(\theta))\right)^H\mathbf{E}_n\mathbf{E}_n^H\left((\mathbf{D}^T\mathbf{a}(\theta)) \otimes (\mathbf{V}_{\text{RF}}^H\mathbf{a}(\theta))\right)} \tag{8.73}$$

After searching the K largest peaks of $P_{\text{MUSIC}}(\theta)$, we can obtain the DOAs of the K targets.

It is worth mentioning that the above DOA estimation algorithm is in fact one-dimensional beamspace MUSIC by expressing $\tilde{\mathbf{a}}(\theta)$ as $(\mathbf{D}^T \otimes \mathbf{V}_{\mathrm{RF}}^H)(\mathbf{a}(\theta) \otimes \mathbf{a}(\theta))$. Following the analysis in [58,59], it is known that the ambiguity can be avoided if $\mathbf{D}$ and $\mathbf{V}_{\mathrm{RF}}$ are designed such that the steering matrix $[\tilde{\mathbf{a}}(\theta_1), \cdots, \tilde{\mathbf{a}}(\theta_{K+1})]$ is full column rank for any distinct angles $\theta_1, \cdots, \theta_{K+1}$. However, it would be quite challenging to take this design consideration into account in the QoS-aware hybrid beamforming design problem (8.45). Even though it is difficult to theoretically prove that the designed beamformers will not cause ambiguity, this issue is not observed in our extensive simulations.

8.3.3.2 Cramér–Rao bound on DOA estimation

Based on the signal model in (8.67), we note that the $\mathbf{h}_{\mathrm{all}}(l)$ obeys the statistical model $\mathcal{CN}(\mathbf{0}, \mathbf{R}_{\mathrm{vir}}(\boldsymbol{\theta}))$. As a consequence, the CRB expressions for estimating the DOAs can be expressed as (see [60], Section 8.2.3)

$$\mathrm{CRB}(\theta_k) = [\mathbf{J}^{-1}]_{k,k} \tag{8.74}$$

where the (i,j)-th entry of the matrix $\mathbf{J}$ is given by

$$\mathbf{J}[i,j] = \mathrm{Tr}\left\{\mathbf{R}_{\mathrm{vir}}^{-1}(\boldsymbol{\theta})\frac{\partial \mathbf{R}_{\mathrm{vir}}(\boldsymbol{\theta})}{\partial \theta_i}\mathbf{R}_{\mathrm{vir}}^{-1}(\boldsymbol{\theta})\frac{\partial \mathbf{R}_{\mathrm{vir}}(\boldsymbol{\theta})}{\partial \theta_j}\right\}, \forall i,j \tag{8.75}$$

where $\frac{\partial \mathbf{R}_{\mathrm{vir}}(\boldsymbol{\theta})}{\partial \theta_i}$ is defined as

$$\frac{\partial \mathbf{R}_{\mathrm{vir}}(\boldsymbol{\theta})}{\partial \theta_i} = \left(\frac{E}{UN}\right)^2 \sigma_{\alpha,k}^2 \left(\tilde{\mathbf{a}}'(\theta_i)\tilde{\mathbf{a}}^H(\theta_i) + \tilde{\mathbf{a}}(\theta_i)\tilde{\mathbf{a}}'^H(\theta_i)\right) \tag{8.76}$$

with

$$\tilde{\mathbf{a}}'(\theta_i) = \left(\mathbf{D}^T\mathbf{a}'(\theta_i)\right) \otimes (\mathbf{V}_{\mathrm{RF}}^H\mathbf{a}(\theta_i)) + \left(\mathbf{D}^T\mathbf{a}(\theta_i)\right) \otimes (\mathbf{V}_{\mathrm{RF}}^H\mathbf{a}'(\theta_i))$$

and

$$\mathbf{a}'(\theta_i) = -\jmath \cdot \mathrm{diag}\left([0, \pi\cos\theta_i, \cdots, \pi(M_t-1)\cos\theta_i]\right)\mathbf{a}(\theta_i)$$

8.3.4 *Simulation results*

In this section, various numerical simulations are provided to examine the performance of the proposed DFRC system with HBF architecture. We first assess the performance of the HBF with the proposed consensus-ADMM algorithm. Then, the MUSIC-based DOA estimation performance of the designed HBF is considered.

Unless otherwise specified, in all simulations, we assume a narrowband DFRC system with $M_t = 32$ transmit antennas sends $N = 8$ symbols within one pulse to U users, all users are equipped with the same number of antennas as $N_{cu} = 4, \forall u$. The symbol width is $\Delta t = 10$ ms, and thus the transmission pulse $T_0 = N \cdot \Delta t = 80$ ms. We adopt the same channel model as mentioned in Section 8.2. We set the intended communication rates for all users to be same, i.e., $\gamma_1 = \cdots = \gamma_U = \gamma$, and set the penalty parameter for the orthogonality constraint to be $\kappa = 0.1$. The total transmit energy is $E = 50$. We assume the noises at all users and radar receive antenna are the

Gaussian white noises with the covariances of $\sigma_z^2 = 0.001$ and $\sigma_{zr}^2 = 0.01$. Consider a desired spatial spectrum with two beams, which is expressed as

$$P(\theta) = \begin{cases} 1, & \theta \in [-30^\circ, \ -10^\circ] \cup [30^\circ, \ 45^\circ] \\ 0, & \text{otherwise} \end{cases}$$

For the consensus-ADMM method, the maximum iteration number is set to $N_{\text{ADMM}}^{\max} = 250$, $\mathbf{V}_D(0) = \frac{\sqrt{U}}{\sqrt{N_{\text{RF}} NU}} \mathbf{1}_{N_{\text{RF}} \times NU}$, $\mathbf{V}_{\text{RF}}(0) = \frac{1}{\sqrt{M_t}} \mathbf{1}_{M_t \times N_{\text{RF}}}$, $\mathbf{T}(0) = \mathbf{V}_{\text{RF}}(0)\mathbf{V}_D(0)$, $Q_{u,n}^{u,m}(0) = 0, \forall u, n, m$, $\varpi_{u,n}(0) = 0 \forall u, n$, and $\lambda_{u,n}^{u,m}(0) = 0, \forall u, n, m$. We also set the convergence tolerance of the RMO approach to be $\varepsilon_{\text{RMO}} = 10^{-6}$.

8.3.4.1 The performance of the proposed consensus-ADMM algorithm

In the first example, we examine the convergence performance of the proposed consensus-ADMM method for solving problem (8.51). We consider the DFRC system with $N_{\text{RF}} = 8$ RF chains serves $U = 3$ users, and the intended achievable rate $\gamma = 1$. The penalty parameters are set to be $\rho_1 = 20, \rho_2 = 10$. The left-hand of Figure 8.7 shows the objective function values in (8.51) versus the iteration number. It is observed that the consensus-ADMM can converge to a sub-optimal value with the iteration number increasing. Apart from this, the primal residuals $\sum_u \|\mathbf{T}_u - \mathbf{V}_{\text{rf}}\mathbf{V}_{Du}\|_F^2$ and $\sum_{u,n,m} \left|Q_{u,n}^{u,m} - \zeta \mathbf{w}_{u,n}^H \mathbf{H}_u \mathbf{t}_{u,m}\right|^2$ versus the iteration number are plotted in top and bottom right-hand in Figure 8.7, the results reveal that the primal residuals approach zeros as the iterations progress. This illustrates the effectiveness of the proposed consensus-ADMM.

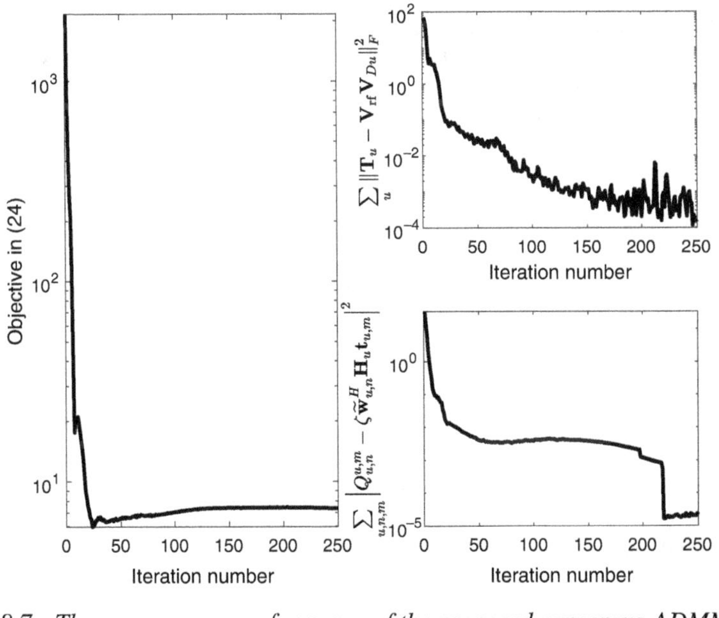

Figure 8.7 The convergence performance of the proposed consensus-ADMM when $N_{\text{RF}} = 8$, $U = 3$ and $\gamma = 1$

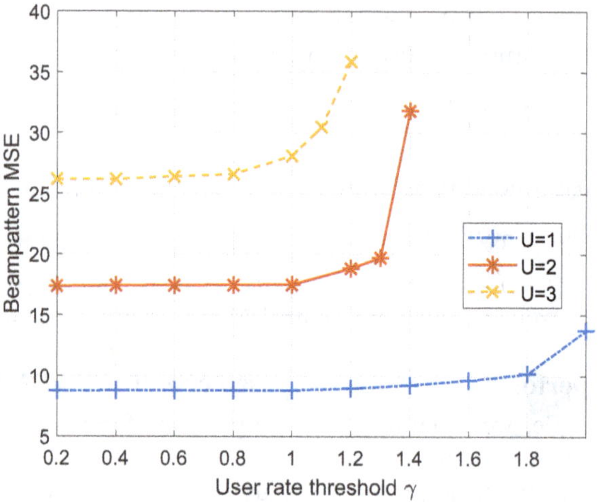

Figure 8.8 The beampattern MSE versus the intended rate threshold γ for different U when considering $N_{\mathrm{RF}} = 8$

Figure 8.8 plots the beampattern MSE as a function of the intended rate threshold γ and number of users U when considering $N_{\mathrm{RF}} = 8$. From the figure, we observe that as the number of users increases, the beampattern MSE property gets worse and worse, the reason for this phenomenon is that larger U implies more constraints should be taken into account in the optimization problem (8.51).

Figure 8.9 shows the resultant beampattern behaviors of different numbers of users when considering $\gamma = 0.6$, $N_{\mathrm{RF}} = 8$. It is interesting to note that as the number of users increases, the sidelobe of the synthesized beampattern becomes higher and higher, this is because the larger U, the less the remaining DoFs can be used to optimize the beampattern.

8.3.4.2 MUSIC-based DOA estimation performance

In this subsection, we assess the MUSIC-based DOA performance of the designed HBF. The number of sample sizes and the reflection powers of all targets are set to $L = 64$ and $\sigma^2_{\alpha,k} = 0$ dB, $\forall k$. In the following examples, the root mean square error (RMSE) and the probability of source resolution are obtained by running $\mathcal{I} = 200$ Monte-Carlo trials for each point.

Figure 8.10 shows the MUSIC spectrum of the HBFs with different number of users U when considering two spatially close sources with the angles $(\theta_1, \theta_2) = (-17°, -15°)$. From the figure, we observe that as the U increases, the DOAs of the two targets are estimated better and better, this result agrees with our expectation, since we can obtain UN_{RF} DoFs at the radar receiver, and the larger U means the larger virtual aperture achieved at DOA estimation.

Further, Figure 8.11 shows the probability of source resolution versus SNR (which is defined as $\mathrm{SNR} = E/\sigma^2_{zr}$) for different U. It is noticed that the two targets are estimated successfully if there are at least two peaks in the MUSIC spectrum

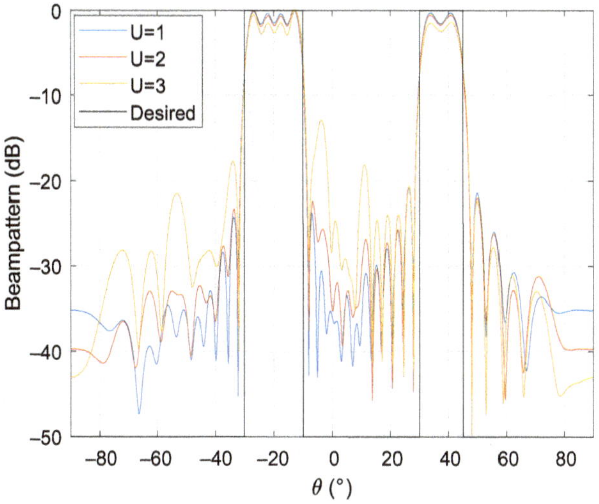

Figure 8.9 The beampattern behaviors of different numbers of users when considering $\gamma = 0.6$, $N_{\mathrm{RF}} = 8$

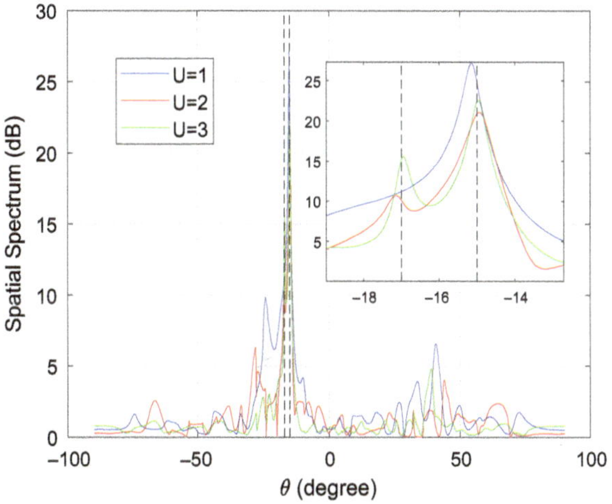

Figure 8.10 The MUSIC spectrums of the HBFs with different number of users U when considering two spatially close sources with the angles $(\theta_1, \theta_2) = (-17^\circ, -15^\circ)$, $N_{\mathrm{RF}} = 4$

and the following condition is met [60]: $\left|\hat{\theta}_i - \theta_i\right| \leq \frac{\Delta\theta}{2}$, $i = 1, 2$ where $\hat{\theta}_i$ and θ_i separately represent the estimated and actual DOA estimations of the i-th target, and $\Delta\theta = \theta_1 - \theta_2$. The result reveals that as the SNR increases, the probability of source resolution starts rising for each U, and when SNR equals some high SNR values, the beamformers with different U exhibit absolutely correct source resolutions. Moreover, we also find that the SNR threshold, which is used to describe the SNR level at which this transition happens, becomes lower as the U increases, this result is aligned with the analysis in Figure 8.10.

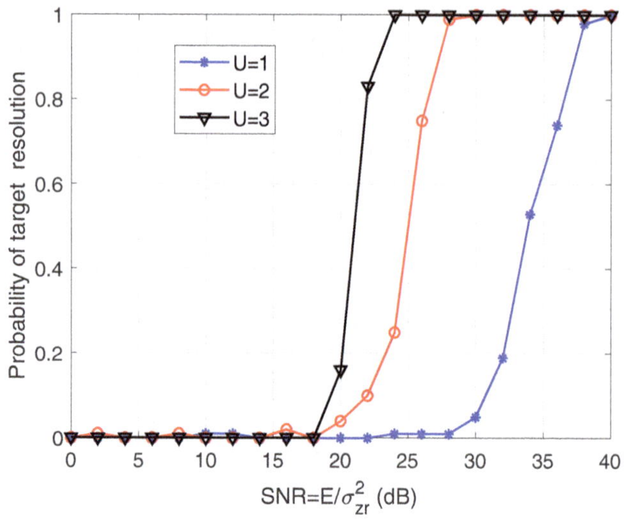

Figure 8.11 *The probability of source resolution versus SNR for* $(\theta_1, \theta_2) = (-17°, -15°)$

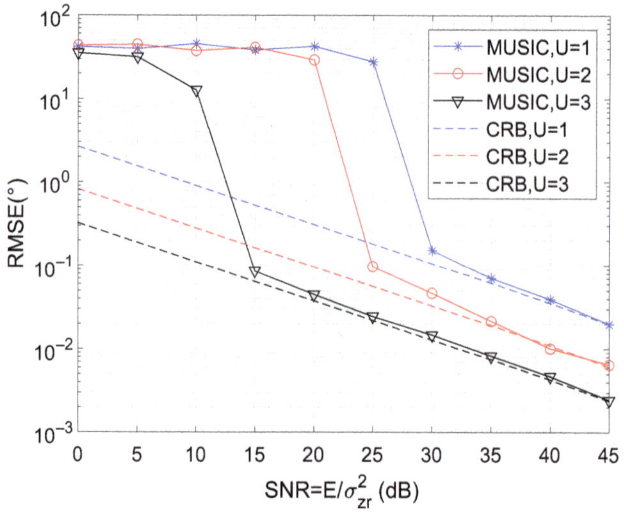

Figure 8.12 *The RMSEs of the MUSIC-based DOA estimators versus SNR for different numbers of users U,* $N_{RF} = 8$

Finally, we analyze the DOA estimation performance of the designed HBF by using the RMSE [60], which is defined as $\text{RMSE} = \sqrt{\frac{1}{\mathscr{I}} \sum_{i=1}^{\mathscr{I}} \left(\hat{\theta}_i - \theta_i\right)^2}$. The RMSEs of the MUSIC-based DOA estimators versus SNR for different cases of U is displayed in Figure 8.12, in which $U = 3$, $\gamma = 0.6$ and only one source with the angle $\theta_1 = -15°$ are considered. The results in the figure show that the HBF with the $U = 3$ case has the best RMSE performance at the high SNR region, and the smaller U will lead to the worst RMSE performance, which implies that the effect of having large effective aperture (i.e., UN_{RF}) is prominent at the high SNR region, while at the low SNR region, the effect is mild. In addition, it is also seen that the performance of the proposed MUSIC-based estimators are close to those of the CRBs at high SNR regions.

8.4 Conclusion

Due to the recent advances in signal processing and hardware design, many novel approaches have been developed to integrate radar sensing and communication over the last half-decade, making DFRC a hot research topic. In this chapter, the problem of HBF design for mmWave DFRC system was studied.

- We consider the HBF design for the single-carrier mmWave DFRC system, in which an HBF design based on the subarray connection architecture is proposed in the presence of extended targets and clutters. We formulate the HBF design problem as the SE maximization subject to the radar SINR and power constraints and propose the THEREON algorithm to achieve HBF design.
- We further study the HBF design for the multi-carrier mmWave DFRC system. In the designed system, the HBF is optimized to focus the transmitted energy within the spatial sectors of interest by taking the QoS requirement for MUs into account. A novel and low-complexity algorithm based on consensus-ADMM framework is devised to achieve HBF design.

Abundant numerical simulations are provided to demonstrate the efficacy of the proposed algorithms and show the superiority of the DFRC system with HBF architecture.

References

[1] Andrews JG, Buzzi S, Choi W, *et al.* What will 5G be? *IEEE Journal on Selected Areas in Communications*. 2014;32(6):1065–1082.

[2] Liu F, Masouros C, Petropulu AP, *et al.* Joint radar and communication design: Applications, state-of-the-art, and the road ahead. *IEEE Transactions on Communications*. 2020;68(6):3834–3862.

[3] Hessar F and Roy S. Spectrum sharing between a surveillance radar and secondary Wi-Fi networks. *IEEE Transactions on Aerospace and Electronic Systems*. 2016;52(3):1434–1448.

[4] Mehrnoush M and Roy S. Coexistence of WLAN network with radar: Detection and interference mitigation. *IEEE Transactions on Cognitive Communications and Networking*. 2017;3(4):655–667.

[5] Zheng L, Lops M, Eldar YC, *et al.* Radar and communication coexistence: An overview: A review of recent methods. *IEEE Signal Processing Magazine*. 2019;36(5):85–99.

[6] Chiriyath AR, Paul B, Jacyna GM, *et al.* Inner bounds on performance of radar and communications co-existence. *IEEE Transactions on Signal Processing*. 2015;64(2):464–474.
[7] Nunn C and Moyer LR. Spectrally-compliant waveforms for wideband radar. *IEEE Aerospace and Electronic Systems Magazine*. 2012;27(8):11–15.
[8] Aubry A, De Maio A, Piezzo M, *et al.* Radar waveform design in a spectrally crowded environment via nonconvex quadratic optimization. *IEEE Transactions on Aerospace and Electronic Systems*. 2014;50(2):1138–1152.
[9] Aubry A, Carotenuto V, and De Maio A. Forcing multiple spectral compatibility constraints in radar waveforms. *IEEE Signal Processing Letters*. 2016;23(4):483–487.
[10] Cheng Z, Liao B, He Z, *et al.* Spectrally compatible waveform design for MIMO radar in the presence of multiple targets. *IEEE Transactions on Signal Processing*. 2018;66(13):3543–3555.
[11] Sodagari S, Khawar A, Clancy TC, *et al.* A projection based approach for radar and telecommunication systems coexistence. In: 2012 *IEEE Global Communications Conference (GLOBECOM)*. IEEE; 2012. p. 5010–5014.
[12] Li B, Petropulu AP, and Trappe W. Optimum co-design for spectrum sharing between matrix completion based MIMO radars and a MIMO communication system. *IEEE Transactions on Signal Processing*. 2016;64(17):4562–4575.
[13] Li B and Petropulu AP. Joint transmit designs for coexistence of MIMO wireless communications and sparse sensing radars in clutter. *IEEE Transactions on Aerospace and Electronic Systems*. 2017;53(6):2846–2864.
[14] Liu F, Masouros C, Li A, *et al.* MU-MIMO communications with MIMO radar: From co-existence to joint transmission. *IEEE Transactions on Wireless Communications*. 2018;17(4):2755–2770.
[15] Mahal JA, Khawar A, Abdelhadi A, *et al.* Spectral coexistence of MIMO radar and MIMO cellular system. *IEEE Transactions on Aerospace and Electronic Systems*. 2017;53(2):655–668.
[16] Zheng L, Lops M, Wang X, *et al.* Joint design of overlaid communication systems and pulsed radars. *IEEE Transactions on Signal Processing*. 2017;66(1):139–154.
[17] Cheng Z, Liao B, Shi S, *et al.* Co-design for overlaid MIMO radar and downlink MISO communication systems via Cramér–Rao bound minimization. *IEEE Transactions on Signal Processing*. 2019;67(24):6227–6240.
[18] Blunt SD, Cook MR, and Stiles J. Embedding information into radar emissions via waveform implementation. In: *2010 International Waveform Diversity and Design Conference*. Piscataway, NJ: IEEE; 2010. p. 000195–000199.
[19] Blunt SD, Metcalf JG, Biggs CR, *et al.* Performance characteristics and metrics for intra-pulse radar-embedded communication. *IEEE Journal on Selected Areas in Communications*. 2011;29(10):2057–2066.
[20] Blunt SD, Yatham P and Stiles J. Intrapulse radar-embedded communications. *IEEE Transactions on Aerospace and Electronic Systems*. 2010;46(3): 1185–1200.
[21] Hassanien A, Amin MG, Zhang YD, *et al.* Dual-function radar-communications: Information embedding using sidelobe control and

waveform diversity. *IEEE Transactions on Signal Processing*. 2015;64(8):2168–2181.

[22] Hassanien A, Amin MG, Zhang YD, *et al.* Phase-modulation based dual-function radar-communications. *IET Radar, Sonar & Navigation*. 2016;10(8):1411–1421.

[23] Hassanien A, Himed B, and Rigling BD. A dual-function MIMO radar-communications system using frequency-hopping waveforms. In: *2017 IEEE Radar Conference*. Piscataway, NJ: IEEE; 2017. p. 1721–1725.

[24] Liu F, Zhou L, Masouros C, *et al.* Toward dual-functional radar-communication systems: Optimal waveform design. *IEEE Transactions on Signal Processing*. 2018;66(16):4264–4279.

[25] Cheng Z, Shi S, He Z, *et al.* Transmit sequence design for dual-function radar-communication system with one-bit DACs. *IEEE Transactions on Wireless Communications*. 2021;1(1):1–15.

[26] Zhang E and Huang C. On achieving optimal rate of digital precoder by RF-baseband codesign for MIMO systems. In: *2014 IEEE 80th Vehicular Technology Conference (VTC2014-Fall)*. Piscataway, NJ: IEEE; 2014. p. 1–5.

[27] Yu X, Shen JC, Zhang J, *et al.* Alternating minimization algorithms for hybrid precoding in millimeter wave MIMO systems. *IEEE Journal of Selected Topics in Signal Processing*. 2016;10(3):485–500.

[28] Han S, Chih-Lin I, Xu Z, *et al.* Large-scale antenna systems with hybrid analog and digital beamforming for millimeter wave 5G. *IEEE Communications Magazine*. 2015;53(1):186–194.

[29] Sohrabi F and Yu W. Hybrid digital and analog beamforming design for large-scale antenna arrays. *IEEE Journal of Selected Topics in Signal Processing*. 2016;10(3):501–513.

[30] Wang Z, Li M, Liu Q, *et al.* Hybrid precoder and combiner design with low-resolution phase shifters in mmWave MIMO systems. *IEEE Journal of Selected Topics in Signal Processing*. 2018;12(2):256–269.

[31] Mo J, Alkhateeb A, Abu-Surra S, *et al.* Hybrid architectures with few-bit ADC receivers: Achievable rates and energy-rate tradeoffs. *IEEE Transactions on Wireless Communications*. 2017;16(4):2274–2287.

[32] Cheng Z, Wu L, Wang B, *et al.* *Hybrid beamforming in mmWave dual-function radar-communication systems: Models, technologies, and challenges*. arXiv preprint arXiv:220904656. 2022.

[33] Cheng Z, Wu L, Wang B, *et al.* Double-phase-shifter based hybrid beamforming for mmWave DFRC in the presence of extended target and clutters. *IEEE Transactions on Wireless Communications*. 2023;22(6):3671–3686.

[34] Cheng Z, He Z, and Liao B. Hybrid beamforming for multi-carrier dual-function radar-communication system. *IEEE Transactions on Cognitive Communications and Networking*. 2021;7(3):1002–1015.

[35] Cheng Z, He Z, and Liao B. Hybrid beamforming design for OFDM dual-function radar-communication system. *IEEE Journal of Selected Topics in Signal Processing*. 2021;15(6):1455–1467.

[36] Cheng Z and Liao B. QoS-aware hybrid beamforming and DOA estimation in multi-carrier dual-function radar-communication systems. *IEEE Journal on Selected Areas in Communications*. 2022;40(6):1890–1905.

[37] Wang B, Li H, and Cheng Z. Dynamic hybrid beamforming design for dual-function radar-communication systems. *IEEE Transactions on Vehicular Technology*. 2024;73(2):2842–2847.

[38] Zeng J and Liao B. Transmit and receive hybrid beamforming design for OFDM dual-function radar-communication systems. *EURASIP Journal on Advances in Signal Processing*. 2023;2023(1):37.

[39] Yin H, Gesbert D, Filippou M, and Liu Y. A coordinated approach to channel estimation in large-scale multiple-antenna systems. *IEEE Journal on Selected Areas in Communications*. 2013;31(2):264–273.

[40] Simeone O, Bar-Ness Y, and Spagnolini U. Pilot-based channel estimation for OFDM systems by tracking the delay-subspace. *IEEE Transactions on Wireless Communications*. 2004;3(1):315–325.

[41] Chen CY and Vaidyanathan PP. MIMO radar waveform optimization with prior information of the extended target and clutter. *IEEE Transactions on Signal Processing*. 2009;57(9):3533–3544.

[42] Leshem A, Naparstek O, and Nehorai A. Information theoretic adaptive radar waveform design for multiple extended targets. *IEEE Journal of Selected Topics in Signal Processing*. 2007;1(1):42–55.

[43] Boyd S and Vandenberghe L. *Convex Optimization*. Cambridge: Cambridge University Press; 2004.

[44] Boyd S, Parikh N, Chu E, *et al.* Distributed optimization and statistical learning via the alternating direction method of multipliers. *Foundations and Trends® in Machine learning*. 2011;3(1):1–122.

[45] Dahlman E, Parkvall S, and Skold J. *5G NR: The Next Generation Wireless Access Technology*. New York: Academic Press; 2020.

[46] Yu X, Zhang J, and Letaief KB. Doubling phase shifters for efficient hybrid precoder design in millimeter-wave communication systems. *Journal of Communications and Information Networks*. 2019;4(2):51–67.

[47] Rappaport TS. *Wireless Communications: Principles and Practice*. vol. 2. Englewood Cliffs, NJ: Prentice-Hall PTR; 1996.

[48] Skolnik MI. *Radar Handbook*. New York: McGraw-Hill Education; 2008.

[49] Schmidt R and Schmidt RO. Multiple emitter location and signal parameter estimation. *IEEE Transactions on Antennas & Propagation*. 1986;34(3): 276–280.

[50] Hassanien A and Vorobyov SA. Transmit energy focusing for DOA estimation in MIMO radar with colocated antennas. *IEEE Transactions on Signal Processing*. 2011;59(6):2669–2682.

[51] Cheng Z, He Z, Zhang S, *et al.* Constant modulus waveform design for MIMO radar transmit beampattern. *IEEE Transactions on Signal Processing*. 2017;65(18):4912–4923.

[52] Shi Q, Razaviyayn M, Luo ZQ, *et al.* An iteratively weighted MMSE approach to distributed sum-utility maximization for a MIMO

interfering broadcast channel. *IEEE Transactions on Signal Processing*. 2011;59(9):4331–4340.

[53] Stoica P, Hao H, and Jian L. New algorithms for designing unimodular sequences with good correlation properties. *IEEE Transactions on Signal Processing*. 2009;57(4):1415–1425.

[54] Hao H, Stoica P, and Jian L. Designing unimodular sequence sets with good correlations—including an application to MIMO radar. *IEEE Transactions on Signal Processing*. 2009;57(11):4391–4405.

[55] Boumal N. *An Introduction to Optimization on Smooth Manifolds*. Cambridge: Cambridge University Press; 2023.

[56] Absil PA, Mahony R, and Sepulchre R. *Optimization Algorithms on Matrix Manifolds*. Princeton, NJ: Princeton University Press; 2009.

[57] Udriste C. *Convex Functions and Optimization Methods on Riemannian Manifolds*. vol. 297. Berlin: Springer Science & Business Media; 2013.

[58] Amini AN and Georgiou TT. Avoiding ambiguity in beamspace processing. *IEEE Signal Processing Letters*. 2005;12(5):372–375.

[59] Guo Z, Wang X, and Heng W. Millimeter-wave channel estimation based on 2-D beamspace MUSIC method. *IEEE Transactions on Wireless Communications*. 2017;16(8):5384–5394.

[60] Van Trees HL. *Detection, Estimation, and Modulation Theory, Part I: Detection, Estimation, and Linear Modulation Theory*. New York: John Wiley & Sons; 2004.

Chapter 9
Robust beamforming design for dual-function radar-communication system

Bin Liao[1], *Hao Liang*[1] *and Zhi Quan*[1]

This chapter considers the problem of robust beamforming for multiple-input multiple-output (MIMO) dual-function radar and communication (DFRC) system, when there exist radar look direction mismatch and communication channel state information (CSI) error. To address these system imperfections, a region-of-interest (ROI) is considered for radar sensing, and the communication CSI is assumed to be available with a bounded unknown error. With these considerations, we design a robust dual-function beamformer by optimizing the radar sensing performance in terms of the energy radiated toward the ROI, while guaranteeing the communication quality-of-service (QoS) measured by the per-user signal-to-interference-plus-noise ratio (SINR) as required. For this robust design, a max–min optimization problem is formulated and a semidefinite programming (SDP) reformulation based on the S-procedure is presented. The popular semidefinite relaxation (SDR) technique is employed to solve the resulting non-convex problem, and the tightness of SDR is discussed.

9.1 Introduction

Radar and communications represent fundamentally important use of the electromagnetic (EM) spectrum. Because of the dramatically growing demand for higher radar sensing capability and communication data rates, greater bandwidth is needed for radar and communications, as well as many other services. This has led to the congestion of the EM spectrum, which is known as a precious resource. For example, the spectrum allocated to WiMAX and LTE (2500–2690 MHz band) is adjacent to that of air traffic control radar (2700–2900 MHz) [1]. Moreover, the millimeter-wave band used by automotive radar and high-resolution imaging radar [2] is shared with 5G communications [3]. Therefore, dual-function radar-communication (DFRC) systems (also named as RadCom, JCAS, and ISAC in the literature), which share the hardware platforms and frequency bands, have become the main objectives guiding

[1]Guangdong Key Laboratory of Intelligent Information Processing, College of Electronics and Information Engineering, Shenzhen University, China

active sensing and wireless technologies [4–6]. Many efforts have been devoted by Zhang, Liu, Masouros, *et al.* to developing emerging DFRC technologies [7].

It is argued in an early work [8] that DFRC systems can allow both environmental sensing and V2X communications to be performed more efficiently. This promising potential lies in designing suitable waveforms and beamformers for simultaneous data transmission and radar sensing. A unique waveform design concept has been presented in [9] to allow for simultaneously performing both digital beamforming radar and multiple-input multiple-output (MIMO) communication operations. As a crucial measure of the waveform for MIMO-DFRC systems, the mutual information (MI) is maximized to design optimal signal by taking into account the training and data symbols in [10]. Besides, the criterion of multi-user interference (MUI) minimization is considered to design orthogonal waveform and constant modulus waveform in [11].

Instead of designing the DFRC waveforms, one can also optimize the transmit beamformer to achieve a desired shape of beampattern suitable for radar sensing and data transmission. With this concept, in [12] the DFRC beamformer is designed by matching the obtained beampattern to a certain desired beampattern, while guaranteeing the signal-to-interference-plus-noise ratio (SINR) of each communication user. Furthermore, the cross-correlation pattern is considered in [13], and the DFRC transmit beamformer is designed by minimizing a weighted sum of beampattern mismatch error and mean-squared cross-correlation pattern. Different from [13], a weighted sum of the radar beampattern mismatch error and communication spectral efficiency is used for hybrid beamforming design in orthogonal frequency-division multiplexing (OFDM) DFRC systems in [14]. In [15], the Kullback-Leibler divergence (KLD) is adopted as the performance metric to formulate the DFRC beamforming problem, while the Cramér-Rao bound (CRB) of direction-of-arrival (DOA) estimation is used for DFRC system design in [16–18].

It should be noticed that most of the existing studies on DFRC beamforming designs assume perfect CSI and prior knowledge of the environment (e.g., angles of target and interferers [19]) at the DFRC system. Necessary information on the radar target and communication channel must be determined before performing the design. Naturally, the estimation error will degrade the performances of radar sensing and communication, as illustrated by numerical results in [20]. The imperfect channel state information (CSI) is considered for physical layer security in DFRC with an ideal radar beampattern in [21]. In general, although robust beamforming has been widely studied for both communication-only and radar-only scenarios over the past few decades (see, e.g., [22] and references therein), this issue is less addressed for DFRC systems.

Motivated by the aforementioned facts, this Chapter aims to investigate the problem of robust transmit beamforming design for MIMO-DFRC systems. Two types of CSI imperfection models are considered: one is modeled by assuming that the true channel vector lies inside a spherical uncertainty set with the center being the CSI estimate, and the other assumes the CSI error follows a complex Gaussian distribution. For the first case, the problem is formulated as maximizing the minimum beampattern response level within the potential target angular range, i.e., the region-of-interest (ROI), under the SINR constraints [23]. For the second case, the problem is formulated as beampattern matching under the outage probability constraints. Optimization approaches such as semidefinite relaxation (SDR) and alternating direction method of multipliers (ADMM) are employed to tackle the resulting nonconvex problems.

The remainder of this chapter is organized as follows. Some preliminaries including the signal model are first given in Section 9.2. According to different channel error models, the SINR-constrained robust beamforming design is presented in Section 9.3, while the outage-constrained robust beamforming design is detailed in Section 9.4, respectively. Finally, the conclusions are drawn in Section 9.6.

Notations: In this chapter, $(\cdot)^*$, $(\cdot)^\mathsf{T}$, and $(\cdot)^\mathsf{H}$ denote the operators of conjugate, transpose, and Hermitian transpose, respectively. $\|\cdot\|_2$ and $\|\cdot\|_\mathsf{F}$ represent the ℓ_2 and Frobenius norms, respectively. $\mathrm{tr}(\mathbf{X})$, $\mathrm{rank}(\mathbf{X})$, $\lambda_{\max}(\mathbf{X})$, and $\mathscr{P}(\mathbf{X})$ are the trace, rank, maximum eigenvalue, and principal eigenvector of the matrix, respectively. $\mathbf{X} \succeq \mathbf{0}$ indicates that the matrix is positive semidefinite. $\mathbf{I}_M$ is the $M \times M$ identity matrix and $\mathbb{E}[\cdot]$ denotes the statistical expectation.

9.2 Signal model of MIMO-DFRC system

Consider a MIMO-DFRC system simultaneously sending symbols to multiple ($K \geq 2$) downlink users and sensing a spatial ROI (denoted as $\mathbf{\Theta}_{\mathrm{ROI}}$) for target detection and localization through the same emitted signal. Assume that the system is equipped with N transmit antennas, while each user equipment (UE) has a single antenna. Let $\mathbf{s}(l) = [s_1(l), \cdots, s_K(l)]^\mathsf{T} \in \mathbb{C}^K$ be the vector containing K data symbols for the users at time instant l and further assume that $\mathbb{E}[\mathbf{s}(l)\mathbf{s}^\mathsf{H}(l)] = \mathbf{I}_K$. In order to jointly perform wireless communication and radar sensing, a suitable dual-function transmit beamforming matrix should be designed:

$$\mathbf{W} = [\mathbf{w}_1, \cdots, \mathbf{w}_K] \in \mathbb{C}^{N \times K} \tag{9.1}$$

where the k-th column, i.e., $\mathbf{w}_k \in \mathbb{C}^N$, denotes the beamformer for user k. The emitted signal $\mathbf{x}(l) \in \mathbb{C}^N$ of the DFRC transmitter is constructed by combining the data symbols with $\mathbf{W}$ as

$$\mathbf{x}(l) = \mathbf{W}\mathbf{s}(l) = \sum_{k=1}^{K} \mathbf{w}_k s_k(l) \tag{9.2}$$

whose covariance matrix is given by

$$\mathbf{R} = \mathbb{E}[\mathbf{x}(l)\mathbf{x}^\mathsf{H}(l)] = \mathbf{W}\mathbf{W}^\mathsf{H} = \sum_{k=1}^{K} \mathbf{R}_k \tag{9.3}$$

where $\mathbf{R}_k = \mathbf{w}_k \mathbf{w}_k^\mathsf{H} \in \mathbb{C}^{N \times N}$.

For the downlink communication, denote by $\mathbf{H} = [\mathbf{h}_1, \cdots, \mathbf{h}_K] \in \mathbb{C}^{N\times K}$ the channel matrix with $\mathbf{h}_k \in \mathbb{C}^N$ being the channel vector of user k, then the received signal by user k is given by

$$y_k(l) = \mathbf{h}_k^{\mathsf{H}}\mathbf{x}(l) + n_k(l) = \mathbf{h}_k^{\mathsf{H}}\mathbf{w}_k s_k(l) + \sum\nolimits_{i\neq k} \mathbf{h}_k^{\mathsf{H}}\mathbf{w}_i s_i(l) + n_k(l) \tag{9.4}$$

where $n_k(l)$ denotes the zero-mean additive white Gaussian noise (AWGN) with variance σ_k^2. Recalling the assumption $\mathbb{E}[\mathbf{s}(l)\mathbf{s}^{\mathsf{H}}(l)] = \mathbf{I}_K$, the SINR of user k can be readily written as

$$\mathrm{SINR}_k(\mathbf{W}) = \frac{\left|\mathbf{h}_k^{\mathsf{H}}\mathbf{w}_k\right|^2}{\sum_{i\neq k}\left|\mathbf{h}_k^{\mathsf{H}}\mathbf{w}_i\right|^2 + \sigma_k^2} \tag{9.5}$$

which indicates the communication quality-of-service (QoS).

In practice, the knowledge of the CSI is usually imperfect, due to the channel estimation errors. Therefore, the CSI should be modeled by taking the errors into account to make the designs robust. Let $\boldsymbol{\delta}_k \in \mathbb{C}^N$ be the channel estimation error, then one gets

$$\mathbf{h}_k = \bar{\mathbf{h}}_k + \boldsymbol{\delta}_k \tag{9.6}$$

where $\bar{\mathbf{h}}_k$ denotes the channel estimate of user k. In order to derive robust beamformers against the CSI uncertainties, two models will be considered in this Chapter. The first one assumes the channel vector $\mathbf{h}_k$ lies in a ball with known radius ε_k around the estimated channel vector $\bar{\mathbf{h}}_k$, i.e.,

$$\|\boldsymbol{\delta}_k\|_2 \leq \varepsilon_k. \tag{9.7}$$

In the second CSI error model, it is assumed that $\boldsymbol{\delta}_k$ follows a complex zero-mean Gaussian distribution as

$$\boldsymbol{\delta}_k \sim \mathcal{CN}(\mathbf{0}, \zeta_k^2\mathbf{I}_{N_t}) \tag{9.8}$$

where ζ_k^2 corresponds to the variance of the estimates.

For radar sensing, the signal seen at a spatial angle θ (where the target is located) can be expressed as $\mathbf{a}^{\mathsf{T}}(\theta)\mathbf{x}(l)$, where $\mathbf{a}(\theta) \in \mathbb{C}^N$ represents the transmit steering vector of the DFRC system. Therefore, the energy radiated toward the target can be expressed as

$$\begin{aligned} B(\mathbf{W},\theta) &= \mathbb{E}[|\mathbf{a}^{\mathsf{T}}(\theta)\mathbf{x}(l)|^2] \\ &= \mathbf{a}^{\mathsf{T}}(\theta)\mathbf{W}\mathbf{W}^{\mathsf{H}}\mathbf{a}^*(\theta) \\ &= \mathbf{a}^{\mathsf{T}}(\theta)\mathbf{R}\mathbf{a}^*(\theta) \\ &= \sum_{k=1}^{K}\mathbf{a}^{\mathsf{T}}(\theta)\mathbf{R}_k\mathbf{a}^*(\theta) \end{aligned} \tag{9.9}$$

which in fact represents the spatial power spectrum (also termed as transmit beampattern). It is known that for radar sensing, a beam should be steered to the direction of the target to make sure that the reflected signal is sufficiently strong. Meanwhile, beampattern nulls toward the angles of interferences are needed.

9.3 SINR-constrained robust beamforming design

9.3.1 Problem formulation

This Section introduces a robust beamforming design for the DFRC system, which can ensure that the users' data are delivered with a specified QoS in terms of the SINR under the CSI error model 9.7 and that a sufficiently strong probing signal can be radiated toward the target direction.

For the aforementioned purpose, the SINR lower bound of user k is assumed to be γ_k, i.e.,

$$\mathrm{SINR}_k(\mathbf{W}) \geq \gamma_k, \forall k. \tag{9.10}$$

In practice, γ_k can be prescribed according to the priority of the user. According to (9.5), the above SINR condition can be rewritten as

$$\gamma_k^{-1}\mathbf{h}_k^{\mathsf{H}}\mathbf{R}_k\mathbf{h}_k - \sum\nolimits_{i \neq k} \mathbf{h}_k^{\mathsf{H}}\mathbf{R}_i\mathbf{h}_k \geq \sigma_k^2, \forall k. \tag{9.11}$$

Then, inserting $\mathbf{h}_k = \bar{\mathbf{h}}_k + \boldsymbol{\delta}_k$ into (9.11) yields

$$-\boldsymbol{\delta}_k^{\mathsf{H}}\mathbf{Y}_k\boldsymbol{\delta}_k - 2\Re\{\bar{\mathbf{h}}_k^{\mathsf{H}}\mathbf{Y}_k\boldsymbol{\delta}_k\} - \bar{\mathbf{h}}_k^{\mathsf{H}}\mathbf{Y}_k\bar{\mathbf{h}}_k + \sigma_k^2 \leq 0 \tag{9.12}$$

where $\mathbf{Y}_k \triangleq \gamma_k^{-1}\mathbf{R}_k - \sum_{i \neq k}\mathbf{R}_i \in \mathbb{C}^{N \times N}$ and $\Re\{\cdot\}$ denotes the real part of a complex value.

On the other hand, in order to detect the target accurately, the energy radiated toward θ should be as large as possible to make sure that the reflected signal is strong enough. However, the design may be less effective or robust if the beamformer is only optimized for a single angle. For practical implementations, it is more suitable to consider a region for the target. Thus, it is proposed herein to maximize the power over an ROI $\boldsymbol{\Theta}_{\mathrm{ROI}}$. This can be achieved by maximizing the minimum beampattern response level within $\boldsymbol{\Theta}_{\mathrm{ROI}}$. Mathematically, it can be expressed as

$$\begin{aligned} &\max_{\mathbf{W}} \; z \\ &\text{ s.t. } B(\mathbf{W}, \theta) \geq z, \; \theta \in \boldsymbol{\Theta}_{\mathrm{ROI}} \end{aligned} \tag{9.13}$$

Consequently, considering the aforementioned design objectives, the robust DFRC beamforming problem is formulated as the following optimization problem:

$$\max_{\mathbf{W}} \; z$$

$$\text{s.t. } B(\mathbf{W}, \theta) \geq z, \; \theta \in \boldsymbol{\Theta}_{\mathrm{ROI}}, \tag{9.14a}$$

$$-\boldsymbol{\delta}_k^{\mathsf{H}}\mathbf{Y}_k\boldsymbol{\delta}_k - 2\Re\{\bar{\mathbf{h}}_k^{\mathsf{H}}\mathbf{Y}_k\boldsymbol{\delta}_k\} - \bar{\mathbf{h}}_k^{\mathsf{H}}\mathbf{Y}_k\bar{\mathbf{h}}_k + \sigma_k^2 \leq 0, \forall k, \tag{9.14b}$$

$$\|\boldsymbol{\delta}_k\|_2 \leq \varepsilon_k, \forall k, \tag{9.14c}$$

$$\mathbf{Y}_k \triangleq \gamma_k^{-1}\mathbf{R}_k - \sum\nolimits_{i \neq k}\mathbf{R}_i \in \mathbb{C}^{N \times N}, \forall k, \tag{9.14d}$$

$$\gamma_k \geq 0, \forall k, \tag{9.14e}$$

$$\mathbf{R}_k = \mathbf{w}_k\mathbf{w}_k^{\mathsf{H}}, \forall k, \tag{9.14f}$$

$$\|\mathbf{W}\|_{\mathsf{F}}^2 \leq P, \; \mathbf{W} = [\mathbf{w}_1, \cdots, \mathbf{w}_K] \tag{9.14g}$$

where the last constraint in (9.14g) is imposed according to the system power budget and P is the upper bound of transmit power. It can be derived from the fact that

$$\mathrm{tr}(\mathbf{R}) = \mathrm{tr}(\mathbf{W}\mathbf{W}^{\mathsf{H}}) = \sum_{k=1}^{K} \mathrm{tr}(\mathbf{R}_k) = \|\mathbf{W}\|_{\mathsf{F}}^2 \le P. \tag{9.15}$$

The problem (9.14) is nonconvex and difficult to solve. Toward this end, a semidefinite programming (SDP) reformulation is derived in the next subsection for the ease of problem solving.

9.3.2 SDP reformulation

To begin with, the ROI $\boldsymbol{\Theta}_{\mathrm{ROI}}$ is uniformly discretized into M angles as $\{\theta_1, \cdots, \theta_M\}$ to make the problem tractable. By doing so, the constraint in (9.14a) can be relaxed as $z \le B(\mathbf{W}, \theta_m), \forall m$, or equivalently

$$\mathrm{tr}(\mathbf{A}_m\mathbf{R}) \ge z,\ m = 1, \cdots, M \tag{9.16}$$

where $\mathbf{A}_m \triangleq \mathbf{a}_m^* \mathbf{a}_m^{\mathsf{T}} \in \mathbb{C}^{N\times N}$ and $\mathbf{a}_m$ represents $\mathbf{a}(\theta_m)$ for notational simplicity. Furthermore, in order to deal with constraint in (9.14b), the S-procedure (see e.g., [24]) is employed. More specifically, the implication $\|\boldsymbol{\delta}_k\|_2 \le \varepsilon_k$ (or $\boldsymbol{\delta}_k^{\mathsf{H}}\mathbf{I}_N\boldsymbol{\delta}_k - \varepsilon_k^2 \le 0$) $\Longrightarrow$ (9.12) holds if and only if there exists a nonnegative λ_k such that

$$\lambda_k \begin{bmatrix} \mathbf{I}_N & \mathbf{0} \\ \mathbf{0} & -\varepsilon_k^2 \end{bmatrix} - \begin{bmatrix} -\mathbf{Y}_k & -\mathbf{Y}_k\bar{\mathbf{h}}_k \\ -\bar{\mathbf{h}}_k^{\mathsf{H}}\mathbf{Y}_k^{\mathsf{H}} & -\bar{\mathbf{h}}_k^{\mathsf{H}}\mathbf{Y}_k\bar{\mathbf{h}}_k + \sigma_k^2 \end{bmatrix} \succeq \mathbf{0}. \tag{9.17}$$

This also implies that the SINR constraint with channel vector error can be equivalently expressed in a semidefinite matrix form. Consequently, the originally formulated robust DFRC robust beamforming problem (9.14) can be recast as an SDP problem, which is convexified via SDR [25], namely, discarding the rank-one constraint. This yields the following problem:

$$\min_{\{\mathbf{R}_k\}_{k=1}^K, \boldsymbol{\lambda}, z} \quad -z$$

$$\text{s.t.} \ \sum_{k=1}^{K} \mathrm{tr}(\mathbf{A}_m\mathbf{R}_k) \ge z,\ \forall m, \tag{9.18a}$$

$$\begin{bmatrix} \mathbf{Y}_k + \lambda_k\mathbf{I}_N & \mathbf{Y}_k\bar{\mathbf{h}}_k \\ \bar{\mathbf{h}}_k^{\mathsf{H}}\mathbf{Y}_k^{\mathsf{H}} & \bar{\mathbf{h}}_k^{\mathsf{H}}\mathbf{Y}_k\bar{\mathbf{h}}_k - \sigma_k^2 - \lambda_k\varepsilon_k^2 \end{bmatrix} \succeq \mathbf{0},\ \forall k, \tag{9.18b}$$

$$\mathbf{Y}_k = \gamma_k^{-1}\mathbf{R}_k - \sum\nolimits_{i\neq k} \mathbf{R}_i,\ \forall k, \tag{9.18c}$$

$$\lambda_k \ge 0,\ \forall k, \tag{9.18d}$$

$$\sum_{k=1}^{K} \mathrm{tr}(\mathbf{R}_k) \le P, \tag{9.18e}$$

$$\mathbf{R}_k \succeq \mathbf{0},\ \forall k \tag{9.18f}$$

where $\boldsymbol{\lambda}$ denotes the collection of $\{\lambda_k\}_{k=1}^K$.

9.3.3 Solution of problem (9.18)

It is known that the tightness of SDR is usually not easy to prove, even though the mystery of a rank-one solution was elegantly unraveled for some communication-only robust beamforming problems (e.g., [26]) and non-robust DFRC beamforming problem (e.g., [18]). For the proposed robust beamforming problem, it is also challenging to specify a general condition to guarantee rank-one solution, although it is experimentally observed that the solution is rank-one in many common scenarios. In what follows, we will give some insights into the SDR tightness from the perspective of duality.

Let α_k, $\mathbf{U}_k$, μ_P, and $\mathbf{Z}_k$ be the Lagrangian dual variables associated with the constraints (9.18a), (9.18b), (9.18e), and (9.18f), respectively, according to [23], the dual problem of (9.18) can be written out as

$$\max_{\boldsymbol{\alpha},\mu_P,\{\mathbf{U}_k\}_{k=1}^K} \sum_{k=1}^{K} \sigma_k^2 \mu_k - P\mu_P$$

$$\text{s.t.} \sum_{m=1}^{M} \alpha_m = 1,\ \alpha_m \geq 0, \forall m, \tag{9.19a}$$

$$\mu_P \mathbf{I}_N - \sum_{m=1}^{M} \alpha_m \mathbf{A}_m + \sum_{i \neq k} \mathbf{J}_i \mathbf{U}_i \mathbf{J}_i^{\mathsf{H}} - \gamma_k^{-1} \mathbf{J}_k \mathbf{U}_k \mathbf{J}_k^{\mathsf{H}} \succeq \mathbf{0}, \forall k, \tag{9.19b}$$

$$\text{tr}(\mathbf{U}_k) \leq (1+\varepsilon_k^2)\mu_k,\ \mathbf{U}_k \succeq \mathbf{0},\ \forall k, \tag{9.19c}$$

$$\mu_P \geq 0, \tag{9.19d}$$

where $\mathbf{J}_k = [\mathbf{I}_N\ \bar{\mathbf{h}}_k] \in \mathbb{C}^{N\times(N+1)}$, $\boldsymbol{\alpha}$ is the collection of dual variables $\{\alpha_m\}_{m=1}^M$, and $\mu_k \geq 0$ denotes the $(N+1)$th diagonal entry of $\mathbf{U}_k$.

Property 1: Suppose that the primal problem (9.18) is feasible, then strong duality holds for the primal problem and dual problem (9.19).

Proof. In fact, for any $\mathbf{U}_k \succ \mathbf{0}, \text{tr}(\mathbf{U}_k) < (1+\varepsilon_k^2)\mu_k, \alpha_m > 0$, one can always find a sufficiently large $\mu_P > 0$, such that $\mu_P \mathbf{I}_N - \sum_{m=1}^{M} \alpha_m \mathbf{A}_m + \sum_{i\neq k} \mathbf{J}_i \mathbf{U}_i \mathbf{J}_i^{\mathsf{H}} - \gamma_k^{-1} \mathbf{J}_k \mathbf{U}_k \mathbf{J}_k^{\mathsf{H}} \succ \mathbf{0}$. This implies that the dual problem is strictly feasible and strong duality holds. □

According to the Karush–Kuhn–Tucker (KKT) condition $\mathbf{Z}_k \mathbf{R}_k = \mathbf{0}$ with $\mathbf{Z}_k \succeq \mathbf{0}$ and $\text{rank}(\mathbf{Z}_k) \leq N-1$, it is known that if and only if the dual variable

$$\mathbf{Z}_k = \mu_P \mathbf{I}_N - \sum_{m=1}^{M} \alpha_m \mathbf{A}_m + \sum_{i\neq k} \mathbf{J}_i \mathbf{U}_i \mathbf{J}_i^{\mathsf{H}} - \gamma_k^{-1} \mathbf{J}_k \mathbf{U}_k \mathbf{J}_k^{\mathsf{H}} \tag{9.20}$$

has a rank of $N-1$ for $k = 1, \cdots, K$, then the solution of problem (9.18) is rank-one. However, it is unclear how the rank relates to the variables. To shed light on this issue, the rank-one conditions for some specific cases are discussed.

Corollary 1: Consider $M = 1$ and let $(\mu_P, \{\mu_k\}_{k=1}^K, \{\mathbf{U}_k\}_{k=1}^K)$ be the optimal solution of the dual problem (9.19), if

$$\mu_P > \frac{\mu_k}{\gamma_k} \left(\|\bar{\mathbf{h}}_k\|_2 + \varepsilon_k \right)^2, \ \forall k \tag{9.21}$$

then the solution of the primal problem (9.18) is rank-one.

Proof. To begin with, let us express the dual variable $\mathbf{U}_k$ as

$$\mathbf{U}_k = \begin{bmatrix} \mathbf{D}_k & \mathbf{d}_k \\ \mathbf{d}_k^{\mathsf{H}} & \mu_k \end{bmatrix} \succeq \mathbf{0}, \;\; \mu_k \geq 0 \tag{9.22}$$

and perform the proof via two cases, i.e., $\mu_k = 0$ and $\mu_k > 0$.

If $\mu_k = 0$, according to (9.19c), we have $\mathbf{U}_k = \mathbf{0}$. In this case, if $M = 1$, then $\mathbf{Z}_k = \mu_P\mathbf{I}_N - \mathbf{A}_1 + \sum_{i\neq k}\mathbf{J}_i\mathbf{U}_i\mathbf{J}_i^{\mathsf{H}}$. Since $\mu_P\mathbf{I}_N + \sum_{i\neq k}\mathbf{J}_i\mathbf{U}_i\mathbf{J}_i^{\mathsf{H}} \succ \mathbf{0}$ and $\text{rank}(\mathbf{A}_1) = 1$, we have $\text{rank}(\mathbf{Z}_k) \geq N-1$. Recalling the fact $\text{rank}(\mathbf{Z}_k) \leq N-1$ yields

$$\text{rank}(\mathbf{Z}_k) = \text{rank}\left(\mu_P\mathbf{I}_N - \mathbf{A}_1 + \sum\nolimits_{i\neq k}\mathbf{J}_i\mathbf{U}_i\mathbf{J}_i^{\mathsf{H}}\right) = N-1. \tag{9.23}$$

As $\mathbf{Z}_k\mathbf{R}_k = \mathbf{0}$, we have $\text{rank}(\mathbf{R}_k) = 1$. Note that the dual variable $\mu_P > 0$, and hence, this case is consistent with the condition (9.21).

If $\mu_k > 0$, according to (9.20), it is known that if $M = 1$ and

$$\mu_P\mathbf{I}_N - \gamma_k^{-1}\mathbf{J}_k\mathbf{U}_k\mathbf{J}_k^{\mathsf{H}} \succ \mathbf{0} \tag{9.24}$$

then $\mu_P\mathbf{I}_N + \sum_{i\neq k}\mathbf{J}_i\mathbf{U}_i\mathbf{J}_i^{\mathsf{H}} - \gamma_k^{-1}\mathbf{J}_k\mathbf{U}_k\mathbf{J}_k^{\mathsf{H}} \succ \mathbf{0}$ and the semidefinite matrix $\mathbf{Z}_k = \mu_P\mathbf{I}_N - \mathbf{A}_1 + \sum_{i\neq k}\mathbf{J}_i\mathbf{U}_i\mathbf{J}_i^{\mathsf{H}} - \gamma_k^{-1}\mathbf{J}_k\mathbf{U}_k\mathbf{J}_k^{\mathsf{H}}$ must be of rank $N-1$ (because $\text{rank}(\mathbf{A}_1) = 1$). Thus, $\text{rank}(\mathbf{R}_k) = 1$. It is noted that the above condition is satisfied if $\mu_P > \gamma_k^{-1}\text{tr}(\mathbf{J}_k\mathbf{U}_k\mathbf{J}_k^{\mathsf{H}})$. Following the definition (9.22), it can be shown that

$$\begin{aligned}
\text{tr}(\mathbf{J}_k\mathbf{U}_k\mathbf{J}_k^{\mathsf{H}}) &= \text{tr}(\mathbf{D}_k) + \mu_k\|\bar{\mathbf{h}}_k\|_2^2 + 2\Re(\bar{\mathbf{h}}_k^{\mathsf{H}}\mathbf{d}_k) && (9.25\text{a})\\
&\leq \mu_k\varepsilon_k^2 + \mu_k\|\mathbf{h}_k\|_2^2 + 2\|\bar{\mathbf{h}}_k\|_2\|\mathbf{d}_k\|_2 && (9.25\text{b})\\
&\leq \mu_k\varepsilon_k^2 + \mu_k\|\bar{\mathbf{h}}_k\|_2^2 + 2\mu_k\varepsilon_k\|\bar{\mathbf{h}}_k\|_2 && (9.25\text{c})\\
&= \mu_k(\|\bar{\mathbf{h}}_k\|_2 + \varepsilon_k)^2 && (9.25\text{d})
\end{aligned}$$

where we have applied the Cauchy–Schwarz inequality for (9.25b) and the derivation of (9.25c) relies on the fact that $\mu_k\mathbf{D}_k \succeq \mathbf{d}_k\mathbf{d}_k^{\mathsf{H}}$ and

$$\|\mathbf{d}_k\|_2^2 = \text{tr}(\mathbf{d}_k\mathbf{d}_k^{\mathsf{H}}) \leq \mu_k\text{tr}(\mathbf{D}_k) = \mu_k(\text{tr}(\mathbf{U}_k) - \mu_k) \leq \mu_k^2\varepsilon_k^2. \tag{9.26}$$

From (9.25), we know that if $\mu_P > \frac{\mu_k}{\gamma_k}(\|\mathbf{h}_k\|_2 + \varepsilon_k)^2$, then $\mu_P > \gamma_k^{-1}\text{tr}(\mathbf{J}_k\mathbf{U}_k\mathbf{J}_k^{\mathsf{H}})$ and condition (9.24) is satisfied to ensure $\text{rank}(\mathbf{R}_k) = 1$. The proof is complete. □

Note that Corollary 1 only provides a sufficient condition to ensure the rank-one solution of the problem (9.18). However, its generalization to $M \geq 2$ has not been adequately discovered yet, and it is not clear enough how μ_P and μ_k are scaled with the system parameters. It is also noted that the proposed design is straightforwardly applicable to a special case of $\varepsilon_k = 0$, i.e., no CSI error. Accordingly, the following conclusion can be obtained.

Corollary 2: For the case of $M \leq 2$ and $\varepsilon_k = 0$, the solution of the problem (9.18) is rank-one if (1) $M = 1$, or (2) $M = 2$ and $\mathbf{H}^{\mathsf{H}}[\mathbf{a}_1^*, \mathbf{a}_2^*]$ is of full column rank.

Proof. When $M = 1$ and $\varepsilon_k = 0$, the proposed method reduces to a separable SDP, which has a rank-one solution as proved in [27]. When $M = 2$ and $\varepsilon_k = 0$, the dual variable $\mathbf{Z}_k$ in (9.20) becomes

$$\mathbf{Z}_k = \mu_P \mathbf{I}_N - \mathbf{V} + \sum_{i \neq k} \mu_i \mathbf{Q}_i - \mu_k \gamma_k^{-1} \mathbf{Q}_k \tag{9.27}$$

where $\mathbf{Q}_k \triangleq \mathbf{h}_k \mathbf{h}_k^{\mathsf{H}}$ and $\mathbf{V} \triangleq \alpha_1 \mathbf{A}_1 + \alpha_2 \mathbf{A}_2 = \alpha_1 \mathbf{a}^*(\theta_1)\mathbf{a}^{\mathsf{T}}(\theta_1) + \alpha_2 \mathbf{a}^*(\theta_2)\mathbf{a}^{\mathsf{T}}(\theta_2)$. Since α_1 and α_2 are nonnegative and $\alpha_1 + \alpha_2 = 1$, we have $\text{rank}(\mathbf{V}) \leq 2$. More specifically, the case of $\text{rank}(\mathbf{V}) = 1$ is the same as that of $M = 1$. If $\text{rank}(\mathbf{V}) = 2$, the two non-zero eigenvalues of $\mathbf{V}$ can be computed as

$$\frac{\alpha_1 \|\mathbf{a}_1\|_2^2 + \alpha_2 \|\mathbf{a}_2\|_2^2 \pm \sqrt{\left(\alpha_1 \|\mathbf{a}_1\|_2^2 - \alpha_2 \|\mathbf{a}_2\|_2^2\right)^2 + 4\alpha_1\alpha_2 |\mathbf{a}_1^{\mathsf{T}} \mathbf{a}_2^*|^2}}{2} \tag{9.28}$$

which are unequal since the steering vectors for different angles are non-orthogonal and $|\mathbf{a}_1^{\mathsf{T}} \mathbf{a}_2^*|^2 \neq 0$. Therefore, analogous to the proof of Theorem 2 in [18], it can be concluded that if $\mathbf{H}^{\mathsf{H}}[\mathbf{a}_1^*, \mathbf{a}_1^*]$ has a full column rank, then $\text{rank}(\mathbf{R}_k) = 1$. This completes the proof. □

In summary, even though the tightness of SDR associated with the robust DFRC beamforming design in this Chapter holds in many common scenarios, much more efforts are still required to unravel this mystery in future work. If the solution appears to be rank-one, we can use the following formula to extract the optimal beamformer as

$$\mathbf{w}_k^\star = \sqrt{\lambda_{\max}(\mathbf{R}_k^\star)}\mathscr{P}(\mathbf{R}_k^\star), \forall k \tag{9.29}$$

where $\mathbf{R}_k^\star$ denotes the solution of the relaxed problem (9.18). Otherwise, it only provides an approximation. Finally, we mention that the considered problem has K matrix variables of size $N \times N$ and $\tilde{M} = M + K + 1$ linear constraints. The worst-case complexity is $O(K^3N^6 + \tilde{M}KN^2)$ in each iteration of interior point methods and is usually much less.

9.3.4 Extensions of the design

Generally speaking, the response of the beampattern at the ROI designed by (9.18) has fluctuation, which increases the dependency of the signal-to-noise ratio (SNR) of radar echo on the angle, and thus target detection performance. Inspired by the concept of [28], the following constraint is utilized to enforce a flat main beam response:

$$\beta \max\{B(\theta)\} - \min\{B(\theta)\} \leq 0, \;\; \theta \in \Theta_{\text{ROI}} \tag{9.30}$$

where $\beta \in [0, 1)$ is a user-defined parameter. Obviously, the larger the value assigned to β, the smaller the fluctuation (ripple) of the main beam is. If β is set to be too small, the constraint may be ineffective. Thus, the modified design is given by

$$
\begin{aligned}
&\min_{\{\mathbf{R}_k\}_{k=1}^K, z, \boldsymbol{\lambda}} \quad -z \\
&\text{s.t.} \sum_{k=1}^{K} \operatorname{tr}(\mathbf{A}_m \mathbf{R}_k) \geq z, \ \forall m, \\
&\begin{bmatrix} \mathbf{Y}_k + \lambda_k \mathbf{I}_N & \mathbf{Y}_k \bar{\mathbf{h}}_k, \\ \bar{\mathbf{h}}_k^{\mathsf{H}} \mathbf{Y}_k^{\mathsf{H}} & \bar{\mathbf{h}}_k^{\mathsf{H}} \mathbf{Y}_k \bar{\mathbf{h}}_k - \sigma_k^2 - \lambda_k \varepsilon_k^2, \end{bmatrix} \succeq \mathbf{0}, \forall k, \\
&\mathbf{Y}_k \triangleq \gamma_k^{-1} \mathbf{R}_k - \sum\nolimits_{i \neq k} \mathbf{R}_i, \\
&\lambda_k \geq 0, \forall k, \\
&\beta \max\{B(\theta_m)\} - \min\{B(\theta_m)\} \leq 0, \ \forall m, \\
&B(\theta_m) = \sum_{k=1}^{K} \mathbf{a}^{\mathsf{T}}(\theta_m) \mathbf{R}_k \mathbf{a}^*(\theta_m), \ \forall m, \\
&\sum_{k=1}^{K} \operatorname{tr}(\mathbf{R}_k) \leq P, \\
&\mathbf{R}_k \succeq \mathbf{0}, \forall k.
\end{aligned}
\tag{9.31}
$$

Solving the above-modified formulation with a suitably selected β ensures that the main beam has a relatively flat response. It is worth mentioning that when set $\beta = 0$, the problem (9.31) reduces to the one in (9.18).

In the previous discussions, we focused on the case of a single target located in the ROI, which is then discretized into a set of consecutive angles. The robust beamformer is thus designed to radiate the energy toward the given region. As a matter of fact, there is no restriction that only one target can be located in the ROI. Instead, the beamforming design can be straightforwardly applied to the scenario where there are multiple targets located in the ROI. Moreover, if the targets are not closely located, the proposed approach can be directly applied to design a beamformer such that the energy can be radiated toward multiple angles (ROIs), by letting the ROI be composed of multiple mildly separated angles or sub-regions.

9.4 Outage-constrained robust beamforming design

9.4.1 Problem formulation

Different from Section 9.3 where the CSI error is modeled as a hypersphere, in this Section, it is assumed that the CSI error follows a complex zero-mean Gaussian distribution $\boldsymbol{\delta}_k \sim \mathcal{CN}(0, \zeta_k^2 \mathbf{I}_N)$. Accordingly, unlike the constraint in (9.10), the communication QoS is quantified by the outage probability as

$$\Pr\{\mathrm{SINR}_k(\mathbf{W}) \geq \gamma_k\} \geq 1 - \nu_k, \ \forall k, \tag{9.32}$$

where $\Pr\{\cdot\}$ denotes the probability, ν_k is the user-defined outage probability for user k. Moreover, for radar sensing, we not only consider the ROI range of the target $\boldsymbol{\theta}_{\text{ROI}}$ but also take the potential angles of interferences, denoted by the set $\boldsymbol{\Omega}$, into account.

We discretize the combined set $\boldsymbol{\theta}_{\text{ROI}} \cup \boldsymbol{\Omega}$ into M angles, then the problem is formulated as minimizing the mean square error (MSE) of the radar beampattern subject to the constraints of SINR outage probability of each user, i.e.,

$$\begin{aligned}
\min_{\mathbf{W}} \ & \sum_{m=1}^{M} \omega_m |\mathbf{a}_m^{\mathsf{T}} \mathbf{W}\mathbf{W}^{\mathsf{H}} \mathbf{a}_m^* - d_m|^2 \\
\text{s.t.} \ & \Pr\{\text{SINR}_k(\mathbf{W}) \geq \gamma_k\} \geq 1 - \nu_k, \ \forall k, && (9.33\text{a}) \\
& \|\mathbf{W}\|_{\mathsf{F}}^2 \leq P, && (9.33\text{b})
\end{aligned}$$

where ω represents the weights at different angles and d_m denotes the desired response level at θ_m.

The optimization problem in (9.33) is non-convex due to the outage probability constraint. To the end, in the following, we shall introduce an alternating optimization procedure based on the ADMM algorithm. First, an auxiliary variable $\mathbf{G} \in \mathbb{C}^{N \times K}$ is used to separate the constraints of problem (9.33). The problem (9.33) can be reformulated as

$$\begin{aligned}
\min_{\mathbf{W},\mathbf{G}} \ & \sum_{m=1}^{M} \omega_m |\mathbf{a}_m^{\mathsf{T}} \mathbf{W}\mathbf{W}^{\mathsf{H}} \mathbf{a}_m^* - d_m|^2 \\
\text{s.t.} \ & \Pr\{\text{SINR}_k(\mathbf{W}) \geq \gamma_k\} \geq 1 - \nu_k, \ \forall k, && (9.34\text{a}) \\
& \|\mathbf{G}\|_{\mathsf{F}}^2 \leq P, && (9.34\text{b}) \\
& \mathbf{G} = \mathbf{W}. && (9.34\text{c})
\end{aligned}$$

Furthermore, as an effective but simple way to make the fourth-order function more tractable, following the previous works (see e.g., [29] and [30]), the objective function in (9.34) can be approximated, by introducing variables $\mathbf{p}_m \in \mathbb{C}^{K \times 1}$, $m = 1, \cdots, M$, as

$$\begin{aligned}
\min_{\mathbf{p}_m} \ & \sum_{m=1}^{M} \omega_m \|\mathbf{W}^{\mathsf{H}} \mathbf{a}_m - \mathbf{p}_m\|_2^2 \\
\text{s.t.} \ & \|\mathbf{p}_m\|_2^2 = d_m, \forall m && (9.35)
\end{aligned}$$

Next, to deal with the outage probability constraints in (9.34a), the outage constraint for the k-th user is rewritten as

$$\Pr\{\text{SINR}_k(\mathbf{W}) \geq \gamma_k\} = \Pr\{f_k(\boldsymbol{\delta}_k) \geq \gamma_k\} = 1 - \nu_k, \qquad (9.36)$$

where $f_k(\boldsymbol{\delta}_k)$ is defined as

$$\begin{aligned}
f_k(\boldsymbol{\delta}_k) = \ & \boldsymbol{\delta}_k^{\mathsf{H}} \mathbf{W}\mathbf{T}_k\mathbf{W}^{\mathsf{H}} \boldsymbol{\delta}_k + \bar{\mathbf{h}}_k^{\mathsf{H}} \mathbf{W}\mathbf{T}_k\mathbf{W}^{\mathsf{H}} \bar{\mathbf{h}}_k \\
& + 2\Re\{\boldsymbol{\delta}_k^{\mathsf{H}} \mathbf{W}\mathbf{T}_k\mathbf{W}^{\mathsf{H}} \bar{\mathbf{h}}_k\}, && (9.37)
\end{aligned}$$

and $\mathbf{T}_k$ is a $K \times K$ diagonal matrix with the k-th diagonal entry being γ_k^{-1} and the rest being -1. It can be derived that the mean of $f(\boldsymbol{\delta}_k)$ is given by

$$\mathbb{E}\{f_k(\boldsymbol{\delta}_k)\} = \mathrm{tr}\{(\bar{\mathbf{H}}_k + \zeta_k^2 \mathbf{I}_N)\mathbf{W}\mathbf{T}_k\mathbf{W}^{\mathsf{H}}\} - \sigma_k^2, \tag{9.38}$$

where $\bar{\mathbf{H}}_k \triangleq \bar{\mathbf{h}}_k \bar{\mathbf{h}}_k^{\mathsf{H}} \in \mathbb{C}^{N\times N}$. Moreover, the variance of $f_k(\boldsymbol{\delta}_k)$ can be derived as

$$\begin{aligned}\mathrm{Var}\{f_k(\bar{\boldsymbol{\delta}}_k)\} &= \mathrm{tr}\{(2\zeta_k^2 \bar{\mathbf{h}}_k \bar{\mathbf{h}}_k^{\mathsf{H}} + \zeta_k^4 \mathbf{I}_N)(\mathbf{W}\mathbf{T}_k\mathbf{W}^{\mathsf{H}})^2\} \\ &\approx 2\zeta_k^2 \|\bar{\mathbf{h}}_k^{\mathsf{H}} \mathbf{W}\mathbf{T}_k\mathbf{W}^{\mathsf{H}}\|_2^2,\end{aligned} \tag{9.39}$$

where the approximation in (9.39) is obtained by assuming that the variance of channel error ζ_k^2 is small, such that the term dependent on ζ_k^4 will be much smaller than the other term and is negligible.

In order to convert the probabilistic constraints into deterministic but more conservative constraints, the Cantelli's inequality [31] is applied. According to this principle, it is known that if the following condition is satisfied:

$$\mathbb{E}\{f(\boldsymbol{\delta}_k)\} \geq \sqrt{(1/\nu_k - 1)\mathrm{Var}\{f(\boldsymbol{\delta}_k)\}} \tag{9.40}$$

then the constraint (9.34a) can be satisfied. Thus, inserting (9.38) and (9.39) into (9.40) yields

$$\mathrm{tr}\{(\bar{\mathbf{H}}_k + \zeta_k^2 \mathbf{I}_N)\mathbf{F}_k\} - \sigma_k^2 \geq \upsilon_k \|\bar{\mathbf{h}}_k^{\mathsf{H}} \mathbf{F}_k\|_2 \tag{9.41}$$

where $\mathbf{F}_k = \mathbf{W}\mathbf{T}_k\mathbf{W}^{\mathsf{H}}$ and $\upsilon_k \triangleq \sqrt{2\zeta_k^2(1/\nu_k - 1)}$.

According to the above transformations and approximations, the problem (9.33) is recast as

$$\begin{aligned}\min_{\boldsymbol{\Pi}} \ & \sum_{m=1}^{M} \omega_m \|\mathbf{W}^{\mathsf{H}}\mathbf{a}_m - \mathbf{p}_m\|_2^2 && (9.42\text{a}) \\ \text{s.t.} \ & \|\mathbf{G}\|_{\mathsf{F}}^2 \leq P, && (9.42\text{b}) \\ & \|\mathbf{p}_m\|_2^2 = d_m, \ \forall m && (9.42\text{c}) \\ & \mathbf{G} = \mathbf{W}, && (9.42\text{d}) \\ & \mathrm{tr}\{(\bar{\mathbf{H}}_k + \rho_k^2 \mathbf{I}_N)\mathbf{F}_k\} - \sigma_k^2 \geq \upsilon_k \|\bar{\mathbf{h}}_k^{\mathsf{H}}\mathbf{F}_k\|_2, && (9.42\text{e}) \\ & \mathbf{F}_k = \mathbf{G}\mathbf{T}_k\mathbf{W}^{\mathsf{H}}, \ \forall k && (9.42\text{f})\end{aligned}$$

where $\boldsymbol{\Pi}$ denotes the set of variables as

$$\boldsymbol{\Pi} = \{\mathbf{G}, \mathbf{W}, \{\mathbf{F}_k\}_{k=1}^{K}, \{\mathbf{p}_m\}_{m=1}^{M}\}. \tag{9.43}$$

In what follows, the non-convex problem is solved under the ADMM framework.

9.4.2 ADMM-based solution

Considering all the equality constraints problem (9.42), the augmented Lagrangian function can be expressed as

$$\mathscr{L} = \sum_{m=1}^{M} \omega_m \|\mathbf{W}^{\mathsf{H}}\mathbf{a}_m - \mathbf{p}_m\|_2^2 + \frac{\rho_1}{2}\sum_{k=1}^{K} \|\mathbf{F}_k - \mathbf{G}\mathbf{T}_k\mathbf{W}^{\mathsf{H}} + \mathbf{B}_{1,k}\|_{\mathsf{F}}^2 + \frac{\rho_2}{2}\| \mathbf{G} - \mathbf{W} + \mathbf{B}_2\|_{\mathsf{F}}^2 \tag{9.44}$$

where $\{\mathbf{B}_{1,k}\} \in \mathbb{C}^{N\times N}$, $\mathbf{B}_2 \in \mathbb{C}^{N\times K}$ are dual variables and $\rho_1, \rho_2 > 0$ are corresponding penalty parameters, respectively. According to the ADMM framework, we can iteratively update all variables.

(1) Optimization of $\{\mathbf{p}_m\}_{m=1}^{M}$*:* By fixing all the other variables, the minimization of $\mathscr{L}$ with respect to $\mathbf{p}_1, \cdots, \mathbf{p}_M$ can be solved by a set of sub-problems in parallel as

$$\begin{aligned} \min_{\mathbf{p}_m} \quad & \|\mathbf{W}^{\mathsf{H}}\mathbf{a}_m - \mathbf{p}_m\|_2^2 \\ \text{s.t.} \quad & \|\mathbf{p}_m\|_2^2 = d_m, \end{aligned} \tag{9.45}$$

whose solution can be analytically obtained as

$$\mathbf{p}_m = \frac{\sqrt{d_m}}{\|\mathbf{W}^{\mathsf{H}}\mathbf{a}_m\|_2}\mathbf{W}^{\mathsf{H}}\mathbf{a}_m. \tag{9.46}$$

(2) Optimization of $\{\mathbf{F}_k\}_{k=1}^{K}$*:* From the augmented Lagrangian function $\mathscr{L}$ in (9.44), the problem of updating $\mathbf{F}_k$ is equivalent to

$$\begin{aligned} \min_{\{\mathbf{F}_k\}} \quad & \sum_{k=1}^{K} \|\mathbf{F}_k - \mathbf{G}\mathbf{T}_k\mathbf{W}^{\mathsf{H}} + \mathbf{B}_{1,k}\|_{\mathsf{F}}^2 \\ \text{s.t.} \quad & \mathrm{tr}\{(\bar{\mathbf{H}}_k + \rho_k^2\mathbf{I}_N)\mathbf{F}_k\} - \sigma_k^2 \geq \upsilon_k\|\bar{\mathbf{h}}_k^{\mathsf{H}}\mathbf{F}_k\|_2, \ \forall k. \end{aligned} \tag{9.47}$$

which is a second-order cone program and can be optimally solved by interior-point methods-based solvers such as CVX toolbox [32]. In what follows, a more computationally efficient algorithm is presented to solve this problem based on the KKT conditions.

It is seen that the variables are not coupled with the constraints. This implies that we can decompose the problem into K sub-problems and solve them in parallel. First, we vectorize each matrix term in (9.47) and the following equivalent problem can be obtained:

$$\begin{aligned} \min_{\mathbf{f}_k} \ & (\mathbf{f}_k - \mathbf{c}_k)^{\mathsf{H}}(\mathbf{f}_k - \mathbf{c}_k) \\ \text{s.t.} \ \ & \Re\{\mathbf{e}_k^{\mathsf{H}}\mathbf{f}_k\} - \sigma_k^2 \geq \upsilon_k\|\mathbf{L}_k^{1/2}\mathbf{f}_k\|_2 \end{aligned} \tag{9.48}$$

where $\mathbf{f}_k = \mathrm{vec}(\mathbf{F}_k)$, $\mathbf{c}_k = \mathrm{vec}(\mathbf{C}_k)$, $\mathbf{C}_k = \mathbf{G}\mathbf{T}_k\mathbf{W}^{\mathsf{H}} - \mathbf{B}_{2,k}$ $\mathbf{e}_k = \mathrm{vec}(\mathbf{E}_k)$, $\mathbf{E}_k^{\mathsf{H}} = \bar{\mathbf{H}}_k + \rho_k^2\mathbf{I}_N$ and $\mathbf{L}_k = \mathbf{I}_N \otimes \bar{\mathbf{H}}_k$. It should be mentioned that if problem (9.48) has a feasible solution, it should satisfy an implicit condition $\Re\{\mathbf{e}_k^{\mathsf{H}}\mathbf{f}_k\} - \sigma_k^2 \geq 0$. We temporarily assume this constraint is already satisfied, and after obtaining the optimal solution, we then substitute it back into the constraint to verify its feasibility. Proceeding in this way, the problem (9.48) can be written as

$$\begin{aligned} &\min_{\mathbf{f}_k} \ (\mathbf{f}_k - \mathbf{c}_k)^{\mathsf{H}}(\mathbf{f}_k - \mathbf{c}_k) \\ &\text{s.t.} \quad (\Re\{\mathbf{e}_k^{\mathsf{H}}\mathbf{f}_k\} - \sigma_k^2)^2 \geq \upsilon_k^2 \mathbf{f}_k^{\mathsf{H}}\mathbf{L}_k\mathbf{f}_k \end{aligned} \tag{9.49}$$

which can be solved by KKT conditions.

By introducing dual variable $\tau \geq 0$, the Lagrangian function of this problem can be written as

$$\mathscr{L}(\mathbf{f}_k, \tau) = (\mathbf{f}_k - \mathbf{c}_k)^{\mathsf{H}}(\mathbf{f}_k - \mathbf{c}_k) + \tau\left(\upsilon_k^2 \mathbf{f}_k^{\mathsf{H}}\mathbf{L}_k\mathbf{f}_k - (\Re\{\mathbf{e}_k^{\mathsf{H}}\mathbf{f}_k\} - \sigma_k^2)^2\right) \tag{9.50}$$

and the KKT conditions are given by

$$\mathbf{f}_k = (\mathbf{I} + \tau\upsilon_k^2\mathbf{L}_k)^{-1}(\mathbf{c}_k + \tau\Re\{\mathbf{e}_k^{\mathsf{H}}\mathbf{f}_k\}\mathbf{e}_k - \tau\sigma_k^2\mathbf{e}_k), \tag{9.51a}$$

$$\tau\left(\upsilon_k^2\mathbf{f}_k^{\mathsf{H}}\mathbf{L}_k\mathbf{f}_k - (\Re\{\mathbf{e}_k^{\mathsf{H}}\mathbf{f}_k\} - \sigma_k^2)^2\right) = 0. \tag{9.51b}$$

where (9.51a) is obtained by setting the gradient $\frac{\partial \mathscr{L}(\mathbf{f}_k, \tau)}{\partial \mathbf{f}_k}$ to be zero. The optimal solution of problem (9.49), can be discussed in two cases, i.e., $\tau = 0$ and $\tau > 0$.

Denote by $\mathbf{f}_k^\star$ the optimal solution, if $\tau = 0$, then we have

$$\mathbf{f}_k^\star = \mathbf{c}_k, \tag{9.52}$$

and $\mathbf{f}_k^\star$ must satisfy $(\Re\{\mathbf{e}_k^{\mathsf{H}}\mathbf{f}_k\} - \sigma_k^2)^2 > \upsilon_k^2\mathbf{f}_k^{\mathsf{H}}\mathbf{L}_k\mathbf{f}_k$. To derive the optimal solution when $\lambda > 0$, let us define

$$\begin{aligned} \mathbf{M}_k &\triangleq (\mathbf{I}_{N^2} + \tau\upsilon_k^2\mathbf{L}_k)^{-1} \\ &= (\mathbf{I}_{N^2} + \tau\upsilon_k^2\mathbf{I}_N \otimes \bar{\mathbf{H}}_k)^{-1} \\ &= \left(\mathbf{I}_N \otimes (\mathbf{I}_N + \tau\upsilon_k^2\bar{\mathbf{H}}_k)\right)^{-1} \\ &= \mathbf{I}_N \otimes (\mathbf{I}_N + \tau\upsilon_k^2\bar{\mathbf{H}}_k)^{-1} \\ &= \mathbf{I}_N \otimes \left(\mathbf{I}_N - \frac{\tau\upsilon_k^2}{1 + \tau\upsilon_k^2\|\bar{\mathbf{h}}_k\|_2^2}\bar{\mathbf{H}}_k\right) \\ &= \mathbf{I}_{N^2} - \frac{\tau\upsilon_k^2}{1 + \tau\upsilon_k^2\|\bar{\mathbf{h}}_k\|_2^2}\mathbf{I}_N \otimes \bar{\mathbf{H}}_k \\ &= \mathbf{I}_{N^2} - \frac{\tau\upsilon_k^2}{1 + \tau\upsilon_k^2\|\bar{\mathbf{h}}_k\|_2^2}\mathbf{L}_k \end{aligned} \tag{9.53}$$

Then, multiplying (9.51a) by $\mathbf{e}_k^{\mathsf{H}}$ and taking the real part yields

$$\Re\{\mathbf{e}_k^{\mathsf{H}}\mathbf{f}_k\} = \frac{\Re\{\mathbf{e}_k^{\mathsf{H}}\mathbf{M}_k\mathbf{c}_k\} - \tau\sigma_k^2\mathbf{e}_k^{\mathsf{H}}\mathbf{M}_k\mathbf{e}_k}{1 - \tau\mathbf{e}_k^{\mathsf{H}}\mathbf{M}_k\mathbf{e}_k} \tag{9.54}$$

and

$$\mathbf{f}_k = \mathbf{M}_k \left(\mathbf{c}_k + \tau \frac{\Re\{\mathbf{e}_k^{\mathsf{H}}\mathbf{M}_k\mathbf{c}_k\} - \tau\sigma_k^2\mathbf{e}_k^{\mathsf{H}}\mathbf{M}_k\mathbf{e}_k}{1 - \tau\mathbf{e}_k^{\mathsf{H}}\mathbf{M}_k\mathbf{e}_k}\mathbf{e}_k - \tau\sigma_k^2\mathbf{e}_k \right) \tag{9.55}$$

Furthermore, it can be derived that

$$\Re\{\mathbf{e}_k^{\mathsf{H}}\mathbf{M}_k\mathbf{c}_k\} = \Re\{\mathbf{e}_k^{\mathsf{H}}\mathbf{c}_k\} - \frac{\tau\upsilon_k^2}{1 + \tau\upsilon_k^2\|\bar{\mathbf{h}}_k\|_2^2}\Re\{\mathbf{e}_k^{\mathsf{H}}\mathbf{L}_k\mathbf{c}_k\} \tag{9.56}$$

$$\mathbf{e}_k^{\mathsf{H}}\mathbf{M}_k\mathbf{e}_k = \|\mathbf{e}_k\|_2^2 - \frac{\tau\upsilon_k^2}{1 + \tau\upsilon_k^2\|\bar{\mathbf{h}}_k\|_2^2}\mathbf{e}_k^{\mathsf{H}}\mathbf{L}_k\mathbf{e}_k. \tag{9.57}$$

Using the above notations and recalling the fact that when $\tau > 0$, the equality $\Re\{\mathbf{e}_k^{\mathsf{H}}\mathbf{f}_k\} - \sigma_k^2 = \upsilon_k\|\mathbf{L}_k^{\frac{1}{2}}\mathbf{f}_k\|_2$ must be satisfied, then inserting (9.54) and (9.55) into this equality yields an equality with respect to τ. More precisely, a polynomial with respect to τ can be obtained. In this work, we adopt the bisection method to find the positive root, denoted as $\tau^\star$. Then, the optimal solution to the problem (9.48) can be obtained by inserting $\tau^\star$ into (9.55).

(3) Optimization of $\{\mathbf{W}\}$*:* The sub-problem concerning $\mathbf{W}$ can be formulated as follows:

$$\min_{\mathbf{W}} \sum_{m=1}^{M} \omega_m\|\mathbf{W}^{\mathsf{H}}\mathbf{a}_m - \mathbf{p}_m\|_2^2 + \frac{\rho_1}{2}\sum_{k=1}^{K}\|\mathbf{F}_k - \mathbf{G}\mathbf{T}_k\mathbf{W}^{\mathsf{H}} + \mathbf{B}_{1,k}\|_{\mathsf{F}}^2 + \frac{\rho_2}{2}\| \mathbf{G} - \mathbf{W} + \mathbf{B}_2\|_{\mathsf{F}}^2 \tag{9.58}$$

Setting the first-order derivative to be zero yields a Sylvester equation

$$\mathbf{X}_1\mathbf{W}^\star + \mathbf{W}^\star\mathbf{X}_2 = \mathbf{X}_3 \tag{9.59}$$

where

$$\mathbf{X}_1 = 2\sum_{m=1}^{M}\omega_m\mathbf{a}_m\mathbf{a}_m^{\mathsf{H}} \tag{9.60a}$$

$$\mathbf{X}_2 = \rho_1\sum_{k=1}^{K}\mathbf{T}_k\mathbf{G}^{\mathsf{H}}\mathbf{G}\mathbf{T}_k + \rho_2\mathbf{I}_K \tag{9.60b}$$

$$\mathbf{X}_3 = 2\sum_{m=1}^{M}\omega_m\mathbf{a}_m\mathbf{p}_m^{\mathsf{H}} + \rho_2(\mathbf{G} + \mathbf{B}_2) + \rho_1\sum_{k=1}^{K}(\mathbf{F}_k^{\mathsf{H}} + \mathbf{B}_{1,k}^{\mathsf{H}})\mathbf{G}\mathbf{T}_k \tag{9.60c}$$

When $\mathbf{X}_1$ and $-\mathbf{X}_2$ do not share common eigenvalues, (9.59) has a unique solution.

(4) Optimization of $\{\mathbf{G}\}$*:* The sub-problem w.r.t. $\mathbf{G}$ is given by

$$\min_{\mathbf{G}} \frac{\rho_1}{2}\sum_{k=1}^{K}\|\mathbf{F}_k - \mathbf{G}\mathbf{T}_k\mathbf{W}^{\mathsf{H}} + \mathbf{B}_{1,k}\|_{\mathsf{F}}^2 + \frac{\rho_2}{2}\|\mathbf{G} - \mathbf{W} + \mathbf{B}_2\|_{\mathsf{F}}^2 \tag{9.61a}$$

$$\text{s.t. } \|\mathbf{G}\|_F^2 \leq P. \tag{9.61b}$$

which is a convex problem and can be solved using the KKT condition. By introducing a parameter $\kappa > 0$, the corresponding KKT condition can be expressed as

$$\mathbf{G}^{\star}(\kappa) = \mathbf{\Psi}(2\kappa\mathbf{I}_K + \mathbf{\Phi})^{-1}, \tag{9.62a}$$

$$\kappa(\|\mathbf{G}^{\star}(\kappa)\|_F^2 - P) = 0, \tag{9.62b}$$

where

$$\begin{aligned}\mathbf{\Psi} &= \rho_1 \sum_{k=1}^{K} \{(\mathbf{F}_k + \mathbf{B}_{1,k})\mathbf{W}\mathbf{T}_k + \rho_2(\mathbf{W} - \mathbf{B}_2)\}\mathbf{\Phi} \\ &= \rho_1 \sum_{k=1}^{K} \mathbf{T}_k\mathbf{W}^{\mathsf{H}}\mathbf{W}\mathbf{T}_k + \rho_2\mathbf{I}_K \end{aligned} \tag{9.63}$$

According to the KKT conditions framework, we can obtain the optimal value of $\mathbf{G}^{\star}$ in the following two conditions: (1) when $\kappa = 0$, we can get the $\mathbf{G}^{\star}$ by calculating

$$\mathbf{G}^{\star} = \mathbf{\Psi}\mathbf{\Phi}^{-1} \tag{9.64}$$

and (2) when $\kappa > 0$, considering the eigenvalue decomposition $\mathbf{\Phi} = \mathbf{P}\mathbf{\Sigma}\mathbf{P}^{\mathsf{H}}$, we can obtain

$$\begin{aligned}\|\mathbf{G}(\kappa)\|_{\mathsf{F}}^2 &= \mathrm{tr}\{\mathbf{\Psi}\mathbf{P}(\mathbf{\Sigma} + 2\kappa\mathbf{I}_K)^{-2}\mathbf{P}^{\mathsf{H}}\mathbf{\Psi}^{\mathsf{H}}\} \\ &= \sum_{n=1}^{N_t} \frac{[\mathbf{P}^{\mathsf{H}}\mathbf{\Psi}^{\mathsf{H}}\mathbf{\Psi}\mathbf{P}]_{n,n}}{([\mathbf{\Sigma}]_{n,n} + 2\kappa)^2}\end{aligned} \tag{9.65}$$

Plugging (9.65) into (9.62b) yields an equality with respect to κ, and κ can be obtained by the bisection method. Then, substituting κ into (9.62a) to get the value of $\mathbf{G}^{\star}$.

(5) Optimization of $\{\{\mathbf{B}_{1,k}\}, \mathbf{B}_2\}$*:* The dual variables are updated according to the following rules:

$$\mathbf{B}_{1,k} = \mathbf{F}_k - \mathbf{W}\mathbf{T}_k\mathbf{W}^{\mathsf{H}} + \mathbf{B}_{1,k}, \ \forall k \tag{9.66a}$$

$$\mathbf{B}_2 = \mathbf{G} - \mathbf{W} + \mathbf{B}_2. \tag{9.66b}$$

Finally, the proposed robust beamforming algorithm for the DFRC system is summarized in Algorithm 9.1. In each iteration, the main computations are caused by (9.54) and (9.55) with complexity $O(N^4)$. It is worth mentioning that the problem considered in this work is nonconvex, and the general convergence of such kinds of optimizations is still an open issue. Nevertheless, it can be clearly seen from the simulations that the objective function is convergent and the residuals of the proposed method approach zero with increasing iterations.

9.5 Simulation results

9.5.1 *Results of SINR constrained design*

The transmitter deploys a ULA with half-wavelength inter-element spacing. The power budget is $P = 1$ (30 dBm), and UE noise covariances are

Algorithm 9.1 ADMM-based robust beamforming algorithm for MIMO-DFRC system

1: **Input:** Antenna number N, user number K, noise variances σ_k^2, SINR thresholds γ_k, channel vector estimates $\bar{\mathbf{h}}_k$, steering vectors $\mathbf{a}_m$, outage probability ν_k, transmit power P, channel error variances ζ_k^2.
2: **Initialization:** Randomly generate $\mathbf{W}^{(0)}$, $\mathbf{G}^{(0)}$, $\{\mathbf{F}_k^{(0)}\}_{k=1}^K$, $\{\mathbf{p}_m^{(0)}\}_{m=1}^M$, $\{\mathbf{B}_{1,k}^{(0)}\}_{k=1}^K$, $\mathbf{B}_2^{(0)}$.
3: Set $t = 1$.
4: **repeat**
5: For given $\mathbf{W}^{(t-1)}$, obtain the solutions of $\{\mathbf{p}_m^{(t)}\}_{m=1}^M$ based on (9.46).
6: For given $\mathbf{W}^{(t-1)}$, $\{\mathbf{B}_{1,k}^{(t-1)}\}$, $\mathbf{G}^{(t-1)}$ solve problems (9.47) in parallel to obtain the solutions of $\{\mathbf{F}_k^{(t)}\}$.
7: For given $\mathbf{G}^{(t-1)}$, $\{\mathbf{F}_k^{(t)}\}_{k=1}^K$, $\{\mathbf{p}_m^{(t)}\}_{m=1}^M$, $\{\mathbf{B}_{1,k}^{(t-1)}\}_{k=1}^K$, $\mathbf{B}_2^{(t-1)}$, update $\mathbf{W}^{(t)}$ by solving the problem (9.59).
8: For given $\mathbf{W}^{(t)}$, $\{\mathbf{F}_k^{(t)}\}_{k=1}^K$, $\{\mathbf{B}_{1,k}^{(t-1)}\}_{k=1}^K$, $\mathbf{B}_2^{(t-1)}$ update $\mathbf{G}^{(t)}$ base on (9.61).
9: Update $\{\mathbf{B}_{1,k}^{(t)}\}_{k=1}^K$, $\mathbf{B}_2^{(t)}$ according to (9.66).
10: **until** $\sum_{k=1}^K \|\mathbf{B}_{1,k}^{(t)} - \mathbf{B}_{1,k}^{(t-1)}\|_\mathsf{F}^2 + \|\mathbf{B}_2^{(t)} - \mathbf{B}_2^{(t-1)}\|_\mathsf{F}^2 \leq \varepsilon$ or the maximum iteration number is reached.

$\sigma_1^2 = \cdots = \sigma_K^2 = 10^{-3}$ (0 dBm). The Rayleigh fading is adopted for communication, and the channel vector estimate $\bar{\mathbf{h}}_k$ is Gaussian distributed with zero-mean and covariance matrix $\mathbf{I}_N$, i.e., $\bar{\mathbf{h}}_k \sim \mathscr{CN}(\mathbf{0}, \mathbf{I}_N), \forall k$. The CSI error vector $\boldsymbol{\delta}_k$ is randomly generated with $\varepsilon_1 = \cdots = \varepsilon_K = \varepsilon$. For the communication SINR thresholds, we assume $\gamma_1 = \cdots = \gamma_K = \gamma$ for simplicity. Throughout the simulations, the CVX toolbox is used to solve the SDP problems, and we assume $N = 15$, $K = 4$, unless otherwise specified.

First, we set $\boldsymbol{\Theta}_{\text{ROI}} = [-10°, 10°]$ with a discretization stepsize of 2°. To examine the tightness of relaxation in (9.18), let $e(\mathbf{R}_k^\star)$ be the vector composed of sorted (in descending order) eigenvalues of $\mathbf{R}_k^\star$ and we compute the summation of all the eigenvalues as $\mathbf{d} = \sum_{k=1}^K e(\mathbf{R}_k^\star)$. When the second largest eigenvalue to largest eigenvalue ratio, i.e., $\mathbf{d}(2)/\mathbf{d}(1)$, e.g., is less than 10^{-5}, then we can say that $\mathbf{R}_k$ is rank-one and the solution is tight. For each (γ, ε), 100 independent experiments are run (in each experiment, the channel vectors are randomly generated) to compute the percentage of rank-one solutions for different CSI error bounds and SINR thresholds. The results are shown in Figure 9.1.

From Figure 9.1, it is seen that the tightness of relaxation of the proposed robust DFRC beamforming problem relies on several aspects including the per-user SINR threshold, CSI, and error bound. In fact, the tightness also relates to the number of angles M (as discussed in Section III-C) and the number of users K. Corresponding illustrations are not shown here due to space limitations. An interesting observation from Figure 9.1 is that when the SINR threshold is low, e.g., less than 5 dB, the tightness does not hold even when there is no CSI error. The fact is that when γ is set to be very small, the SINR constraints are ineffective, and hence, $\mathbf{U}_k = \mathbf{0}$

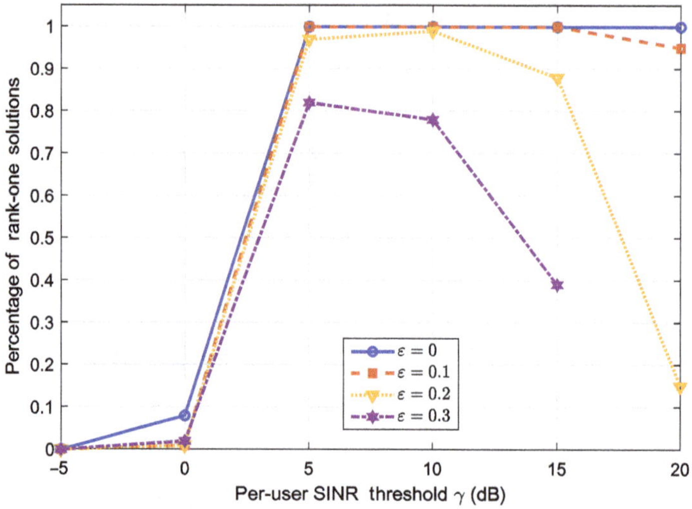

Figure 9.1 Percentage of rank-one solutions for different CSI error bounds and SINR thresholds. $\boldsymbol{\Theta}_{\text{ROI}} = [-10°, 10°]$

and $\mathbf{Z}_k = \mu_P \mathbf{I}_N - \sum_{m=1}^{M} \alpha_m \mathbf{A}_m$. Only if $M \leq 2$, we can ensure rank($\mathbf{Z}_k$) $= N - 1$ and rank($\mathbf{R}_k$) $= 1$. Note that a small γ is less meaningful for communication in practice and hence will not be considered in the following examples. When $\gamma \geq 5$ dB, a large CSI error could decrease the possibility of a rank-one solution, and if the error is sufficiently small, the tightness can be well guaranteed. Overall, the tightness of relaxation for the proposed DFRC beamforming design remains to be an open problem, and more efforts are needed to unravel the rank-one solution mystery.

Now, we examine the transmit beampattern behavior of the proposed robust DFRC beamforming approach. Two kinds of ROIs, i.e., $[-10°, 10°]$ and $[-5°, 5°]$, are considered. For comparison, we also test the non-robust beamforming approach [18], of which the number of receive antennas is 17 and the presumed target angle is 0°. The SINR threshold is set to be 10 dB for both methods. The transmit beampatterns corresponding to the three cases are drawn in Figure 9.2. As expected, a broadened main beam can be obtained by the proposed method. This allows us to guarantee the radar sensing performance for a large region, rather than a specific angle. Such a property makes the radar sensing robust against look direction mismatch. It should be mentioned that the results are obtained from a single trial. For different channel vectors, the resulting beampatterns (sidelobes) are not exactly identical, but the overall performances are the same.

In order to further show the superiority of robust DFRC beamforming, the performance of the designed transmit beamformer for DOA estimation is evaluated and compared to that of a non-robust approach [18]. For this purpose, a ULA with 17 antennas is considered for the receiver, the target is located at $\theta_0 = 6°$, and the output of the receiver can thus be expressed as

$$\mathbf{r}(l) = \xi \mathbf{b}(\theta_0) \mathbf{a}^{\mathsf{T}}(\theta_0) \mathbf{W} \mathbf{s}(l) + \mathbf{v}(l) \tag{9.67}$$

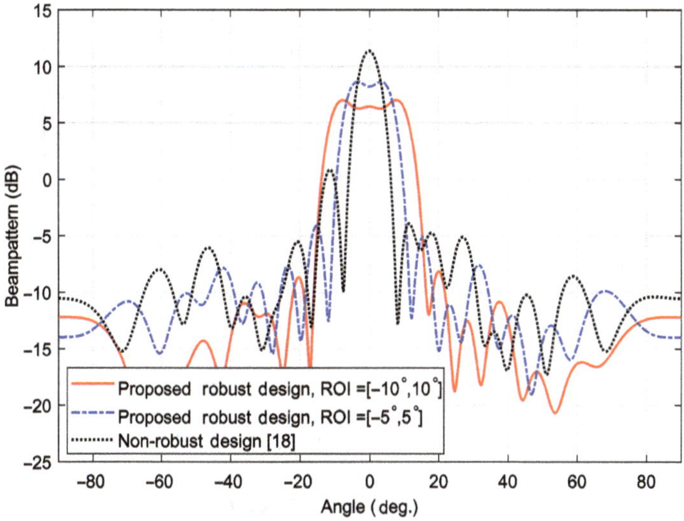

Figure 9.2 Comparison of beampatterns of the proposed robust DFRC beamformer and non-robust beamformer [18]. $\varepsilon = 0.1$ *and* $\gamma = 10$ dB

where $\mathbf{b}(\theta)$ is the receive array steering vector, ξ is the target reflection coefficient which is constant within a radar pulse (communication frame) following the Swerling II target model, $\mathbf{v}(l)$ is the vector of zero-mean AWGN with covariance matrix $\sigma_v^2\mathbf{I}_{N_r}$. Accordingly, the SNR of radar echo is defined as $|\xi|^2/\sigma_v^2$. For illustration, the DOA is estimated by root-MUSIC and the performance is measured in terms of root mean square error (RMSE). For the proposed robust method, we set $\boldsymbol{\Theta}_{\text{ROI}} = [-10°, 10°]$, while for the non-robust method [18], the presumed target angle is 0° and no CSI error is assumed.

Let the number of snapshots be $L = 100$, and the results of RMSE, obtained from 2000 independent trials, are shown in Figure 9.3, where the CRB is determined by the energy radiated toward the target, i.e., $\mathbf{a}^{\mathsf{T}}(\theta_0)\mathbf{W}\mathbf{W}^{\mathsf{H}}\mathbf{a}^*(\theta_0)$. We observe that the proposed approach achieves a better DOA estimation performance when the actual target angle deviates a lot from the nominal one, by designing the transmit beamformer for an ROI rather than a single angle. However, it is still worth mentioning that the non-robust method [18] is preferred when the angle mismatch is small, since a larger array gain can be achieved at the target angle, as shown in Figure 9.2.

We proceed to show how the communication can benefit from the proposed robust DFRC beamforming design. Specifically, the robust transmit beamformer is designed by assuming $\boldsymbol{\Theta}_{\text{ROI}} = [-10°, 10°]$, $\varepsilon = 0.1$, and $\gamma = 10$ dB, while the non-robust beamformer [18] assumes no CSI error and target angle 0°. The beamformers are optimized based on $\bar{\mathbf{h}}_k$, and the achievable SINR of each user is computed with the actual CSI $\mathbf{h}_k = \bar{\mathbf{h}}_k + \boldsymbol{\delta}_k$. 100 independent trials are run, and the results are compared and shown in Figure 9.4. Obviously, it is noted that the SINRs of all users achieved by the proposed robust method are no less than the threshold $\gamma = 10$ dB. On the contrary, the beamformer designed by the non-robust method cannot always ensure the per-user SINR constraint.

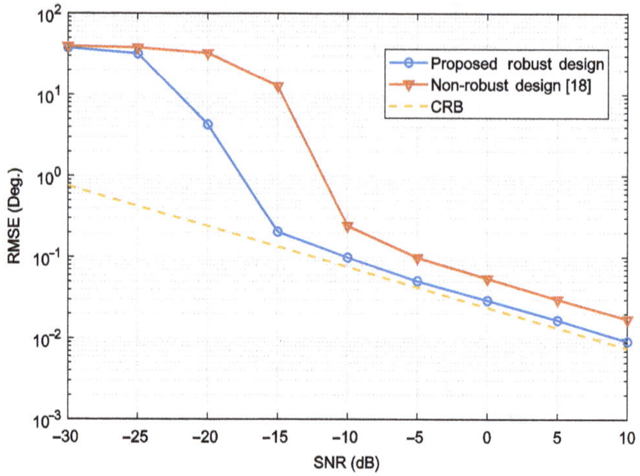

Figure 9.3 RMSE of estimation versus SNR of radar echo. $\varepsilon = 0.1$ *and* $\Theta_{\mathrm{ROI}} = [-10^\circ, 10^\circ]$

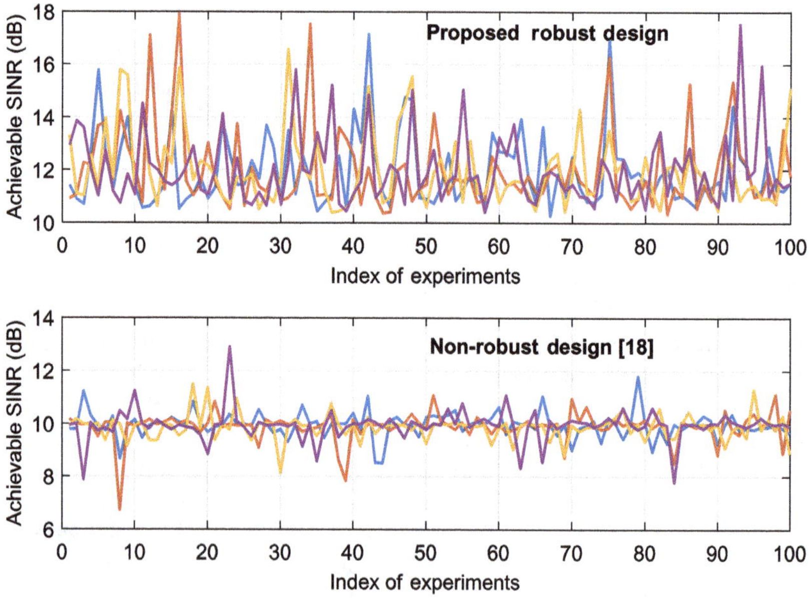

Figure 9.4 Comparison of achievable SINRs for 100 experiments. $\gamma = 10$ dB, $\varepsilon = 0.1$, *and* $\Theta_{\mathrm{ROI}} = [-10^\circ, 10^\circ]$

In the last example, we briefly show the extensions of the proposed robust beamformer to the cases of flat beams as well as multiple beams for sensing multiple targets. More concretely, the ROI is considered to be $\Theta_{\mathrm{ROI}} = [-30^\circ, -10^\circ] \cup 30^\circ$, i.e., targets may locate within $[-30^\circ, -10^\circ]$ or at 30°. The CSI error bound is

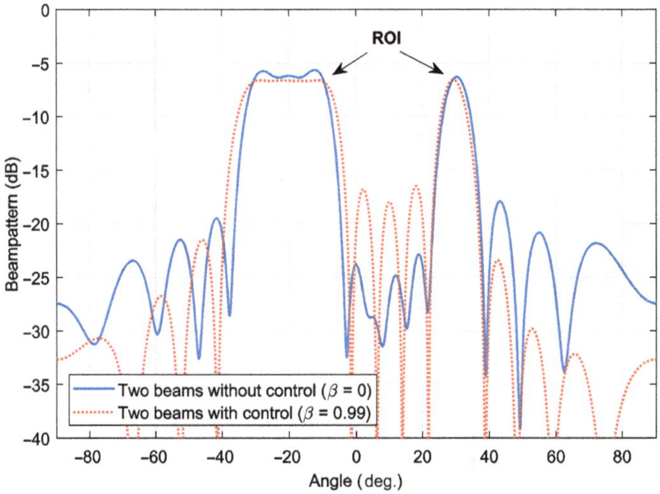

Figure 9.5 *Designed beamformer with multiple beams and beam control.* $\gamma = 10$ dB, $\varepsilon = 0.1$, *and* $\boldsymbol{\Theta}_{\text{ROI}} = [-30°, -10°] \cup 30°$

assumed to be 0.1. The resulting beampatterns are shown in Figure 9.5. It is seen that if we set $\beta = 0.99$, a relatively flat beam toward the region $[-10°, 10°]$ can be obtained. Moreover, a beam toward 30° is formed. The results indicate that the proposed method can flexibly design a beamformer for multiple target sensing.

9.5.2 *Results of outage-constrained design*

Suppose that the BS with $N = 16$ antennas serves $K = 3$ communication users. The estimated channel vector $\bar{\mathbf{h}}_k$ is Gaussian distributed with zero mean and covariance matrix $\mathbf{I}_N$. The CSI error vector $\boldsymbol{\delta}_k$ is randomly generated from complex Gaussian distribution with zero mean and variances $\zeta_k^2 = 0.008$. We further set the communication SINR thresholds as $\gamma_1 = \gamma_2 = \cdots = \gamma_K = \gamma = 10$ dB, and set the outage probability thresholds as $\nu_1 = \nu_2 = \cdots = \nu_K = \nu = 0.0005$. Moreover, we give the objective radar beam band $\boldsymbol{\Theta}_{\text{ROI}} = [-10°, 10°]$, $\boldsymbol{\Omega} = [-60°]$. In order to combat the interference and improve the performance of sensing, we set the weight ω to 2 and 30 at 0° and $-60°$, respectively. Based on this, we calculate the gain of radar with $\frac{P}{2 \sin 10°}$. The total transmit power P is set to 30 dBm, and the noise power is $\sigma_1^2 = \sigma_2^2 = \cdots = \sigma_K^2 = 0$ dBm.

Using the parameters mentioned above, we first compared the beampatterns under different processing methods, as illustrated in Figure 9.6. The non-robust method involves setting ζ_k^2 to a small value (e.g., 10^{-5}) to obtain the results. The Bernstein method represents another classical algorithm for outage probability constraint using Bernstein-type inequality. From Figure 9.6, it can be observed that the proposed method achieves similar performance as the non-robust method, which can simultaneously transmit enough energy to meet the detection requirements and suppress the interference target. Since the Bernstein method is a more conservative algorithm compared to the proposed method, it allows for a more safe fulfillment of

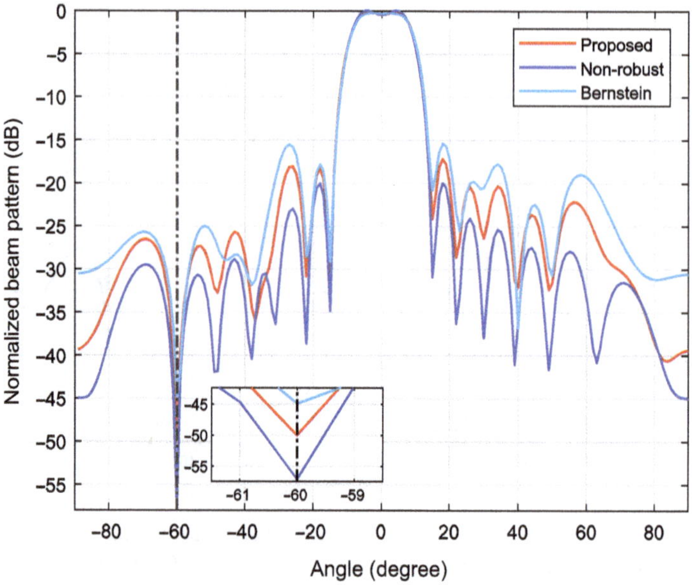

Figure 9.6 Comparison of the transmit beampattern of different methods where $\boldsymbol{\Theta}_{\mathrm{ROI}} = [-10^\circ, 10^\circ]$ *and* $\boldsymbol{\Omega} = [-60^\circ]$.

communication requirements. In the present scenario, both robust processing methods demonstrate comparable performance (Bernstein is slightly worse at suppressing jamming targets).

Figures 9.7 and 9.8 demonstrate the MSE of beampattern which is defined as $\frac{1}{M}\sum_{m=1}^{M}\omega_m\,|\mathbf{a}_m^{\mathsf{T}}\mathbf{W}\mathbf{W}^{\mathsf{H}}\mathbf{a}_m^* - d_m|^2$. The results indicate that an increase in ζ^2, a higher user requirement SINR γ, and a reduction in the outage probability μ all lead to an escalation in the Radar MSE. This implies a greater deterioration in radar performance. Besides, under the same parameters, the results obtained from the proposed method demonstrate better radar performance; however, the corresponding user's actual SINR tends to be lower.

In the next experiment, we briefly show the extensions of the proposed robust beamformer to the multiple beams bandwidth for sensing multiple targets. More concretely, the ROI is considered to be $\boldsymbol{\Theta}_{\mathrm{ROI}} = [-20^\circ, 0^\circ] \cup [25^\circ, 35^\circ]$, and the interference region is considered to be $\boldsymbol{\Omega} = [-60^\circ] \cup [60^\circ]$. The results shown in Figure 9.9 indicate that the channel error does not have a great impact on the main lobe, but when the channel error or the SINR threshold increases, the side lobe becomes larger and the suppression ability of the interference target turns to be weaker.

In Figure 9.10, we generated 1000 random channel errors and calculated the actual SINR for each user in each experience (only 200 selected runs are displayed in the figure). The results indicate that the non robust design fails to meet the users' expected SINR requirement. Conversely, both the proposed method and the Bernstein method are able to satisfy this criterion. Note that the SINR achievable with the proposed method tends to be closer to 10 dB compared to the Bernstein

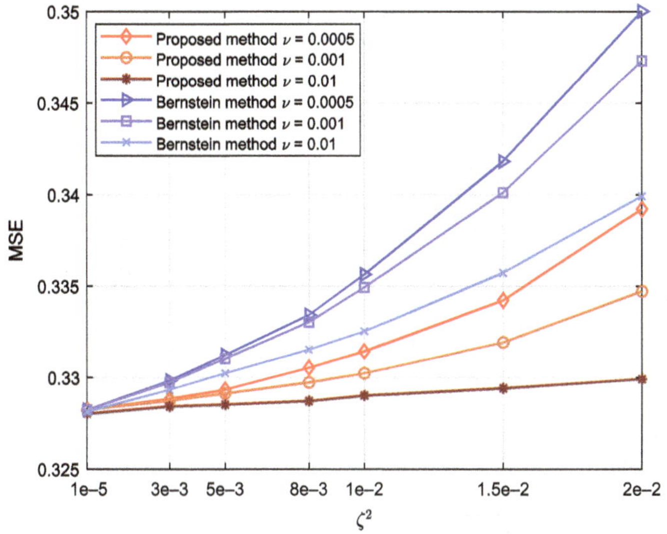

Figure 9.7 *Beampattern MSE under different channel error when changing outage probability thresholds with fixed* $\gamma = 10$ *dB*

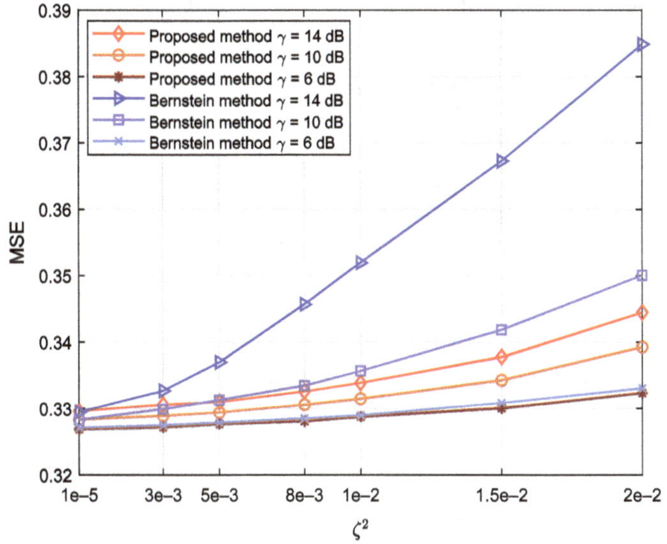

Figure 9.8 *Beampattern MSE under different channel error when changing SINR thresholds with fixed* $\nu = 0.0005$

method, suggesting potentially smaller radar performance loss. However, in practical scenarios, there may be a consideration to disregard these losses and prioritize superior communication performance.

In order to assess the impact of the approximation in (9.39) and compare the actual achievable interruption probabilities for users, we conducted experiments

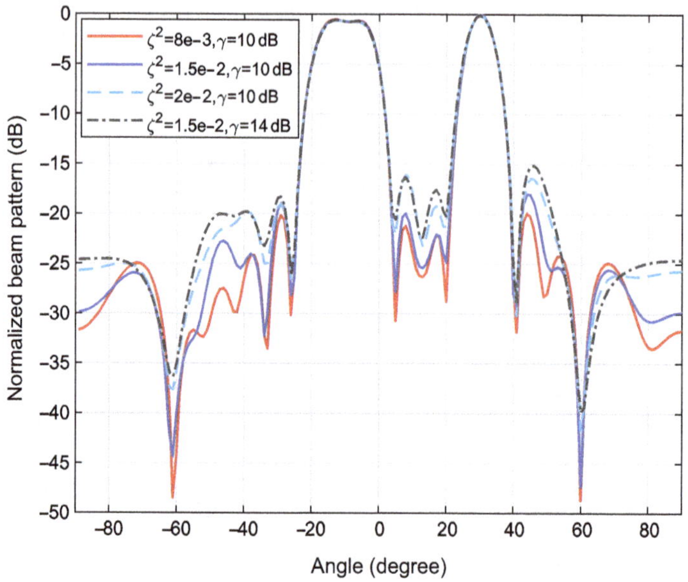

Figure 9.9 Designed beamformer with multiple beams where $\boldsymbol{\Theta}_{\mathrm{ROI}} = [-20^\circ, 0^\circ] \cup [25^\circ, 35^\circ]$, $\boldsymbol{\Omega} = [-60^\circ] \cup [60^\circ]$, *and* $\nu = 0.0005$

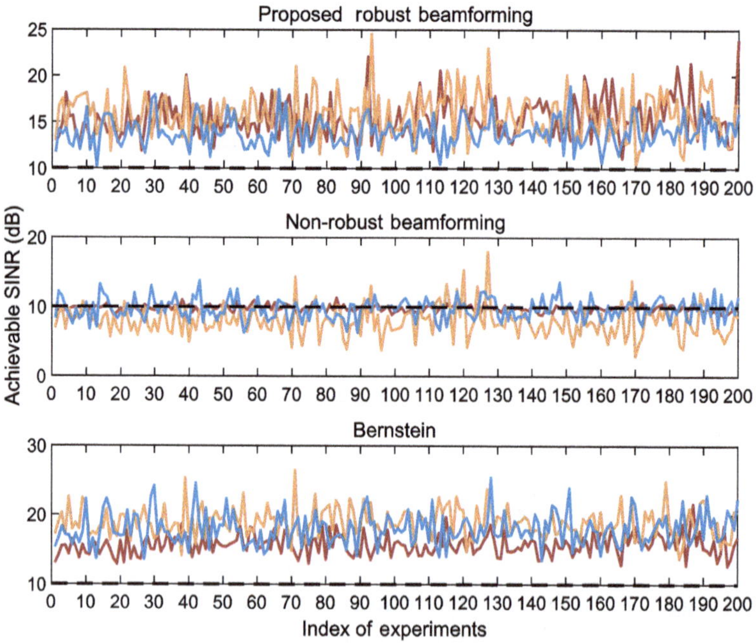

Figure 9.10 Comparison of achievable SINRs of different methods, where $\zeta^2 = 0.008$, $\nu = 0.0005$, *and* $\gamma = 10$ *dB*

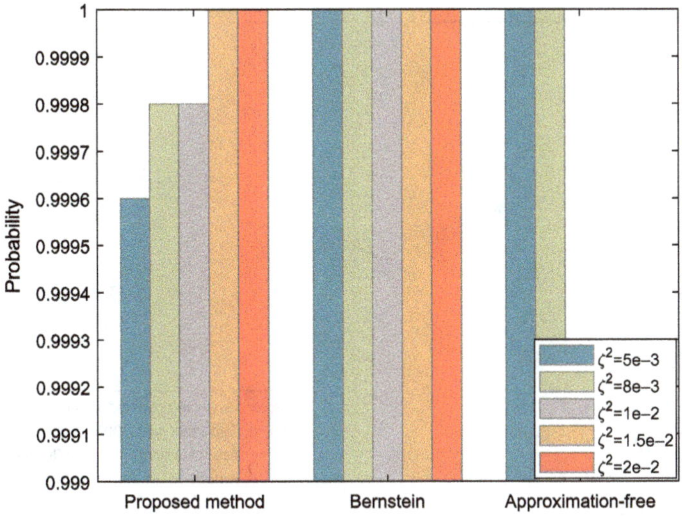

Figure 9.11 The probability of meeting users' communication requirements under different ζ^2, where ν = 0.0005 and γ = 10 dB

as depicted in Figure 9.11. The method without approximation (denoted as "approximation-free method") represents the results obtained without neglecting the fourth order term regarding ζ_k. As the approximation-free method shares similar intermediate processes and equation structures with the proposed algorithm, we only need to substitute the corresponding values in the proposed algorithm (e.g., setting $\mathbf{L}_k = \mathbf{I}_N \otimes (2\bar{\mathbf{H}}_k + \zeta_k^2 \mathbf{I}_N)$ and $\upsilon_k = \sqrt{\zeta_k^2(1/\nu_k - 1)}$), maintaining consistency in the procedure of the algorithm. Results indicate that the proposed method can satisfy communication constraints. Meanwhile, the Bernstein method, due to its conservative nature, ensures that the outage probability obtained in any situation is always below the threshold. However, the non-approximated methods, being overly conservative, may lead to situations where no solution is found, especially as channel errors continue to increase.

In Figure 9.12, the average SINRs of users obtained by different methods are illustrated. As previously described, the results obtained from the approximation-free method tend to be more conservative, and in scenarios with significant channel errors, it may lead to instances where the primal problem is infeasible. On the other hand, the proposed algorithm and the Bernstein method, being relatively less conservative, demonstrate adaptability to situations with larger channel errors.

Figure 9.13 shows the convergence of radar MSE, where $[\rho_1, \rho_2] = [80, 100]$. It can be seen that the radar MSE gradually flattens with iterations. Likewise, Figure 9.14 presents that the raw residuals and dual residuals tend to zero as the iterative process goes on. This verifies the effectiveness of the proposed algorithm for the MIMO-DFRC system. Finally, we compare the algorithm execution time for 1000 iterations, as depicted in Figure 9.15. It is evident that the proposed method has a notable speed advantage compared with those using solving with the CVX solver

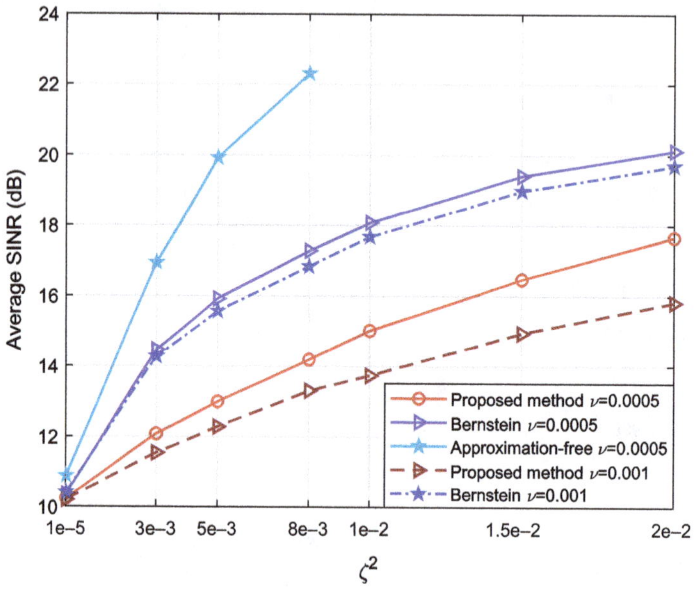

Figure 9.12 Average SINR attainable by users under different ζ^2

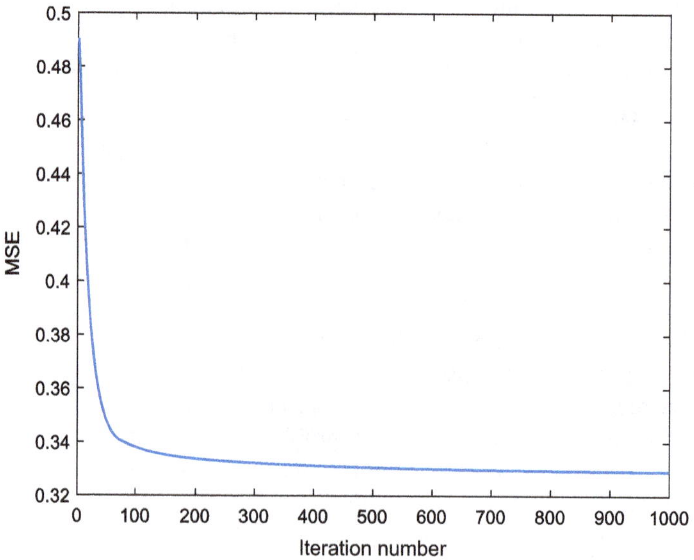

Figure 9.13 The convergence of the objective function value with respect to the number of iterations, at $\zeta^2 = 0.008$, $\nu = 0.0005$, $\gamma = 10$ dB

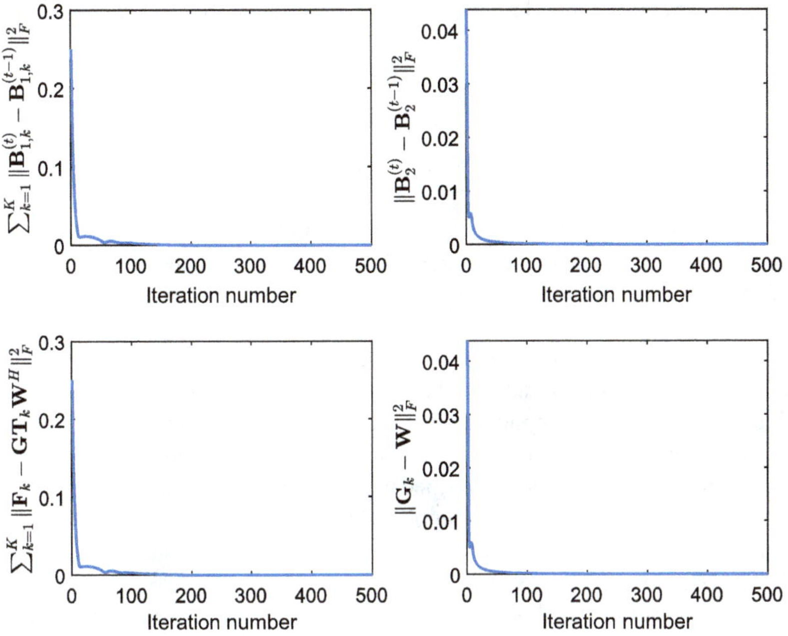

Figure 9.14 *The convergence of raw residuals and dual residuals with respect to the number of iterations, at* $\zeta^2 = 0.008$, $\nu = 0.0005$, $\gamma = 10$ *dB*

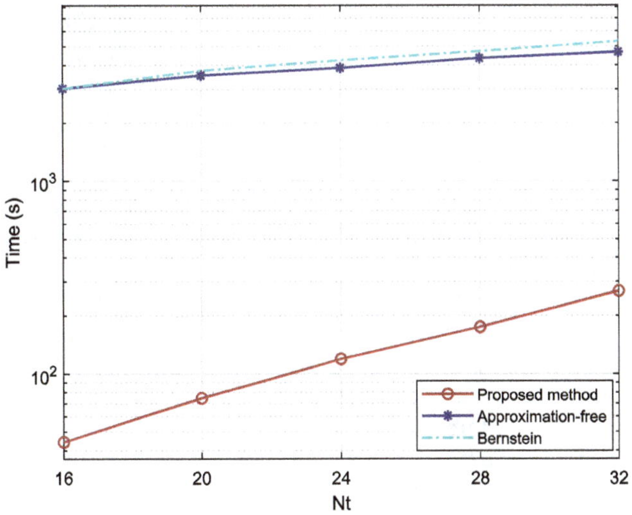

Figure 9.15 *Comparison of execution times for different algorithms with varying numbers of antennas*

and the Bernstein method. The speed of the proposed algorithm decreases as the number of antennas increases.

9.6 Conclusion

This chapter considers the problem of robust beamforming design for the MIMO-DFRC system. Specifically, with the modeling of hypersphere CSI errors, the problem is formulated as the maximization of minimum energy radiated toward the ROI where the target lies in, subject to the restriction on the transmit power and requirement of per-user SINR. The SDR scheme is sought to recast the problem as an SDP problem. This approach can flexibly design a robust beamformer without the need for an ideal radar beampattern. Furthermore, it is shown that a constraint on the magnitude response of the beampattern over the ROI can be imposed to form a flat beam, and multiple beams can be formed to multiple target sensing.

When the CSI error follows a complex Gaussian distribution, the beamformer is designed by synthesizing a desired transmit power pattern for radar sensing under the outage probability constraints on the SINR of each user. To deal with the problem, an iterative algorithm based on the ADMM framework is introduced. First, a set of auxiliary variables are utilized to decouple the original constraints. Then, Cantelli's inequality is introduced to approximate the outage probability constraint. It is shown that each subproblem can be solved in a low-complexity manner, which improves the efficiency of the algorithm.

References

[1] Griffiths H, Cohen L, Watts S, *et al.* Radar spectrum engineering and management: Technical and regulatory issues. *Proc IEEE*. 2015;103(1):85–102.

[2] Choi J, Va V, Gonzalez-Prelcic N, *et al.* Millimeter-wave vehicular communication to support massive automotive sensing. *IEEE Commun Mag*. 2016;54(12):160–167.

[3] Roh W, Seol JY, Park J, *et al.* Millimeter-wave beamforming as an enabling technology for 5G cellular communications: Theoretical feasibility and prototype results. *IEEE Commun Mag*. 2014;52(2):106–113.

[4] Hassanien A, Amin MG, Zhang YD, *et al.* Dual-function radar-communications: Information embedding using sidelobe control and waveform diversity. *IEEE Trans Signal Process*. 2016;64(8):2168–2181.

[5] Ma S, Sheng H, Yang R, *et al.* Covert beamforming design for integrated radar sensing and communication systems. *IEEE Trans Wirel Commun*. 2022;p. 1–1.

[6] Zhao N, Wang Y, Zhang Z, *et al.* Joint transmit and receive beamforming design for integrated sensing and communication. *IEEE Commun Lett*. 2022;26(3):662–666.

[7] Zhang JA, Liu F, Masouros C, *et al.* An overview of signal processing techniques for joint communication and radar sensing. *IEEE J Sel Top Signal Process.* 2021;15(6):1295–1315.
[8] Sturm C and Wiesbeck W. Waveform design and signal processing aspects for fusion of wireless communications and radar sensing. *Proc IEEE.* 2011;99(7):1236–1259.
[9] Sturm C and Wiesbeck W. Waveform design for joint digital beamforming radar and MIMO communications operability. In: *Principles of Waveform Diversity and Design.* SciTech Publishing; 2011.
[10] Yuan X, Feng Z, Zhang JA, *et al.* Spatio-temporal power optimization for MIMO joint communication and radio sensing systems with training overhead. *IEEE Trans Veh Technol.* 2021;70(1):514–528.
[11] Liu F, Zhou L, Masouros C, *et al.* Toward dual-functional radar-communication systems: Optimal waveform design. *IEEE Trans Signal Process.* 2018;66(16):4264–4279.
[12] Liu F, Masouros C, Li A, *et al.* MU-MIMO communications with MIMO radar: From co-existence to joint transmission. *IEEE Trans Wirel Commun.* 2018;17(4):2755–2770.
[13] Liu X, Huang T, Shlezinger N, *et al.* Joint transmit beamforming for multiuser MIMO communications and MIMO radar. *IEEE Trans Signal Process.* 2020;68:3929–3944.
[14] Cheng Z, He Z, and Liao B. Hybrid beamforming design for OFDM dual-function radar-communication system. *IEEE J Sel Top Signal Process.* 2021;15(6):1455–1467.
[15] Tian T, Zhang T, Kong L, *et al.* Transmit/receive beamforming for MIMO-OFDM based dual-function radar and communication. *IEEE Trans Veh Technol.* 2021;70(5):4693–4708.
[16] Cheng Z, Liao B, Shi S, *et al.* Co-design for overlaid MIMO radar and downlink MISO communication systems via Cramér-Rao bound minimization. *IEEE Trans Signal Process.* 2019;67(24):6227–6240.
[17] Keskin MF, Koivunen V, and Wymeersch H. Limited feedforward waveform design for OFDM dual-functional radar-communications. *IEEE Trans Signal Process.* 2021;69:2955–2970.
[18] Liu F, Liu YF, Li A, *et al.* Cramér-Rao bound optimization for joint radar-communication beamforming. *IEEE Trans Signal Process.* 2022;70: 240–253.
[19] Tsinos CG, Arora A, Chatzinotas S, *et al.* Joint transmit waveform and receive filter design for dual-function radar-communication systems. *IEEE J Sel Top Signal Process.* 2021;15(6):1378–1392.
[20] Liu F, Masouros C, Petropulu AP, *et al.* Joint radar and communication design: Applications, state-of-the-art, and the road ahead. *IEEE Trans Commun.* 2020;68(6):3834–3862.
[21] Su N, Liu F, and Masouros C. Secure radar-communication systems with malicious targets: Integrating radar, communications and jamming functionalities. *IEEE Trans Wirel Commun.* 2021;20(1):83–95.

[22] Huang Y, Vorobyov SA, and Luo ZQ. Quadratic matrix inequality approach to robust adaptive beamforming for general-rank signal model. *IEEE Trans Signal Process*. 2020;68:2244–2255.
[23] Liao B, Xiong X, and Quan Z. Robust beamforming design for dual-function radar-communication system. *IEEE Trans Veh Technol*. 2023;72(6): 7508–7516.
[24] Song E, Shi Q, Sanjabi M, *et al*. Robust SINR-constrained MISO downlink beamforming: When is semidefinite programming relaxation tight? *J Wirel Commun Netw*. 2012;243:1–11.
[25] Luo Z, Ma W, So AM, *et al*. Semidefinite relaxation of quadratic optimization problems. *IEEE Signal Process Mag*. 2010;27(3):20–34.
[26] Ma WK, Pan J, So AMC, *et al*. Unraveling the rank-one solution mystery of robust MISO downlink transmit optimization: A verifiable sufficient condition via a new duality result. *IEEE Trans Signal Process*. 2017;65(7):1909–1924.
[27] Huang Y and Palomar DP. Rank-constrained separable semidefinite programming with applications to optimal beamforming. *IEEE Trans Signal Process*. 2010;58(2):664–678.
[28] Liao B, Guo C, Huang L, *et al*. Robust adaptive beamforming with precise main beam control. *IEEE Trans Aerosp Electron Syst*. 2017;53(1):345–356.
[29] Cheng Z and Liao B. QoS-aware hybrid beamforming and DOA estimation in multi-carrier dual-function radar-communication systems. *IEEE J Sel Areas Commun*. 2022;40(6):1890–1905.
[30] Cheng Z, Liao B, He Z, *et al*. Transmit signal design for large-scale MIMO system with 1-bit DACs. *IEEE Trans Wirel Commun*. 2019;18(9):4466–4478.
[31] Medra M, Huang Y, and Davidson TN. Offset-based beamforming: A new approach to robust downlink transmission. *IEEE Trans Signal Process*. 2018;67(1):70–82.
[32] Grant M and Boyd S. CVX: Matlab software for disciplined convex programming, version 2.1; 2014. http://cvxr.com/cvx.

Chapter 10

Optimization of sparse MIMO array for enhanced sensing

Xiangrong Wang[1], Weitong Zhai[1] and Xianghua Wang[2]

This chapter considers the joint sparse optimization of multiple-input multiple-output (MIMO) radar transmitter and receiver arrays for improved beamforming while reducing the system overhead. We examine the active sparse array design enabling the maximum signal to interference plus noise ratio (MaxSINR) in two cases with and without the prior environmental information. In the latter case, a cognitive-driven optimization scheme is proposed to adaptively update both MIMO array configuration and beamforming weights via a "perception-action" cycle, which eliminates the prerequisite for prior information.

10.1 Introduction

The radar angular resolution relies on the effective array aperture, which represents the physical aperture normalized by the carrier signal wavelength [1]. This aperture is directly proportional to the number of antennas, particularly when employing a Uniform Linear Array (ULA) with an inter-element spacing equivalent to half-wavelength [2]. High-resolution direction of arrival (DOA) estimation necessitates a large-scale ULA, which can be impractical due to its costliness [3]. Employing multiple-input multiple-output (MIMO) radar systems operating in MIMO mode with multiple transmit and receive antennas offers a solution to achieve higher angular resolution compared to phased array configurations. Each MIMO radar transmitter emits an orthogonal waveform, and each MIMO receiver receives the radar returns. This waveform diversity characteristic leads to the creation of a virtual large apertured array, thereby potentially achieving higher angular resolution [4].

Categorized based on the distance between transmit and receive antennas, MIMO radars fall into two groups: distributed MIMO radars and colocated MIMO radars. Unlike conventional multistatic radar systems, distributed MIMO radars leverage spatial diversity and enlarge the inter-element spacing of the transmit array to enable each transmit antenna to detect radar targets from various directions. In the

[1]School of Electronic and Information Engineering, Beihang University, China
[2]School of Intelligent Engineering and Automation, Beijing University of Posts and Telecommunications, China

context of MIMO radar systems, the received echo signal results from the combination of independent fading signals. Aggregating these echo signals from distinct observation paths yields an almost constant radar cross section (RCS) of the target, thereby mitigating RCS flicker and achieving substantial spatial diversity gain. In colocated MIMO radar setups, the inter-element spacing tends to be relatively small. In typical setups, the transmit array and receive array are either closely spaced or utilize the same antennas. Unlike phased array radar systems, colocated MIMO arrays allow each transmit antenna to convey arbitrary waveform signals. Phased array radar systems can be considered as a particular instance of colocated MIMO radar systems in terms of transmit signals.

In this chapter, the colocated MIMO array is considered. The transmit waveform of the colocated MIMO array can be categorized into two types: orthogonal and partially correlated. When the transmit waveforms are orthogonal, an omni-directional transmit beampattern is achieved. In this scenario, high resolution can be obtained by separating signals at the receiving end to realize virtual aperture expansion. When the transmit waveforms are partially correlated, the transmit beampattern can be optimized to any desired shape and direction. In this scenario, transmit beampatterns can be designed to allocate transmit energy reasonably to the target area of interest.

Taking a ULA as an example, considering that the transmitter of the MIMO radar consists of a ULA comprising M antennas, while the receiver comprises another ULA consisting of N antennas. The transmit signal in the direction θ of the MIMO array can be expressed as

$$y(t) = \mathbf{a}_T^T(\theta)\mathbf{s}(t), \tag{10.1}$$

where $(.)^T$ represents transpose, $\mathbf{a}_T(\theta)$ is the transmit steering vector and $\mathbf{s}(t)$ is the vector composed of the signal transmitted by each antenna. Then, the transmit beampattern is given by

$$\begin{aligned} B(\theta) &= E\{y(t)y^*(t)\}, \\ &= E\{\mathbf{a}_T^T(\theta)\mathbf{s}(t)\mathbf{s}^H(t)\mathbf{a}_T^*(\theta)\}, \\ &= \mathbf{a}_T^T(\theta)E\{\mathbf{s}(t)\mathbf{s}^H(t)\}\mathbf{a}_T^*(\theta), \end{aligned} \tag{10.2}$$

where $E\{\cdot\}$ means taking expectation. We define the autocorrelation matrix of the transmit waveform as

$$\mathbf{R}_s = E\{\mathbf{s}(t)\mathbf{s}^H(t)\}. \tag{10.3}$$

where $(.)^H$ represents conjugate transpose. When the transmit signals are orthogonal, $\mathbf{R}_s$ is an identity matrix, that is $\mathbf{R}_s = \mathbf{I}$. Accordingly,

$$B(\theta) = \mathbf{a}_T^T(\theta)\mathbf{I}\mathbf{a}_T^*(\theta) = M, \tag{10.4}$$

where $(.)^*$ represents taking conjugation. It is indicated by Equation (10.4) that the radiation power is uniform in all directions, defining the omnidirectional detection scenario.

When the transmit signals are partially correlated, the matrix $\mathbf{R}_s$ ceases to be an identity matrix. By optimizing the matrix $\mathbf{R}_s$, we can synthesize transmit beampatterns of arbitrary shapes. A special instance is the multi-carrier waveform, where the transmit signal of each antenna can be seen as a linear combination of a set of orthogonal signals. In such cases, the transmit signal vector can be expressed as

$$\mathrm{s}(t) = \mathbf{W}\mathrm{s}_o(t), \tag{10.5}$$

where $\mathbf{W}$ is the weighting matrix and $\mathrm{s}_o(t)$ is the vector of M' orthogonal signals. Substituting (10.5) into (10.2), the transmit beampattern becomes

$$\begin{aligned} B(\theta) &= E\{\mathbf{a}_T^T(\theta)\mathbf{s}(t)\mathbf{s}^H(t)\mathbf{a}_T^*(\theta)\}, \\ &= \mathbf{a}_T^T(\theta)\mathbf{W}E\{\mathbf{s}_o(t)\mathbf{s}_o^*(t)\}\mathbf{W}^H\mathbf{a}_T^*(\theta), \\ &= \mathbf{a}_T^T(\theta)\mathbf{W}\mathbf{W}^H\mathbf{a}_T^*(\theta), \\ &= \mathbf{a}_T^T(\theta)\mathbf{R}_W\mathbf{a}_T^*(\theta), \end{aligned} \tag{10.6}$$

where $\mathbf{R}_W = \mathbf{W}\mathbf{W}^H$ is the weight autocorrelation matrix. The optimization problem of the transmit beamforming, as expressed in (10.6), can be restated as transforming the optimization into either the beamforming weight matrix $\mathbf{W}$ or the autocorrelation matrix $\mathbf{R}_W$.

At the receiving end, the received signal vector can be expressed as

$$\begin{aligned} \mathbf{x}(t) &= \eta\mathbf{a}_R(\theta)x(t), \\ &= \eta\mathbf{a}_R(\theta)\mathbf{a}_T^T(\theta)\mathbf{s}(t), \end{aligned} \tag{10.7}$$

where η represents the target RCS and the two-way path loss, which is assumed to follow the Swerling 0 model, $\mathbf{a}_R(\theta)$ is the receive steering vector. Applying the multi-carrier model, the received signal vector can be formed into a matrix upon matched filtering,

$$\begin{aligned} \mathbf{X} &= \int_T \mathbf{x}(t)\mathbf{s}_o^H(t)\mathrm{d}t, \\ &= \eta\mathbf{a}_R(\theta)\mathbf{a}_T^T(\theta)\mathbf{W}. \end{aligned} \tag{10.8}$$

When $\mathbf{W} = \mathbf{I}$, the transmit signals are orthogonal. In this case, $\mathbf{X}$ can be regarded as the received signal of $M \times N$ virtual antennas derived from the transmit and receive arrays. When $\mathbf{W} \neq \mathbf{I}$, the transmit signals are correlated, and $\mathbf{X}$ becomes the received signal of $M' \times N$ virtual antennas derived from the virtual transmit array (whose steering vector is $\bar{\mathrm{a}}(\theta) = \mathbf{W}^T\mathbf{a}_T(\theta)$) and the receive array. Equation (10.8) indicates mathematically that the MIMO array can provide more spatial DoFs than just $M + N$.

The desire for enhanced angular resolution in radar applications has driven the adoption of larger arrays. However, augmenting the number of antenna elements results in escalated hardware expenses, complexity, and power consumption. Consequently, several approaches have emerged to leverage the inherent redundancy in fully populated arrays to curtail the count of RF front-ends while upholding the array's desirable performance characteristics. Leveraging emerging fast antenna switching technologies, sparse array design can decrease overall system costs by minimizing the number of expensive front-end processing channels. Employing sparse arrays (SAs) synthesized with MIMO radar technology presents an effective means to further decrease costs while maintaining high angular resolution [5,6]. MIMO radar systems, utilizing sparse configurations, achieve reductions in the number of transmit and receive antennas by selectively choosing a subset of antennas from a ULA. This reduction in antennas preserves array aperture while diminishing the overall antenna count.

The conventional MIMO radar features a sparse transceiver array, with the receive array having an inter-element spacing of half wavelength and the transmit array having multiple wavelengths. Consequently, the MIMO sum coarray forms a compact ULA with a substantial virtual aperture, facilitating high spatial resolution [7–9]. However, this configuration might not achieve optimal beamforming for maximizing the output SINR (MaxSINR) in a given environment [10,11]. Sparse MIMO array beamforming design can involve optimizing the transmit sensor locations and the corresponding waveform correlation matrix [12–14]. The optimal design of sparse MIMO transceivers can be categorized into two types: structured and unstructured sparse arrays.

The structured sparse array design, which is independent of the environment, aims to increase the number of spatial autocorrelation lags and maximize the contiguous segment of the coarray aperture with a limited number of sensors. The primary objective is to facilitate DOA estimation involving more sources than the physical sensors [15–18]. In Section 10.2, the design of unstructured sparse MIMO array transceivers in terms of estimation accuracy through antenna selection is introduced. Although commonly used structured SAs, such as minimum redundant array (MRA) [19], nested array [16], and coprime array [17,20], enable DOA estimation of more signals than the number of physical sensors, they do not necessarily provide the most accurate angular estimates.

The environment-dependent, unstructured sparse receive beamformer, which aims to achieve MaxSINR, has the potential to offer sparse configurations that enhance target detection and estimation accuracy for the operating environment. Conversely, the environment-independent design criteria for beamforming applications typically aim to attain desirable beampattern characteristics, such as a broad main lobe, minimum sidelobe levels, and frequency-invariant beampatterns for wideband designs [1,13,21–32].

10.2 Sparse MIMO transceiver design for enhanced DOA estimation

In this section, accurate sensing with enhanced DOA estimation is explored by optimizing the configurations of the sparse MIMO transceiver to minimize the CRB. A collocated MIMO radar system is considered, wherein the transmitter and receiver are closely separated compared to the distance to the far-field sources. The ULA of

M antennas constitutes the transmitter, with each antenna emitting an orthogonal signal, while the ULA of N antennas forms the receiver. The inter-element spacing of the transmit and receive arrays is denoted by d_t and d_r, respectively. As mentioned in Section 10.1, the sum-coarray of a co-located orthogonal MIMO array is a large virtual array with MN virtual antennas, and thus the number of estimated sources can be dramatically increased to MN. Without losing generality, we set $d_t = Nd_r$, then the design of the MIMO array transceiver can generate an MN-element linear uniform virtual array, in turn enabling various kinds of DOA estimation algorithms.

10.2.1 Cramer–Rao bound of multi-source DOA estimation

In a collocated MIMO radar system, the transmitter and receiver are closely spaced relative to the distance to the far-field sources. The transmitter consists of a ULA of M antennas, and the receiver comprises a ULA of N antennas. Let d_t and d_r denote the inter-element spacing of the transmit and receive arrays, respectively. The MIMO radar transmits a set of M linearly independent signals, $\mathbf{s}(t) = \{s_1(t), \cdots, s_M(t)\}$, with a covariance matrix satisfying the orthogonality condition, that is,

$$\mathbf{R}_s = \int_0^{T_p} \mathbf{s}(t)\mathbf{s}^H(t) = \mathbf{I}, \tag{10.9}$$

where T_p is the pulse duration. The superscript "H" represents conjugate transpose.

Assume that K far-field narrow-band sources are located in directions $\theta = [\theta_1, \ldots, \theta_K]$. The received data vector can be written as

$$\mathbf{x}(t,\tau) = \mathbf{A}_r \mathrm{diag}[\beta(\tau)]\mathbf{A}_t^T \mathbf{s}(t) + \mathbf{v}(t,\tau), \tag{10.10}$$

where t denotes the fast-time index (i.e., time within the radar pulse), and τ represents the slow-time index (i.e., pulse number). $\beta(\tau) = [\beta_1(\tau), \cdots, \beta_K(\tau)]^T$ is the reflection coefficient which is rapidly changing and may not satisfy the Swirling II model. The noise vector $\mathbf{v}(t,\tau) = [v_1(t,\tau), \cdots, v_N(t,\tau)]^T$ is modeled as a zero-mean, complex white Gaussian process with power σ_n^2. The transmitting array manifold matrix $\mathbf{A}_t$ is given by $\mathbf{A}_t(\theta) = [\mathbf{a}_T(\theta_1), \ldots, \mathbf{a}_T(\theta_K)]$, where the transmitting steering vector is defined by

$$\mathbf{a}_T(\theta) = [1, e^{j2\pi(d_t/\lambda)\cos\theta}, \ldots, e^{j2\pi(M-1)(d_t/\lambda)\cos\theta}]^T. \tag{10.11}$$

Similarly, the receive array manifold $\mathbf{A}_r$ is denoted as $\mathbf{A}_r(\theta) = [\mathbf{a}_R(\theta_1), \ldots, \mathbf{a}_R(\theta_K)]$, and the receive steering vector is defined by

$$\mathbf{a}_R(\theta) = [1, e^{j2\pi(d_r/\lambda)\cos\theta}, \ldots, e^{j2\pi(N-1)(d_r/\lambda)\cos\theta}]^T. \tag{10.12}$$

Upon performing matched-filtering to the received signal vector by the transmitted orthogonal waveforms, we obtain

$$\begin{aligned}\mathbf{X}(\tau) &= \int_0^{T_p} \mathbf{x}(t,\tau)\mathbf{s}^H(t)dt \\ &= \mathbf{A}_r(\theta)\mathrm{diag}[\beta(\tau)]\mathbf{A}_t^T(\theta) + \mathbf{N}(\tau),\end{aligned} \tag{10.13}$$

where the additive noise is $\mathbf{N}(\tau) = \int_0^{T_p} \mathbf{v}(t,\tau)\mathbf{s}^H(t)dt$. The elements of the matrix $\mathbf{N}(\tau)$ are independent and identically, zero-mean Gaussian with variance σ_n^2.

Applying vectorization to (10.13), we obtain

$$\mathbf{y}(\tau) = \text{vec}(\mathbf{X}(\tau)) = \mathbf{A}\beta(\tau) + \mathbf{n}(\tau), \tag{10.14}$$

where $\mathbf{A} = [\mathbf{b}(\theta_1), \cdots, \mathbf{b}(\theta_K)]$ with the vector $\mathbf{b}(\theta_k)$ is defined in (10.28), $\mathbf{n}(\tau) = \text{vec}\{\mathbf{N}(\tau)\}$ is the extended additive Gaussian noise term with zero mean and covariance $\sigma_n^2\mathbf{I}$. As mentioned above, when $d_t = Nd_r$, $\mathbf{b}(\theta_k)$ is the steering vector of a ULA with MN virtual antennas uniformly spaced by d_r.

Based on the signal model in (10.14), the CRB of multi-source DOA estimation is [33]

$$\text{CRB} = \frac{\sigma_n^2}{2J}\left\{\text{Re}[\{\mathbf{D}^H(\mathbf{I} - \mathbf{A}(\mathbf{A}^H\mathbf{A})^{-1}\mathbf{A}^H)\mathbf{D}\} \odot \mathbf{T}^T]\right\}^{-1}, \tag{10.15}$$

where J is the total number of snapshots and Re[·] denotes the real part of a complex entry. The first order derivative of $\mathbf{a}_i$ with respect to $\omega_i = \frac{2\pi}{\lambda}d_r\sin(\theta_i)$ is denoted as

$$\mathbf{d}_i = \frac{\partial \mathbf{a}_i}{\partial \omega_i} = [0, je^{j\omega_i}, \ldots, j(MN-1)e^{j(MN-1)\omega_i}], \tag{10.16}$$

and $\mathbf{D} = [\mathbf{d}_1, \cdots, \mathbf{d}_K]$. The matrix $\mathbf{T}$ is the source covariance matrix, i.e.,

$$\mathbf{T} = E\{\beta(\tau)\beta^H(\tau)\}. \tag{10.17}$$

When sources are uncorrelated, the matrix $\mathbf{T}$ becomes diagonal with the power $P_i, i = 1, \cdots, K$ along the diagonal. Consequently, the CRB is also a diagonal matrix with each diagonal element representing the DOA estimation accuracy of the corresponding detection target.

10.2.2 Sparse MIMO array transceiver design in the metric of CRB

We take the maximum value of CRBs of all estimated DOAs as the objective function for array thinning, i.e.,

$$\begin{aligned} f &= \max\{[\text{CRB}]_{ii, 1\le ii\le K}\}, \\ &= \max_{1\le i\le K}\left\{\frac{\sigma_n^2}{2N}\{P_i\mathbf{d}_i^H(\mathbf{I} - \mathbf{A}(\mathbf{A}^H\mathbf{A})^{-1}\mathbf{A}^H)\mathbf{d}_i\}^{-1}\right\}. \end{aligned} \tag{10.18}$$

Define the selection vector $\mathbf{z} \in \{0,1\}^{N_tN_r}$, where "1" denotes the corresponding virtual antenna is selected and "0" discarded. According to the properties of MIMO transceiver, "$\mathbf{z}$" exhibits the characteristics of group sparsity. From the perspective of antenna selection, (10.18) can be rewritten as

$$f(\mathbf{z}) = \max_{1\le i\le K}\{P_i^{-1}[\mathbf{d}_i^H\text{diag}(\mathbf{z})\mathbf{d}_i - \mathbf{d}_i^H\text{diag}(\mathbf{z})\mathbf{A}(\mathbf{A}^H\text{diag}(\mathbf{z})\mathbf{A})^{-1}\mathbf{A}^H\text{diag}(\mathbf{z})\mathbf{d}_i]^{-1}\} \tag{10.19}$$

The optimization problem is then formulated as

$$\begin{aligned} &\min f(\mathbf{z}), \\ &s.t.\ \mathbf{z} \in \{0,1\}^{N_t N_r}, \\ &\quad (1/N_2)\mathbf{p}_i^T \mathbf{z} \in \{0,1\}, i = 1,\ldots,N_t, \\ &\quad (1/N_1)\mathbf{q}_j^T \mathbf{z} \in \{0,1\}, j = 1,\ldots,N_r, \\ &\quad ||\mathbf{P}\mathbf{z}||_0 = N_1, \\ &\quad ||\mathbf{Q}\mathbf{z}||_0 = N_2. \end{aligned} \tag{10.20}$$

Note that the selection vector $\mathbf{z}$ exhibits a group sparsity structure, we define

$$\mathbf{p}_i = [\underbrace{0,\ldots,0}_{(i-1)N_r\ 0s}, \underbrace{1,\ldots,1}_{N_r\ 1s}, 0,\ldots,0]^T \in \mathbb{R}^{N_t N_r \times 1}, \tag{10.21}$$

$$\mathbf{q}_j = [\ \underbrace{0,\ldots,0,1,0,\ldots,0}_{\substack{\text{all elements are 0 except} \\ \text{for the } j\text{th element is 1} \\ (N_r \text{ elements})}}\ ,\ldots,\ \underbrace{0,\ldots,0,1,0,\ldots,0}_{\substack{\text{all elements are 0 except} \\ \text{for the } j\text{th element is 1} \\ (N_r \text{ elements})}}\ ]^T \in \mathbb{R}^{N_t N_r \times 1}. \tag{10.22}$$

where $\mathbf{p}_i$ and $\mathbf{q}_j$ are the transmit and receive selection vectors of the ith transmit antenna and the jth receive antenna, respectively, which are used to select all virtual matched filters related to the corresponding antenna. $\mathbf{P}$ and $\mathbf{Q}$ are matrices defined as $\mathbf{P} = [\mathbf{p}_1,\ldots,\mathbf{p}_{N_t}]^T$ and $\mathbf{Q} = [\mathbf{q}_1,\ldots,\mathbf{q}_{N_t}]^T$, respectively.

Introducing some convex relaxation strategies, we can transform the original problem, where both the objective function and the cardinality constraints are nonconvex, into a convex form.

As for the nonconvex objective function, we can transform it into the following equivalent linear matrix inequality (LMI) constraint optimization problem utilizing the Schur complement of a block matrix, that is,

$$\min_{\mathbf{z},\delta}\ -\delta, \tag{10.23}$$

$$\text{s.t.} \begin{bmatrix} \mathbf{A}^H \text{diag}(\mathbf{z})\mathbf{A} & \mathbf{A}^H \text{diag}(\mathbf{z})\mathbf{d}_i \\ \mathbf{d}_i^H \text{diag}(\mathbf{z})\mathbf{A} & \mathbf{d}_i^H \text{diag}(\mathbf{z})\mathbf{d}_i - \frac{\delta}{P_i} \end{bmatrix} \succeq 0,\ 1 \le i \le K. \tag{10.23a}$$

The LMI constraint in (10.23a) implies that

$$\begin{aligned} P_i^{-1}\{&\mathbf{d}_i^H \text{diag}(\mathbf{z})\mathbf{d}_i - \mathbf{d}_i^H \text{diag}(\mathbf{z})\mathbf{A}(\mathbf{A}^H \text{diag}(\mathbf{z})\mathbf{A})^{-1} \\ &\mathbf{A}^H \text{diag}(\mathbf{z})\mathbf{d}_i\}^{-1} \le \frac{1}{\delta}, i \in \{1,2,\ldots,K\} \end{aligned} \tag{10.24}$$

As for the nonconvex cardinality constraints, the binary of variable "z" can be promoted via an improved reweighted l_1 iteration method. In this case, for the mth iteration, (10.20) is transformed into

$$\begin{aligned}
\min \ & -\delta + \mu_1 \mathbf{w}_1^{(m)T}\mathbf{z}_1 + \mu_2 \mathbf{w}_2^{(m)T}\mathbf{z}_2 \\
s.t.\ & 0 \le \mathbf{z} \le 1, \\
& (1/N_2)\mathbf{p}_i^T\mathbf{z} \in \{0,1\}, i = 1,\ldots,N_t, \\
& (1/N_1)\mathbf{q}_j^T\mathbf{z} \in \{0,1\}, j = 1,\ldots,N_r, \\
& \mathbf{z}_1 = (1/N_2)\mathbf{P}\mathbf{z},\ 0 \le \mathbf{z}_1 \le 1,\ \mathbf{1}^T\mathbf{z}_1 = N_1, \\
& \mathbf{z}_2 = (1/N_1)\mathbf{Q}\mathbf{z},\ 0 \le \mathbf{z}_2 \le 1,\ \mathbf{1}^T\mathbf{z}_2 = N_2, \\
& \begin{bmatrix} \mathbf{A}^H\mathrm{diag}(\mathbf{z})\mathbf{A} & \mathbf{A}^H\mathrm{diag}(\mathbf{z})\mathbf{d}_i \\ \mathbf{d}_i^H\mathrm{diag}(\mathbf{z})\mathbf{A} & \mathbf{d}_i^H\mathrm{diag}(\mathbf{z})\mathbf{d}_i - \frac{\delta}{P_i} \end{bmatrix} \succeq 0,\ i = 1,\ldots,K.
\end{aligned} \tag{10.25}$$

where μ_1 and μ_2 are manually set trade-off parameters, and $\mathbf{w}_1^{(m)}$ and $\mathbf{w}_2^{(m)}$ are reweighted coefficient vectors of the transmit antennas and receive antennas, respectively. Note that we relax the binary constraint on $\mathbf{z}$ to a box constraint $0 \le \mathbf{z} \le 1$. And then a reweighted l_1-norm is adopted to promote the boolean sparsity of the selection vector. For the mth iteration, the reweighted coefficient of the ith transmit antenna, that is the ith element of $\mathbf{w}_1$, is updated by

$$(\mathbf{w}_1^{(m)})_i = \frac{1 - (\mathbf{z}_1^{(m-1)})_i}{1 - e^{-\beta_0(\mathbf{z}_1^{(m-1)})_i} + \varepsilon} - \left(\frac{1}{\varepsilon}\right)((\mathbf{z}_1^{(m-1)})_i)^{\alpha_0}, \tag{10.26}$$

where $\mathbf{z}_1^{(m-1)}$ is calculated by the optimal solution of the $m-1$th iteration. Similarly, the reweighted coefficient of the jth receive antenna, that is the jth element of $\mathbf{w}_2$, is updated by

$$(\mathbf{w}_2^{(m)})_j = \frac{1 - (\mathbf{z}_2^{(m-1)})_j}{1 - e^{-\beta_0(\mathbf{z}_2^{(m-1)})_j} + \varepsilon} - \left(\frac{1}{\varepsilon}\right)((\mathbf{z}_2^{(m-1)})_j)^{\alpha_0}. \tag{10.27}$$

10.2.3 Simulations

In this section, the DOA estimation performance of the proposed method is verified.

First, an MIMO array consisting of a 10-antenna ULA transmitter with an inter-element spacing of half wavelength and a 10-antenna ULA receiver with an inter-element spacing of five wavelengths is considered. Five antennas are chosen from the transmit array and another five antennas from the receive array to form the sparse transceiver. There are five targets in the radar field of view which are located at $-70°$, $-30°$, $10°$, $40°$, and $70°$, respectively, with SNR all being -20 dB. In the considered scenario, the iterative reweighted algorithm proposed in Section 10.2.2 is employed to optimize the configuration of the sparse MIMO transceiver, aiming to minimize the DOA estimation CRB. The optimal sparse MIMO transceiver, depicted in Figure 10.1, exhibits the selection of the 1st, 2nd, 3rd, 9th, and 10th antennas for both the transmit and receiver arrays, ensuring the largest virtual array aperture.

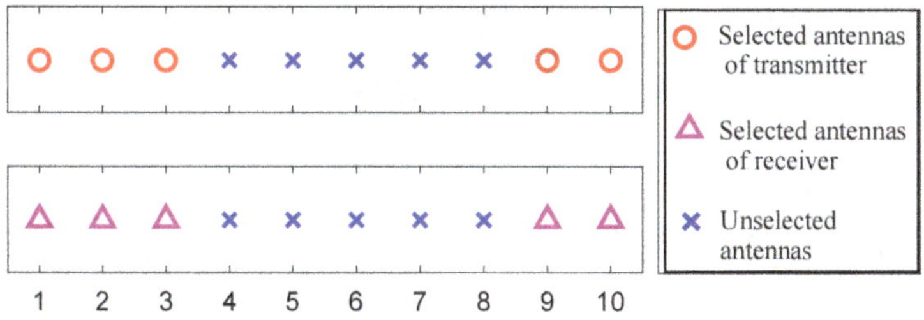

Figure 10.1 *The configurations of selected transmit array (upper figure) and receive array (lower figure) corresponding to the optimal sparse MIMO transceiver*

Figure 10.2 *Virtual array generated by the optimally co-designed sparse MIMO transceiver. Here, "down-triangle" denotes the corresponding virtual antenna selected and "cross" unselected*

Consequently, the generated virtual array, illustrated in Figure 10.2, shows virtual antennas of the same color associated with one transmit antenna and virtual antennas in one row linked to the same receive antenna.

Second, the enhanced DOA estimation accuracy of the optimal transceiver is verified. We select three sparse transceivers, depicted in Figure 10.3, for comparison with the optimal MIMO transceiver obtained by the proposed algorithm. Employing the high-resolution multiple signal classification (MUSIC) algorithms, DOAs of radar targets are estimated, and 10000 Monte-Carlo runs are conducted for statistical performance analysis. The curves plotting estimation mean square errors (MSEs) versus SNRs using different MIMO transceivers are illustrated in Figure 10.4. In Figure 10.4(a), we calculate the MSEs of each target and average them to obtain the averaged estimation accuracy. In Figures 10.4(b)–10.4(f), we depict the MSE versus

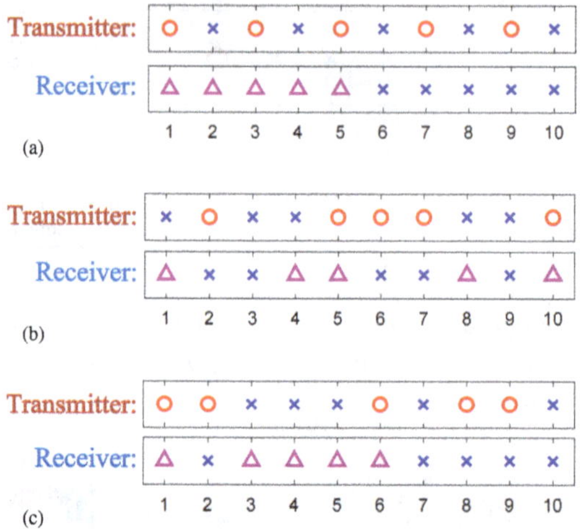

Figure 10.3 The configurations of the selected transmit and receive antennas of uniform, random 1, and random 2: (a) Uniform transceiver, (b) Random sparse transceiver 1 and (c) Random sparse transceiver 2.

SNR curve of each target separately. In Figure 10.4, the CRB is depicted by the red line, while the MSE of the optimal transceiver is represented by the pink line. It can be observed that as the SNR increases, the MSE curves gradually converge toward the CRB. The performance of the scenario where the virtual array forms a ULA, along with its corresponding transceiver configuration depicted in Figure 10.3(a), is indicated by the blue line. The configurations depicted in Figures 10.3(b) and 10.3(c) correspond to the sparse transceivers represented by the sapphire and green lines, respectively. Figure 10.4 illustrates that the DOA estimation accuracy of the optimal transceiver significantly surpasses that of the other arrays, affirming the enhanced DOA estimation achieved by the constructed optimal transceiver. To further illustrate the DOA estimation accuracy of the optimal transceiver, we maintain the SNR of the target echoes at 10 dB and depict the MUSIC spectrum obtained by the optimal transceiver in Figure 10.5. The results reveal that the peak values at the target's directions exceed the sidelobe by more than 20 dB. Thus, the DOAs of the targets can be readily identified, demonstrating the good DOA estimation accuracy of the optimal transceiver.

Finally, we further verify the tightness of the iterative method by depicting the performance of all possible configurations of the transceiver using the exhaustive method, as shown in Figure 10.6. For the sparse transceiver designed for DOA estimation, there are a total of $C_{10}^{5} \times C_{10}^{5} = 63504$ different cases of configurations. The MSEs are calculated under all these transceivers, as shown in Figure 10.6. From the results, it can be seen that the optimal transceiver selected by (10.25) can achieve the optimal DOA estimation accuracy, which proves that the applied relaxation strategies have good tightness.

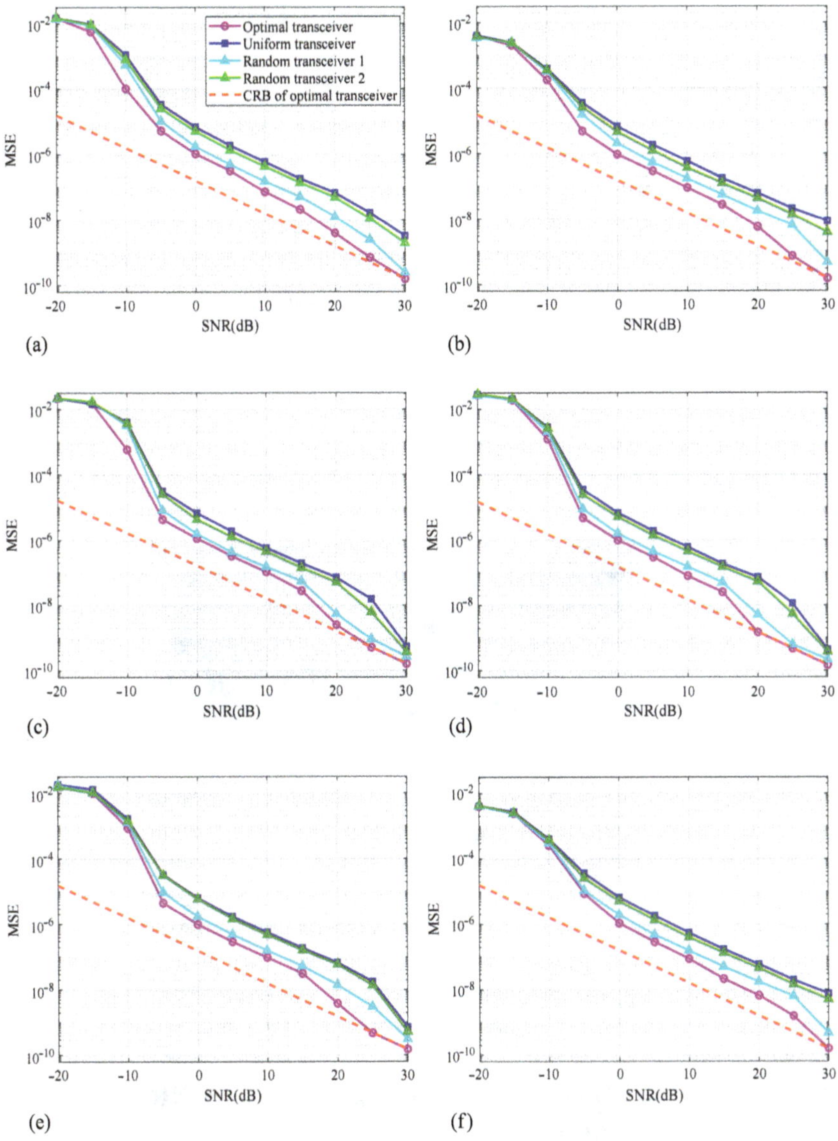

Figure 10.4 The curves of MSE versus SNR for different sparse transceivers: (a) the average of five signals, and (b)–(f) from the first to the fifth signal. (a) Average performance, (b) the target located at −70°, (c) the target located at −30°, (d) the target located at 10°, (e) the target located at 40°, and (f) the target located at 70°.

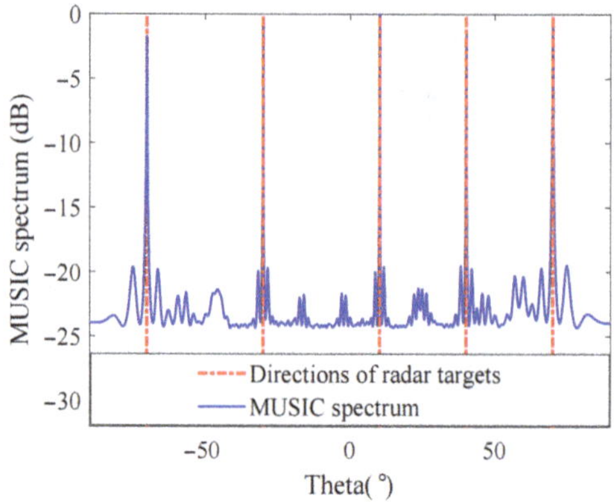

Figure 10.5 The MUSIC spectrum of the optimal transceiver

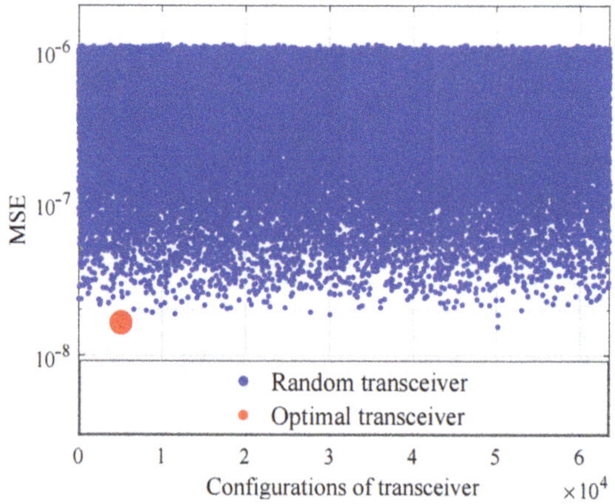

Figure 10.6 The averaged MSE of DOA estimation for all configurations of transceiver

10.3 Sparse MIMO transceiver design for MaxSINR with known environmental information

In this section, we focus on the design of MaxSINR beamformers for MIMO radar systems. This involves jointly selecting the optimal locations for both transmit and receive sensors to efficiently implement a receive beamformer. It is assumed that the transmit sensors emit orthogonal waveforms. However, MIMO radar systems using sensor-based orthogonal transmit waveforms do not benefit from the coherent transmit processing gain achieved when employing directional beamforming. The proposed approach in this context differs fundamentally from parallel sparse array MIMO beamforming designs, which primarily focus on optimizing the transmit sensor locations and the corresponding correlation matrix of the transmit waveform sequence [34–36]. By contrast, the proposed approaches aim to maximize the transmitted signal power toward the target locations while minimizing the cross-correlation of the target returns, thus achieving efficient receive beamforming characteristics. Furthermore, the existing design schemes primarily focus on an uncoupled transmit/receive design, whereas the approach considered in this chapter integrates sparse receiver design alongside transmit array optimization.

The problem of beamforming design for the proposed transceiver essentially involves configuring the transmit/receive array along with the corresponding optimum receiver beamforming weights. In pursuit of this objective, we adopt a data-dependent approach that jointly optimizes the positions of beamforming sensors and their associated weights. For the optimization process, prior knowledge of the operational environment is essential, encompassing the desired locations of potential sources and the covariance matrix of the entire array. We select M_t transmitters optimally from a pool of M options and N_r receivers optimally from a pool of N. Such selection is a binary optimization problem which is NP-hard. In order to avoid extensive computations associated with enumeration of all possible array configurations, we apply the convex relaxation. The design problem is posed as SCA with two-dimensional reweighted mixed $l_{1,2}$-norm penalization to jointly invoke sparsity in the transmit and receive dimensions.

10.3.1 Problem formulation

Considered is a MIMO radar setup where M transmitters and N receivers illuminate the scene using an omni-directional beampattern, typically achieved by emitting orthogonal waveforms across the transmit array. Specifically, we adopt a co-located MIMO arrangement, with the transmit and receive arrays situated in close proximity. Consequently, a far-field source is characterized by equal angles of departure and arrival. The scenario involves K target sources arriving from $\theta_{s,k}$. The received baseband data $\mathbf{x}(n) \in \mathbb{C}^{MN\times 1}$ after matched filtering at the N element uniformly spaced receiver array at time instant n is given by

$$\mathbf{x}(n) = \sum_{k=1}^{K} s_k(n)\mathbf{b}(\theta_{s,k}) + \sum_{l=1}^{L_c} c_l(n)\mathbf{b}(\theta_{j,l}) + \sum_{q=1}^{Q} \mathbf{i}_q(n) + \mathbf{v}(n), \tag{10.28}$$

where $s_k(n) \in \mathbb{C}$ is the kth reflected target signal. The extended steering vector $\mathbf{b}(\theta)$ of the virtual array is $\mathbf{b}(\theta) = \mathbf{a}_T(\theta) \otimes \mathbf{a}_R(\theta)$. In the case of uniform transmit and receive linear arrays with a respective inter-element spacing of d_t and d_r, the transmit and receive steering vectors are given by (10.11) and (10.12), respectively. The variance of additive Gaussian noise $\mathbf{v}(n) \in \mathbb{C}^{MN\times 1}$ is σ_v^2 at the receiver output. There are L_c interferences $c_l(n)$ mimicking target reflected signal and Q narrowband interferences $\mathbf{i}_q(n) \in \mathbb{C}^{MN\times 1}$. The latter is defined as the Kronecker product (which is represented by $\otimes$) of the receiver steering vector $\mathbf{a}_R(\theta_{i,q})$ and the matched filtering output of the interference $\mathbf{j}_q(n)$ such that $\mathbf{i}_q(n) = \mathbf{j}_q(n) \otimes \mathbf{a}_R(\theta_{i,q})$. The received data vector $\mathbf{x}(n)$ is then linearly combined to maximize the output SINR. The output signal $y(n)$ of the optimum beamformer for MaxSINR is given by [37]

$$y(n) = \mathbf{w}^H \mathbf{x}(n), \tag{10.29}$$

where $\mathbf{w}$ is the beamformer weight at the receiver. By solving the optimization problem that aims to minimize the interference power at the receiver output while preserving the desired signal, we can obtain the optimal solution $\mathbf{w}_o$. This problem can be formulated as a constrained minimization problem as follows:

$$\begin{aligned} \underset{\mathbf{w} \in \mathbb{C}^{MN}}{\text{minimize}} \quad & \mathbf{w}^H \mathbf{R}_{\mathbf{x}} \mathbf{w}, \\ \text{s.t.} \quad & \mathbf{w}^H \mathbf{R}_s \mathbf{w} = 1, \end{aligned} \tag{10.30}$$

where the source correlation matrix is $\mathbf{R}_s = \sum_{k=1}^{K} \sigma_{s,k}^2 \mathbf{b}(\theta_{s,k}) \mathbf{b}^H(\theta_{s,k})$, with $\sigma_{s,k}^2 = E\{s_k(n) s_k^H(n)\}$ denoting the average received power from the kth target return. The data correlation matrix, $\mathbf{R}_{\mathbf{x}} \approx (1/T)\mathbf{x}(n)\mathbf{x}(n)^H$, is directly estimated from the T received data snapshots. The solution to the optimum weights in (10.30) is given by $\mathbf{w}_o = \mathscr{P}\{\mathbf{R}_{\mathbf{x}}^{-1}\mathbf{R}_s\}$, with the operator $\mathscr{P}\{.\}$ representing the principal eigenvector of the input matrix. This optimum solution yields the MaxSINR, SINR_o, given by [37]

$$\text{SINR}_o = \Lambda_{max}\{\mathbf{R}_{\mathbf{x}}^{-1}\mathbf{R}_s\}, \tag{10.31}$$

which is the MaxSINR given by the maximum eigenvalue (Λ_{max}) of the product of the two matrices, the inverse of interference plus noise correlation matrix and the desired source correlation matrix. The resulting solution remains consistent regardless of the array configuration, be it uniform or sparse. In the case of sparse arrays, the performance of the MaxSINR beamformer is inherently linked to the array configuration. An explanation of the sparse optimization of the above formulation is provided in Section 10.3.2.

10.3.2 Sparse array transceiver design

The problem of separately selecting sensors for joint transmit and receive design poses a combinatorial optimization challenge and cannot be solved in polynomial time. By leveraging the structure of the received signal model, we formulate the sparse array design problem and solve it using sequential convex approximation. To exploit the sparse structure inherent in the joint selection of transmit and receive sensors, we introduce a two-dimensional $l_{1,2}$-mixed norm regularization to recover group sparse solutions. One dimension corresponds to the sparsity of the transmitters, while the other dimension is associated with the receivers. Moreover, this coupling results in a two-dimensional sparsity pattern where the activation of one transmitter leads to the activation of all N corresponding receivers, and vice versa. Conversely, deactivating one transmitter implies zero received data for all N receivers associated with it, and deactivating one receiver implies zero received data for all M transmitters. The structure of the optimal sparse beamforming weight vector $\mathbf{w} \in \mathbb{C}^{MN}$

is described in (10.32), where $\mathbf{w}(i,j)$ represents the entries of $\mathbf{w}$ indexed from i to j. Additionally, $\checkmark$ indicates an activated sensor location, while $\times$ indicates a deactivated sensor.

$$\overbrace{\underbrace{\begin{pmatrix}\checkmark\\ \times\\ \times\\ \checkmark\\ \vdots\\ \times\end{pmatrix}}_{\text{(Tx 1 active)}}}^{\mathbf{w}(1,N)}\;\overbrace{\underbrace{\begin{pmatrix}\checkmark\\ \times\\ \times\\ \checkmark\\ \vdots\\ \times\end{pmatrix}}_{\text{(Tx 2 active)}}}^{\mathbf{w}(N+1,2N)}\;\overbrace{\underbrace{\begin{pmatrix}\times\\ \times\\ \times\\ \times\\ \vdots\\ \times\end{pmatrix}}_{\text{(Tx 3 inactive)}}}^{\mathbf{w}(2N+1,3N)}\;\cdots\;\overbrace{\underbrace{\begin{pmatrix}\checkmark\\ \times\\ \times\\ \checkmark\\ \vdots\\ \times\end{pmatrix}}_{\text{(Tx }M\text{ active)}}}^{\mathbf{w}(N(M-1)+1,MN)} \tag{10.32}$$

In (10.32), each column vector represents the receive beamformer weights for a fixed transmit location. It's important to note that the optimal sparse beamformer, associated with the active transmit and receive locations, exhibits a group sparse structure along both the transmit and receive dimensions. This entails that when one transmit sensor is excluded, all corresponding N sensors vertically exhibit sparsity, or when one receiver is excluded, all corresponding M sensors horizontally exhibit sparsity. For instance, the absence of a transmit sensor at position 3 results in sparsity across all N entries of $\mathbf{w}$ (denoted by N consecutive $\times$ vertically in (10.32)). Similarly, group sparsity is observed across received signals corresponding to all transmitters (denoted by M consecutive $\times$ horizontally in (10.32)).

10.3.2.1 Group sparse solutions via SCA

The problem in (10.30) can equivalently be rewritten by swapping the objective and constraint functions as follows:

$$\begin{aligned}\underset{\mathbf{w}\in\mathbb{C}^{MN}}{\text{minimize}} \quad & \mathbf{w}^H\bar{\mathbf{R}}_s\mathbf{w}\\ \text{s.t.} \quad & \mathbf{w}^H\mathbf{R}_{\mathbf{x}}\mathbf{w}\leq 1\end{aligned} \tag{10.33}$$

where $\bar{\mathbf{R}}_s = -\mathbf{R}_s$. The sparse MIMO configuration of uniformly spaced receivers and transmitters with a respective inter-element spacing of d_r and $d_t = Nd_r$ is employed for data collection. The covariance matrix $\mathbf{R}_{\mathbf{x}}$ of the fully received virtual array can be then obtained by the matrix completion method proposed in [25].

The beamforming weight vectors are generally complex valued, whereas quadratic functions are real. This observation allows expressing the problem with only real variables which is typically accomplished by replacing the correlation matrix $\bar{\mathbf{R}}_s$ by $\tilde{\mathbf{R}}_s$ and concatenating the beamforming weight vector accordingly [38],

$$\tilde{\mathbf{R}}_s = \begin{bmatrix}\text{real}(\bar{\mathbf{R}}_s) & -\text{imag}(\bar{\mathbf{R}}_s)\\ \text{imag}(\bar{\mathbf{R}}_s) & \text{real}(\bar{\mathbf{R}}_s)\end{bmatrix}, \tag{10.34}$$

$$\tilde{\mathbf{w}} = \begin{bmatrix}\text{real}(\mathbf{w})\\ \text{imag}(\mathbf{w})\end{bmatrix}$$

Similarly, the received data correlation matrix $\mathbf{R_x}$ is replaced by $\tilde{\mathbf{R}}_{\mathbf{x}}$. The problem in (10.33) then becomes

$$\begin{aligned}\underset{\tilde{\mathbf{w}}\in\mathbb{R}^{2MN}}{\text{minimize}} \quad & \tilde{\mathbf{w}}'\tilde{\mathbf{R}}_s\tilde{\mathbf{w}}, \\ \text{s.t.} \quad & \tilde{\mathbf{w}}'\tilde{\mathbf{R}}_{\mathbf{x}}\tilde{\mathbf{w}} \le 1,\end{aligned} \tag{10.35}$$

where $'$ denotes transpose operation. After expressing the constraint in terms of real variables, we convexify the objective function by utilizing the first-order approximation iteratively,

$$\begin{aligned}\underset{\tilde{\mathbf{w}}\in\mathbb{R}^{2MN}}{\text{minimize}} \quad & \mathbf{m}^{(k)\prime}\tilde{\mathbf{w}} + b^{(k)}, \\ \text{s.t.} \quad & \tilde{\mathbf{w}}'\tilde{\mathbf{R}}_{\mathbf{x}}\tilde{\mathbf{w}} \le 1,\end{aligned} \tag{10.36}$$

where $\mathbf{m}$ and b, updated at the $k+1$th iteration, are given by $\mathbf{m}^{(k+1)} = 2\tilde{\mathbf{R}}_s\tilde{\mathbf{w}}^{(k)}$, $b^{(k+1)} = -\tilde{\mathbf{w}}^{(k)\prime}\tilde{\mathbf{R}}_s\tilde{\mathbf{w}}^{(k)}$, respectively. Finally, to invoke sparsity in the beamforming weight vector, the re-weighted mixed $l_{1,2}$-norm is adopted primarily for promoting group sparsity.

$$\min_{\tilde{\mathbf{w}},\mathbf{c},\mathbf{r}} \mathbf{m}^{(\mathbf{k})\prime}\tilde{\mathbf{w}} + b^{(k)} + \alpha(\mathbf{p}'\mathbf{c}) + \beta(\mathbf{q}'\mathbf{r}) \tag{10.37}$$

$$\text{s.t. } \tilde{\mathbf{w}}'\tilde{\mathbf{R}}_{\mathbf{x}}\tilde{\mathbf{w}} \le 1, \tag{10.37a}$$

$$||\mathbf{P}_i \odot \tilde{\mathbf{w}}||_2 \le c_i, \tag{10.37b}$$

$$0 \le c_i \le 1, \quad i = 1, \ldots, M \tag{10.37c}$$

$$||\mathbf{Q}_j \odot \tilde{\mathbf{w}}||_2 \le r_j, \tag{10.37d}$$

$$0 \le r_j \le 1, \quad j = 1, \ldots, N \tag{10.37e}$$

$$\mathbf{1}_M'\mathbf{c} = M_t, \tag{10.37f}$$

$$\mathbf{1}_N'\mathbf{r} = N_r, \tag{10.37g}$$

and

$$\mathbf{P}_i = [\overbrace{0\ldots0\ldots0}^{\substack{N\ elements\\ 1st\ group}}.\overbrace{1\ldots1\ldots1}^{\substack{N\ elements\\ ith\ group}}..\overbrace{1\ldots1\ldots1}^{\substack{N\ elements\\ (M+i)th\ group}}..\overbrace{0\ldots0\ldots0}^{\substack{N\ elements\\ 2Mth\ group}}]', \tag{10.38}$$

$$\mathbf{Q}_j = [\overbrace{\underbrace{0\ \ldots\ 0}_{(j-1)\ 0s}\ 1\ 0\ \ldots\ 0}^{\substack{N\ elements\ of\\ the\ 1st\ group}} \ldots \overbrace{\underbrace{0\ \ldots\ 0}_{(j-1)\ 0s}\ 1\ 0\ \ldots\ 0}^{\substack{N\ elements\ of\\ the\ 2Mth\ group}}]'. \tag{10.39}$$

Here, $\odot$ means the element-wise product, $\mathbf{c} \in [0, 1]^M$ and $\mathbf{r} \in [0, 1]^N$ are two auxiliary selection vectors taking values between 0 and 1, $\mathbf{P}_i \in \{0, 1\}^{2MN}$ in (10.37b)

Algorithm 1 SCA-sparse transmit/receive beamformer design

Input: M, N, M_t, N_r, $\alpha = \beta$, $\alpha_0 = \beta_0$, target direction θ_s, auto-correlation matrix $\mathbf{R}_s$ and $\mathbf{R}_x$

Output: Optimal weight $\mathbf{w}$, location, and number of receivers and transmitters.

Initialization:

Initialize $\mathbf{p}$, $\mathbf{q}$, $\mathbf{m}$ as all one's vectors. Set $\alpha_0 = 1$ and $\beta_0 = 1$. Initialize the tradeoff parameters α and β according to the sparsity requirement, such as $\alpha = \beta = 0.5$.

while $||\tilde{\mathbf{w}}^{(k+1)} - \tilde{\mathbf{w}}^{(k)}||_2 \geq 10^{-5}$ **do**

(a) Convert $\mathbf{w}$, $\mathbf{R}_s$, and $\mathbf{R}_x$ to the real domain to get $\tilde{\mathbf{w}}$, $\tilde{\mathbf{R}}_s$, and $\tilde{\mathbf{R}}_x$ according to (10.34).

(b) Update $\tilde{\mathbf{w}}^{(k+1)}$, $\mathbf{c}^{(k+1)}$, $\mathbf{r}^{(k+1)}$ according to (10.37).

(c) Update $\mathbf{p}^{(k+1)}$ and $\mathbf{q}^{(k+1)}$ according to (10.40) and (10.41).

(d) Convert $\tilde{\mathbf{w}}^{(k+1)}$ back into the complex solution $\mathbf{w}^{(k+1)}$ by $\mathbf{w}^{(k+1)}(i) = \tilde{\mathbf{w}}^{(k+1)}(i) + j\tilde{\mathbf{w}}^{(k+1)}(i + MN)$.

end while

return $\mathbf{w}$

is the transmission selection vector, which is used to select the real and imaginary parts of the N weights corresponding to the ith transmitter, as shown in (10.38). Equations (10.37c) and (10.37f) indicate that we can select M_t transmitters at most. Similarly, $\mathbf{Q}_j$ in (10.37d) is the receiver selection vector, as shown in (10.39). Equations (10.37e) and (10.37g) indicate that we can select N_r receivers at most. The two parameters α and β are used to control the sparsity of the transmitters and the receivers, $\mathbf{p}$ and $\mathbf{q}$ are the reweighting coefficient vectors for the transmitters and receivers, respectively, and the detailed update method is given as follows.

10.3.2.2 Reweighting update

A common method of updating the reweighting coefficient is to take the reciprocal of $|\mathbf{w}|$ [39]. However, this cannot control the number of elements to be selected. Thus, similar to [40], in order to make the number of selected antennas controllable, we update the weights using the following formula:

$$p_i^{(k)} = \frac{1 - c_i^{(k)}}{1 - e^{-\beta_0 c_i^{(k)}} + \varepsilon} - \left(\frac{1}{\varepsilon}\right)(c_i^{(k)})^{\alpha_0} \tag{10.40}$$

$$q_i^{(k)} = \frac{1 - r_i^{(k)}}{1 - e^{-\beta_0 r_i^{(k)}} + \varepsilon} - \left(\frac{1}{\varepsilon}\right)(r_i^{(k)})^{\alpha_0} \tag{10.41}$$

Here, α_0 and β_0 are two parameters that control the shape of the curve, and the parameter ε avoids the unwanted explosive case. The essence of this re-weight update method is to take a large positive penalty for an entry close to zero and a small negative reward for an entry close to 1. As a result, the entries of the two selection vectors $\mathbf{c}$ and $\mathbf{r}$ tend to be either 0 or 1. Through iterative regression, we can select M_t transmitters and N_r receivers. In addition, by controlling the values of α and β, we can obtain different sparsities of transmitters and receivers [40,41]. The proposed algorithm for joint transmit/receive beamformer design is elaborated further in Algorithm 1.

10.3.3 Simulation

In this section, the effectiveness of our proposed sparse MIMO radar is demonstrated from the perspective of output SINR. The performance of the optimal MIMO array transceiver, configured according to the proposed algorithm, is compared with that of randomly configured MIMO arrays. In practice, interference may arise from coexistence within the same bandwidth or deliberate positioning at certain angles, transmitting the same waveform as the targets of interest. Both types of interferences are considered.

10.3.3.1 Example 1

In this example, we fix the direction of the interferences and change the arrival angle of the target from 0° to 90°. We consider a full uniform linear MIMO array consisting of eight transmitters ($M = 8$) and eight receivers ($N = 8$). We select $M_t = 4$ transmitters and $N_r = 4$ receivers. For the receiver, we set the minimum spacing of the sensors to $\lambda/2$, while for the transmitter, we set the minimum spacing of the sensors to $N\lambda/2$. Suppose there are two interferences with an interference to noise-ratio (INR) of 13 dB and arrival angles of $\theta_q = [40°, 70°]$. The SNR of the desired signal is fixed at 20 dB. We simulate co-existing interferences and deliberate interferences. The output SINR versus target angle is shown in Figure 10.7. Case 1 in this figure

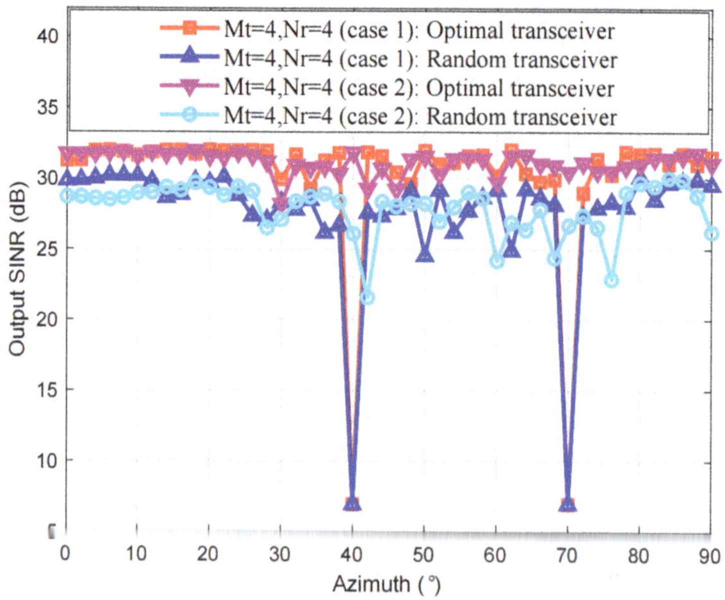

Figure 10.7 Relationship between output SINR and target angle with the directions of interferences being fixed

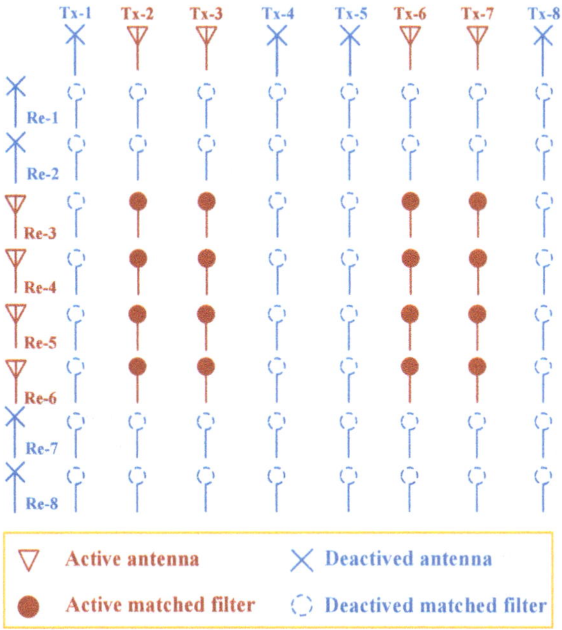

Figure 10.8 Optimal MIMO transceiver configuration when the target is from 80°

represents two deliberate interferences, while case 2 represents two co-existing interferences. It is observable that, in both scenarios, the optimal sparse MIMO arrays outperform randomly selected sparse arrays. Figure 10.8 illustrates the configuration of the optimal MIMO array for a target at 80°.

10.3.3.2 Example 2

In this example, we consider the scenario where the interferences are spatially close to the target which causes a great adverse impact on the array performance. We change the angle of the target from 0° to 90°. For each target angle, two interferences are generated from the proximity of ±5° away from the target. Considering a full uniform linear MIMO array comprising 20 transmitters (M = 20) and 20 receivers (N = 20), we choose five transmitters and five receivers to form the sparse MIMO array. All other simulation parameters remain consistent with those in example 1. Once again, we depict the output SINR against the target angle in Figure 10.9. In case 1, the two interferences are intentional, while in case 2, both interferences coexist. It can be observed that when the interferences are in close spatial proximity to the target, the proposed optimal MIMO sparse array demonstrates greater superiority compared to randomly configured sparse MIMO arrays.

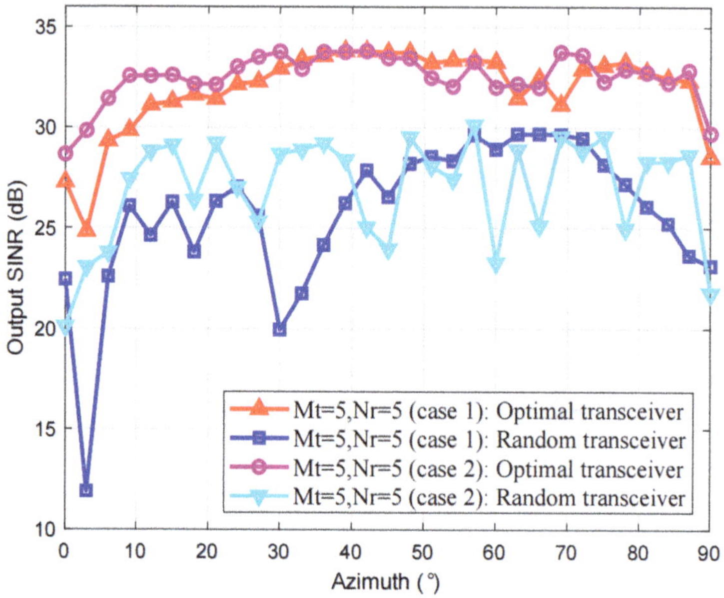

Figure 10.9 Relationship between the output SINR and the target angle when the interferences are closed to the target

10.4 Conclusion

In this chapter, we examined the unstructured sparse MIMO array transceiver design for enhanced sensing.

In Section 10.2, we optimized the configuration of MIMO array transceiver from the perspective of minimizing the CRB of DOA estimation. Applying Schur complement and inequality scaling, we transformed the original problem into an iterative convex optimization form. Simulation results showed that the transceiver optimized by the proposed method can achieve higher DOA estimation accuracy.

In Section 10.3, for a given number of transmitting and receiving antennas, a sparse MIMO array structure in terms of MaxSINR was jointly designed. It was shown by simulations that the proposed sparse MIMO transceiver selection algorithm provides superior interference suppression performance over randomly designed MIMO transceiver arrays.

References

[1] M. G. Amin, X. Wang, Y. D. Zhang, F. Ahmad, and E. Aboutanios, "Sparse arrays and sampling for interference mitigation and DOA estimation in GNSS," *Proceedings of the IEEE*, vol. 104, no. 6, pp. 1302–1317, 2016.

[2] S. Sun and Y. D. Zhang, "Multi-frequency sparse array-based massive MIMO radar for autonomous driving," in *2020 54th Asilomar Conference on Signals, Systems, and Computers*, 2020, pp. 1167–1171.
[3] H. Ameri, A. Attaran, and M. Moghavvemi, "Planning of low-cost 77-GHz radar transceivers for automotive applications," *IEEE Aerospace and Electronic Systems Magazine*, vol. 27, no. 4, pp. 25–31, 2012.
[4] J. Li and P. Stoica, "Mimo radar with colocated antennas," *IEEE Signal Processing Magazine*, vol. 24, no. 5, pp. 106–114, 2007.
[5] S. Sun and Y. D. Zhang, "4d automotive radar sensing for autonomous vehicles: A sparsity-oriented approach," *IEEE Journal of Selected Topics in Signal Processing*, vol. 15, no. 4, pp. 879–891, 2021.
[6] S. Mulleti, C. Saha, H. S. Dhillon, and Y. C. Eldar, "A fast-learning sparse antenna array," in *2020 IEEE Radar Conference (RadarConf20)*, 2020, pp. 1–6.
[7] C.-L. Liu and P. P. Vaidyanathan, "Correlation subspaces: Generalizations and connection to difference coarrays," *IEEE Transactions on Signal Processing*, vol. 65, no. 19, pp. 5006–5020, 2017.
[8] A. Hassanien, M. G. Amin, Y. D. Zhang, and F. Ahmad, "High-resolution single-snapshot DOA estimation in MIMO radar with colocated antennas," in *2015 IEEE Radar Conference (RadarCon)*, 2015, pp. 1134–1138.
[9] A. Hassanien and S. A. Vorobyov, "Transmit energy focusing for DOA estimation in MIMO radar with colocated antennas," *IEEE Transactions on Signal Processing*, vol. 59, no. 6, pp. 2669–2682, 2011.
[10] S. Gogineni and A. Nehorai, "Target estimation using sparse modeling for distributed MIMO radar," *IEEE Transactions on Signal Processing*, vol. 59, no. 11, pp. 5315–5325, 2011.
[11] X. Wang, A. Hassanien, and M. G. Amin, "Dual-function MIMO radar communications system design via sparse array optimization," *IEEE Transactions on Aerospace and Electronic Systems*, vol. 55, no. 3, pp. 1213–1226, 2019.
[12] M. Masood, L. H. Afify, and T. Y. Al-Naffouri, "Efficient coordinated recovery of sparse channels in massive MIMO," *IEEE Transactions on Signal Processing*, vol. 63, no. 1, pp. 104–118, 2015.
[13] X. Wang, C. P. Tan, F. Wu, and J. Wang, "Fault-tolerant attitude control for rigid spacecraft without angular velocity measurements," *IEEE Transactions on Cybernetics*, vol. 51, no. 3, pp. 1216–1229, 2021.
[14] D. S. Kalogerias and A. P. Petropulu, "Matrix completion in colocated MIMO radar: Recoverability, bounds and theoretical guarantees," *IEEE Transactions on Signal Processing*, vol. 62, no. 2, pp. 309–321, 2014.
[15] A. Moffet, "Minimum-redundancy linear arrays," *IEEE Transactions on Antennas and Propagation*, vol. 16, no. 2, pp. 172–175, 1968.
[16] P. Pal and P. P. Vaidyanathan, "Nested arrays: A novel approach to array processing with enhanced degrees of freedom," *IEEE Transactions on Signal Processing*, vol. 58, no. 8, pp. 4167–4181, 2010.

[17] S. Qin, Y. D. Zhang, and M. G. Amin, "Generalized coprime array configurations for direction-of-arrival estimation," *IEEE Transactions on Signal Processing*, vol. 63, no. 6, pp. 1377–1390, 2015.
[18] A. Ahmed, Y. D. Zhang, and J.-K. Zhang, "Coprime array design with minimum lag redundancy," in *ICASSP 2019–2019 IEEE International Conference on Acoustics, Speech and Signal Processing (ICASSP)*, 2019, pp. 4125–4129.
[19] C.-Y. Chen and P. P. Vaidyanathan, "Minimum redundancy MIMO radars," in *2008 IEEE International Symposium on Circuits and Systems (ISCAS)*, 2008, pp. 45–48.
[20] P. P. Vaidyanathan and P. Pal, "Sparse sensing with co-prime samplers and arrays," *IEEE Transactions on Signal Processing*, vol. 59, no. 2, pp. 573–586, 2011.
[21] J.-Y. Lu and J. Greenleaf, "A study of two-dimensional array transducers for limited diffraction beams," *IEEE Transactions on Ultrasonics, Ferroelectrics, and Frequency Control*, vol. 41, no. 5, pp. 724–739, 1994.
[22] J. Doles and F. Benedict, "Broad-band array design using the asymptotic theory of unequally spaced arrays," *IEEE Transactions on Antennas and Propagation*, vol. 36, no. 1, pp. 27–33, 1988.
[23] S. A. Hamza and M. G. Amin, "Sparse array beamforming design for wideband signal models," *IEEE Transactions on Aerospace and Electronic Systems*, vol. 57, no. 2, pp. 1211–1226, 2021.
[24] S. A. Hamza, M. G. Amin, and G. Fabrizio, "Optimum sparse array design for maximizing signal- to-noise ratio in presence of local scatterings," in *2018 IEEE International Conference on Acoustics, Speech and Signal Processing (ICASSP)*, 2018, pp. 3330–3334.
[25] S. A. Hamza and M. G. Amin, "Sparse array design utilizing matrix completion," in *2019 53rd Asilomar Conference on Signals, Systems, and Computers*, 2019, pp. 1207–1211.
[26] X. Wang, M. Amin, and X. Wang, "Optimum sparse array design for multiple beamformers with common receiver," in *2018 IEEE International Conference on Acoustics, Speech and Signal Processing (ICASSP)*, 2018, pp. 3364–3368.
[27] X. Wang, E. Aboutanios, M. Trinkle, and M. G. Amin, "Reconfigurable adaptive array beamforming by antenna selection," *IEEE Transactions on Signal Processing*, vol. 62, no. 9, pp. 2385–2396, 2014.
[28] S. A. Hamza and M. G. Amin, "Optimum sparse array receive beamforming for wideband signal model," in *2018 52nd Asilomar Conference on Signals, Systems, and Computers*, 2018, pp. 89–93.
[29] X. Wang, M. Amin, and X. Cao, "Analysis and design of optimum sparse array configurations for adaptive beamforming," *IEEE Transactions on Signal Processing*, vol. 66, no. 2, pp. 340–351, 2018.
[30] S. A. Hamza and M. G. Amin, "Hybrid sparse array design for underdetermined models," in *ICASSP 2019–2019 IEEE International Conference on Acoustics, Speech and Signal Processing (ICASSP)*, 2019, pp. 4180–4184.

[31] S. A. Hamza and M. G. Amin, "Hybrid sparse array beamforming design for general rank signal models," *IEEE Transactions on Signal Processing*, vol. 67, no. 24, pp. 6215–6226, 2019.
[32] H. Nosrati, E. Aboutanios, and D. Smith, "Multi-stage antenna selection for adaptive beamforming in MIMO radar," *IEEE Transactions on Signal Processing*, vol. 68, pp. 1374–1389, 2020.
[33] P. Stoica and A. Nehorai, "Music, maximum likelihood, and Cramer-Rao bound," *IEEE Transactions on Acoustics, Speech, and Signal Processing*, vol. 37, no. 5, pp. 720–741, 1989.
[34] W. Roberts, L. Xu, J. Li, and P. Stoica, "Sparse antenna array design for MIMO active sensing applications," *IEEE Transactions on Antennas and Propagation*, vol. 59, no. 3, pp. 846–858, 2011.
[35] S. A. Hamza and M. G. Amin, "Sparse array design for transmit beamforming," in *2020 IEEE International Radar Conference (RADAR)*, 2020, pp. 560–565.
[36] Z. Cheng, Y. Lu, Z. He, Yufengli, J. Li, and X. Luo, "Joint optimization of covariance matrix and antenna position for MIMO radar transmit beampattern matching design," in *2018 IEEE Radar Conference (RadarConf18)*, 2018, pp. 1073–1077.
[37] S. Shahbazpanahi, A. Gershman, Z.-Q. Luo, and K. M. Wong, "Robust adaptive beamforming for general-rank signal models," *IEEE Transactions on Signal Processing*, vol. 51, no. 9, pp. 2257–2269, 2003.
[38] M. S. Ibrahim, A. Konar, M. Hong, and N. D. Sidiropoulos, "Mirror-prox SCA algorithm for multicast beamforming and antenna selection," in *2018 IEEE 19th International Workshop on Signal Processing Advances in Wireless Communications (SPAWC)*, 2018, pp. 1–5.
[39] E. Candes, M. Wakin, and S. Boyd, "Enhancing sparsity by reweighted l_1 minimization," *Journal of Fourier Analysis and Applications*, vol. 14, pp. 877–905, 2007.
[40] X. Wang and E. Aboutanios, "Sparse array design for multiple switched beams using iterative antenna selection method," *Digital Signal Processing*, vol. 105, p. 102684, 2020, special issue on Optimum Sparse Arrays and Sensor Placement for Environmental Sensing. [Online]. Available: https://www.sciencedirect.com/science/article/pii/S1051200420300294
[41] O. Mehanna, N. D. Sidiropoulos, and G. B. Giannakis, "Joint multicast beamforming and antenna selection," *IEEE Transactions on Signal Processing*, vol. 61, no. 10, pp. 2660–2674, 2013.

Chapter 11

Transmit–receive beamforming for distributed phased-MIMO radar system

Xianxiang Yu[1], Ruitao Liu[1], Guolong Cui[1] and Lingjiang Kong[1]

This chapter considers the problem of distributed phased multiple-input multiple-output (phased-MIMO) radar system design. The phased-MIMO radar consisting of separated subarrays possesses a high-angle resolution but with sharing high sidelobes. This paper therefore focuses on the design of a virtual low-sidelobe beampattern in distributed phased-MIMO radar. An iterative framework with respect to subarray layout and receive weighting coefficients is proposed to minimize beampattern peak sidelobe level (PSL) accounting for practical position constraints as well as the mainlobe level restriction. In each iteration, an efficient cyclic optimization algorithm based on the coordinate descent (CD) framework is developed to seek the subarray layout by splitting high dimensional layout optimization problem into multiple one-dimensional optimization problems, and the semidefinite relaxation (SDR) technique and rank one approximation are explored to design weighting coefficients. Numerical simulations are provided to assess the performance of the proposed algorithms in terms of the achieved beampattern under different constraints and parameter settings.

11.1 Introduction

The phased array technology has been widely applied in radar, sonar, communication, autonomous driving systems, and various other fields [1–4]. This is primarily attributed to its notable advantages including coherent signal processing capabilities at the transmit/receive arrays [5–7] and efficient management or control of antenna radiation energy distribution in space [8,9].

The emergence of multiple-input multiple-output (MIMO) radar has attracted the attention of researchers [10–23] in the past decade. Unlike conventional phased array radar systems that emit identical waveforms across all antennas, MIMO radar utilizes distinct probing signals, enabling waveform optimization to enhance target

[1]School of Information and Communication Engineering, University of Electronic Science and Technology of China, China

detection, identification, and tracking capabilities [10,12]. The utilization of MIMO radar is extensive in achieving flexible transmit beampattern designs and enhancing the maximum number of uniquely identifiable targets at the receive array owing to its substantial virtual aperture [10,13]. In terms of beampattern and high-resolution applications, waveform design plays a critical role in regulating Sidelobe Level (PSL) for MIMO radar [20–23]. Nevertheless, these benefits provided by MIMO radar are obtained at the cost of compromising the transmit coherent processing gain offered by phased array radar.

To achieve this objective, a novel system known as phased-MIMO radar has been developed in recent years to integrate the advantages of both MIMO radar and phased array radar [24]. The main concept is to replace each transmit/receive antenna of conventional MIMO radar with a phased subarray. Many existing researches focus on colocated phased-MIMO radars [24–31], which divide the input antenna array into subarrays which may be nonoverlapping or fully overlapped. For example, in [24,25], the idea of partitioning transmit array to integrate the phased-array radar into the MIMO radar has been introduced. The advantages of the phased-MIMO radar are analyzed in terms of the corresponding beampattern and signal-to-interference-plus-noise ratio (SINR) expressions as compared to the phased array radar and MIMO radar, respectively. In [26–28], the authors divide the frequency-diverse transmit array into multiple subarrays and each subarray coherently transmits a distinct waveform with a small frequency increment across the array elements. By fine-tuning the frequency increment, the subarrays jointly collectively provide versatile operating modes and range-dependent beamforming. [29] proposes a two-stage design methodology based on phased-MIMO radar to achieve a desired transmit beampattern while incorporating the concept of transmit subarray partitioning. [30] introduces a hybrid phased-MIMO radar with unequal subarrays for steering multiple beams of varying gain and distinct waveforms toward a specific target. The uneven size of subarrays also adds an additional degree of freedom (DoF). [31] deduces an optimal partitioning scheme for phased-MIMO radar at any number of elements. This optimal scheme enhances PSL without compromising directivity compared to other division techniques.

The aforementioned methods do not take into account the scenarios of distributed phased-MIMO radar systems, which consist of sparse subarrays in space. In modern warfare, unmanned aerial vehicles (UAVs) have gained widespread recognition as promising assets for conducting tasks with high flexibility. To enhance target detection and tracking performance in complex environments, it is expected that UAV swarms collaborate to achieve higher resolution and detection range compared to a single UAV [32]. Generally, each UAV is equipped with a phased array radar for signal emission and reception. Consequently, the entire UAV swarm can be regarded as a large distributed array comprising multiple phased subarrays. Therefore, research on phased-MIMO radar based on distributed subarrays holds significant importance for future advancements.

However, due to the sparse distance between subarrays, it is important to note that the distributed array exhibits high sidelobes when achieving high angular resolution. To address this issue, a modified real genetic algorithm for optimizing the element positions under multiple constraints (including number of elements, aperture size, and minimum element spacing) is presented in [33]. In [34], the authors investigate the sparse linear array design problem of coherent unambiguous transmit for the distributed aperture coherence-synthetic radar, considering antenna size constraint and geography constraint. Different from previous studies solely on antenna element position optimization, [35–38] also consider the weighting coefficients to achieve a desired radiation pattern. Moreover, [39], introduces an optimization problem for colocated subarray layouts to generate low sidelobe level. Unlike the existing methods [33–39], this chapter focuses on sparse distributed phased-MIMO radar and considers the joint design of subarray layout and receive weighting coefficients under multiple practical constraints. Our main contributions are highlighted as follows:

- The beampattern model is derived for a sparse phased-MIMO radar system with distributed subarrays, which differs from previous studies [24–31] that focused on the colocated phased-MIMO radar by partitioning the input antenna array into subarrays.
- A hybrid optimization problem is proposed, which focuses on the subarray layout and receive beamforming to minimize beampattern PSL under constraints of subarray layout and mainlobe. To address this optimization problem, a novel joint subarray layout and receive beamforming iterative framework (JSLRBIF) is proposed. In each iteration, the subarray layout design and receive beamforming subproblems are required to be solved separately.
- The introduction of a CD framework enables the subdivision of the subarray layout design subproblem into multiple one-dimensional optimization problems. To obtain receive weighting coefficients, we employ semidefinite relaxation (SDR) and rank one decomposition techniques.
- The convergence and PSL under the different experimental schemes of the proposed technique are assessed in comparison with the existing available approaches. Numerical simulations demonstrate that the proposed technique can gain the lower PSL value.

It is worth emphasizing the definition of distributed phased-MIMO radar in this chapter. Unlike distributed MIMO radar, which utilizes widely separated transmit/receive antennas to capture the spatial diversity of the target radar cross section (RCS) [11,14,15], distributed phased-MIMO radar assumes that the target is located in the far-field. The term "distributed" implies a sparse layout of subarrays with spacings greater than half wavelength. However, for targeting purposes, the subarray layout remains collocated. Therefore, distributed phased-MIMO radar can utilize transmit/receive antennas to coherently form a beam toward a specific direction in space.

The remainder of the chapter is organized as follows. In Section 11.2, we introduce the signal model of distributed phased-MIMO radar from transmit end and receive end, respectively. In Section 11.3, an iterative framework with respect to subarray layout and receive weighting coefficients is proposed to minimize beampattern PSL. The optimization method for each subproblem is put forward accordingly. In Section 11.4, we evaluate the performance of the proposed algorithms. Finally, in Section 11.5, we provide concluding remarks and possible future research tracks.

11.2 Distributed phased-MIMO radar system model

Consider a distributed phased-MIMO radar system with P transmit subarrays and Q receive subarrays, as shown in Figure 11.1. For the sake of convenience, we will focus on a one-dimensional array in this chapter. Each transmit/receive subarray is designed as a uniform linear phased array (ULPA) with half-wavelength separation between two adjacent elements. The transmit and receive subarrays are arranged horizontally in a straight line, respectively. The signals emitted by all transmit subarrays can be stacked into a vector expressed as

$$\mathbf{s}(t) = \left[s_1(t), \ldots, s_p(t), \ldots, s_P(t)\right]^{\mathrm{T}}, \tag{11.1}$$

where $s_p(t)$ denotes the emit waveform of the p-th subarray. t is the time index within the radar pulse. The received signal is initially separated into multi-channel data using a matched filter bank associated with the transmitted signals $s_p(t)$, $p = 1, \ldots, P$ [17]. Subsequently, receive beamforming is applied to enhance useful signals from specific directions while suppressing interference signals and background noises originating from other directions. Detailed descriptions of the transmit/receive signal models, matched-filtering operation, and receive beamforming can be found in the subsequent subsections.

Remark 1: The transmit/receive array of the distributed phased-MIMO radar is assumed to be perfectly calibrated, without considering time and phase synchronization errors between different subarrays. Assuming that the target is a far-field model for the distributed array, the azimuth angles of different subarrays relative to the target are assumed to be identical. Consequently, any received signal delay between the distribution arrays is fully compensated, leaving only the phase difference caused by the wave path between them. For clarity purposes, this problem is described for a one-dimensional distributed array. However, the proposed method can directly extend to a two-dimensional array, enabling beam pattern synthesis over both pitch and azimuth.

11.2.1 Transmit signal model

The layout vector of P transmit subarrays is assumed to be $\mathbf{l}_T = [l_{T_1}, \ldots, l_{T_P}]^{\mathrm{T}}$, where l_{T_p} denotes the position of the first antenna in p-th transmit subarray. The

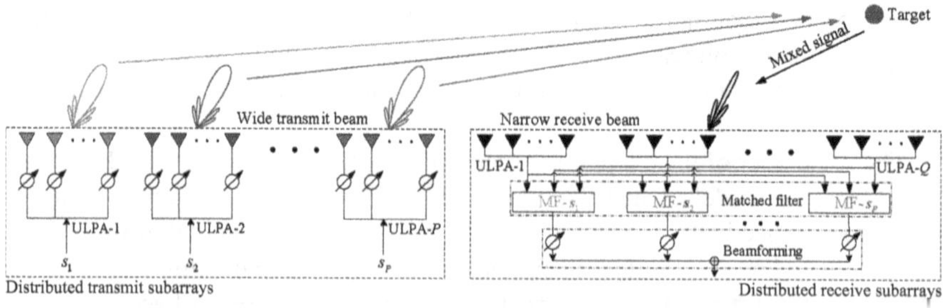

Figure 11.1 Distributed phased-MIMO radar system

phase difference between the p-th subarray and the origin can then be expressed as $\phi_p(\theta) = \frac{2\pi}{\lambda} l_{T_p} \sin\theta$, which is associated with subarray position l_{T_p} and direction θ. $\lambda = c/f_c$ represents the wavelength, where f_c and c denote the carrier frequency and speed of wave propagation, respectively. Therefore, the waveform diversity vector [17,24] between transmit subarrays can be expressed as follows:

$$\mathbf{a}_T(\theta, \mathbf{l}_T) = [e^{j\phi_1(\theta)}, e^{j\phi_2(\theta)}, \ldots, e^{j\phi_P(\theta)}]^{\mathrm{T}}. \tag{11.2}$$

The ULPA consists of N_T isotropic narrowband array elements. Hence, the transmit phased array steering vector within the subarray can be represented as

$$\mathbf{a}_{d_t}(\theta) = [1, e^{j\frac{2\pi d \sin\theta}{\lambda}}, \ldots, e^{j\frac{2\pi(N_T-1)d\sin\theta}{\lambda}}]^{\mathrm{T}}, \tag{11.3}$$

where $d = \lambda/2$ denotes spacing between two adjacent elements of ULPA. Hence, the transmit signal of the p-th subarray is $x_p(\theta, t) = e^{-j\phi_p(\theta)} \mathbf{w}_d^{\dagger} \mathbf{a}_{d_t}(\theta) s_p(t)$, where $\mathbf{w}_d$ is the transmit weighting vector inside of each transmit subarray. Furthermore, the current focus of transmit beampattern synthesis is on utilizing tapers with a constant modulus to maximize amplifier efficiency [40]. Each element of the vector $\mathbf{w}_d$ is enforced to have a constant modulus. For low complexity conventional beamformers [24], $\mathbf{w}_d$ is assigned as $\mathbf{a}_{d_t}(\theta_m)$, where θ_m represents the target direction.

Thus, the superposition of the transmitted signals from all distributed subarrays at direction θ can be obtained as follows:

$$\begin{aligned} x_T(\theta, t) &= \sum_{p=1}^{P} x_p(\theta, t) \\ &= \sum_{p=1}^{P} e^{-j\phi_p(\theta)} \mathbf{w}_d^{\dagger} \mathbf{a}_{d_t}(\theta) s_p(t) \\ &= \mathbf{w}_d^{\dagger} \mathbf{a}_{d_t}(\theta) \sum_{p=1}^{P} e^{-j\phi_p(\theta)} s_p(t) \\ &= \mathbf{w}_d^{\dagger} \mathbf{a}_{d_t}(\theta) \mathbf{a}_T^{\dagger}(\theta, \mathbf{l}_T) \mathbf{s}(t). \end{aligned} \tag{11.4}$$

The power of the detection signal at direction θ can be written as follows:

$$\begin{aligned} Po(\theta) &= \mathbb{E}\{|x_T(\theta, t)|^2\} \\ &= (\mathbf{w}_d^{\dagger} \mathbf{a}_{d_t}(\theta) \mathbf{a}_T^{\dagger}(\theta, \mathbf{l}_T)) \mathbb{E}\{\mathbf{s}(t) \mathbf{s}^{\dagger}(t)\} (\mathbf{w}_d^{\dagger} \mathbf{a}_{d_t}(\theta) \mathbf{a}_T^{\dagger}(\theta, \mathbf{l}_T))^{\dagger} \\ &= (\mathbf{w}_d^{\dagger} \mathbf{a}_{d_t}(\theta) \mathbf{a}_T^{\dagger}(\theta, \mathbf{l}_T)) \mathbf{R} (\mathbf{w}_d^{\dagger} \mathbf{a}_{d_t}(\theta) \mathbf{a}_T^{\dagger}(\theta, \mathbf{l}_T))^{\dagger} \\ &= (\mathbf{w}_d^{\dagger} \mathbf{a}_{d_t}(\theta))^2 (\mathbf{a}_T^{\dagger}(\theta, \mathbf{l}_T) \mathbf{R} \mathbf{a}_T(\theta, \mathbf{l}_T)) \\ &= P_d(\theta) P_T(\theta), \end{aligned} \tag{11.5}$$

where $P_T(\theta) = \mathbf{a}_T^{\dagger}(\theta, \mathbf{l}_T) \mathbf{R} \mathbf{a}_T(\theta, \mathbf{l}_T)$ and $P_d(\theta) = (\mathbf{w}_d^{\dagger} \mathbf{a}_{d_t}(\theta))^2$ are transmit beampattern of distributed MIMO radar and beampattern of ULPA within the subarray, respectively. $\mathbf{R} = \mathbb{E}\{\mathbf{s}(t) \mathbf{s}^{\dagger}(t)\} = \int_{\tau} \mathbf{s}(t) \mathbf{s}^{\dagger}(t)\, dt$ is the covariance matrix of the transmit waveform, where τ is the radar pulse width. The transmit beampattern of distributed phased-MIMO radar is essentially the product of the two aforementioned beampatterns. Specifically, when the transmit signals of subarrays satisfy the orthogonality condition [19]:

$$\int_{\tau} s_i(t) s_j^{\dagger}(t)\, dt = \begin{cases} 1, & i = j \\ 0, & i \neq j \end{cases}, \tag{11.6}$$

where $\mathbf{R} = \mathbf{I}_P$ and $P_T(\theta) = P$ can be easily derived. Consequently, the transmit beampattern of distributed phased-MIMO radar degenerates into the beampattern of ULPA.

11.2.2 Receive signal model

Assume that the layout vector of Q receive subarrays is $\mathbf{l}_R = [l_{R_1}, \ldots, l_{R_Q}]^{\mathrm{T}}$, where l_{R_q} stands for the position of the first antenna in the q-th receive subarray. Then the phase difference between the q-th subarray and the origin can be expressed as $\phi_q(\theta) = \frac{2\pi}{\lambda} l_{R_q} \sin\theta$, which is associated with both the q-th subarray position l_{R_q} and θ. Therefore, we can define the steering vector between receive subarrays as follows:

$$\mathbf{a}_{Ro}(\theta, \mathbf{l}_R) = [e^{j\phi_1(\theta)}, e^{j\phi_2(\theta)}, \ldots, e^{j\phi_Q(\theta)}]^{\mathrm{T}}. \tag{11.7}$$

The ULPA within each receive subarray is assumed to consist of N_R isotropic narrowband array elements. Consequently, the formulation of the receive phased array steering vector for each subarray can be expressed as follows:

$$\mathbf{a}_{d_r}(\theta) = [1, e^{j\frac{2\pi d \sin\theta}{\lambda}}, \ldots, e^{j\frac{2\pi(N_R-1)d\sin\theta}{\lambda}}]^{\mathrm{T}}. \tag{11.8}$$

The real receive steering vector of all receive array elements can be represented as $\mathbf{a}_R(\theta, \mathbf{l}_R) = \mathbf{a}_{Ro}(\theta, \mathbf{l}_R) \otimes \mathbf{a}_{d_r}(\theta)$. Subsequently, the receive signal of the distributed phased-MIMO radar can be formulated as follows:

$$\begin{aligned} \mathbf{y}_R(t) &= \mathbf{a}_R(\theta, \mathbf{l}_R)\, x_T(\theta, t) \\ &= (\mathbf{a}_{Ro}(\theta, \mathbf{l}_R) \otimes \mathbf{a}_{d_r}(\theta))\, \mathbf{w}_d^{\dagger} \mathbf{a}_{d_t}(\theta) \mathbf{a}_T^{\dagger}(\theta, \mathbf{l}_T)\, \mathbf{s}(t). \end{aligned} \tag{11.9}$$

11.2.3 Matched-filtering operation and receive beamforming

The receive signals from each antenna in the distributed phased-MIMO radar are separated into multi-channel echoes at the receiving end, and a receive beamforming technique is designed to achieve the desired processing gain toward the target direction while suppressing interference signals and background noises originating from other directions.

The echo signals are initially filtered through a matched-filter bank MF-s_p, $p = 1, \ldots, P$. Here, MF-s_p represents the matched filter associated with the p-th transmit signal. The response function of the matched filters is represented by $\mathbf{s}^{\dagger}$, enabling each antenna to obtain a $P \times 1$ dimension virtual data vector. Similarly, each receive subarray can acquire the $N_R P \times 1$ dimension virtual data vector. The outcomes of matched-filtering on the receive signals from Q receive subarrays are as follows:

$$\begin{aligned} \mathbf{Y} &= \int_{\mathcal{T}} \mathbf{y}_R(t)\, \mathbf{s}^{\dagger}(t)\, dt = \int_{\mathcal{T}} \mathbf{a}_R(\theta, \mathbf{l}_R)\, x_T(\theta, t)\, \mathbf{s}^{\dagger}(t)\, dt \\ &= \int_{\mathcal{T}} \mathbf{a}_R(\theta, \mathbf{l}_R)\, \mathbf{w}_d^{\dagger} \mathbf{a}_{d_t}(\theta) \mathbf{a}_T^{\dagger}(\theta, \mathbf{l}_T)\, \mathbf{s}(t)\, \mathbf{s}^{\dagger}(t)\, dt \\ &= \mathbf{a}_R(\theta, \mathbf{l}_R)\, \mathbf{w}_d^{\dagger} \mathbf{a}_{d_t}(\theta) \mathbf{a}_T^{\dagger}(\theta, \mathbf{l}_T) \int_{\mathcal{T}} \mathbf{s}(t)\, \mathbf{s}^{\dagger}(t)\, dt \\ &= \mathbf{a}_R(\theta, \mathbf{l}_R)\, \mathbf{w}_d^{\dagger} \mathbf{a}_{d_t}(\theta) \mathbf{a}_T^{\dagger}(\theta, \mathbf{l}_T)\, \mathbf{R}. \end{aligned} \tag{11.10}$$

Typically, the data matrix **Y** is transformed into a vector format to derive a virtual steering vector,

$$\begin{aligned} &\mathbf{y}(\theta, \mathbf{R}, \mathbf{l}_T, \mathbf{l}_R) = \text{vec}(\mathbf{Y}) \\ &= (\mathbf{a}_R(\theta, \mathbf{l}_R)\, \mathbf{w}_d^{\dagger}\mathbf{a}_{d_t}(\theta)) \otimes (\mathbf{R}\mathbf{a}_T(\theta, \mathbf{l}_T)) \in \mathbb{C}^{QN_RP\times 1}. \end{aligned} \tag{11.11}$$

where vec(·) denotes vectorization operation. The receive weighting coefficients **w** can be expressed as follows:

$$\mathbf{w} = [w_{1,1}, w_{2,1}, \ldots, w_{P,QN_R}]^{\mathrm{T}} \in \mathbb{C}^{QN_RP\times 1}. \tag{11.12}$$

The virtual beamforming of distributed phased-MIMO radar is thus formulated as follows:

$$\begin{aligned} &F(\theta, \mathbf{R}, \mathbf{w}, \mathbf{l}_T, \mathbf{l}_R) = \mathbf{w}^{\dagger}\mathbf{y}(\theta, \mathbf{R}, \mathbf{l}_T, \mathbf{l}_R) \\ &\quad = \mathbf{w}^{\dagger}((\mathbf{a}_R(\theta, \mathbf{l}_R)\, \mathbf{w}_d^{\dagger}\mathbf{a}_{d_t}(\theta)) \otimes (\mathbf{R}\mathbf{a}_T(\theta, \mathbf{l}_T))). \end{aligned} \tag{11.13}$$

For the purpose of facilitating analysis, we decompose the receive weight into two parts, namely, $\mathbf{w} = \mathbf{w}_R \otimes \mathbf{w}_T \in \mathbb{C}^{QN_RP\times 1}$, where $\mathbf{w}_R = [w_R^1, \ldots, w_R^{QN_R}]^{\mathrm{T}} \in \mathbb{C}^{QN_R\times 1}$ denotes real receive weighting coefficients, $\mathbf{w}_T = [w_T^1, \ldots, w_T^P]^{\mathrm{T}} \in \mathbb{C}^{P\times 1}$ denotes virtual transmit weighting coefficients. Note that transmit subarray layout vector $\mathbf{l}_T$ and receive subarray layout vector $\mathbf{l}_R$ are coupled in the virtual steering vector and weighting coefficients. Using matrix algebra for matrices **A**, **B**, **U**, and **V**, we have $(\mathbf{A} \otimes \mathbf{B})^{\dagger} = \mathbf{A}^{\dagger} \otimes \mathbf{B}^{\dagger}$ and $(\mathbf{A} \otimes \mathbf{B})(\mathbf{U} \otimes \mathbf{V}) = (\mathbf{AU} \otimes \mathbf{BV})$ [17]. Therefore (11.13) can be simplified as follows:

$$\begin{aligned} &F(\theta, \mathbf{R}, \mathbf{w}_R, \mathbf{w}_T, \mathbf{l}_T, \mathbf{l}_R) = \mathbf{w}^{\dagger}\mathbf{y}(\theta, \mathbf{R}, \mathbf{l}_T, \mathbf{l}_R) \\ &= (\mathbf{w}_R \otimes \mathbf{w}_T)^{\dagger}((\mathbf{a}_R(\theta, \mathbf{l}_R)\, \mathbf{w}_d^{\dagger}\mathbf{a}_{d_t}(\theta)) \otimes (\mathbf{R}\mathbf{a}_T(\theta, \mathbf{l}_T))) \\ &= (\mathbf{w}_R^{\dagger} \otimes \mathbf{w}_T^{\dagger})((\mathbf{a}_R(\theta, \mathbf{l}_R)\, \mathbf{w}_d^{\dagger}\mathbf{a}_{d_t}(\theta)) \otimes (\mathbf{R}\mathbf{a}_T(\theta, \mathbf{l}_T))) \\ &= (\mathbf{w}_R^{\dagger}(\mathbf{a}_R(\theta, \mathbf{l}_R)\, \mathbf{w}_d^{\dagger}\mathbf{a}_{d_t}(\theta))) \otimes (\mathbf{w}_T^{\dagger}(\mathbf{R}\mathbf{a}_T(\theta, \mathbf{l}_T))) \\ &= (\mathbf{w}_R^{\dagger}\mathbf{a}_R(\theta, \mathbf{l}_R)\, \mathbf{w}_d^{\dagger}\mathbf{a}_{d_t}(\theta)) \otimes (\mathbf{w}_T^{\dagger}\mathbf{R}\mathbf{a}_T(\theta, \mathbf{l}_T)). \end{aligned} \tag{11.14}$$

Since the two parts in (11.14) are both complex numbers, the Kronecker product operation can be replaced by multiplication. Consequently, (11.14) can be further simplified as follows:

$$\begin{aligned} &F(\theta, \mathbf{R}, \mathbf{w}_R, \mathbf{w}_T, \mathbf{l}_T, \mathbf{l}_R) = \mathbf{w}_R^{\dagger}\mathbf{a}_R(\theta, \mathbf{l}_R)\, \mathbf{w}_d^{\dagger}\mathbf{a}_{d_t}(\theta)\mathbf{w}_T^{\dagger}\mathbf{R}\mathbf{a}_T(\theta, \mathbf{l}_T) \\ &= G_R(\theta) \times G_d(\theta) \times G_T(\theta), \end{aligned} \tag{11.15}$$

where $G_R(\theta) = \mathbf{w}_R^{\dagger}\mathbf{a}_R(\theta, \mathbf{l}_R)$ is the real receive beampattern of distributed phased-MIMO radar, $G_d(\theta) = \mathbf{w}_d^{\dagger}\mathbf{a}_{d_t}(\theta)$ is the transmit ULPA beampattern within the subarray, $G_T(\theta) = \mathbf{w}_T^{\dagger}\mathbf{R}\mathbf{a}_T(\theta, \mathbf{l}_T)$ is the waveform diversity beampattern [24]. The overall virtual beampattern of the phased-MIMO radar can be regarded as the multiplication of three individual beampatterns. Now, let us consider three special scenarios for distributed phased-MIMO radar beampattern.

- $N_T = N_R = 1$, each transmit subarray contains only one antenna, $\mathbf{a}_{d_t}(\theta) = 1$. In this scenario, the distributed phased-MIMO radar transforms into a conventional sparse MIMO radar.

- $P = Q = 1$, there is only one transmit subarray while the antennas within the subarray emit identical waveforms to achieve coherent gain. As a result, the signal model can be simplified to that of a conventional phased array radar.
- When all subarrays transmit coherent signals, the distributed phased-MIMO array will be transformed into a large aperture phased array radar composed of several sparse ULPA subarrays.

The large separation between subarrays in distributed sparse-phased arrays can result in higher sidelobes in the transmit beampattern, which may introduce angle ambiguity and impact target detection performance. Moreover, it is desirable for the radar's transmitter to rapidly cover a wide beam over the region of interest. However, distributed sparse phased arrays inherently possess extremely narrow beams that are not suitable for scanning.

The distributed phased-MIMO radar proposed in this chapter not only synthesizes a wide beam at the transmit end for target detection but also achieves high angular resolution at the receive end. Additionally, it simultaneously incorporates all the advantages of MIMO radar and phased array radar. Unfortunately, due to the large spacing between subarrays, the beampattern of the distributed phased-MIMO radar exhibits significant sidelobes. To overcome this issue, we will employ an iterative technique below.

Remark 2: The proposed distributed phased-MIMO radar model can be effectively employed for collaborative operations among a swarm of distributed UAVs in real combat scenarios. By optimizing the spatial positioning of UAV nodes and corresponding weighting coefficients, we can synthesize an optimized beampattern to achieve highly precise target detection performance. The spatial layout and weighting coefficients can be dynamically adjusted in real-time to adapt to changing environmental conditions, enabling dynamic cooperative combat capabilities within the UAV swarm. Additionally, this technique is not limited to UAV platforms but can also be utilized by other distributed platforms such as airborne radar, ground radar, communication base stations, etc., facilitating enhanced performance including extended detection range and improved angular resolution. Compared to distributed UAV platforms, ground stationary platforms exhibit smaller errors and pose fewer practical implementation challenges.

11.3 Problem formulation and optimization algorithm

In this section, we present the problem formulation for minimizing the beampattern PSL in distributed phased-MIMO radar systems and demonstrate a detailed optimization procedure to determine subarray layout and receive weighting coefficients. Specifically, two assumptions are made before further analysis.

- The waveforms transmitted between the different subarrays $s_p (p = 1, \ldots, P)$ are orthogonal, thereby ensuring orthogonality in the waveform covariance matrix $\mathbf{R} = \mathbf{I}_P$.
- The receive distributed subarrays are the same as the transmit distributed subarrays, that is $\mathbf{l} = \mathbf{l}_R = \mathbf{l}_T$, $N = N_T = N_R$, and $Q = P$ (the same subarrays).

The notation can be simplified by representing $\mathbf{w}_d^\dagger \mathbf{a}_{d_t}(\theta)$ as $\omega_d(\theta)$. Consequently, the virtual beampattern of distributed phased-MIMO radar can be reformulated as follows:

$$\begin{aligned} F(\theta, \mathbf{w}_R, \mathbf{w}_T, \mathbf{l}) &= (\mathbf{w}_R^\dagger \mathbf{a}_R(\theta, \mathbf{l})(\mathbf{w}_d^\dagger \mathbf{a}_{d_t}(\theta)))(\mathbf{w}_T^\dagger \mathbf{R} \mathbf{a}_T(\theta, \mathbf{l})) \\ &= \omega_d(\theta)\, \mathbf{w}_R^\dagger \mathbf{a}_R(\theta, \mathbf{l})\, \mathbf{w}_T^\dagger \mathbf{a}_T(\theta, \mathbf{l}). \end{aligned} \tag{11.16}$$

The angle region of interest $\theta \in [-90°, 90°]$ is uniformly discretized into a grid of discrete angles. The discrete angle space is divided into the mainlobe direction θ_m, transition band Θ_e [40] and sidelobe region $\Theta_s = \{\theta_1, \ldots, \theta_S\}$. Thus, the objective function in terms of PSL can be written as

$$\gamma = \max \frac{|F(\theta_s, \mathbf{w}_R, \mathbf{w}_T, \mathbf{l})|^2}{|F(\theta_m, \mathbf{w}_R, \mathbf{w}_T, \mathbf{l})|^2}, \theta_s \in \Theta_s, \tag{11.17}$$

where γ represents the PSL at a specific sidelobe region Θ_s.

The distance between adjacent distributed subarrays is typically much larger than half the wavelength, resulting in numerous high sidelobes in the virtual beampattern of distributed phased-MIMO radar. To tackle the challenging problem, we aim to reduce the sidelobe level of the virtual beampattern from two perspectives: the subarray layout vector $\mathbf{l}$ and the receive weighting coefficients $\mathbf{w}$. It is easy to verify that the virtual beampattern coupled between $\mathbf{w}_R$, $\mathbf{w}_T$, and $\mathbf{l}$, the likes of which is very difficult and complex to solve directly.

Remark 3: The complex problem can generally be tackled via employing a Cyclic Alternate (CA) optimization approach. To decrease the sidelobe level, we can alternately iterate $\mathbf{l}$, $\mathbf{w}_T$, and $\mathbf{w}_R$. Unfortunately, this optimization framework only converges after one iteration as depicted in Figure 11.10. Note that there is no performance improvement after one iteration for this optimization framework. This performance behavior implies that after the first iteration, the updated $\mathbf{w}_T$ and $\mathbf{w}_R$ make the subsequent optimization of the subarray layout $\mathbf{l}$ falling into local optimum, and the objective function will not decrease as the iteration progresses, leading to the failure of the CA algorithm.

To address this issue, a novel modified alternate iterative optimization framework called JSLRBIF is proposed, and the overall flowchart is shown in Figure 11.2. At k-th iteration, $\eta^{(k)}$ represents the upper bound level of the sidelobe region, and the detailed procedure is outlined as follows:

- *Step 1:* The subarray layout $\mathbf{l}^{(k)}$ is updated by minimizing the PSL in the presence of the fixed conventional beamforming technique $\mathbf{w}_T = \mathbf{a}_T(\theta_m)$ and $\mathbf{w}_R = \mathbf{a}_R(\theta_m)$. Then let $\mathbf{w}_R^{(k,0)} = \mathbf{a}_R(\theta_m)$.
- *Step 2:* The virtual transmit weight $\mathbf{w}_T^{(k,h)}$ and the actual receive weight $\mathbf{w}_R^{(k,h)}$ are updated H times alternately, assuming a fixed $\mathbf{l}^{(k)}$.
- *Step 3:* Computing the k-th objective function $\eta^{(k)}$ with updated $\mathbf{l}^{(k)}$, $\mathbf{w}_T^{(k,H)}$, and $\mathbf{w}_R^{(k,H)}$.
- *Step 4:* If $\eta^{(k)} < \eta^{(k-1)}$, update $\mathbf{l}_{opt}$, $\mathbf{w}_{T_{opt}}$, and $\mathbf{w}_{R_{opt}}$. Otherwise, let $\eta^{(k)} = \eta^{(k-1)}$ and $\mathbf{l}^{(k)}$ is re-initialized.

All the above operations are repeated K times to find the joint optimization result of $\mathbf{l}_{opt}$, $\mathbf{w}_{T_{opt}}$, and $\mathbf{w}_{R_{opt}}$.

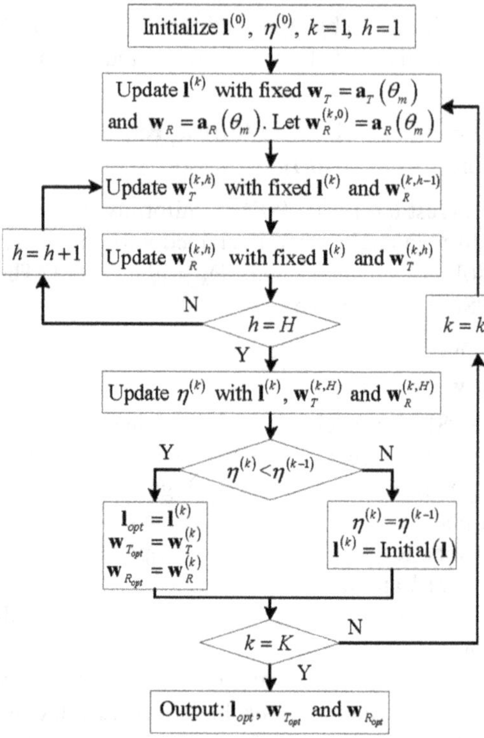

Figure 11.2 Flowchart of JSLRBIF

Remark 4: We employ the conventional beamforming technique to receive weighting coefficients during the updating $\mathbf{l}^{(k)}$ to avoid the local optimal. Additionally, by fixed $\mathbf{l}^{(k)}$, an alternate optimization of $\mathbf{w}_T$ and $\mathbf{w}_R$ can yield optimal weighting coefficients under the current subarray layout. The initialization operation helps to achieve a superior optimal solution of the subarray layout. Numerical simulations show that a single alternation between updating $\mathbf{w}_T$ and $\mathbf{w}_R$ suffices, i.e., $H = 1$.

The following section presents the optimization problem regarding subarray layout, real receive, and virtual transmit weighting coefficients, respectively. Subsequently, a detailed description of the solving process for each subproblem is provided.

11.3.1 Design of $\mathbf{l}^{(k)}$

The weighting coefficients $\mathbf{w}_R$ and $\mathbf{w}_T$ are fixed as a low complexity solution, namely conventional beamforming, when considering the optimization of subarray layout. The virtual beampattern of distributed phased-MIMO radar can be simplified as follows:

$$F(\theta, \mathbf{l}^{(k)}) = \omega_d(\theta)\, \mathbf{w}_R^{\dagger}\mathbf{a}_R(\theta, \mathbf{l}^{(k)})\mathbf{w}_T^{\dagger}\mathbf{a}_T(\theta, \mathbf{l}^{(k)}), \tag{11.18}$$

where $\mathbf{w}_R = \mathbf{a}_R(\theta_m, \mathbf{l}^{(k)})$, $\mathbf{w}_T = \mathbf{a}_T(\theta_m, \mathbf{l}^{(k)})$.

11.3.1.1 Practical constraints

The assumption is made that the phased-MIMO subarrays in space are distributed linearly, and we take into account the following three practical constraints regarding the positioning of subarrays.

Constraint 1: In practice, the length of each subarray itself is $D = (N-1)d$, and the actual interval between adjacent subarrays (UAV) should not be too small. Therefore, the spacing between arbitrary adjacent subarrays should satisfy $\Delta l \geq l_{\min} + D$, where Δl represents the spacing between adjacent subarrays and $l_{\min}$ denotes the minimum distance constraint. The distance constraint between adjacent subarrays is illustrated in Figure 11.3.

Constraint 2: The positions of the first and last subarrays are fixed to ensure the aperture of the distributed array, i.e., $l_1 = 0, l_P = L - D$, where D and L represent the aperture of the last subarray and the entire distributed radar system respectively. Therefore, only $P-2$ subarrays can be effectively optimized.

Constraint 3: The available position region of the $P-2$ subarray is restricted to an isometric area to achieve rapid and real-time position change of distribution subarrays in practical applications.

11.3.1.2 Subarray layout transform and inverse transform

The practical constraints being complex and difficult to solve directly, we initially perform a subarray layout transform operation on $\mathbf{l}$ to eliminate this inequality constraint [34]. Let $\tilde{D} = l_{\min} + D$ represents the minimal distance of l_{p-1} and l_p. Specifically, the detailed procedure is illustrated in Figure 11.4.

- *Step 1:* the position of the first and last subarrays, l_1 and l_P, in the distributed array with aperture $[0, L]$ are fixed to ensure the array aperture.
- *Step 2:* the minimum spacing constraint of $P-2$ subarrays is subtracted, $\tilde{\mathbf{l}} = \mathbf{l} - [0, \tilde{D}, 2\tilde{D}, \ldots, (P-2)\tilde{D}, 0]^{\mathrm{T}}$. The vector $\tilde{\mathbf{l}} = [l_1, \tilde{l}_2, \ldots, \tilde{l}_{P-1}, l_P]^{\mathrm{T}}$ represents the subarray layout within the remaining available aperture. By doing so, the original inequality constraint $\Delta l \geq \tilde{D}$ can be eliminated.

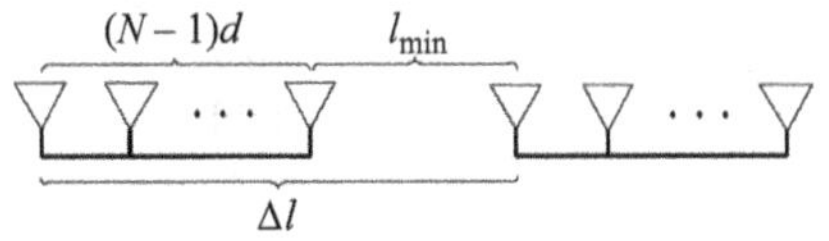

Figure 11.3 Position constraint 1

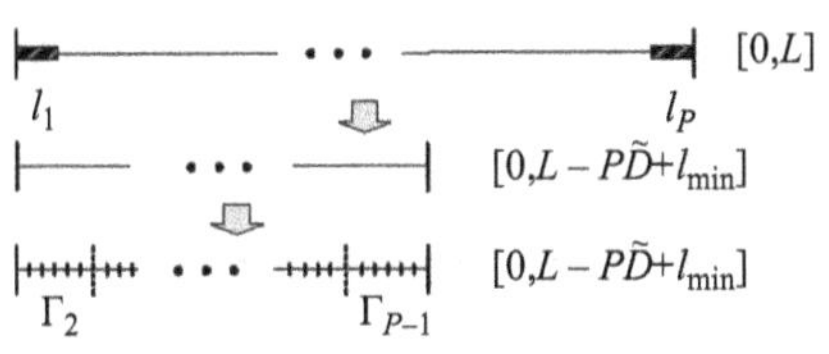

Figure 11.4 Subarray layout transform

- *Step 3:* the remaining aperture is discretized into Z discrete points. These discrete positions are then uniformly divided into $P-2$ parts, denoted as $\Gamma_2, \ldots, \Gamma_{P-1}$, while satisfying $\tilde{l}_p \in \Gamma_p$. Each part contains Z_P positions, i.e., $\Gamma_p = [\Gamma_p^1, \ldots, \Gamma_p^{Z_P}]$, where $Z_P = \frac{Z}{P-2}$.

Remark 5: The value of Z is primarily determined by the positional control accuracy of the UAV. After performing subarray layout transformation to eliminate the inequality constraint $\Delta l \geq l_{\min} + D$, we can obtain the remaining available aperture $L - PD - (P-1)\,l_{\min}$ without any constraints. Subsequently, this remaining aperture is discretized into Z discrete points, with each point representing a position interval of $\frac{L-PD-(P-1)l_{\min}}{Z}$ for the UAV. The larger the value of Z, the greater the precision in controlling the UAV.

The real subarray layout must be decoded to calculate the PSL of the virtual beampattern. Similar to the process of transforming subarray layout, we can obtain the real layout vector of subarrays $\mathbf{l}$ through inverse transform $\mathbf{l} = \tilde{\mathbf{l}} + [0, \tilde{D}, 2\tilde{D}, \ldots, (P-2)\,\tilde{D}, 0]^{\mathrm{T}}$. The process of inverse transformation for subarray layout is illustrated in Figure 11.5.

The design of the subarray layout vector can be formulated as a constrained optimization problem, based on previous considerations and discussions.

$$\begin{cases} \min\limits_{\mathbf{l}} \ \gamma \\ \text{s.t.} \ \ \gamma = \max \frac{|F(\theta_s,\mathbf{l})|^2}{|F(\theta_m,\mathbf{l})|^2}, \theta_s \in \Theta_s \\ \qquad l_1 = 0, l_P = L - D \\ \qquad \mathbf{l} = \tilde{\mathbf{l}} + [0, \tilde{D}, 2\tilde{D}, \ldots, (P-2)\,\tilde{D}, 0]^{\mathrm{T}} \\ \qquad \tilde{l}_p \in \Gamma_p, p = 2, \ldots, P-1 \end{cases} \tag{11.19}$$

Considering the NP-hard nature of problem (11.19) and the complexity in handling the constraints of $\mathbf{l}$, in the following, we propose a CD algorithm to monotonically decrease the objective value.

11.3.1.3 CD algorithm

We decompose the high-dimensional optimization problem into multiple one-dimensional problems by iteratively optimizing the position variables $l_2, \ldots, l_{P-1}$ in a cyclic manner. Each variable is sequentially optimized while keeping the remaining variables fixed. Additionally, we separate the variables associated with l_p to simplify the computation of the objective function. At the kth iteration, it is assumed that the previous $(P-2)$ subarray positions have been updated when solving the l_p,

$$\mathbf{l} = [l_1, l_2^{(k)}, \ldots, l_{p-1}^{(k)}, l_p, l_{p+1}^{(k-1)}, \ldots, l_{P-1}^{(k-1)}, l_P]^{\mathrm{T}}. \tag{11.20}$$

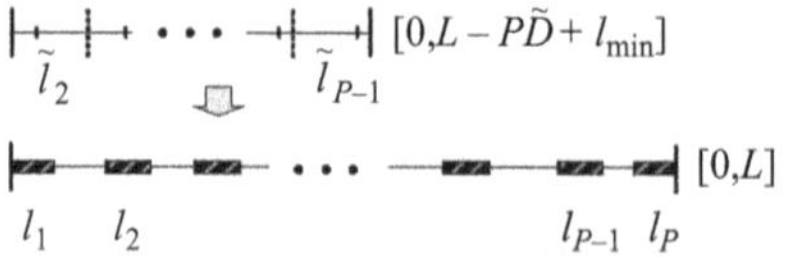

Figure 11.5 Subarray layout inverse transform

The virtual aperture beampattern associated with θ and l_p can be expressed as follows:

$$\begin{aligned}\tilde{F}\left(\theta, l_p\right) &= \omega_d\left(\theta\right)\tilde{\mathbf{w}}_R^{\dagger}\tilde{\mathbf{a}}_R\left(\theta, l_p\right)\tilde{\mathbf{w}}_T^{\dagger}\tilde{\mathbf{a}}_T\left(\theta, l_p\right)\\ &= \omega_d\left(\theta\right)(\mathbf{w}_{R_0} + \mathbf{w}_{R_a})^{\dagger}(\mathbf{a}_{R_0} + \mathbf{a}_{R_a})(\mathbf{w}_{T_0} + \mathbf{w}_{T_a})^{\dagger}(\mathbf{a}_{T_0} + \mathbf{a}_{T_a}),\end{aligned} \tag{11.21}$$

where $\tilde{\mathbf{w}}_R = \mathbf{w}_{R_0} + \mathbf{w}_{R_a}$, $\tilde{\mathbf{a}}_R(\theta, l_p) = \mathbf{a}_{R_0} + \mathbf{a}_{R_a}$, $\tilde{\mathbf{w}}_T = \mathbf{w}_{T_0} + \mathbf{w}_{T_a}$, and $\tilde{\mathbf{a}}_T(\theta, l_p) = \mathbf{a}_{T_0} + \mathbf{a}_{T_a}$, $\mathbf{a}_{T_0}$, $\mathbf{a}_{T_a}$, $\mathbf{a}_{R_0}$, $\mathbf{a}_{R_a}$, $\mathbf{w}_{T_0}$, $\mathbf{w}_{T_a}$, $\mathbf{w}_{R_0}$, and $\mathbf{w}_{R_a}$ are defined as follows:

$$\mathbf{a}_{T_0} = [0, \ldots, e^{j\frac{2\pi}{\lambda}l_p \sin\theta}, \ldots, 0]^{\mathrm{T}}, \mathbf{a}_{R_0} = \mathbf{a}_{T_0} \otimes \mathbf{a}_{d_r}(\theta) \tag{11.22a}$$

$$\mathbf{w}_{T_0} = [0, \ldots, e^{j\frac{2\pi}{\lambda}l_p \sin\theta_m}, \ldots, 0]^{\mathrm{T}}, \mathbf{w}_{R_0} = \mathbf{w}_{T_0} \otimes \mathbf{a}_{d_r}(\theta_m) \tag{11.22b}$$

$$\mathbf{a}_{T_a} = [e^{j\frac{2\pi}{\lambda}l_1^{(k+1)}\sin\theta}, \ldots, e^{j\frac{2\pi}{\lambda}l_{p-1}^{(k+1)}\sin\theta}, 0, e^{j\frac{2\pi}{\lambda}l_{p+1}^{(k)}\sin\theta}, \ldots, e^{j\frac{2\pi}{\lambda}l_P^{(k)}\sin\theta}]^{\mathrm{T}} \tag{11.22c}$$

$$\mathbf{w}_{T_a} = [e^{j\frac{2\pi}{\lambda}l_1^{(k+1)}\sin\theta_m}, \ldots, e^{j\frac{2\pi}{\lambda}l_{p-1}^{(k+1)}\sin\theta_m}, 0, e^{j\frac{2\pi}{\lambda}l_{p+1}^{(k)}\sin\theta_m}, \ldots, e^{j\frac{2\pi}{\lambda}l_P^{(k)}\sin\theta_m}]^{\mathrm{T}} \tag{11.22d}$$

$$\mathbf{a}_{R_a} = \mathbf{a}_{T_a} \otimes \mathbf{a}_{d_r}(\theta), \mathbf{w}_{R_a} = \mathbf{w}_{T_a} \otimes \mathbf{a}_{d_r}(\theta_m) \tag{11.22e}$$

Exploiting the fact that $\mathbf{w}_{R_0}^{\dagger}\mathbf{a}_{R_a} = \mathbf{w}_{R_a}^{\dagger}\mathbf{a}_{R_0} = \mathbf{w}_{T_0}^{\dagger}\mathbf{a}_{T_a} = \mathbf{w}_{T_a}^{\dagger}\mathbf{a}_{T_0} = 0$, the virtual aperture beampattern can be rewritten as follows:

$$\begin{aligned}\tilde{F}\left(\theta, l_p\right) &= \omega_d\left(\theta\right)(\mathbf{w}_{R_0}^{\dagger}\mathbf{a}_{R_0} + \mathbf{w}_{R_a}^{\dagger}\mathbf{a}_{R_a})(\mathbf{w}_{T_0}^{\dagger}\mathbf{a}_{T_0} + \mathbf{w}_{T_a}^{\dagger}\mathbf{a}_{T_a})\\ &= \omega_d(\theta)(\mathbf{w}_{R_0}^{\dagger}\mathbf{a}_{R_0}\mathbf{w}_{T_0}^{\dagger}\mathbf{a}_{T_0} + \alpha\mathbf{w}_{T_0}^{\dagger}\mathbf{a}_{T_0} + \mathbf{w}_{R_0}^{\dagger}\mathbf{a}_{R_0}\beta + \alpha\beta),\end{aligned} \tag{11.23}$$

where $\alpha = \mathbf{w}_{R_a}^{\dagger}\mathbf{a}_{R_a}$, $\beta = \mathbf{w}_{T_a}^{\dagger}\mathbf{a}_{T_a}$. $\mathbf{w}_{T_a}$, $\mathbf{w}_{R_a}$, $\mathbf{a}_{T_a}$, and $\mathbf{a}_{R_a}$ are variables independent of l_p. Therefore, α and β are constant in the process of optimizing l_p. The optimization subproblem of l_p can be further simplified as follows:

$$\begin{cases}\min\limits_{l_p} \gamma\\ \text{s.t. } \gamma = \max \frac{\left|\tilde{F}(\theta_s, l_p)\right|^2}{\left|\tilde{F}(\theta_m, l_p)\right|^2}, \theta_s \in \Theta_s\\ \quad l_p = \tilde{l}_p + (p-1)\tilde{D}\\ \quad \tilde{l}_p \in \Gamma_p = [\Gamma_p^1, \ldots, \Gamma_p^{Z_P}]\end{cases} \tag{11.24}$$

The current solution of l_p is obtained by employing an exhaustive search for each element of Γ_p. To reduce the computational burden on the beampattern, α and β can be pre-calculated when utilizing an exhaustive search for the optimal solution of l_p to reduce the amount of calculation of the beampattern. Based on these considerations, Algorithm 11.1 summarizes the proposed method for optimizing the subarray layout vector.

11.3.1.4 Computational complexity and convergence analysis

The computational complexity of the proposed algorithm in each iteration is determined by the number of subarrays and discrete positions for $\Gamma_p(p = 2, \ldots, P-1)$, resulting in a computational complexity of $O(PZ_P)$.

Next, we evaluate the convergence of the CD algorithm for solving $\mathbf{l}^{(k)}$, where each iteration employs the proposed fast iterative algorithm to solve the l_p.

Algorithm 11.1 Subarray layout optimization via CD algorithm at the k-th iteration

---

Input: $\mathbf{l}^{(k-1)}, f_c, N, d, l_{\min}, \theta_m, P$;
Output: $\mathbf{l}^{(k)}$;
1: $\mathbf{l}^{(k)} = \mathbf{l}^{(k-1)}$;
2: $p = 1$;
3: $p = p + 1$;
4: Calculate α and β;
5: **for** $z = 1, \ldots, Z_P$
    $l_p^z = \Gamma_p^z + (P-1)\tilde{D}$;
    Calculate $\mathbf{w}_{R_0}^\dagger \mathbf{a}_{R_0}$ and $\mathbf{w}_{T_0}^\dagger \mathbf{a}_{T_0}$;
    Calculate $\tilde{F}\left(\theta, l_p^z\right)$ for θ_m and θ_s by (11.23);
    Calculate the PSL $\gamma_{p_z}^{(k)}$ of virtual beampattern;
  end
6: Find the label $z_{\min}$ and position $l_p^{z_{\min}}$ for minimum PSL and then obtain the corresponding objective function value $\gamma_{p_{z_{\min}}}^{(k)}$;
7: Calculate objective function value $\gamma_p^{(k)}$ of $\mathbf{l}^{(k)}$;
8: If $\gamma_{p_{z_{\min}}}^{(k)} < \gamma_p^{(k)}$, update $l_p^{(k)} = l_p^{z_{\min}}$ for $\mathbf{l}^{(k)}$.
9: If $p = P - 1$, output $\mathbf{l}^{(k)}$.
  Otherwise, go to Step 3;

Specifically, $\gamma_p^{(k)}$ represents the objective function value after updating the p-th entry of l_p at the k-th iteration. Based on the above discussion,

$$\gamma^{(k-1)} \geq \gamma_2^{(k)} \geq \gamma_3^{(k)} \cdots \geq \gamma_{P-1}^{(k)} = \gamma^{(k)}. \tag{11.25}$$

Thus, we can derive $\gamma^{(k-1)} \geq \gamma^{(k)}$. Additionally, it is easy to obtain $\gamma^{(k)} \geq 0$. As a consequence, we have the result of $\gamma^{(k-1)} \geq \gamma^{(k)} \geq 0$ showing that the obtained $\gamma^{(k)}$ decreases monotonously and shares a lower bound guaranteeing to converge finite values.

11.3.2 Design of $\mathbf{w}_T^{(k)}$

Next, we consider to optimize the waveform diversity beampattern $\mathbf{w}_T^\dagger \mathbf{a}_T(\theta, \mathbf{l}^{(k)})$.

11.3.2.1 Problem formulation

The subarray layout vector $\mathbf{l}^{(k)}$ and receive weighting coefficients $\mathbf{w}_R = \mathbf{a}_R(\theta_m, \mathbf{l}^{(k)})$ remain fixed during the optimization of the virtual transmit weighting coefficients $\mathbf{w}_T$ for all subarrays. By defining $\omega_r(\theta) = \mathbf{w}_R^\dagger \mathbf{a}_R(\theta, \mathbf{l}^{(k)})$, the virtual beampattern can be equivalently transformed to

$$\begin{aligned} F(\theta, \mathbf{w}_T) &= \omega_d(\theta)\, \mathbf{w}_R^\dagger \mathbf{a}_R(\theta, \mathbf{l}^{(k)}) \mathbf{w}_T^\dagger \mathbf{a}_T(\theta, \mathbf{l}^{(k)}) \\ &= \omega_d(\theta)\, \omega_r(\theta)\, \mathbf{w}_T^\dagger \mathbf{a}_T(\theta, \mathbf{l}^{(k)}) \\ &= \mathbf{w}_T^\dagger \hat{\mathbf{a}}_T(\theta), \end{aligned} \tag{11.26}$$

where $\hat{\mathbf{a}}_T(\theta) = \omega_d(\theta)\,\omega_r(\theta)\,\mathbf{a}_T(\theta, \mathbf{l}^{(k)})$ represents the reconstructed steering vector. Then the optimization problem with respect to $\mathbf{w}_T$ is as follows:

$$\mathcal{P}_{\mathbf{w}_T^{(k)}} \begin{cases} \min\limits_{\mathbf{w}_T,\eta} \ \eta \\ \text{s.t.} \quad |\mathbf{w}_T^{\dagger}\hat{\mathbf{a}}_T(\theta_s)|^2 \leq \eta, \theta_s \in \Theta_s \\ \qquad \chi - \varepsilon \leq |\mathbf{w}_T^{\dagger}\hat{\mathbf{a}}_T(\theta_m)|^2 \leq \chi + \varepsilon \end{cases} \tag{11.27}$$

where $\chi - \varepsilon$ and $\chi + \varepsilon$ represent the lower and upper power bound level, respectively, of the mainlobe θ_m [41,42].

11.3.2.2 Optimization method

$\mathcal{P}_{\mathbf{w}_T^{(k)}}$ can be simplified as follows:

$$\begin{cases} \min\limits_{\mathbf{w}_T,\eta} \ \eta \\ \text{s.t.} \quad \mathbf{w}_T^{\dagger}\mathbf{R}_s\mathbf{w}_T \leq \eta, \theta_s \in \Theta_s \\ \qquad \mathbf{w}_T^{\dagger}\mathbf{R}_m\mathbf{w}_T \leq \chi + \varepsilon \\ \qquad \chi - \varepsilon \leq \mathbf{w}_T^{\dagger}\mathbf{R}_m\mathbf{w}_T \end{cases} \tag{11.28}$$

where $\mathbf{R}_s = \hat{\mathbf{a}}_T(\theta_s)\hat{\mathbf{a}}_T^{\dagger}(\theta_s)$, $s = 1, \ldots, S$, $\mathbf{R}_m = \hat{\mathbf{a}}_T(\theta_m)\hat{\mathbf{a}}_T^{\dagger}(\theta_m)$.

Further, we convert the problem (11.28) into the real domain for implementation [41].

$$\begin{cases} \min\limits_{\tilde{\mathbf{w}}_T,\eta} \ \eta \\ \text{s.t.} \quad \tilde{\mathbf{w}}_T^{\dagger}\tilde{\mathbf{R}}_s\tilde{\mathbf{w}}_T \leq \eta, \theta_s \in \Theta_s \\ \qquad \tilde{\mathbf{w}}_T^{\dagger}\tilde{\mathbf{R}}_m\tilde{\mathbf{w}}_T \leq \chi + \varepsilon \\ \qquad \chi - \varepsilon \leq \tilde{\mathbf{w}}_T^{\dagger}\tilde{\mathbf{R}}_m\tilde{\mathbf{w}}_T \end{cases} \tag{11.29}$$

where

$$\tilde{\mathbf{R}}_s = \begin{bmatrix} \Re\{\mathbf{R}_s\} & -\Im\{\mathbf{R}_s\} \\ \Im\{\mathbf{R}_s\} & \Re\{\mathbf{R}_s\} \end{bmatrix} \in \mathbb{R}^{2P\times 2P} \tag{11.30a}$$

$$\tilde{\mathbf{R}}_m = \begin{bmatrix} \Re\{\mathbf{R}_m\} & -\Im\{\mathbf{R}_m\} \\ \Im\{\mathbf{R}_m\} & \Re\{\mathbf{R}_m\} \end{bmatrix} \in \mathbb{R}^{2P\times 2P} \tag{11.30b}$$

$$\tilde{\mathbf{w}}_T = \begin{bmatrix} \Re\left\{\mathbf{w}_T^{\mathrm{T}}\right\} & \Im\left\{\mathbf{w}_T^{\mathrm{T}}\right\} \end{bmatrix}^{\mathrm{T}} \in \mathbb{R}^{2P\times 1} \tag{11.30c}$$

Applying the technique SDR [43] to relax the problem, the optimization problem can be transformed as follows:

$$\begin{cases} \min\limits_{\mathbf{W}_T,\eta} \ \eta \\ \text{s.t.} \quad \mathrm{Tr}\left(\tilde{\mathbf{R}}_s\mathbf{W}_T\right) \leq \eta, \theta_s \in \Theta_s \\ \qquad \mathrm{Tr}\left(\tilde{\mathbf{R}}_m\mathbf{W}_T\right) \leq \chi + \varepsilon \\ \qquad \mathrm{Tr}\left(\tilde{\mathbf{R}}_m\mathbf{W}_T\right) \geq \chi - \varepsilon \\ \qquad \mathbf{W}_T \geq 0 \end{cases} \tag{11.31}$$

where $\mathbf{W}_T = \tilde{\mathbf{w}}_T\tilde{\mathbf{w}}_T^{\dagger}$. The formulation in (11.31) can be efficiently and reliably solved using the convex optimization toolbox CVX [44].

The computational complexity of the SDR technique is $O(P^{3.5})$. However, simply applying rank-one approximation [41,42] may not yield a feasible solution to problem (11.28). To ensure that the beampattern obtained after the rank-one approximation satisfies the mainlobe constraints of the original problem, an overall scaling of $\mathbf{w}_T$ is performed based on the changes in mainlobe before and after the rank-one approximation.

11.3.3 Design of $\mathbf{w}_R^{(k)}$

The focus of this section is to minimize the sidelobe level of the virtual beampattern by considering real receive weighting coefficients.

11.3.3.1 Problem formulation

Similarly, the subarray layout vector $\mathbf{l}^{(k)}$ and virtual transmit weighting coefficients $\mathbf{w}_T^{(k)}$ is fixed. Next, we consider to optimize the real receive beampattern $\mathbf{w}_R^{\dagger}\mathbf{a}_R(\theta, \mathbf{l}^{(k)})$. Letting $\omega_t(\theta) = \mathbf{w}_T^{(k)\dagger}\mathbf{a}_T(\theta, \mathbf{l}^{(k)})$, the virtual beampattern for $\mathbf{w}_R$ can be recast as follows:

$$\begin{aligned} F(\theta, \mathbf{w}_R) &= \omega_d(\theta)\, \mathbf{w}_R^{\dagger}\mathbf{a}_R(\theta, \mathbf{l})\, \mathbf{w}_T^{\dagger}\mathbf{a}_T(\theta, \mathbf{l}^{(k)}) \\ &= \omega_d(\theta)\, \omega_t(\theta)\, \mathbf{w}_R^{\dagger}\mathbf{a}_R(\theta, \mathbf{l}^{(k)}) \\ &= \mathbf{w}_R^{\dagger}\hat{\mathbf{a}}_R(\theta), \end{aligned} \tag{11.32}$$

where $\hat{\mathbf{a}}_R(\theta) = \omega_d(\theta)\, \omega_t(\theta)\, \mathbf{a}_R(\theta, \mathbf{l}^{(k)})$ represents the reconstructed steering vector, then the optimization problem with respect to $\mathbf{w}_R$ is constructed as follows:

$$\mathcal{P}_{\mathbf{w}_R^{(k)}} \begin{cases} \min\limits_{\mathbf{w}_R, \eta} \ \eta \\ \text{s.t.} \quad |\mathbf{w}_R^{\dagger}\hat{\mathbf{a}}_R(\theta_s)|^2 \le \eta, \theta_s \in \Theta_s \\ \qquad \chi - \varepsilon \le |\mathbf{w}_R^{\dagger}\hat{\mathbf{a}}_R(\theta_m)|^2 \le \chi + \varepsilon \end{cases} \tag{11.33}$$

11.3.3.2 Optimization method

Similar operation as in the previous subsection, we can simplify $\mathcal{P}_{\mathbf{w}_R^{(k)}}$ to

$$\begin{cases} \min\limits_{\mathbf{w}_R, \eta} \ \eta \\ \text{s.t.} \quad \mathbf{w}_R^{\dagger}\Xi_s\mathbf{w}_R \le \eta, \theta_s \in \Theta_s \\ \qquad \mathbf{w}_R^{\dagger}\Xi_m\mathbf{w}_R \le \chi + \varepsilon \\ \qquad \chi - \varepsilon \le \mathbf{w}_R^{\dagger}\Xi_m\mathbf{w}_R \end{cases} \tag{11.34}$$

where $\Xi_s = \hat{\mathbf{a}}_R(\theta_s)\, \hat{\mathbf{a}}_R^{\dagger}(\theta_s)$, $s = 1, \ldots, S$, $\Xi_m = \hat{\mathbf{a}}_R(\theta_m)\, \hat{\mathbf{a}}_R^{\dagger}(\theta_m)$. After further derivation and simplification, we can get

$$\begin{aligned} \Xi_s &= \hat{\mathbf{a}}_R(\theta_s)\, \hat{\mathbf{a}}_R^{\dagger}(\theta_s) \\ &= (\omega_d(\theta_s)\, \omega_t(\theta_s)\, \mathbf{a}_R(\theta_s, \mathbf{l}))^{\dagger} (\omega_d(\theta_s)\, \omega_t(\theta_s)\, \mathbf{a}_R(\theta_s, \mathbf{l})) \\ &= \omega_d^2(\theta_s)\, \omega_t^2(\theta_s)\, \mathbf{a}_R(\theta_s, \mathbf{l})\, \mathbf{a}_R^{\dagger}(\theta_s, \mathbf{l}) \\ &= \omega_d^2(\theta_s)\, (\mathbf{w}_T^{\dagger}\mathbf{a}_T(\theta_s, \mathbf{l})\, \mathbf{a}_T^{\dagger}(\theta_s, \mathbf{l})\, \mathbf{w}_T)\mathbf{R}_{rs} \\ &= \omega_d^2(\theta_s)\, \mathrm{Tr}(\mathbf{a}_T(\theta_s, \mathbf{l})\, \mathbf{a}_T^{\dagger}(\theta_s, \mathbf{l})\, \mathbf{w}_T^{\dagger}\mathbf{w}_T)\mathbf{R}_{rs}, \end{aligned} \tag{11.35}$$

where $\mathbf{R}_{ts} = \mathbf{a}_T(\theta_s, \mathbf{l})\,\mathbf{a}_T^\dagger(\theta_s, \mathbf{l})$, $\mathbf{R}_{rs} = \mathbf{a}_R(\theta_s, \mathbf{l})\,\mathbf{a}_R^\dagger(\theta_s, \mathbf{l})$.

$$\begin{aligned}\Xi_m &= \hat{\mathbf{a}}_R(\theta_m)\,\hat{\mathbf{a}}_R^\dagger(\theta_m)\\ &= (\omega_d(\theta_m)\,\omega_t(\theta_m)\,\mathbf{a}_R(\theta_m, \mathbf{l}))^\dagger\,(\omega_d(\theta_m)\,\omega_t(\theta_m)\,\mathbf{a}_R(\theta_m, \mathbf{l}))\\ &= \omega_d^2(\theta_m)\,\omega_t^2(\theta_m)\,\mathbf{a}_R(\theta_m, \mathbf{l})\,\mathbf{a}_R^\dagger(\theta_m, \mathbf{l})\\ &= \omega_d^2(\theta_m)\,(\mathbf{w}_T^\dagger \mathbf{a}_T(\theta_m, \mathbf{l})\,\mathbf{a}_T^\dagger(\theta_m, \mathbf{l})\,\mathbf{w}_T)\mathbf{R}_{rm}\\ &= \omega_d^2(\theta_m)\,\mathrm{Tr}(\mathbf{a}_T(\theta_m, \mathbf{l})\,\mathbf{a}_T^\dagger(\theta_m, \mathbf{l})\,\mathbf{w}_T^\dagger \mathbf{w}_T)\mathbf{R}_{rm},\end{aligned} \tag{11.36}$$

where $\mathbf{R}_{tm} = \mathbf{a}_T(\theta_m, \mathbf{l})\,\mathbf{a}_T^\dagger(\theta_m, \mathbf{l})$, $\mathbf{R}_{rm} = \mathbf{a}_R(\theta_m, \mathbf{l})\,\mathbf{a}_R^\dagger(\theta_m, \mathbf{l})$.

The problem (11.34) is further transformed into the real domain, which is equivalent to updating $\mathbf{w}_T$. Subsequently, substituting the previous optimized $\mathbf{W}_T$, we obtain

$$\begin{cases}\min\limits_{\tilde{\mathbf{w}}_R, \eta} \ \eta\\ \text{s.t.} \quad \tilde{\mathbf{w}}_R^\dagger \tilde{\Xi}_s \tilde{\mathbf{w}}_R \le \eta, \theta_s \in \Theta_s\\ \qquad\ \ \tilde{\mathbf{w}}_R^\dagger \tilde{\Xi}_m \tilde{\mathbf{w}}_R \le \chi + \varepsilon\\ \qquad\ \ \chi - \varepsilon \le \tilde{\mathbf{w}}_R^\dagger \tilde{\Xi}_m \tilde{\mathbf{w}}_R\end{cases} \tag{11.37}$$

where

$$\begin{aligned}\tilde{\Xi}_s &= \begin{bmatrix}\Re\{\Xi_s\} & -\Im\{\Xi_s\}\\ \Im\{\Xi_s\} & \Re\{\Xi_s\}\end{bmatrix} \in \mathbb{R}^{2QN\times 2QN}\\ &= \omega_d^2(\theta_s)\,\mathrm{Tr}(\tilde{\mathbf{R}}_{ts}\tilde{\mathbf{w}}_T^\dagger \tilde{\mathbf{w}}_T)\tilde{\mathbf{R}}_{rs}\\ &= \omega_d^2(\theta_s)\,\mathrm{Tr}(\tilde{\mathbf{R}}_{ts}\mathbf{W}_T)\tilde{\mathbf{R}}_{rs}\end{aligned} \tag{11.38a}$$

$$\begin{aligned}\tilde{\Xi}_m &= \begin{bmatrix}\Re\{\Xi_m\} & -\Im\{\Xi_m\}\\ \Im\{\Xi_m\} & \Re\{\Xi_m\}\end{bmatrix} \in \mathbb{R}^{2QN\times 2QN}\\ &= \omega_d^2(\theta_m)\,\mathrm{Tr}(\tilde{\mathbf{R}}_{tm}\tilde{\mathbf{w}}_T^\dagger \tilde{\mathbf{w}}_T)\tilde{\mathbf{R}}_{rm}\\ &= \omega_d^2(\theta_m)\,\mathrm{Tr}(\tilde{\mathbf{R}}_{tm}\mathbf{W}_T)\tilde{\mathbf{R}}_{rm}\end{aligned} \tag{11.38b}$$

$$\tilde{\mathbf{w}}_R = \begin{bmatrix}\Re\left\{\mathbf{w}_R^{\mathrm{T}}\right\} & \Im\left\{\mathbf{w}_R^{\mathrm{T}}\right\}\end{bmatrix}^{\mathrm{T}} \in \mathbb{R}^{2QN\times 1} \tag{11.38c}$$

where $\tilde{\mathbf{R}}_{ts}$, $\tilde{\mathbf{R}}_{rs}$, $\tilde{\mathbf{R}}_{tm}$, and $\tilde{\mathbf{R}}_{rm}$ are real domain of $\mathbf{R}_{ts}$, $\mathbf{R}_{rs}$, $\mathbf{R}_{tm}$, and $\mathbf{R}_{rm}$, respectively, similarly with (11.30). Exploiting the technique SDR to relax the problem, the optimization problem can be recast as follows:

$$\begin{cases}\min\limits_{\mathbf{W}_R, \eta} \ \eta\\ \text{s.t.} \quad \mathrm{Tr}(\tilde{\Xi}_s \mathbf{W}_R) \le \eta, \theta_s \in \Theta_s\\ \qquad\ \ \mathrm{Tr}(\tilde{\Xi}_m \mathbf{W}_R) \le \chi + \varepsilon\\ \qquad\ \ \mathrm{Tr}(\tilde{\Xi}_m \mathbf{W}_R) \ge \chi - \varepsilon\\ \qquad\ \ \mathbf{W}_R \ge 0\end{cases} \tag{11.39}$$

The optimized $\mathbf{W}_R$ can be obtained using the convex optimization toolbox CVX. However, it should be noted that simply performing rank-one approximation [41,42] may not yield a feasible solution for problem (11.34). To ensure that the resulting beampattern satisfies the mainlobe constraints of the original problem, an overall scaling of $\mathbf{w}_R$ is performed based on the changes in the mainlobe before and after rank one approximation. Additionally, it is worth mentioning that the SDR technique has a polynomial time computational complexity of $O((QN)^{3.5})$.

11.3.4 JSLRBIF algorithm

The procedure of sequentially optimizing $\mathbf{l}$, $\mathbf{w}_T$ and $\mathbf{w}_R$ can be summarized in Algorithm 11.2 based on the aforementioned discussions. Specifically, each iteration of the proposed algorithm requires executing Algorithm 11.1 once.

Remark 6: The weighting coefficients in conventional beamforming are phase-only, ensuring a constant mainlobe level. The objective function for subarray layout optimization is relative PSL, without constraining the mainlobe level of the virtual beampattern. The change of the subarray layout only affects sidelobe performance. Changes in the subarray layout only affect sidelobe performance. However, it is necessary to restrict the mainlobe level while optimizing receive weighting coefficients $\mathbf{w}_T/\mathbf{w}_R$ due to their non-constant modulus. It should be noted that although the optimization purposes for both parts are essentially the same, solutions to these two subproblems are independent of each other. Therefore, in our proposed JSLRBIF algorithm, different objective functions for subarray layout optimization and receive weighting optimization do not impact algorithm convergence. Additionally, to reduce the computational complexity, we do not calculate the corresponding $\mathbf{w}_T/\mathbf{w}_R$ from $\mathbf{W}_T/\mathbf{W}_R$ in the iterative process, but apply the rank one approximation and scaling to the optimal solution $\mathbf{W}_{T_{opt}}$ and $\mathbf{W}_{R_{opt}}$ after the end of the iteration to get the final optimization result $\mathbf{w}_{T_{opt}}$, $\mathbf{w}_{R_{opt}}$.

The subsequent focus lies on the convergence analysis of the JSLRBIF algorithm. Let $\eta^{(k)}$ denote the objective value at the kth iteration. Since there is a judgment regarding η in step 6 of Algorithm 11.2, it can be easily observed that $0 \leq \eta^{(k)} \leq \eta^{(k-1)}$. Consequently, we can conclude that the objective function value of the proposed JSLRBIF algorithm decreases monotonically and has a lower bound, ensuring finite convergence.

Algorithm 11.2 JSLRBIF algorithm

Input: $\mathbf{l}^{(0)}, \eta^{(0)}, f_c, N, d, l_{\min}, \theta_m, P, K, \chi, \varepsilon$;
Output: $\mathbf{l}_{opt}, \mathbf{w}_{T_{opt}}, \mathbf{w}_{R_{opt}}$ and η_{opt};
1: $k = 0$;
2: $k = k + 1$;
3: Update $\mathbf{l}^{(k)}$ by **Algorithm 11.1**;
4: Calculate $\tilde{\mathbf{R}}_s$ and $\tilde{\mathbf{R}}_m$ by $\mathbf{l}^{(k)}$ and apply CVX toolbox to obtain the solution $\mathbf{W}_T^{(k)}$ in (11.31);
5: Calculate $\tilde{\Xi}_s$ and $\tilde{\Xi}_m$ by $\mathbf{l}^{(k)}$ and $\mathbf{W}_T^{(k)}$ and apply CVX toolbox to obtain the solution $\mathbf{W}_R^{(k)}$ in (11.39), update $\eta^{(k)}$;
6: If $\eta^{(k)} < \eta^{(k-1)}$, update $\mathbf{l}_{opt} = \mathbf{l}^{(k)}$, $\mathbf{W}_{T_{opt}} = \mathbf{W}_T^{(k)}$, $\mathbf{W}_{R_{opt}} = \mathbf{W}_R^{(k)}$ and $\eta_{opt} = \eta^{(k)}$.
Otherwise, update $\eta^{(k)} = \eta^{(k-1)}$, $\mathbf{l}^{(k)} = \text{Initial}\,(\mathbf{l})$.
7: If $k = K$, applying the rank one approximation to get $\mathbf{w}_{T_{opt}}$, $\mathbf{w}_{R_{opt}}$ from $\mathbf{W}_{T_{opt}}$, $\mathbf{W}_{R_{opt}}$ and then output $\mathbf{l}_{opt}, \mathbf{w}_{T_{opt}}, \mathbf{w}_{R_{opt}}$ and η_{opt}.
Otherwise, go to Step 2.

11.4 Simulation results

The performance evaluation of the proposed algorithms focuses on a distributed phased-MIMO radar system with 8 subarrays, where $P = Q = 8$. The transmit waveforms for different subarrays are orthogonal frequency division multiplexing linear frequency modulated (OFDM-LFM) signals. The transmit narrowband signal of the p-th subarray can be written as follows:

$$s_p(t) = \mathrm{rect}(\frac{t}{T_{pd}}) \exp\left[j2\pi\left(f_p t + 0.5\mu t^2\right)\right], \tag{11.40}$$

where T_{pd} is pulse duration, $f_p = f_0 + (p-1)\Delta f$ for $p = 1, \ldots, P$ while f_0 denotes initial carrier frequency and Δf is step frequency interval. $\mu = B/T_{pd}$ represents the frequency modulation rate of LFM and B is the signal bandwidth. $\mathrm{rect}(\cdot)$ is the rectangular modulation function as follows:

$$\mathrm{rect}(i) = \begin{cases} 1, & 0 < i \leq 1 \\ 0, & \text{others} \end{cases} \tag{11.41}$$

The carrier frequency is set to $f_c = 1$GHz. Each subarray is equipped with a ULPA consisting of $N = N_T = N_R = 8$ omnidirectional antennas spaced half a wavelength apart from each other. The aperture length of the distributed array is defined as $L =$ 20m. To ensure the desired aperture, we fix the first subarray location $l_1 = 0$ and the last subarray location $l_P = L - (N-1)d$. The beampattern is assumed to steer toward $\theta_m = 20°$. The vector $\mathbf{l}^{(0)}$ is the randomly initialized vector satisfying the location constraints of the distributed subarrays. We select $\Theta_s = [-90°, 19°] \bigcup [21°, 90°]$ as the sidelobe regions uniformly discretized with a grid size 0.1°. The lower and upper power bounds for mainlobe θ_m in virtual beampattern are set as $\chi = (PQN^2)^2$ and $\varepsilon = 6$, respectively. The simulations are performed using Matlab 2016a version on a standard PC with a 2.8 GHz Core i5 CPU and 8 GB RAM.

11.4.1 CD algorithm for solving subarray layout vector

The convergence of the CD algorithm for solving $\mathbf{l}$ is assessed in this section, where each iteration utilizes the proposed fast iterative algorithm to solve the l_p. In general, different initializations have an impact on the algorithm's performance. Therefore, each curve represents an average of 500 randomized and independent trials.

Firstly, we demonstrate the impact of Z on the performance with $l_{\min}$ set at 0.5m.

In Figure 11.6, we present γ versus iteration number k for various discrete position numbers Z. Note that the proposed CD algorithm can rapidly converge to a stable value under different parameters of Z. As expected, larger values of Z result in smaller objective function values after convergence since more discrete points provide higher degrees of freedom within the feasible region. Interestingly, as Z increases, the value of the objective function after convergence will not continue to decrease. This performance behavior implies that the optimization result of the discrete position is close to continuous and will not be affected by the increase of the number of discrete positions Z. Conversely, the running time of the algorithm will significantly increase with the increase of Z as shown in Figure 11.7. Notably,

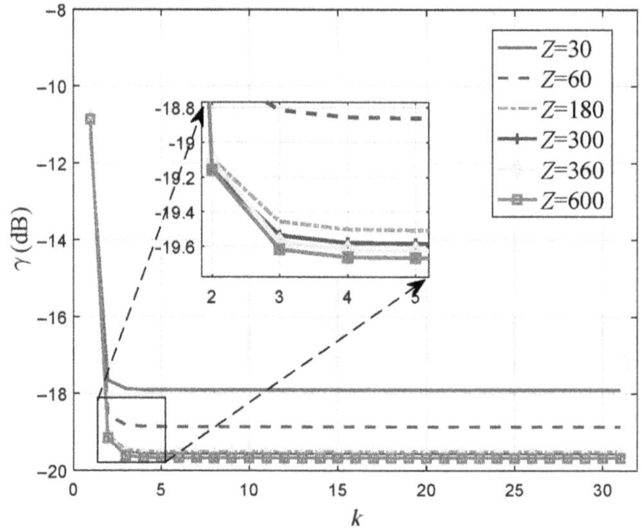

Figure 11.6 γ versus iteration number k for different Z

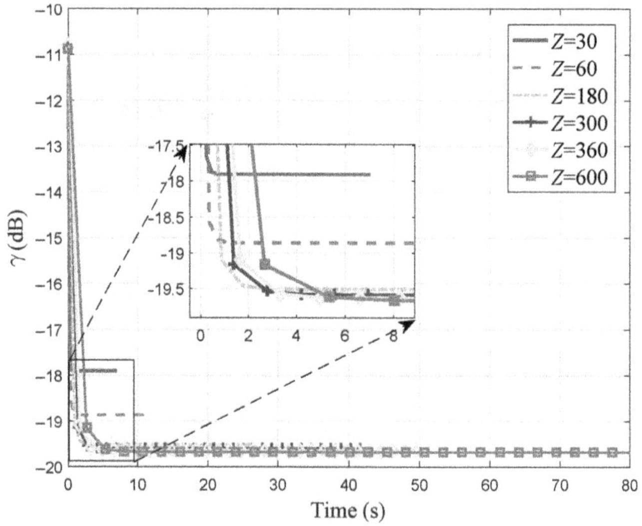

Figure 11.7 γ versus computation time for different Z

selecting an appropriate parameter for Z helps balance objective function PSL and algorithm operation time.

The impact of $l_{\min}$ on performance is evaluated by plotting γ versus iteration number k for different minimum intervals $l_{\min}$ with $Z = 300$, as shown in Figure 11.8. As expected, the objective value decreases as $l_{\min}$ decreases, which can be attributed to the range of feasible set in Problem (11.19). Since the smaller $l_{\min}$ can provide

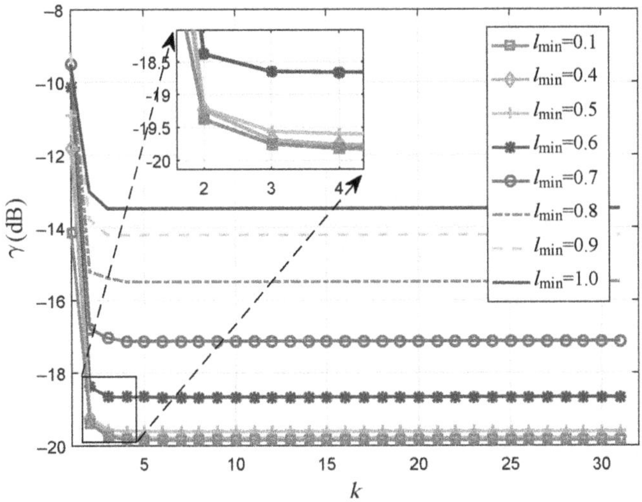

Figure 11.8 γ versus iteration number k for different $l_{\min}$

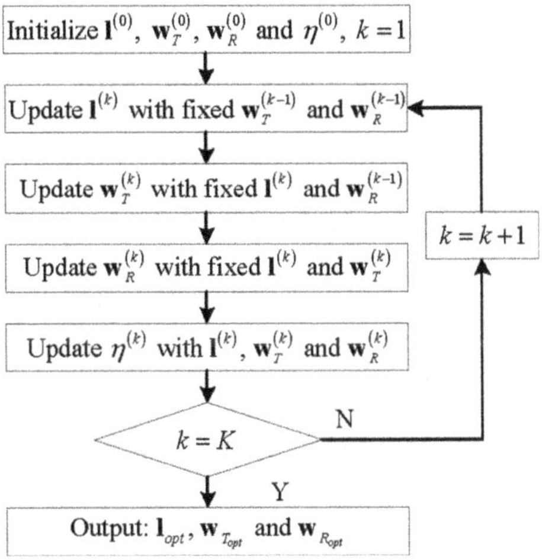

Figure 11.9 Flowchart of CA optimization

the more DoF of $\mathbf{l}$ in the feasible region. Specifically, when $l_{\min} \leq 0.4$, the objective value reaches the lower bound.

11.4.2 CA optimization approach

This subsection is dedicated to verifying the performance of CA optimization approach. Figure 11.9 illustrates the basic flow chart of CA. At kth iteration, $\mathbf{l}^{(k)}$, $\mathbf{w}_T{}^{(k)}$, and $\mathbf{w}_R{}^{(k)}$ are updated separately. The remaining two variables are fixed when

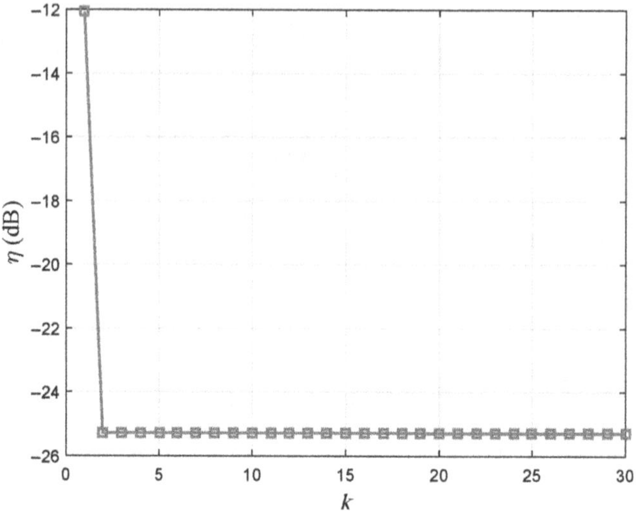

Figure 11.10 η versus iteration number k of the CA optimization approach

solving each variable in turn. Subsequently, the objective function of the current iteration $\eta^{(k)}$ is compared with that of the previous iteration $\eta^{(k-1)}$ to determine whether updating the optimized result is necessary. This loop continues until reaching the exit condition $k = K$. Consequently, this algorithm transforms a complex multivariate optimization problem into several independent subproblems. It should be noted that convergence occurs after only one iteration, as shown in Figure 11.10, which emphasizes that the objective function becomes immediately trapped in a local optimum with respect to $\mathbf{l}$. Under the framework of this algorithm, the joint optimization of $\mathbf{l}$, $\mathbf{w}_T$, and $\mathbf{w}_R$ do not play a role, and the optimization result is not satisfactory.

11.4.3 Joint design of subarray layout and weighting coefficients

In this subsection, we evaluate the performance of the proposed JSLRBIF algorithm. We assume a minimum distance constraint of $l_{\min} = 0.5$ m and set the number of discrete positions as $Z = 300$. For comparison purposes, virtual beampatterns synthesized by conventional beamforming are also considered, with an initialized subarray layout and optimized subarray layout using Algorithm 11.1.

The normalized virtual beampattern performance comparison and corresponding local magnification are shown in Figure 11.11(a) and (b). The proposed JSLRBIF algorithm demonstrates an improvement in the objective values PSL of the I-CBF (initialization layout with conventional beamforming), I-OBF (initialization layout with optimized beamforming), O-CBF (optimized layout with conventional beamforming), and O-OBF (optimized layout with optimized beamforming) and CA.

It is evident that the JSLRBIF algorithm achieves a very low level of −30 dB throughout the sidelobe region. Specifically, the weighting coefficients for I-CBF and O-CBF are low-complexity conventional beamforming. Thus, there exist many high sidelobes around the mainlobe of the virtual beampattern due to the low DoF available at the subarray layout design. Conversely, other optimization methods (I-OBF, I-OBF, CA, and JSLRBIF) involving receive weighting coefficients maintain a flat sidelobe performance. Additionally, Figure 11.12 illustrates η versus the iteration number k of the JSLRBIF algorithm. As expected, the proposed algorithm yields monotonically decreasing sequences for η, while avoiding falling into local optimality compared to CA.

In addition, it is important to emphasize that the initial value $\mathbf{l}^{(0)}$ can be randomly generated, and different $\mathbf{l}^{(0)}$ may lead to varying outcomes. Therefore, it is general to initialize the proposed method with several candidate $\mathbf{l}^{(0)}$ and choose the one that achieves the minimal objective value over the different runs. We also analyze the PSL performance of different initialization of the proposed algorithm as compared

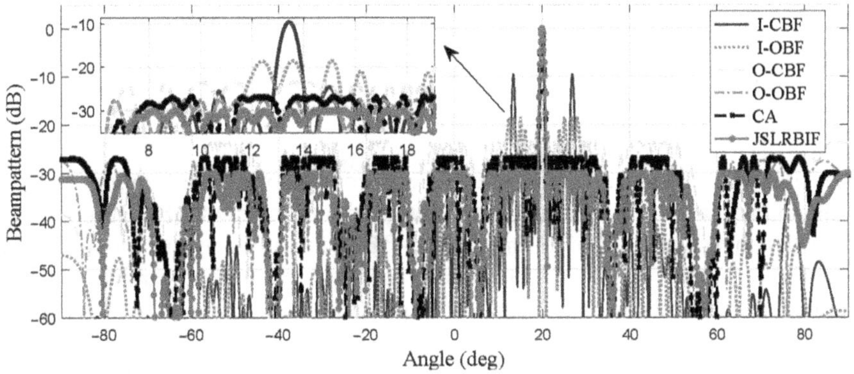

Figure 11.11 Normalized virtual beampattern performance comparison

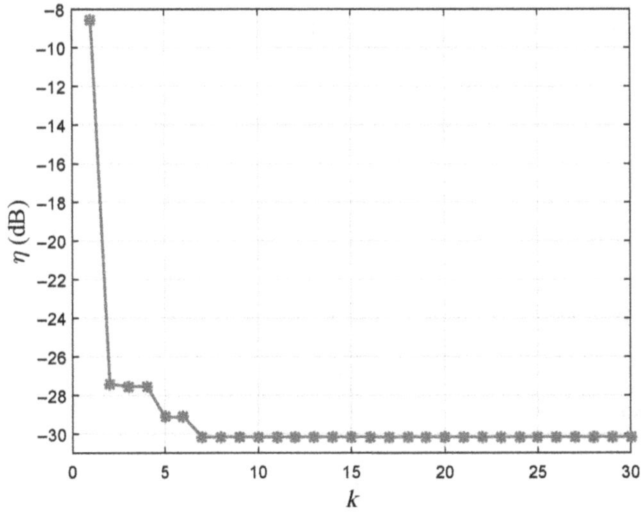

Figure 11.12 η versus iteration number k of the proposed JSLRBIF algorithm

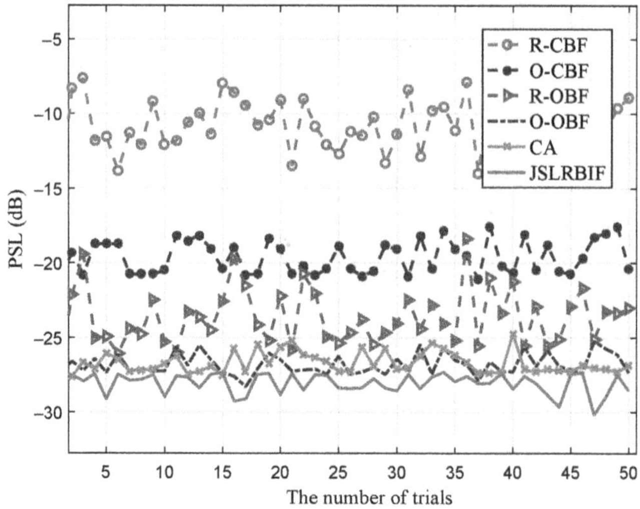

Figure 11.13 PSL of virtual beampattern versus the number of trials

to the R-CBF (random initialization layout with conventional beamforming), R-OBF (random initialization layout with optimized beamforming), O-CBF (optimized layout with conventional beamforming), and O-OBF (optimized layout with optimized beamforming) and CA. The results are averaged with 50 independent Monte Carlo simulations. Specifically, in each independent experiment, $K = 30$ is set for CA and JSLRBIF.

The curves of PSL versus the number of trials are illustrated in Figure 11.13. The proposed JSLRBIF algorithm consistently achieves lower objective values compared to other methods, indicating its superior robustness. Since CA only iterates once and then falls into local optimum, its optimization process is essentially the same as O-OBF, and the results of random experiments are basically consistent with O-OBF, while both of them have better performance than R-CBF, O-CBF, and R-OBF.

11.4.4 Transmit beampattern and virtual beampattern of distributed phased-MIMO radar

Finally, our focus lies on the normalized transmit and virtual beampattern of distributed phased-MIMO radar, as depicted in Figure 11.14. It is worth noting that the phased-MIMO array exhibits an identical transmit beampattern to that of a typical conventional beampattern found in subarray ULPA systems. This behavior is due to the transmit waveforms between the different subarrays that are orthogonal. Consequently, this enables wide area coverage at the transmit end, which holds significant importance for expedited target search.

The virtual beampattern at the receiving end exhibits a significantly narrow mainlobe, thereby enabling high resolution in terms of target detection and tracking. For comparative analysis, we also present the virtual beampattern obtained using conventional beamforming techniques for uniform linear subarrays (ULS) and sparse

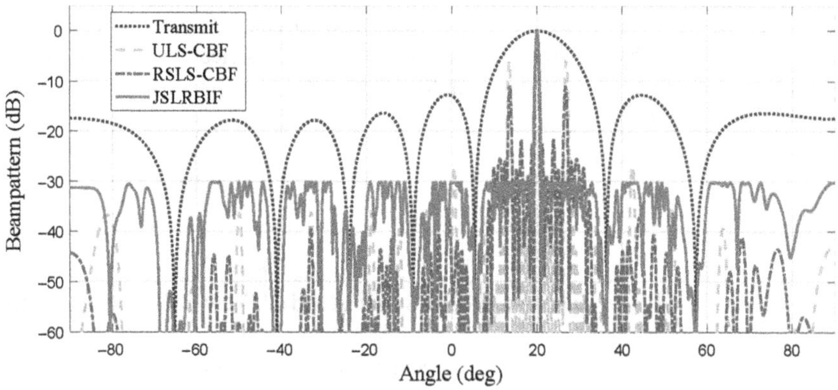

Figure 11.14 Normalized transmit beampattern and virtual beampattern of distributed phased-MIMO array

linear subarrays (SLS). The locations $l_2, \ldots, l_{P-1}$ of SLA are randomly arranged while satisfying the position constraints. Note that the virtual beampattern of ULS-CBF (uniform linear subarrays with conventional beamforming) and RSLS-CBF (random sparse linear subarrays with conventional beamforming) have high sidelobe levels near the mainlobe due to excessively sparse subarray spacing. However, our proposed JSLRBIF algorithm achieves superior performance in terms of achieved PSL in the virtual beampattern.

The optimized virtual beampattern is obtained by increasing the sidelobe level away from the mainlobe region, which can be justified by the principle of energy conservation. Our primary concern lies in minimizing the sidelobe interference within the transmit mainlobe, as it can lead to target detection ambiguity. Therefore, sacrificing some gain in the far-angle area becomes highly worthwhile.

11.5 Conclusions

This chapter has addressed the virtual beampattern optimization problem for distributed phased-MIMO radar. An iterative framework with respect to subarray layout and receive weighting coefficients has been proposed to minimize beampattern PSL while considering practical position constraints and mainlobe level restriction. Compared to the existing CA optimization framework, our JSLRBIF approach can avoid falling into the local optimum of the objective function. To solve the subproblems in each iteration, we have employed the CD algorithm for subarray layout and the SDR algorithm for receiving beamforming. Numerical examples have been presented to validate the effectiveness of our proposed algorithms, demonstrating superior performance in terms of achieved objective values and relatively better robustness.

Possible future research tracks may include the transmit waveform design [21,45], optimum sparse subarray design for beam scanning [4,46,47], and wideband

MIMO system [48]. Additionally, it is important to pay attention to the robustness of distributed phased-MIMO radar models in the presence of array errors [49]. Furthermore, there is a high demand for real-time performance in the physical implementation of this technique.

References

[1] Skolnik M.I. *Introduction to Radar Systems*. 3rd ed. New York: McGraw-Hill; 2001.

[2] Hunter A.J. and Dugelay S. 'Repeat-pass synthetic aperture sonar micronavigation using redundant phase center arrays'. *IEEE Journal of Oceanic Engineering*. 2016;41(4):820–830.

[3] Han G. and Du B. 'A novel hybrid phased array antenna for satellite communication on-the-move in Ku-band'. *IEEE Transactions on Antennas and Propagation*. 2015;63(4):1375–1383.

[4] Dudek M., Nasr I., Bozsik G., Hamouda M., Kissinger D., and Fischer G. 'System analysis of a phased-array radar applying adaptive beam-control for future automotive safety applications'. *IEEE Transactions on Vehicular Technology*. 2015;64(1):34–47.

[5] Van Veen B.D. and Buckley K.M. 'Beamforming: a versatile approach to spatial filtering'. *IEEE ASSP Magazine*. 1988;5(2):4–24.

[6] Lebret H. and Boyd S. 'Antenna array pattern synthesis via convex optimization'. *IEEE Transactions on Signal Processing*. 1997;45(3):526–532.

[7] Van Trees H.L. *Optimum Array Processing*. New York, USA: Wiley; 2002.

[8] Skolnik M.I. *Radar Handbook*. 2nd ed. New York, USA: McGraw-Hill; 1991.

[9] Godara L.C. *Handbook of Antennas in Wireless Communications*. Boca Raton, FL, USA: CRC Press; 2001.

[10] Li J. and Stoica P. 'MIMO radar with colocated antennas'. *IEEE Signal Processing Magazine*. 2007;24(5):106–114.

[11] Haimovich A.M., Blum R.S., and Cimini L.J. 'MIMO radar with widely separated antennas'. *IEEE Signal Processing Magazine*. 2008;25(1):116–129.

[12] Bekkerman I. and Tabrikian J. 'Target detection and localization using MIMO radars and sonars'. *IEEE Transactions on Signal Processing*. 2006;54(10):3873–3883.

[13] Fuhrmann D.R. and San Antonio G. 'Transmit beamforming for MIMO radar systems using signal cross-correlation'. *IEEE Transactions on Aerospace and Electronic Systems*. 2008;44(1):171–186.

[14] Zhang Y., Huo Y., Wang D., Dong X., and You X. 'Channel estimation and hybrid precoding for distributed phased arrays based MIMO wireless communications'. *IEEE Transactions on Vehicular Technology*. 2020;69(11):12921–12937.

[15] Amiri R., Behnia F., and Sadr M.A.M. 'Exact solution for elliptic localization in distributed MIMO radar systems'. *IEEE Transactions on Vehicular Technology*. 2018;67(2):1075–1086.

[16] Yang Y. and Blum R.S. 'MIMO radar waveform design based on mutual information and minimum mean-square error estimation'. *IEEE Transactions on Aerospace and Electronic Systems*. 2007;43(1):330–343.
[17] Ahmed S. and Alouini M. 'MIMO-radar waveform covariance matrix for high SINR and low side-lobe levels'. *IEEE Transactions on Signal Processing*. 2014;62(8):2056–2065.
[18] Yu X., Cui G., Yang J., and Kong L. 'MIMO radar transmit–receive design for moving target detection in signal-dependent clutter'. *IEEE Transactions on Vehicular Technology*. 2020;69(1):522–536.
[19] Cheng Z., Liao B., He Z., Li J., and Xie J. 'Joint design of the transmit and receive beamforming in MIMO radar systems'. *IEEE Transactions on Vehicular Technology*. 2019;68(8):7919–7930.
[20] Aubry A., De Maio A., and Huang Y. 'MIMO radar beampattern design via PSL/ISL optimization'. *IEEE Transactions on Signal Processing*. 2016;64(15):3955–3967.
[21] Fan W., Liang J., and Li J. 'Constant modulus MIMO radar waveform design with minimum peak sidelobe transmit beampattern'. *IEEE Transactions on Signal Processing*. 2018;66(16):4207–4222.
[22] Esmaeili-Najafabadi H., Ataei M., and Sabahi M.F. 'Designing sequence with minimum PSL using chebyshev distance and its application for chaotic MIMO radar waveform design'. *IEEE Transactions on Signal Processing*. 2017;65(3):690–704.
[23] Alaee-Kerahroodi M., Modarres-Hashemi M., and Naghsh M.M. 'Designing sets of binary sequences for MIMO radar systems'. *IEEE Transactions on Signal Processing*. 2019;67(13):3347–3360.
[24] Hassanien A. and Vorobyov S.A. 'Phased-MIMO radar: A tradeoff between phased-array and MIMO radars'. *IEEE Transactions on Signal Processing*. 2010;58(6):3137–3151.
[25] Hassanien A. and Vorobyov S.A. 'Why the phased-MIMO radar outperforms the phased-array and MIMO radars'. *Proceedings of the European Signal Processing Conference*; Aalborg, Denmark, 2010.
[26] Wang W. 'Phased-MIMO radar with frequency diversity for range-dependent beamforming'. *IEEE Sensors Journal*. 2013;13(4):1320–1328.
[27] Wang W. and Zheng Z. 'Hybrid MIMO and phased-array directional modulation for physical layer security in mmwave wireless communications'. *IEEE Journal on Selected Areas in Communications*. 2018;36(7):1383–1396.
[28] Wang W. and So H.C. 'Transmit subaperturing for range and angle estimation in frequency diverse array radar'. *IEEE Transactions on Signal Processing*. 2014;62(8):2000–2011.
[29] Hassanien A., Vorobyov S.A., Yoon Y., and Park J. 'Two-stage based design for phased-MIMO radar with improved coherent transmit processing gain'. *Proceedings of the IEEE 15th International Workshop on Signal Processing Advances in Wireless Communications*; Toronto, ON, Canada, Jun 2014, pp. 45–49.

[30] Khan W., Qureshi I.M., Basit A., and Zubair M. 'Hybrid phased MIMO radar with unequal subarrays'. *IEEE Antennas and Wireless Propagation Letters*. 2015;14:1702–1705.
[31] Alieldin A., Huang Y., and Saad W.M. 'Optimum partitioning of a phased-MIMO radar array antenna'. *IEEE Antennas and Wireless Propagation Letters*. 2017;16:2287–2290.
[32] Fabra F., Zamora W., Reyes P., *et al.* 'MUSCOP: mission-based UAV swarm coordination protocol'. *IEEE Access*. 2020;8:72498–72511.
[33] Chen K., He Z., and Han C. 'A modified real GA for the sparse linear array synthesis with multiple constraints'. *IEEE Transactions on Antennas and Propagation*. 2006;54(7):2169–2173.
[34] Yu X., Cui G., Yang S., Kong L., and Yi W. 'Coherent unambiguous transmit for sparse linear array with geography constraint'. *IET Radar, Sonar & Navigation*. 2017;11(2):386–393.
[35] Liang J., Zhang X., So H.C., and Zhou D. 'Sparse array beampattern synthesis via alternating direction method of multipliers'. *IEEE Transactions on Antennas and Propagation*. 2018;66(5):2333–2345.
[36] Liang J., Fan X., So H.C., and Zhou D. 'Array beampattern synthesis without specifying lobe level masks'. *IEEE Transactions on Antennas and Propagation*. 2020;68(6):4526–4539.
[37] Wang X., Aboutanios E., and Amin M.G. 'Thinned array beampattern synthesis by iterative soft-thresholding-based optimization algorithms'. *IEEE Transactions on Antennas and Propagation*. 2014;62(12):6102–6113.
[38] Wang X., Amin M., Wang X., and Cao X. 'Sparse array quiescent beamformer design combining adaptive and deterministic constraints'. *IEEE Transactions on Antennas and Propagation*. 2017;65(11):5808–5818.
[39] Feng L., Cui G., Yu X., and Kong L. 'Beampattern synthesis via the constrained subarray layout optimization'. *IEEE Transactions on Antennas and Propagation*. 2021;69(1):182–194.
[40] Gemechu A.Y., Cui G., Yu X., and Kong L. 'Beampattern synthesis with sidelobe control and applications'. *IEEE Transactions on Antennas and Propagation*. 2020;68(1):297–310.
[41] Zhang X., He Z., Zhang X., and Peng W. 'High-performance beampattern synthesis via linear fractional semidefinite relaxation and quasi-convex optimization'. *IEEE Transactions on Antennas and Propagation*. 2018;66(7):3421–3431.
[42] Fuchs B. 'Application of convex relaxation to array synthesis problems'. *IEEE Transactions on Antennas and Propagation*. 2014;62(2):634–640.
[43] Luo Z., Ma W., So A.M., Ye Y., and Zhang S. 'Semidefinite relaxation of quadratic optimization problems'. *IEEE Signal Processing Magazine*. 2010;27(3):20–34.
[44] Grant M. and Boyd S. 'CVX package'. http://www.cvxr.com/cvx.r, accessed Feb. 2012.
[45] Wang X., Hassanien A., and Amin M.G. 'Dual-function MIMO radar communications system design via sparse array optimization'. *IEEE Transactions on Aerospace and Electronic Systems*. 2019;55(3):1213–1226.

[46] Wang X., Amin M., and Wang X. 'Optimum sparse array design for multiple beamformers with common receiver'. *Proceedings of the IEEE International Conference on Acoustics, Speech and Signal Processing*; Calgary, AB, Canada, 2018, pp. 3364–3368.

[47] Zhang X., Liang J., Fan X., Yu G., and So H.C. 'Reconfigurable array beampattern synthesis via conceptual sensor network modeling and computation'. *IEEE Transactions on Antennas and Propagation*. 2020;68(6):4512–4525.

[48] Yu X., Cui G., Yang J., Kong L., and Li J. 'Wideband MIMO radar waveform design'. *IEEE Transactions on Signal Processing*. 2019;67(13):3487–3501.

[49] Yu X., Cui G., Kong L., Li J., and Gui G. 'Constrained waveform design for colocated MIMO radar with uncertain steering matrices'. *IEEE Transactions on Aerospace and Electronic Systems*. 2019;55(1):356–370.

Index

www.ingramcontent.com/pod-product-compliance
Lightning Source LLC
Chambersburg PA
CBHW050509190326
41458CB00005B/1481
* 9 7 8 1 8 3 9 5 3 9 3 3 6 *